LEACHABLES AND EXTRACTABLES HANDBOOK

LEACHABLES AND EXTRACTABLES HANDBOOK

Safety Evaluation, Qualification, and Best Practices Applied to Inhalation Drug Products

Edited by

Douglas J. Ball
Pfizer Global Research and Development

Daniel L. Norwood
Boehringer Ingelheim Pharmaceuticals

Cheryl L.M. Stults
Nektar Therapeutics

Lee M. Nagao
Drinker Biddle & Reath LLP

A JOHN WILEY & SONS, INC., PUBLICATION

Library of Congress Cataloging-in-Publication Data:

Leachables and extractables handbook : safety evaluation, qualification, and best practices applied to
inhalation drug products / edited by Douglas J.
Ball . . . [et al.].
      p. ; cm.
   Includes bibliographical references.
     ISBN 978-0-470-17365-7 (hardback)
   I. Ball, Douglas J.
   [DNLM: 1. Drug Delivery Systems–standards.   2. Nebulizers and Vaporizers–standards.
3. Drug Contamination–prevention & control.   4. Drug Packaging–standards.   5. Pharmaceutical
Preparations–administration & dosage.   6. Risk Assessment–standards.   QV 785]
   LC classification not assigned
   615'.6–dc23
                                                2011026193

Printed in the United States of America

10 9 8 7 6 5 4 3 2 1

ROBERT KROES: IN MEMORIAM

This book is dedicated to the memory of Professor Robert Kroes whose scientific contributions played a vital role in developing the concept of the threshold of toxicological concern and the application of that concept to important societal issues including the safety evaluation of inhalable pharmaceutical products.

Robert Kroes, known as Bobby to his friends and colleagues around the world, was a native of The Netherlands. He received his Doctor of Veterinary Medicine in 1964. His training in Veterinary Medicine provided him with a solid scientific basis for a career grounded in comparative medicine, toxicology, and risk assessment, with a focus on the promotion of human health. In 1964, he was appointed Research Scientist at the National Institute of Public Health, which later became the National Institute of Public Health and Environment (known by the Dutch acronym, RIVM) in Bilthoven, The Netherlands. In 1970, he received a PhD in experimental pathology. He became a certified toxicologist in 1988 and a certified laboratory animal pathologist in 1989.

In 1972, he became Head of the Department of Oncology in the National Institute of Public Health. During this time in his career, he made important scientific contributions to understanding carcinogenicity. Moreover, he soon became a key contributor to major scientific committees within The Netherlands and on the international scene including the Benelux, the European Community, the Food and Agricultural Organization of the United Nations, and the World Health Organization. He was a member and, ultimately, Chair of the Dutch Scientific Council on Cancer Research of The Netherlands Academy of Science. He was a key contributor in the development of the first cancer research policy plan (1980–1984) of the Dutch Organization for Cancer Research.

In 1977, he was appointed Deputy Director of the Central Institute for Food and Nutrition Research (CIVO-TNO). In that position, he provided critical leadership for stimulating research in carcinogenesis, toxicology, biochemistry, and nutrition. In 1980, he became Director of the CIVO-TNO Institute for Toxicology and Nutrition. In 1983, he was appointed as a Director of RIVM with responsibility for managing the Institute's toxicology and pharmacology programs. He was also responsible for guiding the institute's advisory mission to the government with respect to the safety of chemicals. In 1988, he developed the Center for Epidemiology, further broadening the scope of RIVM's activities. In 1988, he began a part-time association as a Professor of Biological Toxicology in the Research Institute for Toxicology of the University of Utrecht. In 1989, he became Deputy Director-General of RIVM. In 1995, he retired from his leadership roles at RIVM. In that year, he became the Scientific Director of the Institute for Risk Assessment Sciences (IRAS) of the University of Utrecht. He retired from IRAS in 2005.

The use of the word "retired" certainly did not apply to Bobby's scientific activities. He continued to play a prominent role in many scientific advisory groups in The Netherlands and on the international scene. He had a key role in the National Institute of Toxicology. Of special note are the key roles he played in the International Life Sciences Institute (ILSI) and the related ILSI Risk Science Institute, as well as the International Union of Toxicology. He served the latter organization in multiple roles including service as president-elect and was scheduled to assume the position of president in 2007. Unfortunately, Bobby lost a courageous battle with cancer and died on December 28, 2006.

During his scientific career spanning over four decades, Bobby's many important scientific contributions to the fields of oncology, toxicology, comparative medicine, and risk assessment are well documented in some 200 publications he authored or coauthored. As noteworthy as those contributions are, his most significant contributions came from his ability to rise above the scientific details and understand how to synthesize and integrate science and relate it to important societal health issues. He took a pragmatic view and focused on concepts and solutions to resolving complex issues. He was truly a problem solver.

This pragmatic, science-based approach was exemplified by Professor Kroes championing the use of the concept of "threshold of toxicological concern" (TTC) and its application to the safety of food and pharmaceuticals. The TTC concept refers to the establishment of a generic human exposure threshold for groups of chemicals below which there would be no appreciable risk to human health. He recognized that such a value could be identified for many chemicals, including those of unknown toxicity, by considering their chemical structure and drawing analogies from the known toxicity and modes of action of many chemicals that have been extensively studied. In December 2005, the Product Quality Research Institute organized a workshop to address the use of the TTC concept in evaluating the safety of inhalable pharmaceuticals. The organizers were unanimous in deciding that Professor Kroes should be invited to give an opening presentation to set the stage for the workshop. He gave a marvelous review of the developing field. His presentation served to energize activities that culminated in preparation of this volume. Therefore, it is indeed fitting that this volume be dedicated to the memory of Professor Robert Kroes. In using the science-based concepts championed by Professor Kroes, we celebrate the value of his contributions as a scientist and, for many of us, also have the opportunity to recall a wonderful friend who lived life to its fullest.

Roger O. McClellan, DVM, DSc (Honorary), DABVT, DABT, FATS

CONTENTS

vii

PREFACE

The establishment of data-based safety thresholds for leachables and extractables in orally inhaled and nasal drug products (OINDPs) is an important scientific advancement that helps OINDP manufacturers make knowledge-based safety and risk assessments for extractables and leachables and ensure the safety of their products for patient use. This book describes the development and application of these safety thresholds for OINDP and best practices for the chemical evaluation and management of extractables and leachables throughout the pharmaceutical product life cycle. Although the book addresses OINDP-specific thresholds and best practices, many of the general concepts presented can be applied to extractables and leachables assessments for other drug product types and dosage forms. The purpose of this book is to provide the reader with practical knowledge regarding how and why the thresholds were developed and how they can be applied, as well as practical approaches to management of extractables and leachables. This book is useful to analytical chemists, packaging and device engineers, formulation development scientists, component suppliers, regulatory affairs specialists, and toxicologists, all of whom must work together in the pharmaceutical development process to identify, qualify, and manage extractables and leachables.

Management of extractables and leachables in OINDP is a critical part of the OINDP life cycle. By "management" we mean a thorough understanding of (1) potential and actual extractables from a given container closure system or device material for the purposes of eliminating or limiting the levels of leachables from such materials and (2) potential safety concerns associated with these extractables and/or leachables. These issues highlight the key regulatory and industrial concern regarding leachables in OINDPs as well as other drug products—that of patient safety. Regulatory guidance identifies patient exposure to leachables via OINDPs as an area of high importance in risk assessments for these products. Over the last 30 years, scientific and regulatory thought has evolved on the best ways to approach both chemical and safety assessments of extractables and leachables in the OINDP pharmaceutical development process. A vexing challenge in these assessments has been knowing "how low to go" in determining what concentrations of extractables and leachables should be evaluated for safety assessments; that is, is there a threshold of safety that can be established for the majority of compounds that could be found as leachables or extractables in OINDPs, such that compounds existing at levels below the threshold need not undergo safety evaluation? This question has become increasingly important with the continuous advancement of chemical analysis techniques, which have been, for the past four decades, able to detect chemical compounds at picogram levels and below.

In 2006, the Product Quality Research Institute's (PQRI) Leachables and Extractables Working Group, consisting of scientists from the United States Food and Drug Administration (FDA), academia, and industry, answered this question by developing data-based safety and analytical thresholds for OINDP extractables and leachables, and corresponding best practices for analytical evaluation of these compounds. This book is based on the information contained in the Working Group's recommendations (publicly available through PQRI); but it provides further, more in-depth context and background, case studies, and specific regulatory perspectives and extends the concepts to practices that may be implemented across the industry.

Douglas J. Ball
Daniel L. Norwood
Cheryl L.M. Stults
Lee M. Nagao

ACKNOWLEDGMENTS

We thank the Product Quality Research Institute (PQRI) for supporting the development of this book, and the members of the PQRI Leachables and Extractables (L&E) Working Group, whose efforts formed the basis for this volume. We also thank the International Pharmaceutical Aerosol Consortium on Regulation and Science (IPAC-RS) for initiating the process to develop safety thresholds for inhalation and nasal drug products, for providing the impetus to form the PQRI L&E Working Group, and for giving its ongoing support of collaborative efforts addressing the most challenging aspects of leachables and extractables in inhalation and nasal drug products.

Mr. Ball and Dr. Norwood thank Pfizer, Inc. and Boehringer Ingelheim Pharmaceuticals, Inc., respectively, for supporting their efforts in the PQRI L&E Working Group and in the development of this book. Dr. Stults thanks Novartis Pharmaceuticals Corporation for supporting her efforts in the development of this book and thanks colleagues across the industry for their support in the preparation of this book. We extend a very large thank you to Mr. Duane Van Bergen and Ms. Kara Young of Drinker Biddle & Reath LLP, who worked extremely hard to format, harmonize, and help edit the chapters of this book. Also from Drinker Biddle & Reath LLP, we thank Ms. Mary Devlin Capizzi, Esq. for invaluable guidance on contracts and agreements; Dr. Svetlana Lyapustina and Ms. Melinda Munos for assistance in managing the work of the PQRI L&E Working Group; and Ms. Dede Godstrey and Ms. Kim Rouse for their invaluable assistance in managing and planning the meetings, teleconferences, and administrative details critical in the completion of this book. We thank Mr. Gordon Hansen, Dr. Terrence Tougas, and Ms. Devlin Capizzi for helping to guide the development of this book through the PQRI process. Finally, we thank Dr. Roger McClellan for sharing with us his inhalation toxicology expertise and for helping to facilitate the creation of the PQRI Group's seminar on safety thresholds at the 2007 Society of Toxicology meeting, which lead to the publication of this book.

D.J.B.
D.L.N.
C.L.M.S.
L.M.N.

CONTRIBUTORS

David Alexander, DA Nonclinical Safety Ltd., Cambridgeshire, United Kingdom

Douglas J. Ball, Drug Safety Research & Development, Pfizer Global Research & Development, Groton, CT

William P. Beierschmitt, Drug Safety Research and Development, Pfizer Global Research and Development, Groton, CT

James Blanchard, Preclinical Development, Aradigm Corp, Hayward, CA

James R. Coleman, Boehringer Ingelheim Pharmaceuticals, Inc., Ridgefield, CT

Jason M. Creasey, GlaxoSmithKline, Ware, Hertfordshire, United Kingdom

Tianjing Deng, PPD, Inc., Middleton, WI

Xiaoya Ding, PPD, Inc., Middleton, WI

Barbara Falco, Barbara Falco Pharma Consult, LLC, Bethlehem, PA

Andrew D. Feilden, Smithers Rapra, Shawbury, Shropshire, United Kingdom

Thomas N. Feinberg, Catalent Pharma Solutions, LLC, Research Triangle Park, NC

Cornelia B. Field, Boehringer Ingelheim Pharmaceuticals, Inc., Ridgefield, CT

Alice T. Granger, Boehringer Ingelheim Pharmaceuticals, Inc., Ridgefield, CT

John Hand, Sr., New Rochelle High School, New Rochelle, NY

Alan D. Hendricker, Catalent Pharma Solutions, Morrisville, NC

David Jacobson-Kram, Office of New Drugs, Center for Drug Evaluation and Research, U.S. Food and Drug Administration, Silver Spring, MD

Dennis Jenke, Baxter Healthcare Corporation, Round Lake, IL

Song Klapoetke, PPD, Inc., Middleton, WI

Shuang Li, PPD, Inc., Middleton, WI

Ernest L. Lippert, American Glass Research, Maumee, OH

Timothy J. McGovern, SciLucent, LLC, Herndon, VA

Keith McKellop, Boehringer Ingelheim Pharmaceuticals, Inc., Ridgefield, CT

Kimberly Miller, West Pharmaceutical Services, Lionville, PA

Brian D. Mitchell, American Glass Research, Maumee, OH

James O. Mullis, Boehringer Ingelheim Pharmaceuticals, Inc., Ridgefield, CT

Melinda K. Munos, Drinker Biddle & Reath LLP, Washington, DC

Lee M. Nagao, Drinker Biddle & Reath LLP, Washington, DC

Kumudini Nicholas, Bureau of Pharmaceutical Sciences, Health Canada, Ottawa, Ontario, Canada

Daniel L. Norwood, Boehringer Ingelheim Pharmaceuticals, Inc., Ridgefield, CT

David Olenski, Intertek, Whitehouse, NJ

Diane Paskiet, West Pharmaceutical Services, Lionville, PA

Scott J. Pennino, Boehringer Ingelheim Pharmaceuticals, Inc., Ridgefield, CT

Fenghe Qiu, Boehringer Ingelheim Pharmaceuticals, Inc., Ridgefield, CT

Michelle Raikes, Boehringer Ingelheim Pharmaceuticals, Inc., Ridgefield, CT

Andy Rignall, Analytical Chemistry, AstraZeneca, Loughborough, United Kingdom

Suzette Roan, Pfizer Global Research & Development, Groton, CT

John A. Robson, Boehringer Ingelheim Pharmaceuticals, Inc., Ridgefield, CT

Michael A. Ruberto, Material Needs Consulting, LLC, Montvale, NJ

Arthur J. Shaw, Pfizer Analytical Research and Development, Groton, CT

John A. Smoliga, Boehringer Ingelheim Pharmaceuticals, Inc., Ridgefield, CT

Ronald D. Snyder, Schering-Plough Research Institute, Summit, NJ

Laura Stubbs, West Pharmaceutical Services, Lionville, PA

Cheryl L.M. Stults, Novartis Pharmaceuticals Corporation, San Carlos, CA

Terrence Tougas, Boehringer Ingelheim, Ridgefield, CT

W. Mark Vogel, Drug Safety Research & Development, Pfizer Global Research & Development, Chesterfield, MO

Ronald Wolff, Preclinical Safety Assessment, Novartis Institutes for Biomedical Research, Emeryville, CA

Derek Wood, PPD, Inc., Middleton, WI

Xiaochun Yu, PPD, Inc., Middleton, WI

Diego Zurbriggen, West Analytical Services, Lionville, PA

DEVELOPMENT OF SAFETY THRESHOLDS, SAFETY EVALUATION, AND QUALIFICATION OF EXTRACTABLES AND LEACHABLES IN ORALLY INHALED AND NASAL DRUG PRODUCTS

OVERVIEW OF LEACHABLES AND EXTRACTABLES IN ORALLY INHALED AND NASAL DRUG PRODUCTS

Douglas J. Ball, Daniel L. Norwood, and Lee M. Nagao

1.1 INTRODUCTION

The purpose of this book is to provide a historical perspective on the development and application of safety thresholds in pharmaceutical development, and to discuss the development and implementation of safety thresholds for the qualification of organic leachables, a particular class of drug product impurity, in orally inhaled and nasal drug products (OINDPs). The book will also describe and consider the United States Food and Drug Administration (FDA) and international regulatory perspectives concerning the qualification of organic leachables in OINDP. Although the book is written specifically for OINDP, the principles used in defining safety thresholds could be applied to organic leachables in other drug product types.

Since the environmental movement of the 1970s, analytical chemistry and analytical techniques have become increasingly sophisticated and sensitive, capable of detecting, identifying, and quantifying both organic and inorganic chemical entities at ultratrace (i.e., parts per trillion) levels.[1] However, it is generally accepted that there are levels of many chemicals below which the risks to human health are so negligible as to be of no consequence. This rationale has been a strong impetus for development of safety thresholds for regulating chemicals to which humans are exposed, most notably in the federal regulations for food packaging.[2,3] Safety thresholds have also been developed for application to pharmaceuticals, including organic impurities in drug substances[4] (process and drug related), drug products,[5] and residual solvents in drug substances and drug products.[6] Note that the international regulatory guidance for drug product impurities specifically excludes from consideration "impurities . . . leached from the container closure system."[5]

Leachables and Extractables Handbook: Safety Evaluation, Qualification, and Best Practices Applied to Inhalation Drug Products, First Edition. Edited by Douglas J. Ball, Daniel L. Norwood, Cheryl L.M. Stults, Lee M. Nagao.

Figure 1.1   Patients using metered dose inhaler (top) and dry powder inhaler (bottom) drug products. Note that each patient's mouth is in direct contact with the drug delivery device/container closure system, and that doses of drug formulation are delivered directly into each patient's mouth for inhalation. (Images provided by Bespak, a division of Consort Medical plc; www.bespak.com.)

OINDPs are developed for delivery of active pharmaceutical ingredient (API or drug substance) directly to the respiratory or nasal tract, to treat either a respiratory or nasal condition, or a systemic disease. Examples of OINDP include metered dose inhalers (MDIs), dry powder inhalers (DPIs), solutions/suspensions for nebulization, and nasal sprays (see Figs. 1.1 and 1.2). These drug product types incorporate complex delivery devices and container closure systems whose function and

Figure 1.2   Patient using a nasal spray drug product. Note that the patient's nasal mucosa is in direct contact with the drug delivery device/container closure system, and that doses of drug formulation are delivered directly into the patient's nasal passages for inhalation. (Image provided by Bespak, a division of Consort Medical plc; www.bespak.com.)

performance are critical to the safety and efficacy of the drug product. Components of OINDP delivery systems can be composed of polymers, elastomers, and other materials from which minute quantities of chemicals can migrate (i.e., leach) into the drug product formulation and be delivered to the sensitive surfaces of the respiratory and/or nasal tract along with the therapeutic agent. FDA guidance considers these drug product types high risk for containing leachables, which are delivered to the patient, because of the route of administration and because of the direct interaction of packaging and/or device components with drug formulation.[7] While every effort is usually taken to reduce the levels of leachables, complete removal is neither practical nor desirable as many of these chemical entities perform important functions in container closure system components. Since leachables are non-drug-related impurities, there is an increased concern regarding the human risk associated with inhaling them on a daily basis, often for many years or decades. Historically, acceptable levels of leachables in OINDP have been set by negotiation with regulatory authorities on a case-by-case basis with no standard guidelines available. Recently, however, safety thresholds for risk assessment of organic leachables have been developed through a joint effort of scientists from the FDA, academia, and industry.[8,9] This book will address the concepts, background, historical use, and development of safety thresholds and their utility in qualifying organic leachables in OINDP.

1.2   LEACHABLES IN OINDP: THE ISSUE IN DETAIL

The FDA guidance documents for MDIs/DPIs,[10] and nasal spray, and inhalation solution, suspension and spray drug products[11] state that *leachables* are "compounds that leach from elastomeric, plastic components or coatings of the container and closure system as a result of direct contact with the formulation," and *extractables* are "compounds that can be extracted from elastomeric, plastic components or coatings of the container and closure system when in the presence of an appropriate solvent(s)."

In short, extractables are chemical entities that are derived from container closure and/or device components *under laboratory conditions*. Leachables are chemical entities derived from container closure and/or device components when they are *part of the final drug product and under patient-use conditions*. Leachables are, therefore, either a subset of extractables or can be correlated indirectly with extractables (e.g., via chemical reaction), and all extractables are *potential leachables*. Patients can be exposed to leachables through the normal use of the drug product.

OINDPs are used in the treatment of a variety of lung- and nasal-related conditions such as asthma, chronic obstructive pulmonary disease (COPD, such as emphysema or chronic bronchitis), and allergic rhinitis, as well as systemic diseases such as diabetes. This latter therapeutic application suggests that the inhalation route has potential for wider use in the treatment and management of a variety of disease states.

All OINDP types include a drug product formulation (API along with excipients) in direct contact with areas of the container closure system and parts of the drug product device that facilitate accurate dose delivery for inhalation by the patient and/or protect the integrity of the formulation. Figure 1.3 shows a schematic diagram of an MDI drug product, and Figure 1.4 shows a "cutaway" view of a dose metering valve. The MDI consists of a solution or suspension formulation containing a drug substance (API), chlorofluorocarbon (CFC), or hydrofluoroalkane (HFA) propellant to facilitate aerosol dose delivery, and surfactants, co-solvents and other excipients to help stabilize the formulation. The container closure and device system includes a metal canister to contain the pressurized formulation, a valve to meter the dose to the patient, elastomeric components to seal the valve to the canister, and an actuator/mouthpiece to facilitate patient self-dosing. The formulation and container closure system are closely integrated in the MDI drug product, and leachables may be derived from the elastomeric seals between the valve and metal canister (e.g., gaskets), plastic and other types of polymeric valve components (e.g., metering chamber, valve stem), and organic residues or coatings on the surfaces of the metal canister and metal valve components. As shown in Figure 1.1, the patient's mouth is also in contact with the actuator/mouthpiece during normal use of the drug product.

Although the DPI can be a more complex device/container closure system than the MDI (see Fig. 1.5), the potential for leachables issues is significantly reduced. This is because the drug product formulation in the DPI is by definition a dry powder and, therefore, contains no solvent systems such as the organic propellants and co-solvents in the MDI formulation, which can facilitate leaching. However, DPI doses are usually contained in unit dose blister packs, capsules, and similar packaging systems, which include plastic, foil, and/or laminate overwraps that contact the drug

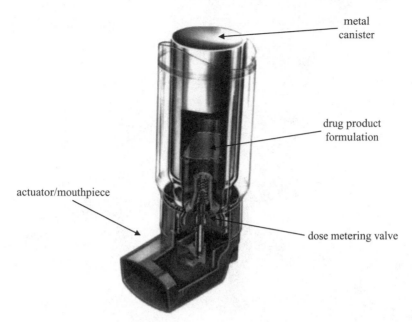

Figure 1.3   Schematic diagram of a metered dose inhaler (MDI) drug product. Note that the elastomeric, plastic, and metal components of the dose metering valve, as well as the metal canister inner surfaces, are capable of leaching chemical entities into the drug product formulation. The actuator/mouthpiece is in contact with the patient's mouth (see Fig. 1.1). (Images provided by Bespak, a division of Consort Medical plc; www.bespak.com.)

product formulation directly during storage. Also, the dry powder can contact certain surfaces of the DPI device during dose delivery, and as with the MDI, the patient's mouth contacts the mouthpiece (Fig. 1.1). Nasal spray and inhalation spray drug products can also include device/container closure system components with leaching potential (i.e., plastic containers and tubes, elastomeric seals); however, these drug product formulations are typically aqueous based and therefore have a generally reduced leaching potential compared with the organic solvent-based MDI drug products. Inhalation solutions are also mostly aqueous based and typically packaged in unit dose plastic containers (e.g., low-density polyethylene). Delivery of inhalation solution drug product to patients is usually accomplished via commercially available nebulizer systems. It is interesting to note that certain types of plastic, such as low-density polyethylene, can allow gaseous chemical substances from the surrounding environment to penetrate into the drug product. As a result of this, many inhalation solutions are stored in secondary packaging systems such as foil pouches.

The variety and complexity of OINDP and the different potentials for container closure system leaching among the various OINDP types should be clear from the above discussion. The organic chemicals that can appear as extractables and leachables represent an additional level of complexity. Extractables and leachables are generally low-molecular-weight organic chemicals either purposefully added to the packaging or device materials during synthesis, compounding, or fabrication (e.g.,

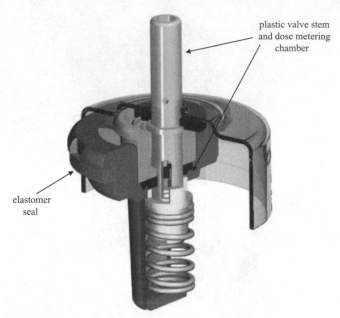

Figure 1.4   Cutaway diagram of a metered dose inhaler (MDI) dose metering valve showing the various metal, plastic and elastomeric components potentially in contact with the drug product formulation. (Images provided by Bespak, a division of Consort Medical plc; www.bespak.com.)

Figure 1.5   Cutaway diagram of a dry powder inhaler (DPI) showing the internal complexity of the device/container closure system and its many components. Many DPI components are plastic or elastomeric and therefore potentially capable of leaching. (Images provided by Valois Pharma.)

polymerization agents, fillers, antioxidants, stabilizers, and processing aids), or present in the materials as a by-product of synthesis, compounding, or fabrication (e.g., oligomers, additive contaminants such as polyaromatic hydrocarbons [PAHs] or polynuclear aromatics [PNAs] and reaction products such as *N*-nitrosamines). All of these chemical entities have the capacity to move from the packaging or device components into the OINDP formulation, and thus be delivered to the patient. Table 1.1 provides examples of potential sources of extractables and leachables from OINDP.[12] Unlike drug-substance-related impurities, leachables can represent a wide variety of chemical types (see some examples in Fig. 1.6) and be present in drug products at widely variable concentration levels, from perhaps several tens of micrograms per canister in the case of named additives to an MDI valve elastomeric seal,

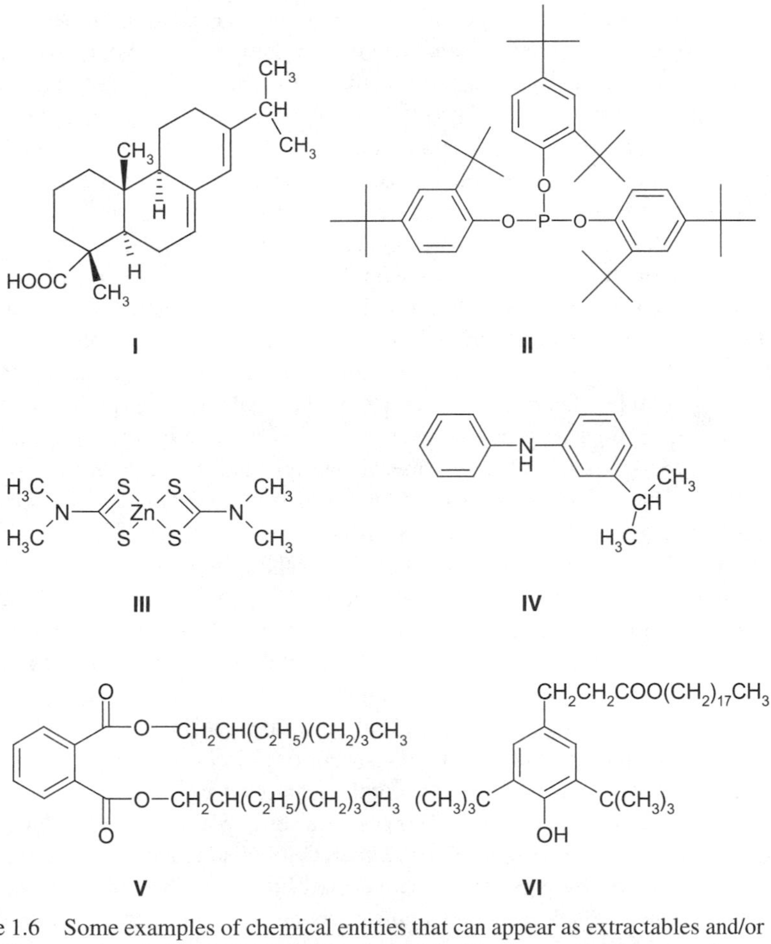

Figure 1.6   Some examples of chemical entities that can appear as extractables and/or leachables associated with OINDP. (I) Abietic acid (a filler for certain elastomers); (II) Irgafos 168 (a phosphite antioxidant); (III) zinc tetramethyldithiocarbamate (an accelerator for certain sulfur-cured elastomers); (IV) isopropyldiphenylamine (an antioxidant); (V) di-2-ethylhexylphthalate (a plasticizer); (VI) Irganox 1076 (an antioxidant).

to several nanograms per canister in the case of a volatile *N*-nitrosamine rubber polymerization by-product. Additional detailed discussions are available regarding the variety and origins of extractables and leachables.[8,12]

1.3   REGULATORY BACKGROUND

The U.S. regulatory history of extractables and leachables in OINDP was summarized and discussed by Dr. Alan Schroeder of the FDA Center for Drug Evaluation and Research (CDER), at a workshop on the topic in 2005.[13] Regulatory attention was focused in two general areas: clinical and quality control. Clinical concerns resulted from the fact that the majority of OINDPs are administered to a sensitive and already compromised patient population, that is, patients with asthma or COPD. It is known that some of these patients can experience a condition known as *para-doxical bronchospasm*. Bronchospasm is defined as a condition in which the airways suddenly narrow, causing coughing or breathing difficulty, like an asthma attack.[14] Paradoxical bronchospasm is a relatively rare event in which a medicine prescribed to treat bronchospasm or the underlying condition, has the effect of causing bron-chospasm, which can be life threatening. Some hypothesized that patient sensitivity to leachables in the drug product could contribute to this condition. Beyond para-doxical bronchospasm, regulators were concerned that OINDPs are often prescribed for chronic use, and therefore, patients would potentially be exposed to leachables over many years. Clinical concerns can be linked to quality control issues, such as control of the OINDP manufacturing process, the consistency of container closure system materials and components, and the control of unintended contaminants.

Schroeder added that regulatory concern and regulation of OINDP leachables have evolved over time as problems were observed in specific drug products and increased knowledge regarding component materials and manufacturing processes was acquired. The first example dates to the mid- to late 1980s and involved the observation of PNAs (PAHs) in extracts of an MDI elastomeric valve component following the detection of PNAs as leachables in the corresponding drug product. The resulting increased awareness and understanding of leachables led FDA to request that MDI manufacturers investigate an additional class of known elastomeric extractables of potential safety concern, the volatile *N*-nitrosamines. *N*-nitrosamines are trace-level reaction by-products of certain sulfur "curing agents" used in rubber vulcanization (cross-linking) processes. *N*-nitrosamines had previously been found in baby bottle rubber nipples at trace (parts per billion) levels, and had been regulated by the FDA as extractables from these components (see Reference 12 for a more detailed discussion and additional references regarding *N*-nitrosamines). Additional concern and investigation centered on 2-mercaptobenzothiazole, another rubber vulcanization reaction by-product and sometimes known rubber additive, again in MDI drug products. As knowledge and understanding built through the 1990s, concern broadened to include other classes of extractables/leachables (Table 1.1), metal component organic residues, as well as the previously mentioned issue of migration of extraneous organics through container walls. For the latter concern, Schroeder described a case study involving the migration of vanillin derived from

TABLE 1.1. Potential Sources of Extractables and Leachables from OINDP[a]

Potential sources	MDI	DPI	Inhalation solutions, suspensions, and sprays	Nasal sprays
Metal components (MDI valve components, canisters, etc.)	■	■	■	
• Residual cleaning agents, organic surface residues	■	■	■	
• Coatings on internal canister surface	■	■	■	
Elastomeric container closure system components (gaskets, seals, etc.)	■	■	■	■
• Antioxidants, stabilizers, plasticizers, and so on	■	■	■	■
• Monomers and oligomers	■	■	■	■
• Secondary reaction products from curing process	■	■	■	■
Plastic container closure system components (plastic MDI valve components, mouthpieces, plastic container material)	■	■	■	■
• Antioxidants, stabilizers, plasticizers, and so on	■	■	■	■
• Monomers and oligomers from the polymeric material	■	■	■	■
• Pigments	■	■	■	■
Processing aids, for example, chemicals applied to surfaces of processing/fabrication machinery, or directly to components	■	■	■	■
• Mold release agents	■	■	■	■
• Lubricants	■	■	■	■
Blisters or capsules containing individual doses of drug product		■		
• Chemical additives		■		
• Adhesives and glues		■		
Labels, for example, paper labels on inhalation solution plastic containers			■	■
• Inks			■	■
• Adhesives/glues			■	■

[a] Shading means that source is relevant for a given dosage form.

cardboard shipping containers through the low-density polyethylene packaging system of an inhalation solution drug product. Vanillin is associated with lignin, which is a major component of wood from which paper is derived.[15]

As knowledge of the identities and origins of extractables and leachables associated with OINDP increased, regulatory interest and concern both increased and broadened. The initial focus on PNAs in MDI drug products has now evolved into a general interest and concern regarding safety and quality control for all leachables and potential leachables in every OINDP type.

1.4 WHY DO WE NEED SAFETY THRESHOLDS?

Modern analytical chemistry has enormous capability for analyzing extractables and leachables in OINDP and other drug product types. Analytical challenges of this general type are best approached as problems in the field of trace organic analysis (TOA).[1] TOA can be defined as the qualitative and/or quantitative analysis of a complex mixture of trace level organic compounds contained within a complex matrix.[16] Solving TOA problems generally requires knowledge of the chemical nature of the analyte mixture; removal or extraction of the analyte mixture from its matrix; separation of the analyte mixture into individual chemical entities; and compound-specific detection of the individual chemical entities.[16] Analytical techniques capable of separating, detecting, identifying, and quantifying individual organic extractables and leachables include gas chromatography/mass spectrometry (GC/MS), (high-performance) liquid chromatography/mass spectrometry (LC/MS or HPLC/MS), and (high-performance) liquid chromatography/diode array detection (LC/DAD or HPLC/DAD). These advanced analytical technologies are now in routine use in pharmaceutical development laboratories (see Fig. 1.7), and have been applied to extractables/leachables problems for almost 20 years (e.g., see Norwood et al.[17] regarding analysis of PNAs in MDI drug products by GC/MS).

A GC/MS extractables "profile" from a laboratory-controlled extraction study[8] conducted on an elastomeric container closure system component material is shown in Figure 1.8. The display in Figure 1.8 is normalized to the most concentrated individual extractable. An expanded view of a similar GC/MS profile is shown in Figure 1.9. The problem faced by the OINDP pharmaceutical development scientist should now be obvious. As Figures 1.8 and 1.9 suggest, a single extractables mixture derived from a single type of container closure system component material and analyzed with a single analytical technique, can result in an extractables profile with perhaps hundreds of individual chemicals to identify and quantify. Under today's typical pharmaceutical development practice, this single mixture would be analyzed by a variety of analytical techniques as described above, resulting in several equally complex extractables profiles. Furthermore, OINDP container closure systems often contain many components with leaching potential (see Fig. 1.10). This consideration does not include the original issues of PNAs, volatile *N*-nitrosamines, and 2-mercaptobenzothiazole, which are still considered as "special case" compounds[8] by the FDA and require special scrutiny by ultrasensitive and specific analytical technologies. Given the enormity of these challenges, it is clear that a more rational approach is needed—one that tells the pharmaceutical development scientist "how low to go" in the search for extractables and leachables.

1.5 SAFETY THRESHOLDS AND THEIR APPLICATION TO LEACHABLES IN OINDP

Safety thresholds for OINDP leachables would provide a means of determining just "how low to go" in their evaluation and management, allowing the pharmaceutical development scientist to confidently identify from the full universe of leachables

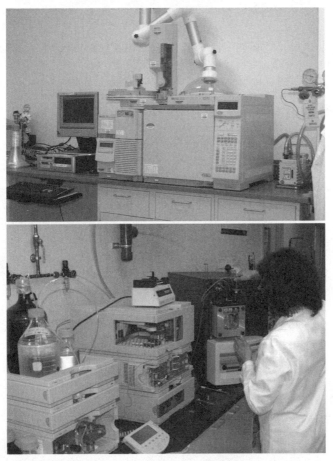

Figure 1.7 Typical GC/MS (top) and LC/MS (bottom) systems in common use in pharmaceutical development laboratories. Such systems are used to identify and quantify drug- and excipient-related impurities and metabolites, as well as extractables and leachables.

only a subset of compounds (i.e., those above a given threshold) that should undergo risk assessment and safety qualification, while still providing an ample margin of assurance that those leachables below the threshold pose no safety concern for patients. Safety thresholds have been developed for other applications where control of human exposure to specific chemicals is important. These include the thresholds for indirect food additives and International Conference on Harmonisation (ICH) thresholds for APIs and residual solvents.[4–6,18] Furthermore, it is well established that there are levels at or below which organic chemical entities in drug product represent no safety concern to patients. Therefore, the establishment of safety thresholds that are protective of patients for OINDP leachables and extractables can be justified and are believed to be necessary to limit unreasonable and extended evaluations of chemicals present at levels that cannot harm patients.

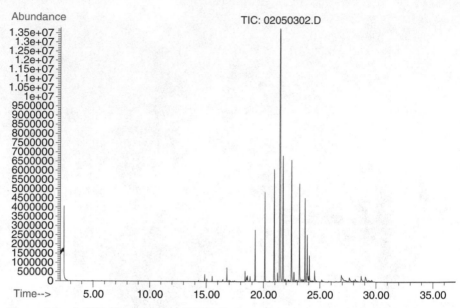

Figure 1.8 A GC/MS extractables "profile" of an elastomer (total ion chromatogram of a solvent extract).

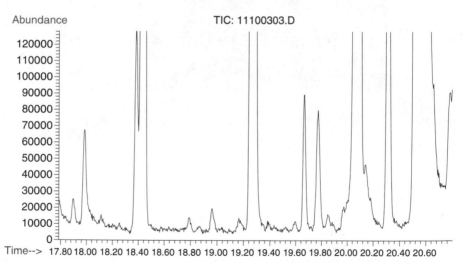

Figure 1.9 Expanded region of a GC/MS extractables "profile" of an elastomer (total ion chromatogram of a solvent extract).

Figure 1.10   Components of the container closure system of an MDI drug product capable of contributing leachables and potential leachables (i.e., extractables).

1.5.1   Context

The first MDI was introduced by Riker Laboratories in the mid-1950s.[19] At that time, there were no regulatory guidance documents that specifically focused on leachables in OINDP. From a safety perspective, however, it is important to note that general guidelines from the federal regulations were available. These explained that drug product is deemed adulterated "if its container is composed, in whole or in part, of any poisonous or deleterious substance which may render the contents injurious to health."[7,20]

As previously mentioned, leachables were treated as common impurities until the 1980s when known leachables issues (e.g., PNAs leached from carbon-black-containing elastomers) raised awareness that MDI container closure system components could affect the overall safety and quality of the drug product. Through the 1990s, the FDA became increasingly concerned about leachables issues in particular drug products. In 1999, the agency issued its guidance on container closure systems,[7] which calls for drug product manufacturers to provide information showing that the proposed container closure system and its component parts are suitable for their intended use. The type and extent of information that should be provided in an application will depend on the dosage form and the route of administration. The guidance also proposed a safety classification based on the type of drug product with the drug products of highest concern having the more stringent safety requirements (Table 1.2).

Shortly thereafter, in 1999 and 2002, FDA issued its specific guidance for pulmonary and nasal products,[10,11] addressing leachables and extractables in detail, stating that

TABLE 1.2.   Safety Characterization of Extractables for Various Routes/Dosage Forms

Route/dosage form	Safety category	Typical safety data provided
Inhalation aerosol Inhalation solution Nasal spray	Case 1s	USP Biological Reactivity Test data, extraction/toxicological evaluation, limits on extractables, batch-to-batch monitoring
Injection Suspension/powder for injection Sterile powders Ophthalmic solution/suspension	Case 2s	USP Biological Reactivity Test data; possibly extraction/toxicological evaluation

Topical delivery system
Topical solution/suspension
Topical and lingual aerosols
Oral solutions/suspensions
	Case 3s	*Aqueous-based solvents*
		    Reference to indirect food additive regulations
		Nonaqueous solvents and co-solvents
		    Reference to indirect food additive regulations
		    Additional suitability information

Topical powders Oral tablets and capsules	Case 4s	Reference to indirect food additive regulations
Inhalation powders	Case 5s	Reference to indirect food additive regulations, USP Biological Reactivity testing for mouthpiece

- the profile of *each critical component extract* should be evaluated both analytically and toxicologically;
- the toxicological evaluation should include appropriate *in vitro* and *in vivo* tests;
- a rationale, based on available toxicological information, should be provided to support acceptance criteria for components in terms of the extractables profile(s);
- safety concerns will usually be satisfied if the components that contact either the patient or the formulation meet food additive regulations and the mouthpiece meets the USP Biological Reactivity Test criteria (USP <87> and <88>); and
- if the components are not recognized as safe for food contact under appropriate regulations, additional safety data may be needed.

In 2001, in response to this guidance, the International Pharmaceutical Aerosol Consortium on Regulation and Science (IPAC-RS) and the Inhalation Technology Focus Group of the American Association of Pharmaceutical Scientists, developed a Points to Consider document proposing safety thresholds for OINDP leachables, as well as a justification for the thresholds, based on human exposure studies of inhaled particulate matter.[21] Specifically, the document proposed that qualification be performed on only those leachables that occur above data-supported thresholds (>0.2 µg total daily intake [TDI]).

1.5.2 Safety Thresholds for OINDP

At the suggestion of the FDA, and with the desire to develop a wider consensus on safety thresholds for OINDP leachables that would include regulators and other stakeholders from industry and the scientific community, IPAC-RS proposed the development of safety thresholds for OINDP as a project for the Product Quality Research Institute (PQRI).

In 2001, PQRI accepted the proposal and commenced this project.[22] At the time, there was no regulatory guidance available for drug products that applied such thresholds. The ICH thresholds for impurities are not applicable to leachables and extractables.[4–6]

The PQRI Leachables and Extractables Working Group, consisting of toxicologists and chemists from industry, FDA, and academia, developed a safety concern threshold (SCT) and a qualification threshold (QT) for leachables; an analytical evaluation threshold (AET) for extractables and leachables; processes for applying these thresholds; and best practices for selecting OINDP container closure system components and conducting controlled extraction studies, leachables studies, and routine extractables testing. These "recommendations" provided, for the first time, data-based safety thresholds for extractables and leachables in OINDP, established with a broad stakeholder consensus.[8] Furthermore, the recommendations provided a comprehensive and rationalized approach to applying these thresholds within the context of the OINDP pharmaceutical development process.

The PQRI SCT was proposed to be 0.15 μg/day, and the QT was 5 μg/day. The SCT is the threshold below which a leachable would have a dose so low as to present negligible safety concerns from carcinogenic and noncarcinogenic toxic effects. The QT is the threshold below which a given noncarcinogenic leachable is not considered for safety qualification (toxicological assessments) unless the leachable presents structure–activity relationship (SAR) concerns. Below the SCT, identification of leachables generally would not be necessary. Below the QT, leachables without structural alerts for carcinogenicity or irritation would not require compound-specific risk assessment.

The recommendations also describe how the SCT can be translated into an AET, using individual product parameters such as dose per day, actuations per canister, and so on. The AET is defined as the threshold at or above which an analytical chemist should begin to identify a particular leachable and/or extractable and report it for potential toxicological assessment. The AET allows the pharmaceutical development scientist to determine, based on safety considerations, "how low to go" in identifying and quantifying peaks in leachables and extractables profiles from OINDP. In 2006, the PQRI recommendations were submitted to the FDA for consideration in the agency's development of regulatory recommendations for OINDP.

1.6 SUMMARY

OINDPs have been available to patients for more that 50 years. Increasingly sophisticated liquid aerosol and DPIs have been developed to provide precise dosing of potent medicines to asthmatic and COPD patients. In parallel, a diverse number of elastomers and polymers have been used in the construction of these inhalers, each with unique extractables and leachables profiles. The application of thresholds such as the SCT, QT, and AET has provided scientifically justified approaches to identifying, reporting, and qualifying extractables and leachables in OINDP.

This book discusses in detail the concepts of safety-based thresholds and their application to leachables in OINDP, extractables from OINDP critical components, and concepts and approaches addressing best practices for management of extractables and leachables from OINDP and OINDP components. Part I of this book addresses development of safety thresholds and their application. Chapter 2 provides the context for safety qualification of extractables and leachables, describing the suitability for intended use requirements for materials used in pharmaceutical products and therefore describing fundamental concepts for understanding extractables and leachables and why evaluation and qualification of these compounds are so important for certain drug products, including OINDP. Background on the development and application of thresholds for various consumer products in general is provided in Chapter 3. Chapter 4 then provides details of the concepts and approaches used to develop safety thresholds for OINDP leachables. Following this, Chapter 5 provides a description of the development and application of the AET for extractables and leachables. Chapter 6 describes the history of safety qualification of OINDP extractables/leachables, from an industry perspective, and also describes, at a high level, how the safety thresholds for OINDP can be applied in the pharmaceutical development process. Chapter 7 provides further detail on the application of safety thresholds, providing case studies on how the chemist and toxicologist can collaborate in the development process to evaluate extractables and leachables, and how in specific cases, thresholds may be applied. Chapter 8 provides a perspective on the FDA's application of safety thresholds in its review of OINDP. Finally, Chapter 9 provides a regulatory perspective from Health Canada on extractables and leachables in drug products as well as the application of safety thresholds. Chapter 10 provides a detailed introduction to Part II of this book, which focuses on the aforementioned best practices.

REFERENCES

1   Hertz, H.S. and Chesler, S.N. *Trace Organic Analysis: A New Frontier in Analytical Chemistry*. NBS Special Publication 519. U.S Department of Commerce/National Bureau of Standards, Washington, DC, 1979.
2   Code of Federal Regulations. Threshold of regulation for substances used in food-contact articles. Part 21, Sec. 170.39, amended September 2000.
3   Federal Register. Volume 60, No. 136, Government Printing Office, 1995, pp. 36581–36596.
4   ICH harmonised tripartite guideline: Q3A(R2) impurities in New Drug Substances. International Conference on Harmonisation of Technical Requirements for Registration of Pharmaceuticals for Human Use, 2006.
5   ICH harmonised tripartite guideline: Q3B(R2) impurities in New Drug Products. International Conference on Harmonisation of Technical Requirements for Registration of Pharmaceuticals for Human Use, 2006.

6   ICH harmonised tripartite guideline: Q3C(R4) residual solvents. International Conference on Harmonisation of Technical Requirements for Registration of Pharmaceuticals for Human Use, 2007.

7   Guidance for industry: Container closure systems for packaging human drugs and biologics. U.S. Department of Health and Human Services, Food and Drug Administration, Center for Drug Evaluation and Research (CDER), Center for Biologics Evaluation and Research (CBER), 1999.

8   Norwood, D.L. and Ball, D. Product Quality Research Institute: Safety thresholds and best practices for extractables and leachables in orally inhaled and nasal drug products. Submitted to the PQRI Drug Product Technical Committee, PQRI Steering Committee, and U.S. Food and Drug Administration by the PQRI Leachables and Extractables Working Group, 2006.

9   Ball, D., Blanchard, J., Jacobson-Kram, D., McClellan, D.R., McGovern, T., Norwood, D.L., Vogel, M., Wolff, R., and Nagao, L. Development of safety qualification thresholds and their use in orally inhaled and nasal drug product evaluation. *Toxicol Sci [Online]* 2007, *97*(2), pp. 226–236.

10 Draft guidance for industry: Metered dose inhaler (MDI) and dry powder inhaler (DPI) drug products. Department of Health and Human Services, Food and Drug Administration, Center for Drug Evaluation and Research (CDER), 1998.

11 Guidance for industry: Nasal spray and inhalation solution, suspension, and spray drug products—Chemistry, manufacturing, and controls documentation. Department of Health and Human Services, Food and Drug Administration, Center for Drug Evaluation and Research (CDER), 2002.

12 Norwood, D.L., Granger, A.T., and Pakiet, D.M. *Encyclopedia of Pharmaceutical Technology*, 3rd ed. Dekker Encyclopedias, New York, 2006; 1693–1711.

13 Schroeder, A.C. Leachables and extractables in OINDP: An FDA perspective. Presented at the PQRI Leachables and Extractables Workshop, Bethesda, Maryland, December 5–6, 2005.

14 Medline Plus. U.S. National Library of Medicine. Available at:http://www.merriam-webster.com/medlineplus/bronchospasm (accessed September 2, 2011).

15 Norwood, D.L. Aqueous halogenation of aquatic humic material: A structural study. PhD Dissertation, University of North Carolina, Chapel Hill, NC, 1985.

16 Norwood, D.L., Nagao, L., Lyapustina, S., and Munos, M. Application of modern analytical technologies to the identification of extractables and leachables. *Am Pharm Rev* 2005, *8*(1), pp. 78–87.

17 Norwood, D.L., Prime, D., Downey, B.P., Creasey, J., Sethi, S.K., and Haywood, P. Analysis of polycyclic aromatic hydrocarbons in metered dose inhaler drug formulations by isotope dilution gas chromatography/mass spectrometry. *J Pharm Biomed Anal* 1995, *13*(3), pp. 293–304.

18 Code of Federal Regulations. Threshold of regulation for substances used in food contact articles. Part 21, Sec. 170.39, amended September 2000.

19 Anderson, P. History of aerosol therapy: Liquid nebulization to MDIs to DPIs. *Respir Care [Online]* 2005, *50*, pp. 1139–1150.

20 United States Food, Drug and Cosmetic Act, Section 501(a)(3). United States Congress, amended through December 31, 2004.

21 ITFG/IPAC-RS CMC Leachables and Extractables Technical Team. Leachables and extractables testing: Points to consider, 2001.

22 Product Quality Research Institute Leachables and Extractables Working Group. Development of scientifically justifiable thresholds for leachables and extractables, 2002.

CHAPTER 2

A GENERAL OVERVIEW OF THE SUITABILITY FOR INTENDED USE REQUIREMENTS FOR MATERIALS USED IN PHARMACEUTICAL SYSTEMS

Dennis Jenke

2.1   INTRODUCTION

Pharmaceutical products are those products that produce a desirable therapeutic outcome when they are administered to a subject to address an issue related to health. To produce the desired therapeutic outcome, pharmaceutical products must be manufactured, stored, and administered (delivered). Systems that accomplish these objectives, such as manufacturing suites, packaging, and devices, have very specific and exacting performance requirements. Such performance requirements are met due to the systems' design and because of their materials of construction. Such performance requirements are, at times, most effectively met by rubber and plastic materials, and thus, it is not surprising that both rubber and plastic materials are widely used in the pharmaceutical industry.

Pharmaceutical products are formulated, and administration regimens are developed, to maximize the therapeutic benefit derived from the product. Any action that modifies the formulation's composition can, either directly or indirectly, adversely impact the derived benefit. One such action is the contact that occurs between the pharmaceutical product and its associated systems while the system is performing its function. Contact between the product and its associated system provides the opportunity for an interaction to occur between the product and the system's materials of construction, including rubber and plastic. The result of such an interaction could be a meaningful change in the product or, less frequently, the

---

Leachables and Extractables Handbook: Safety Evaluation, Qualification, and Best Practices Applied to Inhalation Drug Products, First Edition. Edited by Douglas J. Ball, Daniel L. Norwood, Cheryl L.M. Stults, Lee M. Nagao.

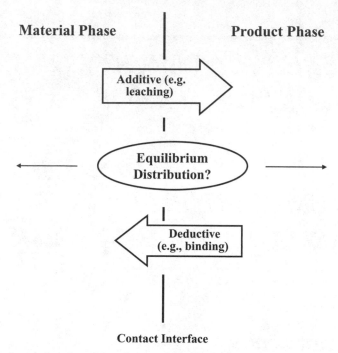

Figure 2.1   Interactions between a therapeutic product and a material (plastic) phase. Such interactions include additive process such as leaching, the migration of material-related components into the product, and deductive processes such as binding, the sorption of product ingredients by the material. Both processes impact the drug product's final composition at its time of use and thus its safety and/or efficacy. Note: The arrows denote the direction of solute movement. The oval represents a solute molecule, which can end up in either phase at equilibrium.

system. While a change in the pharmaceutical product can be manifested in many different ways, in all cases the root cause of the observed effect is that the product's composition has changed as a result of the interaction. This change in the product's composition could impact its ability to produce the desired therapeutic outcome (i.e., its suitability for its intended use). Such a change in the product's composition could be additive, in which case a substance from the system would accumulate in the pharmaceutical product, or it could be deductive, in which case an ingredient in the pharmaceutical product would be taken up by the system (Fig. 2.1). In the case of an additive interaction, the suitability for use issue for the pharmaceutical product is that the added substance could exert an undesirable influence on, or could impart an undesirable characteristic to, the pharmaceutical product. Examples of such undesirable influences or characteristics include the following:

- reduction in product stability,
- alteration of the product's impurity profile,
- formation of extraneous (e.g., particulate) matter,

- inactivation of active ingredients,
- failure to meet established product quality standards,
- development of undesirable aesthetic effects (e.g., smell, taste, discoloration, clarity),
- increase in the risk that product use would adversely affect the health and/or well-being of the user, and
- interference with product testing.

Considering an additive interaction further, one recognizes that an interaction that is additive to the pharmaceutical product is deductive to the contacted systems. Thus, the suitability for use issue for the system is that the loss of its additives may have an undesirable impact on the stability, integrity, and/or performance of the system.

In a deductive interaction, an ingredient of the pharmaceutical product is taken up by the system. If the lost ingredient is the active drug substance, then the relevant suitability for use consideration is the product's potency and efficacy. If the lost ingredient is an excipient (product component that does not produce the therapeutic effect), then the relevant suitability for use consideration is the product's physical or chemical stability.

Both additive and deductive interactions between pharmaceutical products and their associated systems are well documented in the literature. The knowledge that such interactions can and do occur and that they can and do have documented suitability for use consequences has lead to an increased awareness of this issue in the pharmaceutical community and is the driving force behind regulations designed to ensure that suitability for use issues are readily and universally recognized, appropriately investigated, and properly assessed.

2.2 AN OVERVIEW OF THE ISSUE OF SUITABILITY FOR INTENDED USE

The generation of safe and effective products is an obligation for any organization in the pharmaceutical market. To facilitate the industry's effort to live up to this obligation, various government regulatory authorities have provided guidance that enumerates the nature of the issues involved, establishes general and high-level expectations in terms of how the issues are to be assessed, and provides some insights into the strategies and tactics that would be used in such an assessment. Regulatory agencies in the United States and European Union (EU) have issued guidance and guidelines to specifically address packaging (container closure) systems (and their materials of construction) used for pharmaceutical products. The relevant document in the United States is the United States Food and Drug Administration (FDA) Guidance for industry: Container closure systems for packaging human drugs and biologics.[1] In this document, the FDA establishes the concept of "suitable for its intended use." Specifically, in section II.B.1 of the guidance, the FDA noted that "every proposed packaging system should be shown to be suitable

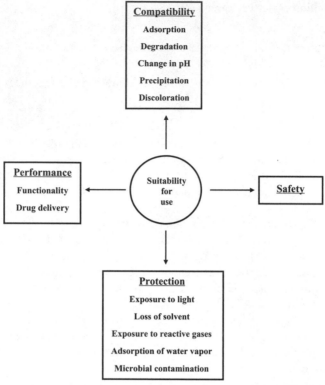

Figure 2.2   Dimensions of suitability of intended use. Abstracted from the FDA Guidance for industry: Container closure systems for packaging human drugs and biologics.[1]

for its intended use." The guidance goes on to establish four aspects of suitability for use (Fig. 2.2):

- protection,
- compatibility,
- safety, and
- performance.

The guidance (and this chapter) considers each of these aspects in somewhat greater detail.

2.2.1   Protection

The guidance notes that "a container closure system should provide the dosage form with adequate protection from factors (e.g., temperature, light) that can cause a degradation in the quality of the dosage form over its shelf-life." Common causes of degradation that are specifically identified in the guidance include exposure to light, loss of solvent, exposure to reactive gases, absorption of water vapor, and microbial contamination. Of these causes, the last four are clearly relevant to closures and seals, and, by inference, to rubber and plastic parts that perform these functions.

2.2.2 Compatibility

A packaging system that is compatible with a dosage form "will not interact sufficiently to cause unacceptable changes in the quality of either the dosage form or the packaging component." Examples of interactions that can change quality include the following:

- loss of potency due to adsorption or absorption of the active drug substance;
- loss of potency due to degradation of the active drug substance induced by a chemical entity leached from the packaging system;
- reduction in the concentration of an excipient due to adsorption, absorption, or leachable-induced degradation;
- precipitation;
- changes in drug product pH;
- discoloration of either the dosage form or the packaging component; and
- increase in brittleness of the packaging component.

One noted that these interactions include the additive and deductive processes discussed previously.

2.2.3 Safety

The guidance notes that "packaging components should be constructed of materials that will not leach harmful or undesirable amounts of substances to which a patient will be exposed when being treated with the drug product." This requirement is very specifically linked to components that have both direct and indirect contact with the drug product and is therefore relevant to rubber and plastic components that are either outside the fluid path of a delivery device or are "protected" from direct solution contact due to the construction or configuration of the packaging system.

2.2.4 Performance

Performance of the container closure system refers to its ability to function in a manner for which it was designed. The guidance identifies two major considerations with respect to performance, system functionality, and drug delivery. System functionality reflects the concept that the system may, due to its design or construction, perform a function other than the obvious. For example, it is obvious that a packaging system must contain the drug product, in which case one would interpret the requirement as "no leakers." However, one could envision a multidose packaging system that includes a component that is designed to count the number of doses that have been delivered. The suitability for use performance requirement in that particular case would be phrased as "the counter provides an accurate assessment of the number of doses delivered."

The second aspect of performance, drug delivery, refers to the ability to deliver the dosage form in the amount, or at the rate, described in the package insert (e.g., a combination of a product description and operating manual that is included with the drug product). For example, consider the case of a syringe with a faulty plunger. If the fault is such that the plunger can only move so far down the barrel, then the amount of drug delivered is less than the total fill volume of the syringe and potentially less than the minimum volume required to produce the desired therapeutic outcome. Another example is a "sticky" plunger. If the contents of the syringe are dispensed via use of a syringe pump, the increased "stickiness" of the plunger may be sufficient that the pump is unable to produce the required plunger movement, once again resulting in the delivery of a suboptimal dose.

The regulatory requirements for the products marketed in the EU are captured in the European Medicines Agency's (EMEA) Guideline on Plastic Immediate Packaging Materials.[2] While there are clear and meaningful differences in the scope and specifics of the U.S. and EU guidance documents, the EU guidelines are very much in line with the suitability for intended use concepts in general and with the four dimensions of suitability for use enumerated in the FDA guidance in particular. The EMEA guidelines deal very specifically with the dimensions of safety and certain aspects of compatibility (primarily drug sorption and altered drug degradation) and consider the dimensions of protection and performance more by inference than substantive text.

2.3   ADDITIVE INTERACTIONS

Although all four dimensions of suitability for use are important, a consideration of all four dimensions of suitability for use and both classes of interactions, (additive and deductive) is beyond the scope of this chapter, which heretofore will focus on additive interactions and their associated suitability dimensions. As noted previously, an additive interaction is one in which the migration of an entity out of the system results in the accumulation of that entity in the therapeutic product. In the simplest case, the entity that migrates out of the system was an intentional ingredient (additive) of the system and the entity that accumulates in the therapeutic product is the same entity that migrated out of the system. However, given the complex and "stressful" processes that occur when either a system is manufactured from its component raw materials or the system and therapeutic product are in contact (and may interact), it is often the case that the relationship between "what was put into the system" and "what is present in the product" is not clear and direct.

The topic of how to perform an efficient, effective, and rigorous impact assessment for additive interactions is one of considerable debate within the pharmaceutical industry and between the industry and its regulators. This is the case because while the statement of the problem is deceptively simple, the mechanics of solving the problem are quite complex. Simply stated, if a substance can only affect a product's suitability for use if it is present in the product, then the most direct and

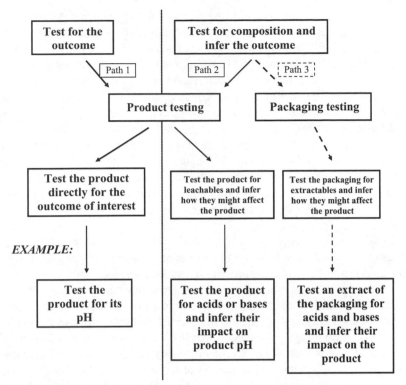

Figure 2.3   Possible means of performing a suitability for use assessment for additive interactions. The example being considered is "Does packaging the drug product affect its pH?"

straightforward means to perform a suitability for use assessment is to contact the product and its system under typical conditions of use and either (1) monitor outcomes (i.e., directly measure the effect that the contact has on the product or its user) or (2) test the potentially affected product directly for added substances and assess the outcome based on the probable impact of the added substances (Fig. 2.3).

Many suitability for use dimensions and aspects are well suited for the "monitor outcomes" approach. Thus, for example, incompatibilities, such as pH change, discoloration, precipitate formation, and issues with protection and performance can readily be assessed by the "*contact* the product and system, and *monitor* the effect" approach. In theory, a contact and monitor approach such as a clinical trial could address several aspects of suitability for use. Because the packaged product is actually used in the clinical setting during such a trial, suitability of use dimensions such as functionality are addressed. Because the product is actually administered to subjects, the subject's responses to the potentially impacted product can, in theory, be observed, measured, and interpreted in the context of suitability for use (specifically safety and efficacy).

The clinical trial approach is rarely, if ever, used as a means of establishing the suitability for use of a packaging system due to practical and economic factors whose discussion is well beyond the scope of this chapter. Although other types of "contact and monitor" studies can be effective in establishing suitability for use aspects such as compatibility, such testing is not diagnostic in the case that an incompatibility is uncovered. Additionally, "contact and monitor" studies carry considerable risks if they are performed with no "up-front insurance" for a positive outcome. That is to say, since contact and monitor studies can be extensive (and expensive), it is prudent to perform such a study only after some information has been obtained up-front that suggests that a positive outcome is likely. Furthermore, if a negative outcome is obtained (e.g., the product is found to be unsuited for its intended use), then it is typically the case that a root cause analysis is performed. "Contact and monitor" studies, while they may reveal an issue, generally produce little, if any, information that would be relevant for root cause analysis and thus additional testing would be required to complete such an analysis.

In those cases where no efficient and/or effective "contact and monitor" methods exist, the only viable means of addressing the suitability for use issue would be to *characterize* the contacted product for added substances and to *interpret* the results in the context of the probable effect of these substances. Testing of the contacted product for added substances is attractive as a means for performing suitability for use assessments because it can directly establish suitability for use, it can provide information with which to diagnose suitability for use failures revealed by other means, and it can provide some degree of "insurance" for successful outcomes in "contact and monitor" studies. The success of testing contacted product depends on the ability to actually accomplish the testing and the ability to interpret the results in the context of potential suitability for use issues. This situation can be understood via a simple example. Let us suppose that an investigator wants to assess the effect of the interaction of the product and the system on the product's pH (an aspect of the compatibility dimension of suitability for use). This can be accomplished by analyzing the contacted drug product for entities that could influence pH (like acids and bases). If the investigator could, in fact, make the required measurements and then correlate the concentrations of the individual acids and bases to product pH, the objective would be realized.

It is readily observed that this example is overly simplistic because a "better" way to approach the issue would be to just measure the product's pH after contact. However, what if the pH is "out of specification?" The "out of specification" result would undoubtedly be investigated, most likely by characterizing the product for acids and bases. In this case, then, the actual pH measurement is only the start of the investigation process. Additionally, what if the suitability for use dimension cannot be readily measured itself? While pH is a relatively simple, straightforward, and inexpensive analytical measurement, similarly simple and inexpensive test methods for other suitability for use dimensions such as safety do not exist. Simply stated, how would one determine the safety of a product that has been contacted by packaging with a test method as simple and straightforward as a pH measurement? It is the author's experience that there are few, if any, biological/biochemical tests that are clearly and definitively demonstrative of product safety and which can be performed on the actual drug product. In this case, the "contact and monitor"

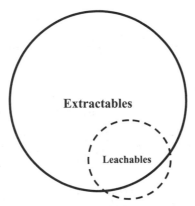

Figure 2.4   The relationship between extractables and leachables. Although these two populations of entities typically share members in common, with leachables being a subset of extractables, there are many reasons and many cases where extractables ≠ leachables and leachables ≠ extractables.

approach is simply not viable and the "characterize and interpret" approach is the only workable option.

The phrase "characterize the contact product for added substances" is misleading in that it implies that this can be accomplished only by testing the product. While it is certainly the case that testing of the product is one way to accomplish this objective, another way can be envisioned if one modifies the statement of the problem. If one changes the statement of the problem from "what is actually in the product" to "what is in the system that could potentially go into the product," then one realizes that characterizing the system for extractable substances is a potential alternative to testing the product for what has leached into it.

At this point in the discussion, it becomes clear that the investigator has two choices in terms of the target of his or her testing, either the product or the packaging, and thus is faced with two populations of potential analytes of interest. These populations are those substances, derived from the packaging, that are present in the product and those substances present in the packaging which could migrate from the packaging and become present in the product. Although these two populations may be closely related (Fig. 2.4), there can be clear differences between them, and thus the terms extractables and leachables were adopted to reflect the populations and emphasize their differences. Working definitions of these two terms follow:

Leachables. Those substances that are present in the therapeutic product due to its contact with a material, component, system, and so on.

Extractables. Those substances that are present in the material, component, system, and so on, that can be extracted from that material by a solvent.

The relationship between an extractable and a leachable is illustrated in Figure 2.5. As the object that is extracted comes closer to the product use system and as

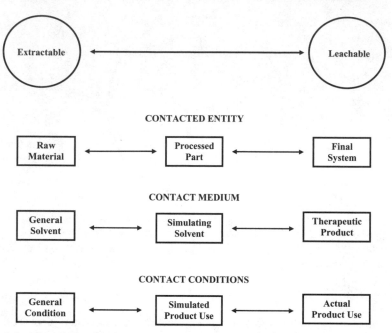

Figure 2.5   The relationship between extractables and leachables. As the contacted entity comes closer to the finished system, as the contact medium comes closer to the therapeutic product, and as the conditions of contract come closer to actual product use, then extractables become closer to leachables.

the extraction conditions used to generate the test sample come closer to the actual conditions of product use (composition of the drug product and actual product use), the population of extractables become closer to being the population of leachables.

On the surface, it seems logical that the "best" and most direct "characterize and interpret" approach is to test the therapeutic product for leachables, as opposed to testing the system for extractables. This is true since testing the product for leachables produces the exact data that needs to be interpreted (i.e., what is actually in the product that could affect its suitability for use), while testing the system for extractables still leaves the question "to what extent will these extractables accumulate in the final product?" Despite this "logic," it is rarely the case that the suitability for use assessment starts with scouting of the finished product for leached substances. The primary reason that this is the case is the complexity of the analytical task involved in such a scouting process. In many cases, the finished drug product contains the active ingredient and multiple formulation components at relatively high concentrations (vs. the leached substances). The finished drug product will also contain impurities and decomposition products associated with these primary ingredients. The analytical challenge in performing a leachables assessment is to uncover, identify, and quan-titate "unknown" leachables (i.e., leachables whose identity cannot be established up-front) in trace quantities in the complex formulation matrix. For organic leachables in particular, such an analytical challenge can only be met with extensive (and expensive) analytical testing, which may, or may not, be successful in terms of meeting its objective.

The analytical challenge of "finding a needle in the haystack when you don't even know what the needle looks like" is greatly simplified if one is given the probable identity of the needle. In the case of leachables testing, this means that it is far easier to determine if a sample contains a known (or suspected) compound than to determine if the sample contains any "unknown" compounds. The key to the approach of "finding knowns" is establishing the list of potential leachables up-front. One means of accomplishing this is to perform extractables testing of the system. In this case, the extractables profile of the system establishes what the probable leachables are. Test methods and procedures can be developed and implemented to specifically determine which of the leachables targets (i.e., extractables) do accumulate in the product in measurable quantities.

Extractables, versus leachables, testing is also relevant in other facets of suitability for use testing. For example, to this point in the discussion, suitability for use testing has been presented as a one-time event, where one establishes the system's suitability for use once, and then it is assumed that the system remains suitable for use throughout its lifetime. It is clear, however, that the system will change over the course of its lifetime, if for no other reason than different lots of its raw materials will be utilized to produce the system over time. It is reasonable to anticipate that there might be circumstances where it is necessary to control, or demonstrate control of, the effective lot-to-lot variation in a system material on the product's leachables profile. This objective can only be met by rigorous batch-to-batch testing. While it makes "logical" sense that such testing should be leachables analysis of the finished product, there is an important practical consideration that makes this logical choice inappropriate. Quality control (QC) by testing the finished product suffers the significant practical issue that the test result is obtained after considerable value has been added to the product. The specter of having to throw away a batch of product because it did not meet a leachables QC specification is an unfortunate one that can be avoided if the QC testing involves extractables testing of incoming raw materials. For example, let us say that a leachable must be present in the finished product at a level less than X parts per million for the product to be suitable for use. If it is possible to quantitatively correlate leachable "a" with extractable "b" from a particular system raw material, then the level of leachable "a" in the finished product can be "controlled" by controlling the concentration of extractable "b" in the raw material. If the QC testing involves testing of incoming raw materials, QC issues are surfaced very early in the manufacturing process, before any value has been added. In this case, the cost of a QC failure is greatly reduced and the time with which to address a QC failure is greatly increased.

Finally, there are certain instances where it is useful, necessary, or required that the system and its components be "characterized." An extractables assessment is a material characterization tool; a leachables assessment is not.

2.4 UTILIZATION OF RUBBER AND PLASTIC MATERIALS IN PHARMACEUTICAL SYSTEMS: OPPORTUNITIES AND ISSUES

The use of elastomers and polymeric materials in the medical industry dates back to the early years of the rubber and plastics industries themselves. The potential utility of elastomers as components of packaging and delivery devices was recognized shortly after the discovery of the vulcanization process. The unique properties of processed rubber, including elasticity, penetrability, resiliency, ability to act as a gas/vapor barrier, and general chemical compatibility, were the driving force behind its ready adoption in early 20th-century pharmaceutical applications (primarily as closures for glass vials), and it is the fact that these properties are largely unmatched by today's polymers and plastics that ensures rubber's continued use in modern pharmaceutical practice (closures, o-rings, plungers, seals, etc.).

Another important application of plastics in the pharmaceutical industry involves primary packaging of solution products. Up until the early 1970s, pharmaceutical solutions (and to some extent powders) were either packaged in glass containers and/or admixed (reconstituted) at time of use. The introduction of flexible container systems for IV products and blood storage revolutionized pharmacy practice, irreversibly changed the pharmaceutical marketplace, and set the stage for innovations in packaging design that continue today, for example, polyolefin-based flexible packaging systems, plastic vials, and prefilled syringes.

In the case of both rubber and plastic materials, it is an unfortunate circumstance that these desirable properties and performance characteristics are not intrinsic to the base materials themselves (e.g., base rubber or plastic monomers and polymers) but rather are imparted to the base materials via their chemical modification. To produce the actual materials that are used in pharmaceutical applications, the base materials are combined and/or reacted with a number of chemical agents (e.g., vulcanizing agents, accelerators, activators, plasticizers, tackifiers, colorants, fillers, antioxidants, lubricants) under "harsh" conditions of high temperature and pressure. These substances, their impurities, and their processing-induced reaction/decomposition products are all potential "participants" in an additive interaction between a system consisting of an elastomeric or plastic part and the pharmaceutical product with which the system is utilized. Considering the conditions of contact, which can include elevated temperatures, long contact times and "aggressive" pharmaceutical products (e.g., products whose composition is such that they are effective "solubilizing agents"), and the nature of the elastomer or plastic (e.g., relatively high amounts of numerous additives, some of which are poorly bound by the base material and some of which are "exposed" to the pharmaceutical product because they have "bloomed" to the contact surface), it not surprising that consequential interactions between elastomeric parts and pharmaceutical products occur with some regularity.

Incompatibility issues associated with the use of rubber closures in pharmaceutical products were observed in the "early days" of rubber utilization. For example, the loss of preservatives and stabilizers such as cresol or phenol from pharmaceutical products was reported as early as 1923 and was the subject of extensive investigation in the 1950s.[3–8] Issues such as haze formation and leaching of zinc from closures and syringe plungers were quantitatively investigated in the mid-1950s as viable analytical methodologies were developed.[9,10] The identification of 2-(methylthio)benzothiazole in water extracts of plungers from disposable syringes was reported in 1965.[11] In a review published in 1966, Capper discussed various types of interactions between rubber and medicants, including the deposition of particulates into the drug product, the adsorption of preservatives and medicants, "yielding into the solution the various materials added as accelerators or antioxidants, or materials derived from vulcanizing agents, and water absorption."[12] Included in this review is the report of the formation of a stearate-containing deposit in eyedrops due to leaching of these substances from the rubber component, and of the "deactivation" of penicillin by rubber tubing by mercaptans leached from the tubing. The utilization of a Teflon liner to retard the leaching of extractives was reported by Lachman et al. in 1964.[13]

The development of "modern" chromatographic and spectroscopic analytical methods facilitated the investigation of rubber materials and plastic packaging systems for organic extractables. The period of time between the early 1970s to the present was one of active research in this area for both rubber materials[14–32] and plastic packaging systems.[33–49]

Given the long and active history of investigation into rubber and product interactions, one might conclude that in the current environment all the issues have been resolved and that rubber and plastic materials used in pharmaceutical applications are inherently and eminently suitable for use. While it is certainly the case that increased awareness of, and knowledge about, suitability for use issues has driven developments in rubber and plastic composition, processing, and utilization, such a desirable state of affairs has not been fully realized. On one hand, the development of the perfectly suitable rubber is limited by the fact that there are limited choices in terms of material composition and processing. The dual requirements of functionality and suitability are, to some extent, inherently mutually exclusive, and thus, there is a limited amount of "wiggle room" in the composition and processing design space that defines a viable product. On the other hand, pharmaceutical products, especially biopharmaceuticals, are becoming more compositionally complex and "sensitive" to perturbations related to material interactions. The juxtaposition of these two trends means that rubber–plastic/product interactions are still an important product design consideration and constraint.

This observation is supported by recently reported adverse events that have been associated with rubber/product interactions. For example, one of the most widely documented instances of an unanticipated incompatibility between a rubber component of a container closure system and a protein drug product is that of EPREX® (epoetinum alfa) and its prefilled syringe packaging system.[50–52] At some point in its product lifetime (ca. 1998), EPREX, a product of recombinant human erythropoietin, was reformulated with polysorbate 80, which replaced human serum albumin as a formulation stabilizer. Shortly after this change, the incidence of antibody-mediated pure red cell aplasia (PCRA) with EPREX used by chronic renal failure patients increased. The cause of PCRA was directly linked to the formation of neutralizing antibodies to both recombinant and endogenous erythropoietin in patients administered EPREX. A considerable, cross-functional technical effort was undertaken to establish the root cause of this phenomenon. One potential root cause involved leached substances. The presence of previously unidentified leachables was suggested as new peaks in the tryptic map of EPREX. Leaching studies determined that the polysorbate 80 extracted low levels of vulcanizing agents (and related substances) from the uncoated rubber components of the container closure system (prefilled syringe). This leaching issue was addressed by replacing the rubber components with components coated with a fluoropolymer. As the fluoropolymer is an effective barrier to migration, the leaching of the rubber's components was greatly reduced. Since the conversion from the uncoated to the coated components, the incidence of PRCA has returned to the baseline rate seen for all marketed epoetin products. This is strong circumstantial evidence that leaching of the vulcanizing agent was, in fact, the root cause of the observed effect.

Additional recent examples of rubber-related product issues include a 2006 recall of Isoket (50-mL vials) by Schwarz Pharmaceuticals in the United Kingdom, where the company said that 50-mL vials of the drug had been contaminated by an impurity from the rubber stopper and the 2008 report by Cubist Pharmaceuticals that mercaptobenzothiazole (MBT) was present in its reconstituted Cubicin injectable antibiotic that had been stored in disposable, single-use ReadyMED infusion pumps.

While the ultimate objective of having cost-effective, broadly applicable, functional, and suitable (inert) pharmaceutical rubber and plastic materials has only been partially realized, ongoing developments and future innovations in rubber and plastic composition and material compounding, processing, and utilization have the potential to close the gap between the utopia of tomorrow and the reality of today.

REFERENCES

1 Guidance for industry: Container closure systems for packaging human drugs and biologics. U.S. Department of Health and Human Services, Food and Drug Administration, Center for Drug Evaluation and Research (CDER), Center for Biologics Evaluation and Research (CBER), 1999.
2 European Medicines Agency Inspections. Guideline on plastic immediate packaging materials. Committee for Medicinal Products for Human Use (CHMP), Committee for Medicinal Products for Veterinary Use. London (CVMP), 2005.
3 Masucci, P. and Moffat, M. The diffusion of phenol and tricresol through rubber. *J Pharm Sci* 1923, *12*, p. 117.
4 McGuire, G. and Falk, K.G. The disappearance of phenols and cresols added to "biological products" on standing. *J Lab Clin Med* 1937, *22*, p. 641.
5 Wiener, S. The interference of rubber with the bacteriostatic action of thiomersalate. *J Pharm Pharmacol* 1955, *7*, p. 118.
6 Berry, H. Pharmaceutical aspects of glass and rubber. *J Pharm Pharmacol [Online]* 1953, *5*, p. 1008.
7 Wing, W.T. An examination of rubber used as closures for containers of injectable solutions (part III). Effect of the chemical composition of the rubber mix on phenol and chlorocresol absorption. *J Pharm Pharmacol* 1956, *8*, p. 738.
8 Lachman, L., Weinstein, S., Hopkins, G., Slack, S., Eisman, P., and Copper, J. Stability of antibacterial preservatives in parenteral solutions I: Factors influencing the loss of antimicrobial agents from solutions in rubber-stoppered containers. *J Pharm Sci* 1962, *51*, p. 224.
9 Reznek, S. Rubber closures for containers of parenteral solutions I: The effect of temperature and pH on the rate of leaching of zinc salts from rubber closures in contact with (acid) solutions. *J Am Pharm Assoc* 1953, *42*, p. 288.

10   Milosovich, G. and Mattocks, A.M. Haze formation of rubber closures for injections. *J Am Pharm Assoc* 1957, *46*, p. 377.

11   Inchiosa, M.A., Jr. Water-soluble extractives of disposable syringes: Nature and significance. *J Pharm Sci* 1965, *54*, p. 1379.

12   Capper, K.R. Interaction of rubber with medicaments. *J Mondial de Pharmacie* 1966, *9*, p. 305.

13   Lachman, L., Pauli, W.A., Sheth, P.B., and Pagliery, M. Lined and unlined rubber stoppers for multiple-dose vial solutions II: Effect of Teflon lining in preservative sorption and leaching of extractives. *J Pharm Sci* 1966, *55*, p. 962.

14   Chrzanowski, F., Niebergall, P.J., Mayock, R., Taubin, J., and Sugita, E. Interference by butyl rubber stoppers in GLC analysis for theophylline. *J Pharm Sci* 1976, *65*, p. 735.

15   Petersen, M.C., Vine, J., Ashley, J.J., and Nation, R.L. Leaching of 2-(2-hydroxyethylmercapto) benzothiazole into contents of disposable syringes. *J Pharm Sci* 1981, *70*, p. 1139.

16   Nation, R.L. Leaching of a contaminant into the contents of disposable syringes. *Aust N Z J Med* 1981, *11*, p. 208.

17   Danielson, J.W., Oxborrow, G.S., and Placencia, A.M. Chemical leaching of rubber stoppers into parenteral solutions. *J Parenter Sci Technol* 1983, *37*, p. 89.

18   Danielson, J.W., Oxborrow, G.S., and Placencia, A.M. Quantitative determination of chemicals leached from rubber stoppers into parenteral solutions. *J Parenter Sci Technol* 1984, *38*, p. 90.

19   Reepmeyer, J.C. and Juhl, Y.H. Contamination of injectable solutions with 2-mercaptobenzothiazole leached from rubber closures. *J Pharm Sci* 1983, *72*, p. 1302.

20   Salmona, G., Assaf, A., Gayte-Sorbier, A., and Airaudo, C.B. Mass spectral identification of benzo-thiazole derivatives leached into injections by disposable syringes. *Biomed Mass Spectrom* 1984, *11*, p. 460.

21   Wells, C.E., Juenge, E.C., and Wolnik, K. Contaminants leached from rubber stoppers into a water-soluble vitamin E intravenous injectable product. *J Pharm Sci* 1986, *75*, p. 724.

22   Jaehnke, R.W.O., Kreuter, J., and Ross, G. Interaction of rubber closures with powders for parenteral administration. *J Parenter Sci Technol* 1990, *44*, p. 282.

23   Jaehnke, R.W.O., Linde, H., Mosandl, A., and Kreuter, J. Contamination of injectable powders by volatile hydrocarbons form rubber stoppers. The C13-oligomer and determination of its structure. *Acta Pharm Technol* 1990, *36*, p. 139.

24   Lasko, J., Jakubik, T., and Michalkova, A. Gas chromatographic-mass spectrometric detection of trace amounts of organic compounds in the intravenous solution Infusio Darrowi. *J Chromatogr* 1992, *603*, p. 294.

25   Danielson, J.W. Toxicity potential of compounds found in parenteral solutions with rubber stoppers. *J Parenter Sci Technol* 1992, *46*, p. 43.

26   Gaind, V.S. and Jedrzajczak, K. HPLC determination of rubber septum contaminants in the iodinated intravenous contrast agent (sodium iothalamate). *J Anal Toxicol* 1993, *17*, p. 34.

27   Norwood, D.L., Prime, D., Downey, B.P., Creasey, J., Sethi, S.K., and Haywood, P. Analysis of polycyclic aromatic hydrocarbons in metered dose inhaler drug formulations by isotope dilution gas chromatography/mass spectrometry. *J Pharm Biomed Anal* 1995, *13*, p. 293.

28   Zhang, X.K., Dutky, R.C., and Fales, H.M. Rubber stoppers as sources of contaminants in electro-spray analysis of peptides and protein. *Anal Chem* 1996, *68*, p. 3288.

29   Paskiet, D.M. Strategy for determining extractables from rubber packaging materials in drug prod-ucts. *PDA J Pharm Sci Technol* 1997, *51*, p. 248.

30   Zhang, F., Chang, A., Karaisz, K., Feng, R., and Cai, J. Structural identification of extractables from rubber closures used for pre-filled semisolid drug applicator by chromatography, mass spectrometry, and organic synthesis. *J Pharm Biomed Anal* 2004, *34*, p. 841.

31   Castner, J., Williams, N., and Bresnick, M. Leachables found in parenteral drug products. *Am Pharm Rev* 2004, *7*(2), p. 70.

32   Xiao, B., Gozo, S.K., and Herz, L. Development and validation of HPLC methods for the determina-tion of potential extractables from elastomeric stoppers in the presence of a complex surfactant vehicle used in the preparation of parenteral drug products. *J Pharm Biomed Anal* 2007, *43*, p. 558.

33   Ulsaker, G.A. and Hoem, R.M. Determination by gas chromatography-single ion monitoring mass spectrometry of phthalate contaminants in intravenous solutions stored in PVC bags. *Analyst* 1978, *103*, p. 1080.

34 Arbin, A. and Ostelius, J. Determination by electron-capture gas chromatography of mono- and di(2-ehtylhexyl)phthalate in intravenous solutions stored in poly(vinyl chloride) bags. *J Chromatogr* 1980, *193*, p. 405.

35 Petersen, M.C., Vine, J.H., Ashley, J.J., and Nation, R.L. Stability and compatibility of 2.5 mg/mL methotrexate solution in plastic syringes over 7 days. *J Pharm Sci* 1981, *70*, p. 1139.

36 Hopkins, J.L., Cohen, K.A., Hatch, F.W., Pitner, T.P., Stevensen, J.M., and Hess, F.K. Pharmaceuticals: Tracking down an unidentified trace level constituent. *Anal Chem* 1987, *59*, p. 784A.

37 Snell, R.P. Capillary GC analysis of compounds leached into parenteral solutions packaged in plastic bags. *J Chromatogr Sci* 1989, *27*, p. 524.

38 Kim-Kang, H. and Gilbert, S.G. Permeation characteristics and extractables from gamma-irradiated and non-irradiated plastic laminates for a unit dosage injection device. *Packag Technol Sci* 1991, *4*, p. 35.

39 Sarbach, C., Yagoubi, N., Sauzieres, J., Renaux, C., Ferrrier, D., and Postaire, E. Migration of impurities from multilayer plastics container into a parenteral infusion fluid. *Int J Pharm* 1996, *140*, p. 169.

40 Jenke, D.R., Martinez, A.V., Cruz, L.A., and Zimmerman, S.R. Accumulation model for solutes leaching from polymeric containers. *J Parenter Sci Technol* 1993, *47*(4), p. 1.

41 Jenke, D.R., Jenne, J.M., Poss, M., Story, J., Tsilipetros, T., Odufu, A., and Terbush, W. Accumulation of extractables in buffer solutions from a polyolefin plastic container. *Int J Pharm* 2005, *297*, p. 120.

42 Jenke, D., Swanson, S., Edgcomb, E., Couch, T., Chacko, M., Garber, M.J., and Fang, L. Strategy for assessing the leachables impact of a material change made to a container/closure system. *PDA J Pharm Sci Technol* 2005, *59*, p. 360.

43 Depaolis, A., Zhy, L., Gunturi, S., Deng, F., Begum, S., Tolman, G., Templeman, T., and Ghobrial, I. Rapid screening of UV absorbing leachables in biologic product placebos. *Am Pharm Rev* 2006, *9*(5), p. 54.

44 Fliszar, K.A., Walker, D., and Allain, L. Profiling of metal ions leached from pharmaceutical packaging materials. *PDA J Pharm Sci Technol* 2006, *60*, p. 337.

45 Fang, X., Cherico, N., Barbacci, D., Harmon, A.M., Piserchio, M., and Perpall, H. Leachable study on solid dosage form. *Am Pharm Rev* 2006, *9*(7), p. 58.

46 Fichtner, S., Giese, U., Pahl, I., and Reif, O.W. Determination of "extractables" on polymer materials by means of HPLC-MS. *PDA J Pharm Sci Technol* 2006, *60*, p. 291.

47 Ito, R., Seshimo, F., Miura, N., Kawaguchi, M., Saito, K., and Nakazawa, H. High-throughput determination of mono- and di(20-ethylhexyl)phthalate migration from PVC tubing to drugs using liquid chromatography-tandem mass spectrometry. *J Pharm Biomed Anal* 2005, *39*, p. 1036.

48 Lennon, J.D., III, Hendricker, A.D., and Feinberg, T.N. Identifying packaging-related drug product impurities. *LC GC* 2007, *25*, p. 710.

49 Pan, C., Harmon, F., Toscano, K., Liu, F., and Vivilecchia, R. Strategy for identification of leachables in packaged pharmaceutical liquid formulations. *J Pharm Biomed Anal* 2008, *46*, p. 520.

50 Sharma, B., Bader, F., Templeman, T., Lisi, P., Ryan, M., and Heavner, G.A. Technical investigations into the cause of the increased incidence of antibody-mediated pure red cell aplasia associated with EPREX®. *Eur J Hosp Pharm* 2004, *5*, p. 86.

51 Boven, K., Knight, J., Bader, F., Rossert, J., Eckardt, K., and Casadevail, N. Epoetin-associated pure red cell aplasia in patients with chronic kidney disease: Solving the mystery. *Nephrol Dial Transplant* 2005, *20*, p. ii30.

52 Pang, J., Blanc, T., Brown, J., Labrenz, S., Villalobos, A., Depaolis, A., Gunturi, S., Grossman, S., Lisi, P., and Heavner, G.A. Recognition and identification of UV-absorbing leachables in EPREX® pre-filled syringes: An unexpected occurrence at a formulation-component interface. *PDA J Pharm Sci Technol* 2007, *61*, p. 423.

CHAPTER 3

CONCEPT AND APPLICATION OF SAFETY THRESHOLDS IN DRUG DEVELOPMENT

David Jacobson-Kram and Ronald D. Snyder

3.1 INTRODUCTION

Over 500 years ago, Paracelsus made the astute observation that "all substances are poisons; there is none which is not a poison. The right dose differentiates a poison from a remedy" (Fig. 3.1).

A corollary to this bit of wisdom is that for most or even perhaps for all toxicological effects, there exist thresholds: a dose below which an exposure imparts no risk. While most toxicologists would likely agree with this principle, the means for calculating the threshold can sometimes be controversial. In addition, for some adverse health end points, that is, mutagenesis and carcinogenesis, most regulatory agencies have assumed a lack of a threshold. Practically, this means that any exposure results in some increased risk for mutation and/or cancer. Some scientists in the field have argued against this assumption. Mammalian cells, after all, have efficient detoxification pathways to prevent DNA damage, efficient DNA repair capabilities should damage occur, and apoptotic options if damage is severe. It has been argued that in the face of all these protective mechanisms, it is unreasonable to assume a lack of threshold. Nevertheless, normal humans show approximately 8 aberrant cells per 1000 in mitogen-stimulated peripheral blood lymphocytes, and the frequency increases with age.[1] Specific locus mutations affecting gene function are seen with a frequency of approximately 10^{-6} per locus and also increase with age.[2] Nearly 1.5 million new cases of cancer are diagnosed each year in the United States.[3] Since it is not possible to empirically determine whether thresholds exist for these adverse health effects, the debate will no doubt continue.

This chapter provides an overview of risk assessment approaches that use the threshold concept and reviews some of the challenges and issues inherent in developing and applying thresholds to chemicals, especially in the case of carcinogens and mutagens.

Leachables and Extractables Handbook: Safety Evaluation, Qualification, and Best Practices Applied to Inhalation Drug Products, First Edition. Edited by Douglas J. Ball, Daniel L. Norwood, Cheryl L.M. Stults, Lee M. Nagao.
© 2012 John Wiley & Sons, Inc. Published 2012 by John Wiley & Sons, Inc.

Dosis facit venenum

„ Was ist das nit Gifft ist? Alle Ding sind Gifft und nichts ohn Gifft. Allein die Dosis macht, das ein Ding kein Gifft ist. "

Paracelsus *(1493-1541)*

Figure 3.1   Paracelsus. Courtesy of the Swiss Society of Toxicology.

3.2   THRESHOLDS AND RISK ASSESSMENT

Two types of methodologies have evolved in risk assessment: one for calculating safe exposure values for health effects thought to have thresholds and a second for calculating "virtually safe" exposure values for health effects thought to lack thresholds. In both cases, the method generally involves extrapolation of agent-induced health effects in animals to human risk.

In the area of drug development, methodologies for threshold-associated exposures are discussed in ICHQ3C(R4) (Impurities: Guideline for Residual Solvents).[4] The guideline discusses a method for calculating a PDE, "permissible daily exposure" to a toxin, in this case a residual solvent. The calculation is based on the use of a "no observed effect level" (NOEL) from an animal toxicology study. The NOEL is divided by a number of safety factors, F1 through F5. F1 is meant to account for variations in species extrapolation and is related to body surface area. For example, F1 = 2 for the dog, 12 for the mouse, and 10 for all other species. A second safety factor, F2 accounts for interindividual variability and is equal to 10. F3 is a third safety factor that compensates for studies of short-term duration. A fourth safety factor, F4 is applied in cases of severe toxicity, nongenotoxic carcinogenicity or teratogenicity. F4 = 1 for fetal toxicity associated with maternal toxicity, while F4 = 10 for teratogenic effect without maternal toxicity. Lastly, F5 is a factor applied if the no-effect level was not established; F5 = 10 when only a lowest-observed effect level (LOEL) is available, and is dependent on the severity of the toxicity. Using this algorithm, one can calculate PDEs for substances whose toxicity is thought to have a threshold.

Calculating PDEs for mutagenic and carcinogenic substances is more complex. With the exception of a few special cases, most regulatory agencies consider these toxicities to lack thresholds. Nevertheless, it is often impossible to reduce human exposures to zero. To calculate PDEs for carcinogens, risk assessors again most often use results from animal studies. However, the concept of a NOEL cannot be used for an effect that lacks a threshold. Instead, risk assessors extrapolate responses, generally tumors, from animal data at the doses used in the studies to much lower exposures that people might experience. Such extrapolation, coupled with a series of assumptions allows regulators to determine "acceptable exposures" often referred to as "virtually safe doses." A virtually safe dose has been defined somewhat differently by different regulatory agencies. However, in general, it refers to lifetime exposures that increase risk of cancer by either 1 in 1 million or 1 in 100,000.

These types of risk assessments are often used to calculate acceptable levels of carcinogens in drinking water, air, and soil at hazardous waste sites. To perform such risk assessments, data from rodent lifetime bioassays are required. However, data from such studies are not always available, and often, it is not cost-effective to perform carcinogenicity studies. This conundrum has led toxicologists to devise the concept of the "threshold of toxicological concern" (TTC).[5] These authors define the TTC as "a level of exposure for all chemicals, whether or not there are chemical-specific toxicity data, below which there would be no appreciable risk to human health."

The TTC concept provides a basis for development of specific exposure levels that can be used in safety qualification considerations for a large number of chemical compounds. An example is the "threshold of regulation" developed by the United States Food and Drug Administration (FDA) Center for Food Safety and Nutrition (CFSAN), established as 1.5 µg/person/day for food contact substances and further standardized by CFSAN in a companion guidance document for food contact substances.[6,7] In general, a food intake exposure level of 1.5 µg/person/day is considered an acceptable threshold below which further qualification for genotoxicity/carcinogenicity concerns would not be required. Substances with no known cause for concern that may migrate into food are exempted from regulation as a food additive if present at daily dietary concentrations at or below 0.5 ppb, corresponding to 1.5 µg/person/day based on a total daily consumption of 3 kg of solid and liquid foods. The threshold is an estimate of daily exposure expected to result in an upper bound lifetime risk of cancer of less than 10^{-6}, considered a "virtually safe dose." The initial CFSAN analysis was based on an assessment of 343 carcinogens from the carcinogenic potency database (CPDB) and was derived from the probability distribution of carcinogenic potencies of those compounds.[8] Subsequent analyses of an expanded database of more than 700 carcinogens further confirmed the threshold.[9] Additional analysis of subsets of highly potent carcinogens suggested that a threshold of 0.15 µg/day, corresponding to a 10^{-6} lifetime risk of cancer, may be more appropriate for chemicals with structural alerts for potential genotoxicity.[5] Some structural groups including aflatoxin-like-, N-nitroso-, and azoxy-compounds were identified to be of extremely high potency and are excluded from the threshold approach.

U.S. federal regulatory agencies such as the United States Environmental Protection Agency (EPA) and FDA typically use a 10^{-6} lifetime risk of cancer to determine "acceptable" risk from chemical exposures, although higher risk levels are accepted under certain circumstances, for example, for active pharmaceutical ingredients in which a benefit may be derived. This level of exposure is expected to produce a negligible increase in carcinogenic risk based on the analysis of the CPDB, and this approach has been proposed for regulating the presence of genotoxic impurities in drugs.[10] Additionally, this risk level is considered to be low enough to ensure that the presence of an unstudied compound that is below the resultant threshold will not significantly alter the risk/benefit ratio of a drug even if the impurity is later shown to be a carcinogen.

3.3   THRESHOLDS AND DRUG DEVELOPMENT

From a drug development perspective, the assumed presence or absence of thresholds can have very practical consequences. For example, it is not uncommon in performing the International Conference on Harmonisation (ICH)-specified genetic toxicology battery to see a positive response for the *in vitro* mammalian cell assay while the other tests in the battery are negative.[11] While there are a number of potential explanations for these results, the most common observation is that one could not achieve plasma levels of the drug *in vivo* that are comparable to the concentrations that induced the positive response *in vitro*. This does not necessarily mean there is a threshold for the response since it can be argued that an effect still occurs at the low exposure, but the test system lacks the sensitivity to demonstrate it.

Kirkland and Müller[12] discussed the concept of thresholds for genotoxic agents and the mechanisms that may be responsible. They also proposed data requirements to demonstrate a threshold. They suggested that mechanisms such as enzyme inhibition, imbalance of DNA precursors, energy depletion, production of active oxygen species, lipid peroxidation, sulfhydryl depletion, nuclease release from lysosomes, inhibition of protein synthesis, protein denaturation, and ionic imbalance are examples of pathways that can give positive responses in genetox assays, especially *in vitro* assays that are expected to yield information that could be used to develop thresholds. As outlined in Table 3.1, a three-step approach was recommended for assessing biological relevance of *in vitro* positives in mammalian cell assays.

While these are reasonable recommendations, they are often not practical in the usual course of drug development. Generally, at the time of an investigational new drug (IND) submission, relatively little is known about absorption, distribution, metabolism, and excretion (ADME) and mechanisms of toxicity of the compound. Also, since an IND generally proposes a first in human study, human exposure data are not available.

TABLE 3.1.   Assessing Biological Relevance of *In Vitro* Positives in Mammalian Cell Assays[12]

1 Provide credible mechanistic/metabolic reasons why positive *in vitro* results are not relevant to *in vivo* exposures.
2 Obtain negative results from "appropriate" *in vivo* genotoxicity assays along with appropriate exposure data.
3 Calculate margins of safety between the likely human exposure and that seen in the positive *in vitro* assay and the negative *in vivo* assay.

As mentioned above, it is not uncommon for sponsors to submit negative results from a bacterial mutation assay, negative results from a rodent bone marrow clastogenicity test, and positive or equivocal data from an *in vitro* assay for chromosomal aberrations or the mouse lymphoma assay. The clinical protocol most often specifies healthy, volunteer subjects. Clearly, in such a situation, there is no risk/benefit balance for the study participants, only risk. For healthy subjects with nothing to gain from the study, risks must be exceedingly low. A sponsor often argues that the positive *in vitro* response is due to "cytotoxicity" and therefore not relevant, and that this observation is negated by the negative *in vivo* assay. While some *in vitro* clastogens show their effects only at relatively high levels of cytoxicity, this alone is not a mechanistic explanation of the result. Many compounds can be highly cytotoxic but not clastogenic. Furthermore, the *in vivo* assay has to be viewed as relatively insensitive; it uses small numbers of animals, and there is significant interanimal variation in background frequencies of micronucleated cells; small increases are not easily detected. Under these circumstances, it is understandable why regulators might be reluctant to allow such trials to proceed.

3.4 THRESHOLDS AND GENOTOXIC EFFECTS

The presence or absence of thresholds for genotoxic effects has been investigated through a variety of approaches. The assessment of chemically induced mutagenicity thresholds *in vitro*, while clearly difficult, is at least made feasible by the relatively straightforward and measurable end point, mutation. Thus, one can plot mutation frequency against, for example, DNA adduct formation in a cell-based system and determine, within the limits of analytical sensitivity, if the resultant curve is linear and extrapolatable through zero or exhibits nonlinearity. Such exercises are informative with respect to how DNA damage is handled at a cellular level, providing insights into DNA repair processes as well as increasing our understanding of the role that specific DNA adducts play in mutagenesis. In a practical sense also, threshold information is important in that it facilitates risk estimation and can provide the basis for mechanism of action (MOA) determinations. The clear demonstration of a threshold in a chemical-induced genotoxic response suggests that cells (and therefore, tissues or organisms) have biological mechanisms in place that limit untoward chemical effects at the low end of the dose spectrum. This means that low exposures to a known mutagen, for example, would not necessarily be expected to lead to mutation.

In theory, at least, the same basic concepts should apply to the *in vitro* and *in vivo* situation, and several studies have been performed to evaluate the existence of chemically induced mutational thresholds *in vivo*. Such studies are inherently more difficult to interpret since, in addition to DNA repair processes (which themselves differ greatly from tissue to tissue), there are questions of biodistribution, metabolism of test article, and pharmacokinetics. Moreover, plasma chemical levels are usually much lower *in vivo* than can be attained *in vitro* experiments and sensitivity of analytical methods often becomes an issue.

Not surprisingly, then, evidence exists both in support of and counter to the existence of *in vivo* thresholds for genotoxicity. Using the *in vivo* HPRT gene mutation assay in lymphocytes in rats treated with ethyl methanesulfonate (EMS) or *N*-ethyl-*N*-nitrosourea (ENU), Jansen et al.[13] concluded that a clear, no-effect level could be seen with EMS, but not with ENU. However, evidence for a threshold for ENU-induced mouse spermatogonial mutations in the specific locus test was reported by Favor et al.[14] Other *in vivo* studies with benzene[15,16] or MeIQx[17-19] Mitomycin C, diepoxybutane,[20] and acrylamide[21] all failed to demonstrate a threshold.

Recently, very compelling evidence was presented for the existence of a threshold for mutation (MutaMouse assay) and micronucleus formation (bone marrow) in mice treated with EMS.[22] These studies were conducted by Roche to assess the risk to patients having taken repeated doses of batches of the antiviral agent, nelfinavir mesylate, inadvertently contaminated with relatively high levels of EMS (such that the worst-case scenario has 0.055 mg/kg EMS ingested per day at the daily dose of 2500 mg nelfinavir). Using ethylvaline adducts in hemoglobin as an internal dosimeter, it was estimated that in mice, both mutation and micronucleus formation were observed only after chronic dosing with EMS at ≥ 25 mg/kg/day. Extrapolation of animal exposure to human exposure (C_{max} analysis) demonstrated that 370 times more EMS would have had to be ingested by patients to pose a significant cancer risk. These values should be applicable to other drugs containing mesylate. Not to be overlooked as a very important additional finding in the Roche study is the fact that ENU, under the same conditions, did not appear to assume a threshold consistent with the findings of Jansen et al.[13] The reason for this may relate to the different spectra of adducts formed by these two alkylating agents, primarily O6 alkylguanine for ENU and N7 alkylguanine for EMS. Since these two adducts are repaired by different enzymes, a reasonable hypothesis might be that ENU repair saturates at low adduct density relative to that of EMS.

3.5 CONCLUSION

The development and use of safety thresholds for use in drug and food safety has matured and reached a certain level of acceptance over the last 15 or so years with the introduction of risk assessment concepts such as those used in the ICH guideline for residual solvents and the TTC. Nevertheless, development and application of safety thresholds to carcinogens and mutagens has been a particular challenge—the weight of the evidence clearly suggests the presence of thresholds for some mutagenic and carcinogenic chemicals, and a lack of a threshold for others. Furthermore, in the case of drug products, genetic toxicity tests can sometimes yield conflicting results, which may require further consideration by the sponsor, possibly with the regulatory reviewer. Thus, the choice of risk assessment methodologies for mutagens and carcinogens will have to continue to be conducted on a case-by-case basis.

REFERENCES

1   Tucker, J.D., Lee, D.A., Ramsey, M.J., Briner, J., and Olsen, L. On the frequency of chromosome exchange in a control population measured by chromosome painting. *Mutat Res* 1994, *313*, pp. 193–202.

2   Curry, J., Karnaukhova, L., Guenette, G.C., and Glickman, B.W. Influence of sex, smoking and age on human hprt mutation frequencies and spectra. *Genetics* 1999, *152*, pp. 1065–1077.

3   Jemal, A., Murray, T., Ward, E., Samuels, A., Tiwari, R.C., Ghatoor, A., Feuer, E.J., and Thun, M.J. Cancer statistics. *Cancer J Clin* 2005, *55*, pp. 10–30.

4   ICH harmonised tripartite guideline: Q3C(R4)residual solvents. International Conference on Harmonisation of Technical Requirements for Registration of Pharmaceuticals for Human Use, 2007.

5   Kroes, R., Renwick, A.G., Cheeseman, M., Kleiner, J., Mangelsdorf, I., Piersma, A., Schilter, B., Schlatter, J., van Schothorst, F., Vos, J.G., and Würtzen, G. Structure-based threshold of toxicological concern (TTC): Guidance for application to substances present at low levels in the diet. *Food Chem Toxicol* 2004, *42*, pp. 65–83.

6   Federal Register. Volume 60, No. 136, Government Printing Office, 1995, pp. 36581–36596.

7   Guidance for industry: Preparation of food contact notifications for food contact substances: Toxicology recommendations. U.S. Food and Drug Administration, Center for Food Safety and Applied Nutrition, Office of Food Additive Safety, 2002.

8   Gold, L.S., Sawyer, C.B., Magaw, R., Backman, G.M., de Veciana, M., Levinson, R., Hooper, N.K., Havender, W.R., Bernstein, L., Peto, R., Pike, M.C., and Ames, B.N. A carcinogenicity potency database of the standardized results of animal bioassays. *Environ Health Perspect* 1984, *58*, pp. 9–319.

9   Fiori, J.M. and Meyerhoff, R.D. Extending the threshold of regulation concept: De minimis limits for carcinogens and mutagens. *Regul Toxicol Pharmacol* 2002, *35*, pp. 209–216.

10 Müller, L., Mauthe, R.J., Riley, C.M., Andino, M.M., DeAntonis, D., Beels, C., DeGeorge, J., De Knaep, A.G.M., Ellison, D., Fagerland, J.A., Frank, R., Fritschel, B., Galloway, S., Harpur, E., Humfrey, C.D.N., Jacks, A.S., Jagota, N., Mackinnon, J., Mohan, G., Ness, D.K., O'Donovan, M.R., Smith, M.D., Vudathala, G., and Yotti, L. A rationale for determining, testing and controlling specific impurities in pharmaceuticals that possess potential for genotoxicity. *Regul Toxicol Pharmacol* 2006, *44*, pp. 198–211.

11 Snyder, R.D. and Green, J.W. A review of the genotoxicity of marketed pharmaceuticals. *Mutat Res* 2001, *488*, pp. 151–169.

12 Kirkland, D.J. and Müller, L. Interpretation of the biological relevance of genotoxicity test results: The importance of thresholds. *Mutat Res* 2000, *464*, pp. 137–147.

13 Jansen, J.G., Vrieling, H., van Teijlingen, C.M., Mohn, G.R., Tates, A.D., and Van Zeeland, A.A. Marked differences in the role of O6 alkylguanine in hprt mutagenesis in T-lymphocytes of rats exposed to ethylmethanesulfonate, N-(2-hydroxyethyl)-N-nitrosourea or N-ethyl-N-nitrosourea. *Cancer Res* 1995, *55*, pp. 1875–1882.

14 Favor, J., Sund, M., Neuhauser-Klaus, A., and Ehling, U.H. A dose response analysis of ethylnitrosourea-induced recessive specific locus mutations in treated spermatogonia of the mouse. *Mutat Res* 1990, *231*, pp. 47–54.

15 McDonald, T.A., Yeowell-O'Connell, K., and Rappaport, S.M. Comparison of proteins adducts of benzene oxide and benzoquinone in the blood and bone marrow of rats and mice exposed to 14C/13C6 benzene. *Cancer Res* 1994, *54*, pp. 4907–4914.

16 Creek, M.R., Mani, C., Vogel, J.S., and Turteltaub, K.W. Tissue distribution and macromolecular binding of extremely low doses of 14C benzene in B6C3F1 mice. *Carcinogenesis* 1997, *18*, pp. 2421–2427.

17 Mauthe, R.J., Dingley, K.H., Leveson, S.H., Freeman, S.P., Turesky, R.J., Garner, R.C., and Turteltaub, K.W. Comparison of DNA adduct and tissue available dose levels of MeIQx in human and rodent colon following administration of a very low dose. *Int J Cancer* 1999, *80*, pp. 539–545.

18 Turteltaub, K.W., Mauthe, R.J., Dingley, K.H., Vogel, J.S., Frantz, C.E., Garner, R.C., and Shen, N. MeIQx-DNA adduct formation in rodent and human tissues at low doses. *Mutat Res* 1997, *376*, pp. 243–252.

19 Hoshi, M., Keiichirou, M., Wanibuchi, H., Wei, M., Okochi, E., Ushijima, T., Takaoka, K., and Fukushima, S. No observed effect levels for carcinogenicity and for in vivo mutagenicity of a genotoxic carcinogen. *Toxicol Sci* 2004, *82*, pp. 273–279.

20 Grawe, J., Abramsson-Zetterberg, L., and Zetterberg, G. Low dose effects of chemicals assessed by the flow cytometric in vivo mouse micronucleus assay. *Mutat Res* 1998, *405*, pp. 199–208.

21 Abramsson-Zetterberg, L. The dose response relationship of very low doses of acrylamide is linear in the flow cytometer-based mouse micronucleus assay. *Mutat Res* 2003, *535*, pp. 215–222.

22 Muller, L., Gocke, E., Larson, P., Lave, T., and Pfister, T. Elevated ethylmethanesulfonte (EMS) in nelfinavir mesylate (Viracept, Roche): Animal studies confirm toxicity threshold and absence of risk to patients. XVII International AIDS Conference, Mexico City, 2008.

CHAPTER 4

THE DEVELOPMENT OF SAFETY THRESHOLDS FOR LEACHABLES IN ORALLY INHALED AND NASAL DRUG PRODUCTS

W. Mark Vogel

4.1 INTRODUCTION

Since the mid-1990s, several regulatory guidelines were issued or drafted by health authorities that address extractables and leachables evaluation.[1-6] However, these guidelines do not provide specific recommendations and rationale for performing safety qualifications of leachables. In particular, they do not address the potential use of safety thresholds for leachables in a drug product.

The Product Quality Research Institute (PQRI) Leachables and Extractables Working Group has developed and recommends a two-tiered qualification strategy for leachables consisting of a safety concern threshold (SCT) of 0.15 µg/day and a qualification threshold (QT) of 5 µg/day for orally inhaled and nasal drug products (OINDPs).[7] At intakes below the SCT, concern for both carcinogenic and noncarcinogenic toxicity is negligible, and identification of leachables below this threshold is not considered routinely necessary. At intakes below the QT, concern for noncarcinogenic toxicity is negligible, and leachables below this threshold without structural alerts for potent low-dose toxicity (e.g., genotoxicity or respiratory tract irritation) should not require compound-specific risk assessment. This chapter describes the derivation and scientific justification for these proposed safety thresholds for OINDP.

Leachables are derived from "critical components" of the OINDP container closure system, as opposed to the drug substance synthetic pathway. Therefore, the proposed SCT and QT are based on microgram per day intake of leachables, unlike International Conference on Harmonisation (ICH) thresholds for drug product impurities, which are linked to the daily dose of the active pharmaceutical ingredient.[8]

Leachables and Extractables Handbook: Safety Evaluation, Qualification, and Best Practices Applied to Inhalation Drug Products, First Edition. Edited by Douglas J. Ball, Daniel L. Norwood, Cheryl L.M. Stults, Lee M. Nagao.

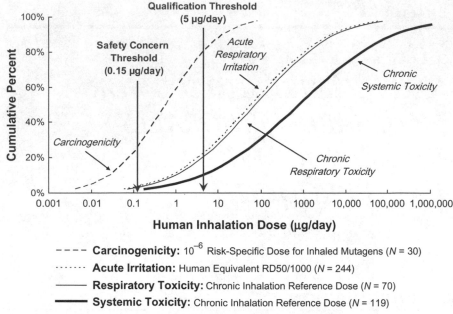

Figure 4.1   Cumulative distributions of estimated safe human exposures for sets of chemicals assessed for different toxicity end points. The vertical axis represents the cumulative percentage of chemicals in a particular data set with an estimated safe human exposure for the indicated toxicity end point less than or equal to the dose on the horizontal axis. Curves shown are the lognormal curve fits for the frequency distributions. N, number of chemicals in each data set; RD50, respiratory irritant dose in mice that reduces respiratory frequency by 50%.

The approach used to define the proposed thresholds is similar to that used by others to define thresholds for substances intentionally or unintentionally ingested orally, such as food additives, pharmaceutical impurities, or household consumer products.[9-18] As illustrated in Figure 4.1, the proposed thresholds were determined in relation to estimated safe human inhalation exposures for sets of chemicals assessed for various toxicity end points. The SCT was determined based on lifetime intakes of genotoxic carcinogens estimated to present acceptably low levels of carcinogenic risk to humans. The QT was determined based on chronic inhalation exposures to known respiratory tract toxicants estimated to present low likelihood for respiratory toxicity. The QT was also benchmarked against intakes of chemicals associated with respiratory irritation and sensitization. The following sections provide a detailed description of the derivation and scientific rationale for the proposed SCT and QT.

4.2   DERIVATION OF THE SCT

Carcinogenicity was used as the basis for the SCT because calculated carcinogenic risks of chemical carcinogens are appreciable at daily intake levels well below the range of no-observed-adverse-effect levels (NOAELs) documented for noncarcinogenic toxicity of a large number of compounds. This was previously demonstrated for orally administered compounds, including those with potent neurotoxicity, reproductive toxicity, or endocrine effects.[14] The PQRI Working Group analysis confirms that genotoxic carcinogenicity is a concern at lower doses than for acute respiratory irritation, and chronic respiratory or systemic toxicity from inhaled compounds.

Data from the carcinogenic potency database (CPDB) were used to determine the SCT.[19] The original analysis was based on data available in 2004; this chapter includes additional data from the most recently published update of the CPDB in 2007. The CPDB is a large, robust database, which was used by the United States Food and Drug Administration (FDA) to set the threshold of regulation for indirect food additives, and by the European Medicines Agency (EMEA) to set limits on genotoxic impurities in human pharmaceutical products.[10,20] The CPDB expresses carcinogenic potency as the TD50, the daily dose inducing a particular tumor type in half of the exposed animals that otherwise would not develop the tumor in a standard lifetime. Human 10^{-6} risk-specific doses were estimated by linear extrapolation from TD50 values, an approach previously used by others.[9,13,15–17,21] A risk-specific dose is the daily dose of a particular carcinogen associated with a specified lifetime excess risk for carcinogenicity, such as 10^{-5} or 10^{-6}. A 10^{-6}, or one-in-a-million, risk-specific dose is sometimes referred to as a "virtually safe dose."[22]

4.2.1 Genotoxic and Nongenotoxic Carcinogens

When available, the CPDB includes results of *Salmonella* bacterial mutagenicity assays (SAL) as an indicator of genetic toxicity. The PQRI Working Group based the SCT on the potencies of SAL-positive rodent (mouse, rat, and hamster) carcinogens in the CPDB. As noted by Cheeseman et al., and illustrated in Figure 4.2, SAL-positive carcinogens are more potent than SAL-negative compounds, and therefore of increased concern.[13] The subset of SAL-positive CPDB compounds was also chosen for analysis because the purpose of the SCT is to establish a threshold for structural identification, and structural alerts are more predictive for SAL-positive carcinogens than for nongenotoxic SAL-negative carcinogens.[23] Furthermore, most known human carcinogens are genotoxic,[24] and the assumption of linear extrapolation of cancer risk is more appropriate for genotoxic compounds than for nongenotoxic compounds, which are likely to exhibit mechanism-based thresholds for tumor induction. Too few inhalation studies with mutagenicity data (26 SAL-negative and 30 SAL-positive compounds as of 2007) are represented in the CPDB to establish a meaningful threshold based solely on inhalation data. However, the potency of the small subset of carcinogens tested by inhalation mirrors that of compounds tested by all routes of administration (Fig. 4.2), suggesting that CPDB data from all routes should be representative of inhalation carcinogens.

Because of the small number of inhalation compounds with mutagenicity data ($N = 56$), a larger set of compounds tested in rodents by inhalation ($N = 81$) with or without available mutagenicity data was also examined. Another reason for evaluating a larger set of compounds is because the results in Figure 4.2 were contrary to the expectation that inhalation might result in more potent carcinogenicity in the

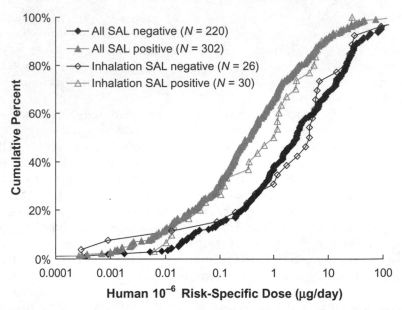

Figure 4.2 Carcinogenic potency of genotoxic (SAL-positive) and nongenotoxic (SAL-negative) carcinogens in the CPDB from studies conducted in rodents by inhalation or all routes combined.

respiratory tract due to delivery of carcinogens directly to the target tissues. A set of 1083 compounds in the CPDB was identified as statistically significant rodent carcinogens ($P < 0.05$ for a two-tailed test that the slope of the dose–response is different from zero); 1002 compounds were tested by an oral route (gavage, diet, or drinking water) and 81 were tested by inhalation. Figure 4.3 shows that the distribution of potencies (TD50 values) was generally similar for these larger sets of rodent carcinogens tested by oral and inhalation routes.

Figure 4.3 also suggests that, at the lower doses, the inhalation TD50s are shifted somewhat leftward toward greater potencies relative to the orally administered compounds (i.e., approximately three- to fourfold lower TD50s). This small shift appears to be due to a greater incidence of respiratory tract tumors occurring at relatively greater potency (lower TD50) by the inhalation route. Table 4.1 shows the distribution of carcinogenic compounds having the most sensitive tumor site in the respiratory tract versus other sites for oral and inhalation routes. A greater percentage of compounds administered by inhalation induced tumors in the respiratory tract as the most sensitive tumor site (25% of inhaled compounds), compared with those administered orally (8% of orally administered compounds). For sites outside the respiratory tract, carcinogenic potency was similar for inhalation and oral routes, with just slightly greater TD50 values on average (i.e., lesser potency) for the inhalation route. For compounds inducing lung tumors as the most sensitive tumor, the TD50 values were about 10-fold lower (i.e., ~10 times greater potency) by inhalation versus oral administration. Curiously, both inhalation and oral administration induced

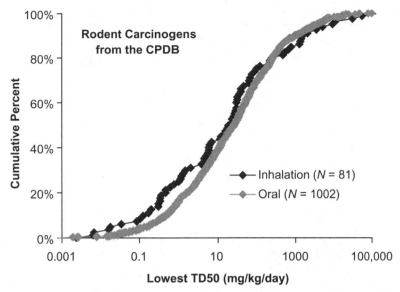

Figure 4.3   Carcinogenic potency of compounds in the CPDB from studies conducted in rodents (mice, rats, and hamsters) by inhalation or oral (gavage, dietary, drinking water) routes. Compounds include SAL positive, SAL negative, and SAL not reported. For compounds with more than one study or tumors in multiple tissues, the TD50 value represents the lowest statistically significant TD50 value among the different studies and tumor sites. For 28 compounds studied by both the oral and inhalation routes, both the lowest inhalation TD50 and the lowest oral TD50 are reported.

nasal cavity tumors at similar potencies, but with a much greater relative incidence by inhalation.

Thus, as expected, inhalation exposure induces respiratory tract tumors at greater incidence and greater potency than by oral administration. However, because respiratory tract tumors still account for only a minority of the most sensitive tumors induced by inhalation, the overall distribution of potencies is similar for carcinogens administered orally and by inhalation. This is consistent with a previous evaluation of compounds tested for carcinogenicity by both the oral and inhalation routes.[25] In that analysis, there was no statistically significant difference in carcinogenic potency between oral and inhalation administration for 14 compounds tested in rats and 9 in mice. In light of these observations, it is considered appropriate to base the SCT on the large set of SAL-positive carcinogens in the CPDB administered by all routes, rather than restricting the analysis solely to compounds administered by inhalation.

A 10^{-6} risk-specific dose was used by the PQRI Working Group as an acceptable carcinogenicity risk, consistent with the FDA threshold of regulation for indirect food additives.[10] In other contexts, regulatory agencies have considered a 10^{-5} carcinogenicity risk acceptable. Examples include the California Environmental Protection Agency (CAL EPA) Proposition 65 No-Significant-Risk Levels, and the EMEA guideline on limits for genotoxic impurities in pharmaceutical products.[20,26]

TABLE 4.1.   Tumor Distribution by Site and Route for the Most Sensitive Tumor Sites

Route		Most sensitive tumor site (lowest TD50)				
		Lungs	Nasal cavity	All respiratory tract	Nonrespiratory tract	All sites respiratory and nonrespiratory
Inhalation	Number of compounds	10	10	20	61	81
	Percent of compounds	12%	12%	25%	75%	100%
	Median TD50 (mg/kg/day)	0.3	0.9	0.4	31	19
	Geometric mean TD50 (mg/kg/day)	0.4	1.5	0.8	36	14
Oral	Number of compounds	71	8	79	923	1002
	Percent of compounds	7%	1%	8%	92%	100%
	Median TD50 (mg/kg/day)	34	0.5	26	26	26
	Geometric mean TD50 (mg/kg/day)	31	0.8	22	21	21

TD50, daily dose in mg/kg inducing a particular tumor type in half of the exposed animals that otherwise would not develop the tumor in a standard lifetime.

Selection of an acceptable risk level is essentially a regulatory rather than a scientific decision. Unlike the threshold for genotoxic impurities addressed in the EMEA guidance, the SCT is linked to an analytical threshold. In this regard, the 10^{-6} risk level may be more appropriate than a 10^{-5} level, especially in the case of mixtures of leachables. For typical drug- or process-related impurities in a drug product, only one or a few have potential genotoxicity issues. However, it is not uncommon in an OINDP for there to be multiple extractables or leachables impurities with potential genotoxicity issues. Moreover, there are many real-world examples of leachables found in OINDP that are potent carcinogens. Examples include nitrosamines, polyaromatic hydrocarbons (PAHs), aromatic amines, 1,3-butadiene, formaldehyde, and styrene. Therefore, it is reasonable to use an analytical threshold linked to the 10^{-6} risk level as a starting point for identification and evaluation of leachables impurities.

4.2.2   Allometric Scaling

The Working Group's estimates of human 10^{-6} risk-specific doses includes allometric scaling factors, based on body weight to the 0.75 power. This is the scaling factor used by the United States Environmental Protection Agency (EPA) to adjust for differences in metabolic rate across animals of different size.[27] The FDA did not use dose-scaling to establish the threshold of regulation for indirect food additives. However, in the absence of toxicokinetic data, dose metrics from rodent carcinogenicity assays are typically scaled to body surface area on a milligram per square meter basis (body weight to the 2/3 power) in approved U.S. pharmaceutical labeling. Data in the CPDB support the dose-scaling approach. For 240 SAL-positive and SAL-negative carcinogens with data from both mice and rats, rats are more sensitive when dose is expressed on a milligram per kilogram basis. The geometric mean ratio of TD50s for mice/rats is 2.4 (95% confidence limits of 2.0–3.0), a dose ratio consistent with similar carcinogenic potencies in mice and rats if dose is scaled to body surface area.

4.2.3   Other Considerations

The Working Group recognized that applying multiple conservative assumptions can unrealistically overestimate carcinogenic risk.[28] Therefore, the Working Group did not include additional conservative assumptions used in some cancer potency risk estimates. Both the EPA slope factors and the FDA estimates for the threshold of regulation for food additives are based on the most sensitive rodent species. Additionally, EPA slope factors are based on the upper 95% limit on slope rather than the central estimate. Both of these conservative approaches are appropriate for estimating the potential risk for an individual regulated chemical. In that case, one wishes to be confident that an estimated risk is likely to be less than some specified level with a high degree of certainty. However, these approaches result in an overestimation of human risk when applied overall to a population of chemicals. It is extraordinarily unlikely that the actual risk for each one in a large set of chemicals would be as great as the upper 95% estimate. Likewise, apart from pharmacokinetic differences that can be addressed by dose scaling, it is also unlikely that, for every carcinogen, humans will always be as sensitive as the most sensitive rodent species. Thus, those assumptions are appropriate for establishing regulatory thresholds for individual chemicals but not for estimating risk parameters for a population of chemicals from a particular data set. To estimate the potency distribution for a population of carcinogens, the Working Group considered it more appropriate to use a central estimate of risk rather than the upper-bound risk estimate, and to use the geometric mean of potencies from rats, mice, or hamsters when data are available from more than one species rather than basing the estimate on the most sensitive species.

Finally, a default human body weight of 70 kg is used by some regulatory agencies such as the EPA. However, a more conservative value of 50 kg is typically used to calculate safety margins relative to human in U.S. pharmaceutical labeling. This 1.4-fold difference is small considering the six to seven orders of magnitude range in carcinogenic potencies. Thus, an assumption of 50 kg versus 70 kg body weight makes relatively little difference in risk estimate, and the Working Group's calculations were based on the more protective 50-kg value.

The distribution of calculated human 10^{-6} risk-specific doses using these assumptions is illustrated in Figure 4.4. The median human equivalent

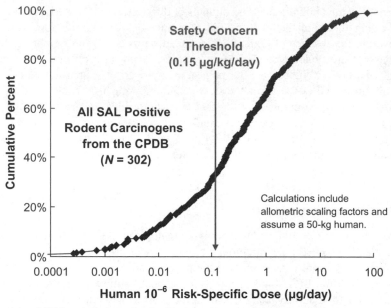

Figure 4.4 SCT in relation to the distribution of calculated human 10^{-6} risk-specific doses for genotoxic (SAL-positive) rodent carcinogens administered by all routes from the CPDB.

10^{-6} risk-specific dose for these 302 SAL-positive carcinogens from the CPDB is 0.36 μg/day, and the median excess cancer risk at the proposed SCT of 0.15 μg/day is 0.41×10^{-6}. If <20% of random chemicals are genotoxic carcinogens,[29,30] <7% of all compounds would exceed 10^{-6} increased cancer risk at intakes <0.15 μg/day lifetime exposure, meeting the criterion that a leachable below the SCT is unlikely to have a lifetime excess cancer risk $>10^{-6}$.

4.3 DERIVATION OF THE QT

The proposed QT for leachables in OINDP is based primarily on reference doses (RfDs) for inhaled chemicals derived by application of standard safety factors to NOAELs for noncarcinogenic toxicity. Such reference values are considered by regulatory agencies to pose negligible human health risks. The Working Group originally analyzed a set of 150 inhaled compounds derived from "chronic RfD" established by the EPA, "minimum risk levels" (MRLs) established by the U.S. Agency for Toxic Substances and Disease Registry, and "reference exposure levels" (RELs) established by CAL EPA. Reference values established by these agencies are available in electronic databases accessible via the Internet.[31–33] Since the original analysis, additional regulatory RfDs have been published by these agencies, and the data presented here are for 189 compounds as of December 2008. The toxic effect upon which the reference values were determined was systemic toxicity for 119

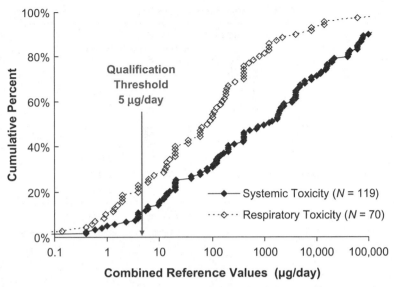

Figure 4.5   Distribution of reference values from combined data set of compounds assessed by the EPA, Agency for Toxic Substances and Disease Registry (ATSDR), and CAL EPA.

TABLE 4.2.   Summary of Inhalation Reference Toxicity Values (µg/day)

Reference value	Respiratory toxicity			Systemic toxicity		
	Median	10th percentile	N	Median	10th percentile	N
EPA RfD	80	1.2	49	1100	8.4	78
ATSDR MRL	64	1.8	20	2065	5.7	45
California REL	60	1.3	34	4000	4.0	46
Combined set	86	1.3	70	1200	4.0	119

Combined values are parameters for the combined data set. The number of compounds in the combined set is less than the sum of compounds in each set because several compounds were assessed by more than one agency.

EPA, United States Environmental Protection Agency; RfD, reference dose; ATSDR, Agency for Toxic Substances and Disease Registry; MRL, minimum risk level; REL, reference exposure level; N, number of substances.

chemicals and respiratory tract toxicity for 70 chemicals. For six chemicals for which no target organ toxicity was identified, the reference values are included in the distributions for both systemic and respiratory tract toxicity. When more than one reference value was available, a combined geometric mean reference value was calculated.

The distribution of reference value is shown in Figure 4.5, with descriptive statistics in Table 4.2. Median and 10th percentile RfDs were similar among the different regulatory agencies. This suggests that the types of chemicals assessed and methods for estimating safe human exposures were comparable, and that it is

TABLE 4.3. Substances with Combined Inhalation Reference Value Less than 5 µg/Day

Substance	CASRN	MRL	RfD	REL	Combined
Hexamethylene diisocyanate, 1,6-	822-060	1.4	0.2		0.5
Chloroacetophenone, 2-	532-27-4		0.6		0.6
Cobalt	7440-48-4	2.0	0.4		0.9
Acrolein	107-02	1.8	0.4	1.2	1.0
Toluene diisocyanate mixture	26471-62-5		1.4	1.4	1.4
Glutaraldehyde	111-30-8			1.6	1.6
Nitroaniline, 2-	88-74-4		2.0		2.0
Hexachlorocyclopentadiene	77-47-4		4.0		4.0
Chlorine	7782-50-5		4.0	4.0	4.0

Combined value is geometric mean of available regulatory reference values.

CASRN, Chemical Abstracts Service Registry Number; MRL, minimum risk level; RfD, reference dose; REL, reference exposure level.

appropriate to combine the reference values into a single set. For the combined data set, the median reference value for chemicals with respiratory toxicity end points was 86 µg/day, with a 10th percentile of 1.3 µg/day, and the median reference value for chemicals with systemic toxicity end points was 1200 µg/day, with a 10th percentile of 4 µg/day. Based on large (~100-fold) safety margins incorporated in the reference values, leachables at intakes <5 µg/day should pose negligible health risks. Substances with respiratory toxicity and inhalation reference values less than 5 µg/day (Table 4.3) are dominated by metals and metal salts, and by reactive chemicals with readily identifiable irritant potential, such as aldehydes and isocyanates.

In addition to chronic respiratory toxicity, the Working Group also considered acute respiratory irritation in relation to the proposed QT, since airway irritation and paradoxical bronchoconstriction is a concern for impurities or excipients in OINDP.[34] A useful metric of sensory irritation is the RD50, the concentration of an irritant that decreases respiratory frequency by 50% in mice.[35] A good correlation was reported ($r^2 = 0.78$) between occupational threshold limit values and the value of 0.03 × RD50.[36] To estimate safe doses for respiratory irritants, the Working Group calculated the microgram dose at the RD50 concentration inhaled for 10 minutes divided by 1000 for a large database of RD50 values.[36] The additional 30-fold safety margin for this metric, compared with the value of 0.03 × RD50, should be sufficient to account for sensitive populations such as asthmatics as illustrated in Table 4.4.

The distribution of RD50/1000 doses was quite similar to the distribution of NOAELs for chronic respiratory toxicity (Fig. 4.1). As with the chronic respiratory toxicants, compounds with RD50/1000 values below the proposed QT of 5 µg/day were predictably irritant compounds such as aldehydes, isocyanates, and nitriles.

The Working Group's analysis of noncarcinogenic toxicity for inhaled compounds indicates that for leachables in OINDP below the proposed QT of 5 µg/day, concern for noncarcinogenic toxicity is very low except for a small proportion of compounds that are strong irritants. Since potent irritancy is predictable based on

TABLE 4.4. Bronchoconstrictor Concentrations in Asthmatics Relative to Occupational Short-Term Exposure Limits (STELs) and RD50 Values

Substance	STEL (mg/m³)	RD50/1000 (μg/m³)	Bronchoconstrictor concentration in asthmatics	Reference
Nitrogen dioxide	9.4	655	None at 753 μg/m³ (12-fold below STEL)	37
Sulfur dioxide	13	523	Range = 666 to 105,000 μg/m³ (20- to 1.2-fold below STEL)	38
Sulfuric acid	3.0	NA	None at 46 μg/m³ (65-fold below STEL) Some at 130 μg/m³ (23-fold below STEL)	39
Formaldehyde	2.45	39	None at 3700 μg/m³ (1.5 × STEL)	40
Toluene diisocyanate	0.14	4.8	Most at >14 μg/m³ (10-fold below STEL) A few at ≤7 μg/m³ (20-fold below STEL)	41

RD50, concentration decreasing respiratory rate by 50% in mice; NA, not available.

chemical structure, leachables compounds below the QT without structural alerts for carcinogenicity or irritation should not require compound-specific risk assessment.

4.4 CONCLUSION

Toxicity threshold approaches have previously been applied or proposed for the regulation of chemicals in various settings such as food additives and pharmaceutical impurities, including genotoxic impurities.[8,10,15,17,18,20] The analysis of the PQRI Leachables and Extractables Working Group demonstrates that a threshold approach to safety qualification of leachables in OINDPs is valid. As demonstrated previously for orally administered compounds, carcinogenicity was shown to be a greater concern at low intakes than other potential toxicities such as acute irritation or chronic toxicity of the respiratory tract. Thus, carcinogenic risk was used as the basis for the SCT. At intakes below the proposed SCT of 0.15 μg/day, potential unidentified leachables impurities are likely to have an acceptable lifetime excess cancer risk <10^{-6}, and have negligible risk for noncarcinogenic systemic or respiratory tract toxicity. Therefore, it is not considered necessary to routinely identify leachables impurities with intakes below the SCT. It is appropriate to identify leachables with intakes greater than the SCT, to determine whether there are structural alerts for carcinogenicity or respiratory tract irritation. Although not routinely needed, it is still appropriate to identify particular agents of concern, such as nitrosamines, at intakes below the SCT when critical components of the OINDP container closure system are likely to include such specific compounds. Because of differences in product configuration (e.g., dose, volume, and delivery characteristics) of different OINDPs, and differences in analytical uncertainty associated with different analytical methods, it is not possible to translate the SCT to a specific concentration cutoff applicable to all OINDPs in general (e.g., in units of parts per million, microgram per canister, or percent of active ingredient). Instead, the SCT is used as the starting point for calculating an analytical evaluation threshold (AET) for any specific OINDP. The AET is a threshold at or above which chemists should begin to identify a particular leachable and/or extractables and report them for potential toxicological assessment. The following chapter describes how the SCT is used to calculate the AET, depending on the specific product characteristics and analytical methods employed. At intakes below the proposed QT of 5 μg/day, identified leachables are very unlikely to exert significant airway irritation (with associated bronchoconstriction), or chronic respiratory or systemic toxicity. Therefore, compounds below the QT without structural alerts for carcinogenicity or irritation should not require compound-specific risk assessment.

REFERENCES

1 Guidance for industry: Container closure systems for packaging human drugs and biologics. U.S. Department of Health and Human Services, Food and Drug Administration, Center for Drug Evaluation and Research (CDER), Center for Biologics Evaluation and Research (CBER), 1999.
2 European Medicines Agency Inspections. Guideline on plastic immediate packaging materials. Committee for Medicinal Products for Human Use (CHMP), Committee for Medicinal Products for Veterinary Use (CVMP). London, 2005.
3 Guidance for industry: Nasal spray and inhalation solution, suspension, and spray drug products—Chemistry, manufacturing, and controls documentation. Department of Health and Human Services, Food and Drug Administration, Center for Drug Evaluation and Research (CDER), 2002.
4 Draft guidance for industry: Metered dose inhaler (MDI) and dry powder inhaler (DPI) drug products. Department of Health and Human Services, Food and Drug Administration, Center for Drug Evaluation and Research (CDER), 1998.
5 Guidance for industry: Pharmaceutical quality of inhalation and nasal products. Health Canada, 2006.
6 European Medicines Agency Inspections (EMEA). Guideline on the pharmaceutical quality of inhalation and nasal products. Committee for Medicinal Products for Human Use (CHMP), 2006.
7 Norwood, D.L. and Ball, D. Product Quality Research Institute: Safety thresholds and best practices for extractables and leachables in orally inhaled and nasal drug products. Submitted to the PQRI Drug Product Technical Committee, PQRI Steering Committee, and U.S. Food and Drug Administration by the PQRI Leachables and Extractables Working Group, 2006.
8 Guidance for industry: Q3B(R2) impurities in new drug products (revision 2). International Conference on Harmonisation of Technical Requirements for Registration of Pharmaceuticals for Human Use, 2006.
9 Rulis, A. Food safety assessment, threshold of regulation: Options for handling minimal risk situations. In: Finley, J.W., Robinson, S.F., and Armstrong, D.J., eds. *American Chemical Society Symposium Series 484*. American Chemical Society, Washington, DC, 1992; 132–139.
10 Code of Federal Regulations. Threshold of regulation for substances used in food-contact articles. Title 21, Sec. 170.39, revised as of April 1, 2011.
11 Federal Register. Volume 60, No. 136, Government Printing Office, 1995, pp. 36581–36596.
12 Munro, I.C., Ford, R.A., Kennepohl, E., Sprenger, J.G., Maier, A., and Dourson, M. Correlation of structural class with no-observed-effect levels: A proposal for establishing a threshold of concern. *Food Chem Toxicol* 1996, *34*, pp. 829–867.

13 Cheeseman, M.A., Machuga, E.J., and Bailey, A.B. A tiered approach to threshold of regulation. *Food Chem Toxicol* 1999, *37*, pp. 387–412.

14 Kroes, R., Galli, C., Munro, I., Schilter, B., Tran, L.A., Walker, R., and Würtzen, G. Threshold of toxicological concern for chemical substances present in the diet: A practical tool for assessing the need for toxicity testing. *Food Chem Toxicol* 2000, *38*, pp. 255–312.

15 Fiori, J.M. and Meyerhoff, R.D. Extending the threshold of regulation concept: De minimis limits for carcinogens and mutagens. *Regul Toxicol Pharmacol* 2002, *35*, pp. 209–216.

16 Kroes, R., Renwick, A.G., Cheeseman, M., Kleiner, J., Mangelsdorf, I., Piersma, A., et al. Structure-based threshold of toxicological concern (TTC): Guidance for application to substances present at low levels in the diet. *Food Chem Toxicol* 2004, *42*, pp. 65–83.

17 Blackburn, K., Stickney, J., Carlson-Lynch, H.L., McGinnis, P.M., Chappell, L., and Felter, S.P. Application of the threshold of toxicological concern approach to ingredients in personal and household care products. *Regul Toxicol Pharmacol* 2005, *43*, pp. 249–259.

18 Dolan, D.G., Naumann, B.D., Sargent, E.V., Maier, A., and Dourson, M. Application of the threshold of toxicological concern concept to pharmaceutical manufacturing operations. *Regul Toxicol Pharmacol* 2005, *43*, pp. 1–9.

19 Gold, L.S. The carcinogenic potency database (CPDB) at the University of California, Berkeley, 2007. Available at: http://potency.berkeley.edu/ (accessed April 27, 2009).

20 European Medicines Agency. CPMP/SWP/5199/02. Committee for Medicinal Products for Human Use: Guideline on the limits of genotoxic impurities, 2006.

21 Krewski, D., Szyszkowicz, M., and Rosenkranz, H. Quantitative factors in chemical carcinogenesis: Variation in carcinogenic potency. *Regul Toxicol Pharmacol* 1990, *12*, pp. 13–29.

22 Gaylor, D.W. and Swirsky, L. Gold regulatory cancer risk assessment based on a quick estimate of a benchmark dose derived from the maximum tolerated dose. *Regul Toxicol Pharmacol* 1998, *28*(3), pp. 222–225.

23 Benigni, R. and Zito, R. The second national toxicology program comparative exercise on the prediction of rodent carcinogenicity: Definitive results. *Mutat Res/Rev Mutat Res* 2004, *566*, pp. 49–63.

24 Bartsch, H. and Malaveille, C. Prevalence of genotoxic chemicals among animal and human carcinogens evaluated in the IARC monograph series. *Cell Biol Toxicol* 1989, *5*, pp. 115–127.

25 Pepelko, W.E. Effect of exposure route on potency of carcinogens. *Regul Toxicol Pharmacol* 1990, *13*, pp. 3–17.

26 Proposition 65: Reproductive and cancer hazard assessment section process for developing safe harbor numbers. Office of Environmental Health Hazard Assessment, California Environmental Protection Agency, 2001.

27 Federal Register. Volume 57, No. 109, United States Environmental Protection Agency, 1992, pp. 24152–24173.

28 Gaylor, D.W., Chen, J.J., and Sheehan, D.M. Uncertainty in cancer risk estimates. *Risk Anal* 1993, *13*, p. 149.

29 Fung, V.A., Barrett, J.C., and Huff, J. The carcinogenesis bioassay in perspective: Application in identifying human cancer hazards. *Environ Health Perspect* 1995, *103*, pp. 680–683.

30 Sawatari, K., Nakanishi, Y., and Matsushima, T. Relationships between chemical structures and mutagenicity: A preliminary survey for a database of mutagenicity test results of new workplace chemicals. *Ind Health* 2001, *39*, pp. 341–345.

31 U.S. Environmental Protection Agency's Integrated Risk Information System (IRIS). Health effects assessment summary tables (HEAST); chemical-specific toxicity values, 2009. Available at: http://rais.ornl.gov/ (accessed April 27, 2009).

32 California Environmental Protection Agency, Office of Environmental Health Hazard Assessment (OEHHA). All chronic reference exposure levels adopted by OEHHA as of February 2005. Available at: http://www.oehha.org/air/chronic_rels/AllChrels.html (accessed April 27, 2009).

33 Agency for Toxic Substances and Disease Registry (ATSDR). Minimal risk levels (MRLs) for hazardous substances, 2008. Available at: http://www.atsdr.cdc.gov/mrls/index.html (accessed April 27, 2009).

34 Shaheen, M.Z., Ayres, J.G., and Benincasa, C. Incidence of acute decreases in peak expiratory flow following the use of metered-dose inhalers in asthmatic patients. *Eur Respir J* 1994, *7*, pp. 2160–2164.

35 Alarie, Y., Kane, L., Barrow, C.S., and Reeves, A.L. Sensory irritation: The use of an animal model to establish acceptable exposure to airborne chemical irritants. In Reeves A., ed. *Toxicology: Principles and Practice 1*, Wiley, New York, 1980, 48–92.

36 Schaper, M. Development of a database for sensory irritants and its use in establishing occupational exposure limits. *Am Ind Hyg Assoc J* 1993, *54*, pp. 488–544.

37 Tunnicliffe, W.S., Burge, P.S., and Ayres, J.G. Effect of domestic concentrations of nitrogen dioxide on airway responses to inhaled allergen in asthmatic patients. *Lancet* 1994, *344*, pp. 1733–1736.

38 Rubinstein, I., Bigby, B.G., Reiss, T.F., and Boushey, H.A. Short term exposure to 0.3 ppm nitrogen dioxide does not potentiate airway responsiveness to sulfur dioxide in asthmatic subjects. *Am Rev Respir Dis* 1990, *141*, pp. 381–385.

39 Avol, E.L., Linn, W.S., Shamoo, D.A., Anderson, K.R., Peng, R.C., and Hackney, J.D. Respiratory responses of young asthmatic volunteers in controlled exposures to sulfuric acid aerosol. *Am Rev Respir Dis* 1990, *142*, pp. 343–348.

40 Sauder, L.R., Green, D.J., Chatham, M.D., and Kulle, T.J. Acute pulmonary response of asthmatics to 3.0 ppm formaldehyde. *Toxicol Ind Health* 1987, *3*, pp. 569–578.

41 O'Brien, I.M., Newman-Taylor, A.J., Burge, P.S., Harries, M.G., Fawcett, I.W., and Pepys, J. Toluene di-isocyanate-induced asthma. II. Inhalation challenge tests and bronchial reactivity studies. *Clin Allergy* 1979, *9*(1), pp. 7–15.

CHAPTER 5

THE ANALYTICAL EVALUATION THRESHOLD (AET) AND ITS RELATIONSHIP TO SAFETY THRESHOLDS

Daniel L. Norwood, James O. Mullis, and Scott J. Pennino

5.1 INTRODUCTION

The proposed safety thresholds for patient exposure to individual organic leachables in orally inhaled and nasal drug products (OINDPs, or "inhalation drug products"), the qualification threshold (QT) (5 µg/day total daily intake), and the safety concern threshold (SCT) (0.15 µg/day total daily intake) represent a significant step forward in increasing the efficiency of the OINDP pharmaceutical development process and ensuring patient safety. However, in order to be of use to analytical chemists concerned with the identification and quantification of leachables and extractables for a particular drug product, these absolute safety thresholds must be converted into thresholds that can be applied to particular leachables/extractables profiles under investigation. The analytical evaluation threshold (AET) concept provides a mechanism for this conversion and defines the levels at which leachables (and extractables) should be identified and evaluated.[1] Therefore, the AET answers the question presented in Chapter 1, which has been repeatedly posed by OINDP pharmaceutical development scientists over the past two decades: *How low do we go?* For additional detailed discussions, the interested reader is referred to the comprehensive recommendation document of the Product Quality Research Institute's (PQRI) Leachables and Extractables Working Group,[1] which was submitted to the United States Food and Drug Administration (FDA) in 2006, as well as two additional scientific publications of the Working Group.[2,3]

The AET for a particular inhalation drug product is derived directly from the SCT (i.e., the lower of the two safety thresholds), which is defined in terms of absolute exposure of a patient to any individual organic leachable contained in the

Leachables and Extractables Handbook: Safety Evaluation, Qualification, and Best Practices Applied to Inhalation Drug Products, First Edition. Edited by Douglas J. Ball, Daniel L. Norwood, Cheryl L.M. Stults, Lee M. Nagao.
© 2012 John Wiley & Sons, Inc. Published 2012 by John Wiley & Sons, Inc.

particular inhalation drug product. The SCT represents the threshold below which a leachable would have a dose so low as to present negligible safety concerns from carcinogenic and noncarcinogenic toxic effects. Deriving the AET for any particular inhalation drug product from the SCT requires consideration of the unique parameters of that particular inhalation drug product, such as the product label claim for total number of individual doses in a single drug product unit and the maximum recommended number of doses per day. Derivation of the AET must also consider the particular analytical technique/method used to establish the leachables/extractables profile under consideration, as well as the uncertainty inherent in that particular analytical technique/method.

Although the SCT applies only to leachables, as patients are exposed to leachables, the AET concept can also be applied to extractables by consideration of the parameters of individual OINDP container closure system critical components associated with a particular drug product (e.g., mass of the particular component). Note that a "critical component" is defined as any component in contact with the drug product formulation, the patient's mouth or nasal mucosa, or that is deemed of particular significance to the functionality of the drug product.[1] Application to extractables profiles considerably expands the utility of the AET, and therefore the SCT, by allowing its use in controlled extraction studies designed to characterize extractables from container closure system critical components, as well as in early-stage development component selection and late-stage development critical component extractables profile quality control and release. Figure 5.1 presents a flowchart representing a typical pharmaceutical development process for leachables and extractables in inhalation drug products, taken from the PQRI recommendations.[1] Note that the AET, and therefore the SCT, is applied to both leachables and extractables in this process.

5.2   LEACHABLES/EXTRACTABLES CHARACTERIZATION AND PROFILING

A leachables or extractables "profile" presents an overall picture of the leachables contained in an inhalation drug product formulation, or the extractables removed from an OINDP critical component in a laboratory investigation. Remember that *leachables* are defined by the FDA as[4] "compounds that leach into the formulation from elastomeric or plastic components of the drug product container closure system," and *extractables* are defined by the FDA as[4] "compounds that can be extracted from elastomeric or plastic components of the container and closure system when in the presence of a solvent."

In practice, leachables and extractables can also originate from organic coatings on container closure system components (e.g., coatings on the interior surface of a metered dose inhaler [MDI] metal canister), organic residues on component surfaces (e.g., heavy drawing oils on MDI canisters and metal valve components), and organic processing aids from component fabrication machinery (e.g., mold release agents used in the injection molding of plastic components).[1,5]

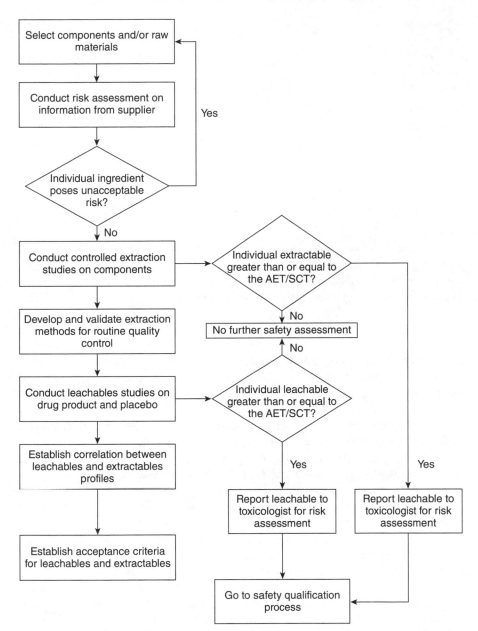

Figure 5.1   Schematic pharmaceutical development process flowchart for leachables and extractables in OINDP (taken from the PQRI Leachables and Extractables Working Group recommendation document to the FDA[1]).

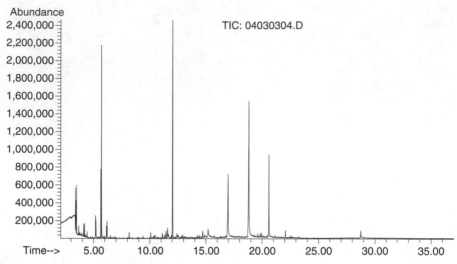

Figure 5.2 A gas chromatography/mass spectrometry (GC/MS) extractables profile of an elastomer (in the form of a total ion chromatogram [TIC]).

Although in principle any analytical technique can produce a leachables/ extractables profile, the most commonly applied techniques are those capable of separating and individually detecting the usually complex mixtures of trace-level organic chemicals. These analytical techniques all involve the combination of chromatography with either compound-specific or nonspecific detection.[5,6] Techniques with compound-specific detection include gas chromatography/mass spectrometry (GC/MS), liquid chromatography/mass spectrometry (LC/MS), and liquid chromatography/diode array detection (LC/DAD). Those with nonspecific detection include gas chromatography/flame ionization detection (GC/FID) and liquid chromatography/ultraviolet (single or variable wavelength) (LC/UV) detection. In addition to separating and detecting leachables and extractables, these analytical techniques are all capable of generating signal levels directly proportional to the amounts of individual organic chemical entities introduced, and can therefore produce quantitative information on individual leachables/extractables.

Figures 5.2–5.5 show extractables "profiles" from GC/MS, LC/MS, GC/FID, and LC/UV. The analytical techniques with compound-specific detection are most useful for leachables/extractables characterization and identification studies (e.g., controlled extraction studies, see Fig. 5.1), and those with nonspecific detection are most useful for more routine studies, such as drug product leachables stability studies and routine extractables testing and release of container closure system critical components for manufacturing. It should again be stated that other analytical techniques can and have been used for leachables/extractables profiling, and the preceding discussion is not intended to preclude any analytical technique. In fact, the AET concept should be applicable to all analytical techniques for leachables/extractables profiling.

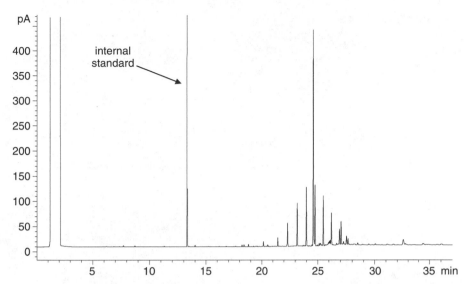

Figure 5.3   A gas chromatography/flame ionization detection (GC/FID) extractables profile of an elastomer.

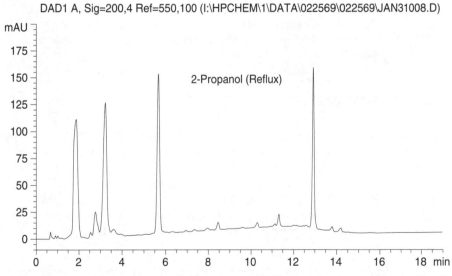

Figure 5.4   A liquid chromatography/ultraviolet (LC/UV) detection extractables profile of a plastic material.

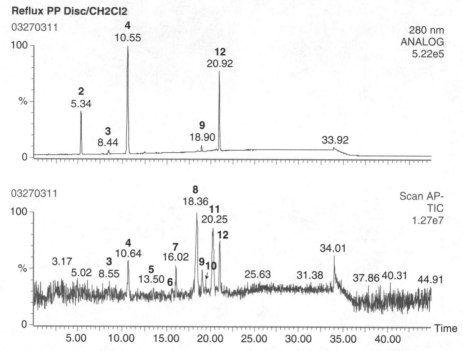

Figure 5.5 A liquid chromatography/mass spectrometry (LC/MS) extractables profile of a plastic. Note that the top trace is an in-line LC/UV chromatogram that is usually acquired simultaneously during LC/MS experiments due to the relatively high chemical background in LC/MS total ion chromatograms (bottom trace; negative ion atmospheric pressure chemical ionization).

5.3 DETERMINATION OF THE AET

The AET is officially defined by PQRI as[1] "the threshold at or above which an OINDP pharmaceutical development team should identify and quantify a particular extractables and/or leachable and report it for potential toxicological assessment."

The overall process for determination of a drug-product-specific AET has also been defined by PQRI[1] and begins with the SCT (0.15 μg/day total daily intake for an individual organic leachable), and an understanding of the individual dosing parameters of the particular inhalation drug product. The steps of the process are as follows[1]:

1. Convert the SCT into an estimated AET (e.g., microgram per canister for an individual organic leachable in an MDI) by considering the dosing parameters of the particular OINDP. Note that only label claims of total doses per drug product unit and maximum recommended doses per day should be considered, and issues such as "overfill" should not be considered.

2. Convert the estimated AET for leachables to an estimated AET for extractables (e.g., microgram per gram elastomer for an individual organic extractable) by

considering the parameters of the particular OINDP container closure system (e.g., weight of elastomer per MDI valve).

3. Locate the estimated AET on a particular leachables or extractables profile.

4. Evaluate the uncertainty of the particular analytical technique/method.

5. Convert the estimated AET to a final AET by considering this analytical uncertainty.

The *estimated AET* is the direct conversion of the SCT to a drug-product-specific analytical threshold, while the *final AET* includes an evaluation of the uncertainty of the particular analytical technique/method that produces a particular leachables/extractables profile.

Obviously, if one is able to accurately quantitate every individual leachable (or extractable) in a particular profile, then the *estimated AET* is exactly equal to the *final AET*. For leachables profiles, this in fact might be the case since comprehensive controlled extraction studies would have been accomplished (see Fig. 5.1), providing identifications of all potential leachables and ample time to develop and validate quantitative leachables methods with all appropriate reference compounds. Given a properly accomplished controlled extraction study and a thorough understanding of manufacturing processes, the detection of a completely unknown leachable during drug product stability studies should be a rare occurrence, although not impossible. During controlled extraction studies, however, where it is not practical to accurately quantitate each and every individual extractable with an authentic reference compound, the *estimated* and *final AETs* are important thresholds that serve to rationalize the overall scope of the study.

The *estimated AET* can be located on a particular extractables/leachables profile (e.g., GC/FID chromatogram, GC/MS total ion chromatogram [TIC], LC/UV chromatogram) relative to the response of an appropriately selected internal standard, or the response(s) of authentic reference compounds representing major extractables/leachables. The *final AET* can then be determined by incorporating into the *estimated AET* a factor that reflects the uncertainty inherent in the particular analytical method. Analytical uncertainty is a result of the differing responses that chemical entities with different molecular structures have with analytical techniques/methods. This analytical uncertainty is of particular significance for leachables and extractables that, as previously discussed in Chapter 1, can represent a wide variety of chemical classes and molecular structure types. The *final AET* is, therefore, dependent on the analytical technique(s)/method(s) used to create the extractables/leachables profile(s) being investigated.

One possible approach to accomplishing an evaluation of analytical uncertainty is through the use of response factors (RFs). An RF is defined as

$$RF = A_a/C_a,$$

where

A_a = response of an individual analyte, for example, chromatographic peak area; and

C_a = concentration (or mass) of the individual analyte.

For a GC/MS method, for example, the chromatographic peak areas for individual analytes, that is, leachables or extractables, as determined from either the TIC or individual mass chromatograms (extracted ion current profiles), are divided by individual analyte concentrations in a known sample of authentic reference compounds. The concentration levels of the authentic reference compounds chosen for RF determination must be within the linear dynamic range of the analytical system. For GC/MS, this means not overloading the GC column or saturating the mass spectrometer's detector. A somewhat more precise uncertainty evaluation can be obtained through the use of relative response factors (RRFs), which are defined as follows:

$$RRF = C_{is}A_a/A_{is}C_a,$$

where

C_{is} = concentration (or mass) of an internal standard,

A_{is} = response of the internal standard,

A_a = response of an individual analyte, and

C_a = concentration of the individual analyte.

The RRF normalizes individual RFs to the RF of an internal standard. The use of internal standards is a well-established procedure for improving the accuracy and precision of trace organic analytical methods.

A summary of the process discussed above as one way to evaluate analytical uncertainty is as follows:

1. Given a particular extractables/leachables profile obtained by a particular analytical technique/method, create a list of individual analytes that have *confirmed* identifications and for which authentic reference compounds are available.

    This analyte list should ideally include chemical entities representing all known ingredients in the appropriate container closure system component(s), and all identified molecular structure classes of extractables/leachables that were not stated explicitly in the ingredients, for example, specific alkanes that constitute the general ingredient "paraffins." A confirmed identification is an identification authenticated with a reference compound, scientific literature, or other authentic reference information.

2. Choose an internal standard appropriate to the particular analytical technique/ method.

    Some characteristics of a good internal standard are the following:

    • It should be compatible with the particular analytical technique.

    • It should be "well behaved" in the particular analytical method. A "well-behaved" internal standard in a GC method, for instance, will not have a significant tailing factor, will not irreversibly adsorb onto the column, and so on.

    • It should be stable in the analytical matrix.

- It should not be interfered with by other analytes or components in the analytical matrix.
- It should possess a response similar to those of other analytes in the particular analytical technique/method.

3. Analyze a mixture(s) of authentic reference compounds with the internal standard using the particular analytical technique/method.

    This analysis should be accomplished according to principles of sound scientific practice, for example, at appropriate concentration levels, with an appropriate number of replicates, with appropriate blanks and controls.

4. Calculate RRFs for all analytes and create an RRF database. An example of an RRF database is presented in Table 5.1.

5. Calculate statistical parameters for the RRF database, including the standard deviation (SD) and % relative SD of RRFs. The analytical uncertainty can then be estimated based on this database and statistical parameters.

The PQRI recommendations state that the analytical uncertainty should be defined as a single % relative SD in an appropriately constituted RF database. However, the alternative of simply cutting the *estimated AET* in half to give the *final AET* was also proposed. The latter option involves less laboratory work and likely provides an equally appropriate uncertainty estimate.

Example:

As an example of AET determination, consider an MDI drug product whose leachables profile, in the form of a GC/MS TIC, is shown in Figure 5.6. Note that the profile contains an internal standard, 1-bromotetradecane at a level of 40 μg/MDI canister, from which the *estimated* and *final AETs* will be extrapolated. This particular drug product was specified to deliver 200 actuations with a maximum daily recommended dose of 8 actuations. The *estimated AET* for leachables can be calculated from the SCT as follows:

$$\text{estimated } AET = (0.15 \text{ μg/day} \div 8 \text{ actuations/day}) \times 200 \text{ actuations/canister}$$

$$\text{estimated } AET = 3.75 \text{ μg/canister}.$$

The calculated *estimated AET* is extrapolated from the peak height (or peak area) of the internal standard and located on the profile as shown in Figure 5.6. The *final AET* in this example is taken to be the *estimated AET* cut in half (see the expanded leachables profile in Fig. 5.7).

Extractables profile AETs for container closure system critical components can be established by consideration of the number of a particular component in an individual MDI canister and the mass of the component. For example, if the above MDI drug product contained a 100-mg elastomeric component whose job is to seal the dose metering valve to the canister, then the *estimated AET* for extractables in a GC/MS TIC extractables profile of that component can be calculated:

$$\text{estimated AET} = (3.75 \text{ μg/canister} \times 1 \text{ canister/valve}) \div 0.1 \text{ g elastomer/valve}$$

$$\text{estimated } AET = 37.5 \text{ μg/g}.$$

TABLE 5.1. Example of a GC/MS Relative Response Factor (RRF) Database for a Group of Organic Chemicals Commonly Observed as Leachables/Extractables (RRFs Are Shown Relative to Two Internal Standards, 2-Fluorobiphenyl and *p*-Terphenyl-d₁₄)

Analyte (leachable/extractable)	RRFs relative to 2-fluorobiphenyl	RRFs relative to *p*-terphenyl-d$_{14}$
α-Methylstyrene	0.563	0.395
Tetramethylthiourea	0.368	0.248
Benzothiazole	0.491	0.336
4-*tert*-butylphenol	0.574	0.372
1,3-Diacetylbenzene	0.383	0.231
2,6-Di-*tert*-butylphenol	0.942	0.639
4'-Hydroxyacetophenone	0.245	0.235
Butylated hydroxytoluene	1.062	0.694
Diphenylamine	0.883	0.634
Phenyl salicylate	0.380	0.256
2,4-Diphenyl-4-methyl-1-pentene	0.824	0.527
2,5-Di-*tert*-butyl hydroquinone	0.422	0.368
Dibutylphthalate	0.843	0.625
Tetramethylthiuram monosulfide	0.071	0.0495
Palmitic acid	0.377	0.274
2,4-Dihydroxybenzophenone	0.245	0.248
N-phenyl-1-naphthylamine	1.211	0.840
Heneicosane	0.565	0.400
Stearic acid	0.385	0.274
Docosane	0.568	0.402
2,2'-Methylene-bis(6-*tert*-butyl-4-methylphenol)	0.519	0.666
Tetracosane	0.584	0.424
Dicyclohexyl phthalate	0.830	0.636
2,2'-Methylene-bis(6-*tert*-butyl-4-ethylphenol)l	0.574	0.639
Bis-2-ethylhexylphthalate	0.870	0.654
Hexacosane	0.542	0.430
Bis-2-ethylhexylisophthalate	0.766	0.630
Di-*n*-octylphthalate	0.822	0.668
Irgafos 168	1.118	0.781
Irganox 1076	0.644	0.668
Tris-phenolic	0.648	0.414
Mean	0.623	0.473
Standard deviation	0.270	0.197
% Relative standard deviation	43.3	41.6

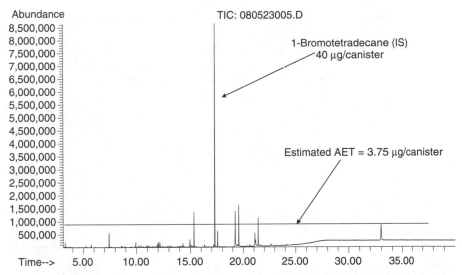

Figure 5.6   A gas chromatography/mass spectrometry (GC/MS) leachables profile of a metered dose inhaler drug product (in the form of a total ion chromatogram [TIC]). Note the location of the estimated AET for this profile.

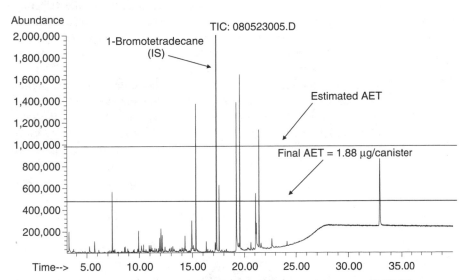

Figure 5.7   Vertically expanded gas chromatography/mass spectrometry (GC/MS) leachables profile of a metered dose inhaler drug product (in the form of a total ion chromatogram [TIC] from Fig. 5.6). Note the locations of both the estimated and final AETs for this profile.

This *estimated AET* could be used to guide controlled extraction studies on this particular elastomeric valve component.

The MDI represents the "worst case scenario" for leachables of all the OINDP; that is, typically all potential leachables (i.e., extractables) detected and identified in controlled extraction studies will be correlated[1] both qualitatively and quantitatively with real leachables observed during drug product stability studies during pharmaceutical development. The MDI is, however, not the worst case scenario with respect to a very low AET. One need only examine the calculations above to discern that the *estimated AET* is a function of the dosing parameters of the particular OINDP. Consider, for example, if the number of doses per drug product unit were one, as with an inhalation solution unit dose nebule. The PQRI recommendation document includes example AET calculations for most OINDP types along with detailed recommendations designed to somewhat ameliorate the problems with extremely low AETs for certain OINDP types.

5.4   ANALYTICAL SENSITIVITY AND THE AET: WHAT IS MODERN ANALYTICAL CHEMISTRY CAPABLE OF?

The AET concept for leachables in inhalation drug products begs the question of the capabilities of modern analytical chemistry, and in particular trace organic analysis.[6] Specifically, are the most commonly utilized analytical techniques for the detection, identification, and quantitation of leachables in inhalation drug products, and extractables in container closure system critical components, capable of producing sufficient compound-specific information to allow identification, and sufficient overall sensitivity to allow quantitation, at AET levels?

To address this issue, consider Figure 5.8, which shows an expanded portion of the GC/MS TIC leachables profile from Figure 5.6. Note the indicated leachables peak (***) whose apex is slightly above the final AET level of 1.88 µg/canister (also note that many additional leachables peaks are clearly visible at significantly lower levels). The electron ionization mass spectrum from this leachables peak is shown in Figure 5.9. An experienced analytical chemist would immediately know from this mass spectrum that the compound is an ethyl ester of an aliphatic acid (characteristic ions at m/z 88 and 101) with molecular weight 256 (molecular ion at m/z 256). A computerized library search confirms the identification as tetradecanoic acid, ethyl ester, or ethyl myristate (Fig. 5.10). Knowledge of the formulation of this particular MDI drug product, along with data from extractables studies of dose metering valve critical components, suggests that ethyl myristate likely formed from a chemical reaction between ethanol in the drug product formulation and myristic acid (a confirmed potential leachable). Ethyl myristate is therefore not a leachable itself, but is qualitatively correlated with an extractable. Clearly, GC/MS has more than adequate sensitivity to facilitate the identification of leachables in this particular drug product at AET levels.

An idea of the ultimate sensitivity of both GC/MS and LC/MS can be obtained through the analysis of authentic reference compounds. Figure 5.11 shows an expanded portion of a TIC (A), a mass chromatogram for m/z 132 (B) and an electron

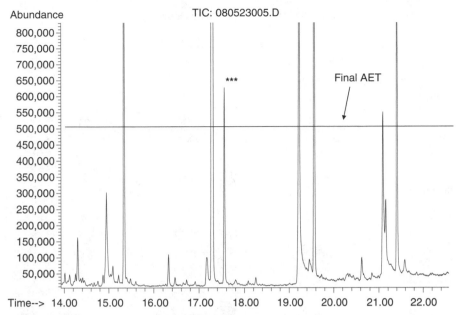

Figure 5.8   Expanded region of a gas chromatography/mass spectrometry (GC/MS) leachables profile of a metered dose inhaler drug product (in the form of a total ion chromatogram [TIC] from Fig. 5.6). Note the location of the final AET for this profile and a leachable of interest (***) just above the final AET.

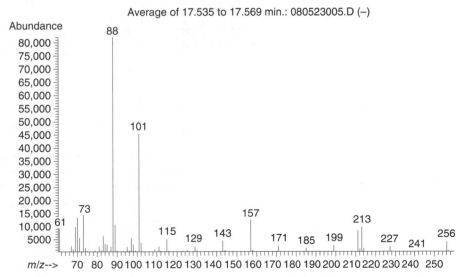

Figure 5.9   Electron ionization (EI) mass spectrum of the leachable of interest (***) from Figure 5.8.

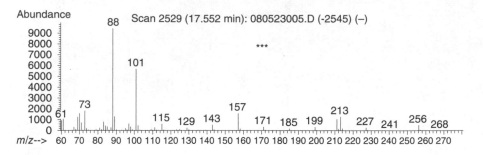

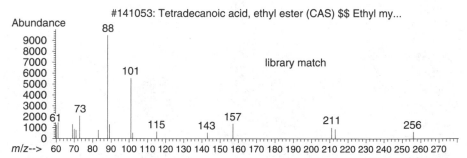

Figure 5.10 Computerized mass spectral library match for the leachable of interest from Figure 5.8. Note the confirmed identification for ethyl myristate (tetradecanoic acid, ethyl ester).

ionization mass spectrum from tetramethylthiourea (C; molecular weight 132; 130 pg on-column).

Tetramethylthiourea

Tetramethylthiourea is often detected as a leachable/extractable when sulfur-cured elastomers are included as MDI dose metering valve critical components. Note that significant fragment ion information is available in this mass spectrum (m/z 88, m/z 72, etc.), and that the overall quality of the spectrum is sufficient to facilitate computerized library searching. The signal-to-noise ratio of the m/z 132 mass chromatogram should allow for accurate quantitation at this level.

Figure 5.12A shows a mass chromatogram (m/z 1175) from the molecular ion of the phenolic antioxidant Irganox® 1010 (Ciba Specialty Chemicals Corporation, Tarrytown, NY) produced by negative ion atmospheric pressure chemical ionization (APCI) LC/MS (1 ng on-column). Irganox 1010 is a commonly used antioxidant in certain types of plastics, which can be used to fabricate many types of OINDP container closure system components (Fig. 5.12B).

Figure 5.12C presents a portion of the negative ion APCI mass spectrum of Irganox 1010, which clearly shows the molecular ion with a high signal-to-noise ratio.

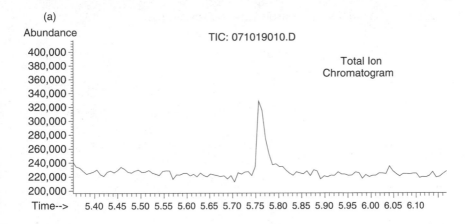

(a)

Abundance

TIC: 071019010.D

Total Ion
Chromatogram

400,000
380,000
360,000
340,000
320,000
300,000
280,000
260,000
240,000
220,000
200,000

Time--> 5.40 5.45 5.50 5.55 5.60 5.65 5.70 5.75 5.80 5.85 5.90 5.95 6.00 6.05 6.10

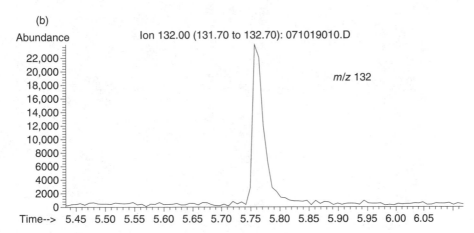

(b)

Abundance

Ion 132.00 (131.70 to 132.70): 071019010.D

m/z 132

22,000
20,000
18,000
16,000
14,000
12,000
10,000
8000
6000
4000
2000
0

Time--> 5.45 5.50 5.55 5.60 5.65 5.70 5.75 5.80 5.85 5.90 5.95 6.00 6.05

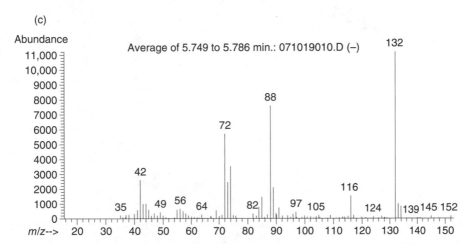

(c)

Abundance

Average of 5.749 to 5.786 min.: 071019010.D (–)

132

11,000
10,000
9000
8000
7000
6000
5000
4000
3000
2000
1000
0

88

72

42

35   49 56 64   82   97 105   116   124   139 145 152

m/z--> 20   30   40   50   60   70   80   90   100 110 120 130 140 150

Figure 5.11   (a) Expanded region of a gas chromatography/mass spectrometry (GC/MS) total ion chromatogram (TIC) showing a peak for 130 pg (on-column) of tetramethylthiourea authentic reference material. (b) Expanded region of a mass chromatogram (m/z 132; molecular ion of tetramethylthiourea) from this GC/MS analysis. (c) Electron ionization (EI) mass spectrum of tetramethylthiourea from this GC/MS analysis.

(a)

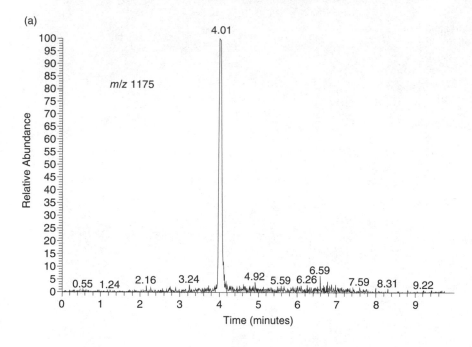

(b)

Figure 5.12   (a) Mass chromatogram (*m/z* 1175) from the atmospheric pressure chemical ionization (APCI) liquid chromatography/mass spectrometry (LC/MS) analysis of Irganox 1010 (1 ng on-column). (b) Irganox 1010. (c) Negative ion APCI mass spectrum of Irganox 1010 (molecular ion region; [M − H]⁻ at *m/z* 1175).

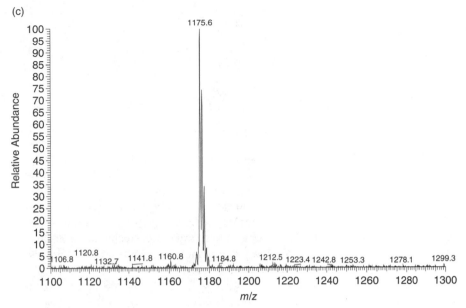

(c)

Figure 5.12   (*Continued*)

The relatively simple examples presented clearly depict the dilemma faced by OINDP pharmaceutical development scientists. Modern analytical techniques and methods are capable of enormous sensitivity and can produce information sufficient to identify and quantitate leachables at very low levels. The AET concept addresses this dilemma by establishing a safety-based threshold and answering the question: *How low do we go?*

5.5 SPECIAL CASE COMPOUNDS

The PQRI Leachables and Extractables Working Group did consider certain leachables to be outside the scope of the threshold concept due to special safety.[1-3] Polycyclic aromatic hydrocarbons (PAHs, or polynuclear aromatics [PNAs]), *N*-nitrosamines, and 2-mercaptobenzothiazole (2-MBT) are considered to be "special case" compounds, requiring special characterization studies using specific analytical techniques/methods and technology-driven thresholds. Table 5.2 lists the PNAs and *N*-nitrosamines that are typically investigated as extractables and leachables in OINDP. Chemical structures of an example PAH (pyrene), an *N*-nitrosamine (*N*-nitrosodimethylamine), and 2-MBT are as follows:

Pyrene

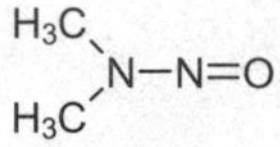

$$H_3C \diagdown N-N=O$$
$$H_3C \diagup$$

N-nitrosodimethylamine

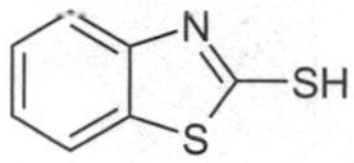

2-MBT

PAHs/PNAs have been associated with carbon black filler used in many types of elastomer, including the sulfur-cured elastomers.[7] Analysis of PAHs/PNAs, either as elastomer extractables or as drug product leachables, usually involves quantitative extraction followed by highly specific and sensitive analysis of the resulting extracts. GC/MS with selected ion monitoring (SIM) has been reported for analysis of target PAHs/PNAs in MDI drug products, for example.[7] Analytical techniques such as GC/MS with SIM are capable of detecting and quantitating PAHs/PNAs at nanogram per canister levels in MDI drug products and low parts per million levels in rubber. N-nitrosamines are reaction products between specific organic precursor molecules, secondary amines (R$_2$NH), and a "nitrosating agent" (NOX).[8] In the compounding

TABLE 5.2.   PAHs/PNAs and N-Nitrosamines Typically Investigated as Extractables and Leachables for OINDP

PAHs/PNAs	N-nitrosamines
Naphthalene	N-nitrosodimethylamine
Acenaphthylene	N-nitrosodiethylamine
Acenaphthene	N-nitrosodi-n-butylamine
Fluorene	N-nitrosomorpholine
Phenanthrene	N-nitrosopiperidine
Anthracene	N-nitrosopyrrolidine
Fluoranthene	
Pyrene	
Benzo(a)anthracene	
Chrysene	
Benzo(b)fluoranthene	
Benzo(k)fluoranthene	
Benzo(e)pyrene	
Benzo(a)pyrene	
Indeno(1,2,3-cd)pyrene	
Dibenzo(ah)anthracene	
Benzo(ghi)perylene	

of rubber, secondary amines are likely formed from certain vulcanization accelerators such as thiurams and dithiocarbamates. For example, tetramethylthiuramdisulfide (**I**) can liberate dimethylamine (**II**), which can then react to form N-nitrosodimethylamine (**III**) as depicted in a simplified reaction sequence below:

I   II   III

Potential nitrosating agents include NO+, N_2O_3, and N_2O_4. Some of these can be formed from commonly used chemicals such as sodium nitrite ($NaNO_2$), which has many industrial uses.[8] The formation of N-nitrosamines in rubber has been extensively studied. The analysis of N-nitrosamines in rubber as potential extractables is accomplished by quantitative extraction followed by analysis of extracts with gas chromatography/thermal energy analysis (GC/TEA®) detection.[5] GC/TEA is based on the phenomenon of chemiluminescence, a complete discussion of which is beyond the scope of this chapter. For a more thorough discussion of N-nitrosamines in rubber and their analysis as extractables, the reader is referred to the previously indicated citations.[5,8] Sensitivities for N-nitrosamine analytical techniques/methods for rubber are in the low parts per billion range, which could correlate with low nanogram per canister levels in MDIs, although at the time of this writing, no literature is available regarding the analysis of N-nitrosamines as leachables in inhalation drug products.

2-MBT is a known ingredient in certain sulfur-cured elastomers and functions as a vulcanization accelerator when used in combination with other vulcanization agents.[9] It is most often analyzed by LC/UV- and LC/MS-based analytical methods.[10–12]

5.6 SUMMARY AND CONCLUSIONS

The safety threshold (i.e., QT and SCT) and AET concepts represent a significant step forward in the pharmaceutical development process for OINDP. However, both the AET concept and the process for AET determination have limitations. For example, while it might be relatively easy to determine both *estimated* and *final AETs* for extractables/leachables profiles acquired by GC/MS, GC/FID, and LC/UV detection, it might not be so simple for a technique like LC/MS, which does not create a readily usable extractables/leachables profile due to relatively high background levels of chemical noise. However, in spite of its limitations, the AET concept represents a significant reduction in the uncertainty associated with the OINDP pharmaceutical development process, which translates to significant gains in pharmaceutical development efficiency.

For fundamental information on modern analytical chemistry (and in particular GC/MS and LC/MS) and applications for trace organic analysis, the interested reader is referred to the following comprehensive treatises.[13–15]

REFERENCES

1   Product Quality Research Institute (PQRI) and Leachables and Extractables Working Group. Safety thresholds and best practices for extractables and leachables in orally inhaled and nasal drug products. Product Quality Research Institute, Arlington, VA, 2006.

2   Ball, D., Blanchard, J., Jacobson-Kram, D., McClellan, R., McGovern, T., Norwood, D.L., Vogel, W.M., Wolff, R., and Nagao, L.M. Development of safety qualification thresholds and their use in orally inhaled and nasal drug product evaluation. *Toxicol Sci* 2007, *97*(2), p. 226.

3   Norwood, D.L., Paskiet, D., Ruberto, M., Feinberg, T., Schroeder, A., Poochikian, G., Wang, Q., Deng, T.J., DeGrazio, F., Munos, M.K., and Nagao, L.M. Best practices for extractables and leachables in orally inhaled and nasal drug products: An overview of the PQRI recommendations. *Pharm Res* 2008, *25*(4), p. 727.

4   Guidance for industry: Container closure systems for packaging human drugs and biologics. Department of Health and Human Services, Food and Drug Administration, 1999.

5   Norwood, D.L., Granger, A.T., and Paskiet, D.M. *Encyclopedia of Pharmaceutical Technology*, 3rd ed. Dekker Encyclopedias (a product line from Taylor and Francis Books), New York, 2006; 1693–1711.

6   Norwood, D.L., Nagao, L., Lyapustina, S., and Munos, M. Application of modern analytical technologies to the identification of extractables and leachables. *Am Pharm Rev* 2005, *8*(1), pp. 78–87.

7   Norwood, D.L., Prime, D., Downey, B.P., Creasey, J., Sethi, S.K., and Haywood, P. Analysis of polycyclic aromatic hydrocarbons in metered dose inhaler drug formulations by isotope dilution gas chromatography/mass spectrometry. *J Pharm Biomed Anal* 1995, *13*(3), p. 293.

8   Willoughby, B.G. and Scott, K.W. *Nitrosamines in Rubber*. Rapra Technology Ltd., Shawbury, UK, 1997.

9   Morton, M. *Rubber Technology*, 3rd ed. Kluwer Academic Publishers, Dordrecht, 1999.

10 Bergendorff, O., Persson, C., and Hansson, C. High-performance liquid chromatography analysis of rubber allergens in protective gloves used in health care. *Contact Dermatitis* 2006, *55*(4), p. 210.

11 Kloepfer, A., Jekel, M., and Reemtsma, T. Determination of benzothiazoles from complex aqueous samples by liquid chromatography-mass spectrometry following solid-phase extraction. *J Chromatogr A* 2004, *1058*(1–2), p. 81.

12 Reemtsma, T. Determination of 2-substituted benzothiazoles of industrial use from water by liquid chromatography/electrospray ionization tandem mass spectrometry. *Rapid Commun Mass Spectrom* 2000, *14*(17), p. 1612.

13 Gross, J.H. *Mass Spectrometry: A Textbook*. Springer-Verlag, Berlin/Heidelberg, 2004.

14 de Hoffmann, E. and Stroobant, V. *Mass Spectrometry: Principles and Applications*, 3rd ed. John Wiley & Sons, Ltd., Chichester, England, 2007.

15 Boyd, R.K., Basic, C., and Bethem, R. *Trace Quantitative Analysis by Mass Spectrometry*. John Wiley & Sons, Ltd., Chichester, England, 2008.

CHAPTER **6**

SAFETY THRESHOLDS IN THE PHARMACEUTICAL DEVELOPMENT PROCESS FOR OINDP: AN INDUSTRY PERSPECTIVE

David Alexander and James Blanchard

6.1 INTRODUCTION

Safety thresholds for leachables and extractables in orally inhaled and nasal drug products (OINDPs) should be used within a rational and cohesive pharmaceutical development process, which also encompasses the safety qualification process. This chapter will describe, from an industry perspective, the use of safety thresholds in OINDP, the benefits of safety thresholds, and how thresholds can be used generally for safety qualification of OINDP within the pharmaceutical development process.

6.2 USE OF SAFETY THRESHOLDS IN OINDP: HISTORY AND BACKGROUND

6.2.1 The Toxicologist's Role: 1950s to 1985

From the first introduction of the metered dose inhaler (MDI) in the 1950s until about the middle of the 1980s, the role of the toxicologist in terms of the safety qualification of leachables and extractables was almost nonexistent.

Although early MDI drug products (MDIDPs) did not undergo a comprehensive extractables and leachables evaluation, they did undergo toxicological assessment and generally showed no adverse toxicity that could be associated with the formulation. Thus, by definition, the elastomers and valve leachables were "qualified." One of the first potential safety issues associated with MDIDP was carbon

Leachables and Extractables Handbook: Safety Evaluation, Qualification, and Best Practices Applied to Inhalation Drug Products, First Edition. Edited by Douglas J. Ball, Daniel L. Norwood, Cheryl L.M. Stults, Lee M. Nagao.
© 2012 John Wiley & Sons, Inc. Published 2012 by John Wiley & Sons, Inc.

TABLE 6.1. Example of Ingredients in a Sulfur-Cured Elastomer Test Article

Ingredient	CAS #	Percent (w/w)
Calcined clay	308063-94-7	8.96
Blanc fixe (barium sulfate)	7727-43-7	25.80
Crepe	9006-04-6	38.22
Brown sub MB (ingredients below)	NA (not available)	16.84
Brown sub loose	NA	33.30
Crepe	9006-04-6	66.70
1722 MB (ingredients below)	NA	2.11
Standard Malaysian rubber (SMR)	NA	60.00
FEF carbon black (low PAH)	1333-86-4	40.00
Zinc oxide	1314-13-2	4.04
2,2′-Methylene-bis(6-*tert*-butyl-4-ethylphenol)	88-24-4	0.56
Coumarone-indene resin	164325-24-0	1.12
	140413-58-7	
	140413-55-4	
	68956-53-6	
	68955-30-6	
Paraffin	8002-74-2	1.12
	308069-08-1	
Tetramethylthiuram monosulfide	97-74-5	0.11
Zinc 2-mercaptobenzothiazole	149-30-4	0.29
	155-04-4	
Sulfur	7704-34-9	0.84

CAS, Chemical Abstracts Service; MB, medium brown; NA, not available; FEF, fast extrusion furnace.

black, which was used as a filler in most valve elastomers. Spraying older MDIs onto a white surface could, with some formulations, produce a black deposit derived from the elastomers, but scant concern was paid to this effect. Although data were not published, some early formulations are rumored to have produced more extraneous material than drug.

Investigation and analysis of these older, carbon black valves showed that they contained multiple extractives with potentially toxic implications including but not limited to polyaromatic hydrocarbons (PAHs) and nitrosamines along with many other leachables. Typical analysis would often identify 100 or more peaks using, what was at the time, advanced analytical techniques.

An example of a sulfur-cured elastomer containing carbon black is shown in Table 6.1. The extractables and/or leachables from older, sulfur-cured elastomers could contain some or all of the following materials (Table 6.2) of known toxicological concern.

Many PAHs and nitrosamines are possible direct-acting carcinogens and have been shown in both humans and rats to induce tumors including bronchial and bronchiolar–alveolar carcinomas and adenomas. Therefore, potentially delivering PAHs and nitrosamines directly to the respiratory tract of patients is not acceptable.

TABLE 6.2.   **Polyaromatic Hydrocarbons (PAHs)
and Nitrosamines**

PAH/PNA[a]	N-nitrosamines
Naphthalene	N-nitrosodimethylamine
Acenaphthylene	N-nitrosodiethylamine
Acenaphthene	N-nitrosodi-n-butylamine
Fluorene	N-nitrosomorpholine
Phenanthrene	N-nitrosopiperidine
Anthracene	N-nitrosopyrrolidine
Fluoranthene	
Pyrene	
Benzo(a)anthracene	
Chrysene	
Benzo(b)fluoranthene	
Benzo(k)fluoranthene	
Benzo(e)pyrene	
Benzo(a)pyrene	
Indeno(1,2,3-cd)pyrene	
Dibenzo(ah)anthracene	
Benzo(ghi)perylene	

[a] PAHs are known also as polynuclear aromatics (PNAs).

One cause for concern in the past was the lack of process control on the source and quality of the elastomers and polymers used to construct the valves. Suppliers could and did change the specification or composition of the input materials throughout the life of the product often with little or no communication of such changes to the pharmaceutical manufacturer. Under most circumstances, there appeared to be little issue with this approach, and for more than 40 years, patients received effective medication delivered via convenient discrete MDIs. As noted in Chapter 1, however, in the mid- to late 1980s, regulators and the industry became more aware of the presence of and potential safety issues associated with PAHs and nitrosamines.

In 1991, Murphy et al.[1] undertook a study that showed that a chlorofluorocarbon (CFC) formulation delivered from MDIs with valves constructed with an acetyl metering chamber was more irritating to the larynx of rats than the same formulation delivered from MDIs constructed with stainless steel metering chambers. The laryngeal changes seen included necrosis of the ventral cartilage, epithelial hyperplasia, epithelial squamous metaplasia and keratinization, the presence of subepithelial inflammation/fibrosis/granulation tissue, and atrophy of the submucosal glands in all animals. In comparison, the rats exposed to the formulation delivered from the MDIs fitted with the stainless steel metering chambers showed minimal focal epithelial hyperplasia and focal squamous metaplasia in just 20% of the exposed rats. These results suggest that some leachable from the acetyl was causing the increased and more extensive laryngeal irritancy.

6.2.2 The Toxicologist's Role: 1985–1999

Toward the end of the 1990s, coupled with concerns over the ozone-depletion potential of the CFC propellants used in the MDI industry, the international regulatory community became more aware of and concerned with the presence of leachables in OINDPs. Their response, in some cases, was to request a toxicological assessment of each extractable and leachable that was detected in a new OINDP. This involved full analysis of fresh and end-of-shelf-life products involving qualitative identification and quantification. The analysis covered not only the more conventional MDI, but also mouthpieces, dry powder inhaler (DPI) formulations, nebulizers/nebule formulations, and solutions for inhalation. As the investigations were undertaken, it became apparent that leachables were not only related to MDI container closure systems or devices. Migration of solvents and dyes was possible across materials that previously had been considered relatively impervious. For example, leachables from glues used to apply labels to the outer wraps of nebule formulations were being found inside the nebule formulation. Plasticizers and other unexpected leachables were being found in minute (e.g., nanogram and picogram) quantities. Multiple complex analyses were being undertaken by the analysts to identify these compounds. Such analyses, especially when conducted following completion of other nonclinical and clinical evaluations and development, consumed a great deal of time and manpower, and could, in some cases, delay the introduction of new products often by a year or more.

At this time, the valve manufacturers, pharmaceutical industry, and international regulatory community had started to recognize the safety implications associated with the inclusion of sulfur and carbon black in the valve elastomers. There resulted a rapid switch to white, peroxide-cured elastomers that removed most polynuclear aromatics (PNAs) and nitrosamines. Additionally, valve manufacturers introduced pre-extraction to the valve components by washing the components in CFC-11. Subsequently, following the phaseout and general unavailability of CFC-11, most valve manufacturers have moved to an ethanol pre-extraction step for the valve components prior to delivery to the pharmaceutical industry. These processes have produced significant reductions in the extractables from the elastomers and polymer components and therefore of the leachables seen in the products.

A full analysis, however, was still required where toxicologists were asked to provide safety assessments for every leachable irrespective of the levels. Rarely, due to time constraints in the development programs, were the final valve assemblies used in the long-term toxicology studies and, if they were, usually the valve assemblies and products were at the beginning of their shelf life.

One major issue with leachables levels is that some will increase over the shelf life of the product, some will decrease, and others will remain fairly constant following an initial leaching period. Thus, toxicological evaluations in long-term studies generally use freshly produced formulations. The levels of leachables seen at the beginning of product shelf life were not necessarily relevant to the product's leachables profile at the end of shelf life.

Other types of recommended safety evaluations included pharmacopeial and International Organization for Standardization (ISO) tests. USP <87> and <88> and/

or ISO 10993[2-4] assessments were generally required for suppliers of OINDP device components. The United States Food and Drug Administration (FDA) also recommends USP <87> and <88> testing for OINDP manufacturers. Brief details of the tests are given below.

USP <87> (USP) determines the biological reactivity of mammalian cells in culture following contact with elastomeric and polymers. Three tests are specified as

- an agar diffusion test in which extracts from elastomeric components are applied to a layer of agar overlying a culture of mammalian fibroblast cells for not less than 24 hours;

- a direct contact test in which materials are placed in direct contact with the fibroblasts for not less than 24 hours; and

- an elution test in which extracts are taken under physiological and nonphysiological temperature from polymeric components and applied to fibroblasts for 48 hours.

USP <88> (USP) determines the biological activity *in vivo*. Three tests that are involved are

- a systemic test in which extracts from the components are injected systemically into mice;

- an intracutaneous test in which extracts from the components are injected intracutaneously into rabbits; and

- an implantation of materials in which strips of polymeric materials are implanted intramuscularly into rabbits.

ISO 10993 is the European Pharmacopeia directive for the biological evaluation of medical devices and defines a range of testing applicable to medical device testing and Conformité Européene (CE) marking that includes cytotoxicity, sensitization, irritation, systemic toxicity, subchronic toxicity, implantation, hemocompatability, chronic toxicity, and carcinogenicity.

In many cases, both OINDP manufacturers as well as regulators, considered these tests too insensitive to detect anything but significant safety effects of the materials used in OINDP container closure systems and devices, and thus inadequate to justify the levels of leachables in drug products. Subsequently, the Product Quality Research Institute (PQRI) Leachables and Extractables Working Group (2006)[5] recommended that the device and/or drug product manufacturers should not be required to undertake these tests when more comprehensive toxicological evaluations have been undertaken.

The use of materials permitted elsewhere in the world, for example, pigments permitted for baby products in Europe, was generally considered unacceptable in the United States unless the individual material was listed for food contact use in the Code of Federal Regulations (CFR) (21 CFR 174-21 CFR 190) documentation. Therefore, toxicologists performed additional studies, often of 90 days' duration, to qualify, for example, a valve leachables profile.

The situation in Europe and the rest of the world generally tended to be less specific. The regulators were taking a "wait-and-see" approach based on developments in the United States. Already the industry had switched from sulfur-cured elastomers with high levels of PAHs and nitrosamines to peroxide-cured prewashed elastomers. Products were showing lower levels of leachables, particularly PAHs or nitrosamines, than historically seen. Provided versions of the valve containing the same or very similar materials of construction had been used in the toxicological assessment; regulators outside the United States tended to be less demanding for individual assessments of each leachable and assessed product on a case-by-case basis.

6.2.3 The Toxicologist's Role: 1994–2006

The years 1994–1996 saw the development of the tripartite International Conference on Harmonisation (ICH) Q3A, B, and C guidelines that included the concept of thresholds.[6–8] The thresholds, typically <0.1% for low-dose OINDP products, considered within Q3A and B were too high to be sensibly applied to potentially more toxic leachables. Additionally, the ICH Q3B (R2) guideline specifically excludes leachables. However, the international regulatory community had started to accept the principle that there is a level of an impurity or degradant that is below that which may cause a toxicological concern.

In November 1998 and June 1999, the FDA issued for comment two draft CMC guidances for the industry[9,10]:

1. MDI and DPI drug products chemistry, and manufacturing and controls documentation; and

2. nasal and inhalation solution, suspension and spray drug products chemistry, and manufacturing and controls documentation.

These draft guidelines defined FDA requirements for extractables and leachables testing, which required the identification and quantification of each peak detected.

Subsequently, a threshold of toxicological concern (TTC) was described by Kroes et al.[11] as "a level of exposure for all chemicals, whether or not there are chemical-specific toxicity data, below which there would be no appreciable risk to human health."

This threshold principle was used in some pharmaceutical company submissions to the FDA. Likely, human daily exposure to specified leachables was compared against known toxic materials including some carcinogens, such as benzopyrenes and other highly toxic compounds. Although many of these early attempts failed, there began a gradual acceptance of the threshold concept for leachables in OINDP. The general level of safety considered acceptable was approximately a 1 in 1 million risk of an additional cancer formation (above the baseline frequency).

As the threshold concept matured, at least one pharmaceutical company adopted the approach of comparing the likely exposures to leachables with the normal daily exposure to environmental pollutants in nonindustrial rural environments, showing that the added exposure to leachables from OINDP was small compared with "normal" day-to-day exposure of a patient's respiratory tract to known and unknown environmental pollutants. These comparisons appeared, in some cases, to be acceptable to regulators.

This approach was supported by the general acceptance in the United States and Europe of the specifications for the alternative propellants HFA 134a and HFA 227ea, both of which contained impurities that in themselves, at certain levels, were toxic. However, the two propellants had undergone extensive nonclinical evaluation, which was sponsored by the International Pharmaceutical Aerosol Consortium on Toxicology Testing (IPACT) I and II consortia for HFA 134a and HFA 227ea, respectively, and determined that the impurities had no toxicological impact at the levels seen in those propellants.[12]

In 2001, the Inhalation Technology Focus Group (ITFG) of the American Association of Pharmaceutical Scientists and the International Pharmaceutical Aerosol Consortium on Regulation and Science (ITFG/IPAC-RS Collaboration) submitted their comments to the FDA draft guidance documents. The ITFG/IPAC-RS group had initiated a scientific, data-driven collaboration to address specific issues in the draft FDA guidance documents and first introduced a proposal for the establishment of reporting and qualification thresholds (QTs) for leachables in OINDP, contained in the Extractables and Leachables Testing: Points to Consider.[13]

Subsequently, at the suggestion of the FDA, a larger group of stakeholders, under the auspices of the PQRI was formed. This more broad-based approach took in the views of the regulators, industry, and academics to develop the IPAC-RS fundamental concepts into a more widely acceptable threshold-based approach. The concept and draft documents went through much iteration, and the final version, published in 2006,[5] described in detail elsewhere in this book, and published by Ball et al.,[14] was agreed to by the PQRI Steering Committee, formally submitted to the FDA, and is currently being successfully applied in submissions to regulatory authorities around the world. It must be noted that, at the time of publication, the PQRI recommendations have not been formally adopted as policy by regulatory authorities.

6.2.4   The Toxicologist's Current Approach

The modern assessment of extractables and leachables starts early in the development of the product. As early as material selection, the formulators, chemists, and toxicologists should be discussing the final expected formulation and packaging, that is, whether the products will be an MDI, DPI, or nebulizer formulation and packaging types.

If an MDI is proposed, then the valve, container, and mouthpiece manufacturers are often consulted to discuss the potential materials of construction. Potential formulation compatibility is one of the first considerations in this process. Material selection is then based on well tried, tested, and wherever possible 21 CFR-approved materials or materials that have been used and thus previously toxicologically qualified. Specifications for input materials, pre-extraction, and construction methods can be selected. Control at the source by the manufacturer of the container closure and packaging materials becomes critical for the provision of controllable and known leachables levels in the final product. Using this prior knowledge and experience makes for easier regulatory passage later in development.

As product development progresses, inevitably there will be changes in formulations, packaging, and the like, but usually by the end of the phase II clinical program, a good indication of the final formulation, delivery device, and materials of construction are available. These can be defined and subjected to stringent change control processes, ensuring that the manufacturers of the container closure systems and packaging supply materials and components of consistent quality and formulation to the pharmaceutical manufacturers who have control over the final product quality and performance.

At this point, the analyst, formulator, and toxicologist should be looking for the unexpected in terms of leachables reaching the product from the container closure system, device, overwrap, inks, labels, adhesives, or other packaging material used for the product. The source of any unexpected peaks seen on the chromatograms/analyses should be identified to ensure that there are no unexpected leachables reaching the product.

The role of the toxicologist is to work with the analysts and formulators to start the extractables and leachables assessments as it can take many months to complete the work. Under normal circumstances, the final formulation with containers/closures/mouthpieces or other product-specific delivery devices where required, are usually defined, and the longer-term toxicology work, that is, chronic toxicity studies, oncogenicity studies, and reproductive toxicology studies should be designed and conducted, wherever possible, to take packaging materials into consideration. A typical example would be to use the mouthpieces for delivery of an MDI formulation to nonrodents.

The analysts undertake full controlled extraction studies using polar and nonpolar solvent systems of the components to determine the extractables profiles. Modern elastomers, such as peroxide-cured elastomers, can show no PAHs or nitrosamines in the extractables profiles; however, usually for MDIs, the FDA recommends the establishment of acceptance criteria for these compounds.

Leachables analysis will include product samples taken from accelerated or real-time stability studies and those used on the longer-term toxicity studies (under ideal circumstances these come from the same manufacturing batches). The liaison between the toxicologist and analysts throughout the pharmaceutical development process (from selection of materials, through controlled extraction studies, and through leachables studies) is crucial to ensure that potential safety issues for extractables/leachables are addressed and understood throughout the process, and that relevant data are collected. Table 6.3 provides examples of the type of data that would be collected.

Chapter 7 describes in more detail the collaboration between the toxicologist, the analyst, and other members of the pharmaceutical development team, as well as the types of data that are obtained and shared in the extractables/leachables evaluation process. Table 6.4 lists examples of analytes that might be found in an extractables and leachables evaluation.

PAHs are considered special case compounds by the FDA and require evaluation by specific analytical techniques.[5] Leachables might also include pigments from the mouthpieces, silicone oil lubricants, and mold release agents used during the manufacturing process. Given the results from these analyses, the toxicologist undertakes a review of the data and conducts a safety qualification of the leachables.

TABLE 6.3. Example Measurements for Extractables and Leachables

Extractables	Leachables
Qualitative analysis of analytes above an analytical evaluation threshold following controlled extraction studies using different solvents, for example, dichloromethane, ethanol, and water	Qualitative analysis looking for the leachables from product prepared fresh and taken from real-time or accelerated storage programs
Quantification of extractable/gram of material	Quantification of leachables above safety thresholds
Duplicate analysis from several batches of input material to give range of results. At least three batches are recommended.	Duplicate analysis from several batches of product to assess range of leachables values seen for each analyte. At least three batches are recommended.
Weight of material in a single valve/closure or device component	Tabulation of methods used, limits of quantification and typical chromatograms, and validation of methods
Calculation of the amount of extractable/ component in the closure/device	Tabulation of levels of leachables seen from several batches of product/devices
List of extractables with quantification, sensitivity of methods, and results for submission	

TABLE 6.4. Examples of Possible Analytes from an MDI Extractables/Leachables Program

Container closure component(s)	Analyte	PAH analyte	Nitrosamine analyte
Valve body, dosage chamber, lower stem, upper stem	Irganox 1010 Irganox 1076	PBT dimer PBT trimer PBT tetramer PBT pentamer	
Ring	Irganox 245 Butylated hyroxytoluene (BHT) Oleamide Irganox 259 Irganox 1010 Irganox 1076 Irgafos 168		

(Continued)

TABLE 6.4. (*Continued*)

Container closure component(s)	Analyte	PAH analyte	Nitrosamine analyte
Gaskets	2,2-Methylene (6-*tert*-butyl-4-methylphenol)	Naphthalene	N-nitososdimethylamine
		Acenaphthylene	N-nitrososmethylethylamine
		Fluorene	N-nitrososdiethylamine
	Dodecanethiol	Anathrancene	N-nitroso-di-*n*-propylamine
	Abietic acid	Phenanthrene	N-nitrosodibutylamine
	1.6-Dichloro-1,5-cyclo-octadiene	Fluoranthene	N-nitrosospiperidine
	Irganox 245	Chrysene	N-nitosospyrrolodine
	Irganox 259	Benzo(a)pyrene	N-nitrososmorpholine
	Irganox 1010	Benz(k) and (b) fluoranthene	N-nitroso-di-phenylamine
	Irganox 1076	Dibenzo(a,h) anthracene	2-Mercaptobenzothiazole (2-MBT)
	Kemamide	Benzo(ghi)perylene	
	Irgafos 168	Indenol (1,2,3-cd) pyrene	
	BHT	Acenaphthene	
	2,2-Methylene-bis(6-*tert*-butyl-4-methylphenol)	Benzo(a)anthracene	
Canister lining	Polytetrafluorocarbon monomers and polymers		

PBT, polybutylene terephthalate.

6.3 QUALIFICATION OF LEACHABLES

This section describes the recommended process to qualify a leachable, particularly when the leachable exceeds the safety concern threshold (SCT) proposed to be 0.15 μg/day or the QT proposed to be 5 μg/day. A key element is the use of a decision tree to guide the selection of appropriate qualification steps according to the threshold and the total daily intake (TDI) of the leachable. The TDI of the leachable is based on the maximum daily dose of the drug product per day, assuming the worst case that the entire inhaled dose is delivered to the lungs.

The qualification steps considered appropriate depend on the patient population, the TDI of the leachable, and the duration of drug administration. The qualification steps consist of two types: (1) risk assessment, including structure–activity relationship (SAR) analysis and literature review; and (2) safety studies, including genotoxicity, general toxicity, and specialized toxicity studies, as needed. The risk assessment is the recommended first step; however, if there are insufficient SAR or literature data to qualify the leachable, then safety studies may be required. In either case, the qualification process may determine that a higher or lower threshold for a leachable is warranted based on scientific rationale, potency, and level of concern. In such a case, an alternate acceptable level may be proposed to regulators and treated on a case-by-case basis.

It is recognized that when the leachable is present at a very low level, it may be problematic to have a confirmed identity for SAR analysis. Instead, there may be sufficient structural data to preclude all but the most closely related compounds, or, if even less structural data are available, the identity may be limited to only the class of molecule. In the latter instance, the risk assessment would be based on the general class of compound.

Safety studies would usually be conducted on the OINDP that contains the leachable; however, studies on the isolated leachable may sometimes be appropriate. In either case, the level of the leachable should represent end-of-shelf-life conditions. Genotoxicity studies would involve a minimum of two *in vitro* tests, such as the bacterial point mutation and mammalian chromosomal aberration. General toxicity studies would compare unqualified with qualified material. Studies would be done in one relevant species. The study duration would usually be 14–90 days, depending on the duration of patient dosing; however, single-dose studies may be appropriate for single-dose drugs. As an example of a specialized toxicity study, if the leachable has a SAR alert for carcinogenicity, then carcinogenicity studies may be needed.

It is noteworthy to reiterate that when drug product manufacturers have access to comprehensive *in vivo* safety data on the leachable, neither the USP biocompatibility tests <87>[2] nor <88>[3] nor the ISO biocompatibility test 10993[4] needs to be performed. However, these tests may be appropriate for suppliers of OINDP device components.

6.4   USE OF THRESHOLDS FOR QUALIFICATION: DECISION TREE

The "decision tree for safety qualification" (Fig. 6.1) describes considerations for the qualification of leachables when thresholds are exceeded. The decision tree was proposed in the PQRI leachables and extractables (L&E) recommendations[5,14] and is used with the proposed PQRI SCT and QT. The format and qualification steps of the decision tree are based on ICH guidelines for impurities in new drug substances (Q3A)[6] and new drug products (Q3B).[7] In some cases, decreasing the level of a leachable to not more than the threshold can be simpler than providing safety data. Alternatively, as mentioned above, adequate data could be available in the scientific literature to qualify a leachable. If neither is the case, additional safety testing should be considered. The studies considered appropriate to qualify a leachable will depend on a number of factors, including the patient population, daily dose, and duration of drug administration. The decision tree is described as follows:

A. If the leachable's TDI ≤ SCT, then it is qualified (i.e., no further action is required).

An exception would be if the leachable is a compound of special toxicological concern (e.g., nitrosamines, PAHs, or mercaptobenzothiazole, which

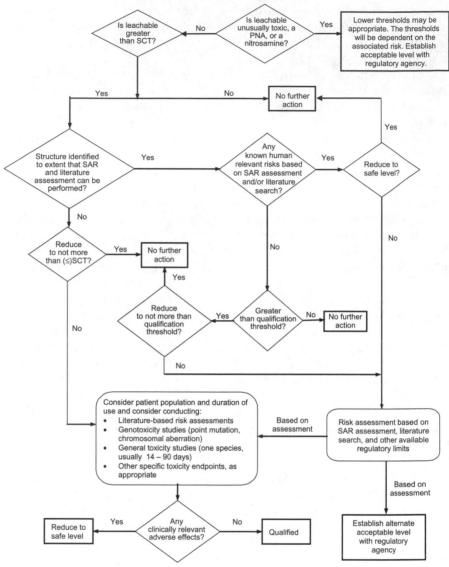

Figure 6.1   Decision tree for safety qualification. If considered desirable, a minimum screen, e.g., genotoxic potential, should be conducted. A study to detect point mutations and one to detect chromosomal aberrations, both *in vitro*, are considered appropriate minimum screens. If general toxicity studies are desirable, one or more studies should be designed to allow comparison of unqualified to qualified material. The study duration should be based on available relevant information and performed in the species most likely to maximize the potential to detect the toxicity of a leachable. On a case-by-case basis, single-dose studies can be appropriate, especially for single-dose drugs. In general, a minimum duration of 14 days and a maximum duration of 90 days would be considered appropriate. For example, do known safety data for this leachable or its structural class preclude human exposure at the concentration present? Reproduced from Ball et al.[14]

are all unusually toxic). In such a situation, a lower acceptable level would be considered by regulators on a case-by-case basis.

B. If the leachable's TDI > SCT but ≤QT, then the options are to

1. reduce the leachables level so it is ≤SCT, or

2. conduct a risk assessment, then either

- qualify the leachable via the risk assessment and/or safety testing, or

- propose a higher acceptable level for the leachable to the regulators.

For example, if the risk assessment determined that the leachable had no structural alert or known class effect for genotoxicity, carcinogenicity, or immediate hypersensitivity, then the leachable would be considered qualified. Similarly, if the risk assessment and/or safety testing determined that there were no clinically relevant adverse effects, then the leachable would be considered qualified.

C. If the leachable's TDI > QT, then the options are to

1. reduce the leachable ≤QT, or

2. conduct a risk assessment, then

   a. qualify the leachable via risk assessment and/or safety testing, or propose a higher acceptable level for the leachable to the regulators;

   b. if considered desirable, a minimum screen, for example, genotoxic potential, should be conducted. A study to detect point mutations and one to detect chromosomal aberrations, both *in vitro*, are considered an appropriate minimum screen;

   c. if general toxicity studies are desirable, one or more studies should be designed to allow comparison of unqualified to qualified material. The study duration should be based on available relevant information and performed in the species most likely to maximize the potential to detect the toxicity of a leachable. On a case-by-case basis, single-dose studies can be appropriate, especially for single-dose drugs. In general, a minimum duration of 14 days and a maximum duration of 90 days would be considered appropriate;

   d. for example, do known safety data for this leachable or its structural class preclude human exposure at the concentration present?

6.5 CONCLUSION

The introduction and acceptance of the principle of thresholds has significantly improved the process of assessing extractables and leachables during OINDP development and has helped to introduce process control in the manufacturing and production of the OINDP container closure, device, and packaging systems. From an industry perspective, implementation of data-based thresholds for OINDP leachables and extractables provides OINDP developers with knowledge-based risk management tools that will assist in streamlining and making more effective extractables/leachables safety assessments, and helping to ensure the safety of patients.

REFERENCES

1 Murphy, D., Schwartz, L., Milosovich, S., Bannerman, M., and Mullins, P. Comparison of laryngeal changes in rats following inhalation of a vehicle formulation generated by metered dose inhalers (MDIs) with two different types of valves. *Tox Path* 1991, *19*(4, part 2), p. 618.

2 U.S. Pharmacopeia (USP). Chapter 87. Biological Reactivity Tests, In Vitro.

3 U.S. Pharmacopeia (USP). Chapter 88. Biological Reactivity Tests, In Vivo.

4 International Organization for Standardization (ISO). TC 194 biological evaluation of medical devices, ISO 10993-1:2009. International Organization for Standardization. Geneva. Switzerland.

5 Norwood, D.L. and Ball, D. Product Quality Research Institute: Safety thresholds and best practices for extractables and leachables in orally inhaled and nasal drug products. Submitted to the PQRI Drug Product Technical Committee, PQRI Steering Committee, and U.S. Food and Drug Administration by the PQRI Leachables and Extractables Working Group, 2006.

6 ICH harmonized tripartite guideline: Q3A(R2) impurities in new drug substances. International Conference on Harmonisation of Technical Requirements for Registration of Pharmaceuticals for Human Use, 2002.

7 ICH harmonised tripartite guideline: Q3B(R2) impurities in new drug products. International Conference on Harmonisation of Technical Requirements for Registration of Pharmaceuticals for Human Use, 2006.

8 ICH harmonised tripartite guideline: Q3C(R4) residual solvents. International Conference on Harmonisation of Technical Requirements for Registration of Pharmaceuticals for Human Use, 2007.

9 Draft guidance for industry: Metered dose inhaler (MDI) and dry powder inhaler (DPI) drug products chemistry, manufacturing and control documentation. U.S. Food and Drug Administration, Center for Drug Evaluation and Research (CDER), 1998.

10 Draft guidance for industry: Nasal and inhalation solution, suspension and spray drug products chemistry, manufacturing and control documentation. U.S. Food and Drug Administration, 1999.

11 Kroes, R., Galli, C., Munro, I., Schilter, B., Tran, L., Walker, R., and Würtzen, G. Threshold of toxicological concern for chemical substances present in the diet: A practical tool for assessing the need for toxicology testing. *Food Chem Toxicol* 2000, *38*, pp. 255–312.

12 Manville, D. Toxicology testing of HFA-134a and HFA-227. The story of IPACT-I and IPACT-II. 1996. Unpublished work. Available from Drinker Biddle & Reath LLP:1500 K Street, NW. Suite 1100, Washington, DC, 2005.

13 ITFG/IPAC-RS Collaboration and CMC Leachables and Extractables Technical Team. Leachables and extractables testing: Points to consider, 2008. IPAC-RS publication.

14 Ball, D., Blanchard, J., Jacobson-Kram, D., McClellan, R.O., McGovern, T., Norwood, D.L., Vogel, W., Wolff, R., and Nagao, L. Development of safety qualification thresholds and their use in orally inhaled and nasal drug products evaluation. *Toxicol Sci* 2007, *97*(2), pp. 226–236.

CHAPTER 7

THE CHEMISTRY AND TOXICOLOGY PARTNERSHIP: EXTRACTABLES AND LEACHABLES INFORMATION SHARING AMONG THE CHEMISTS AND TOXICOLOGISTS

Cheryl L.M. Stults, Ronald Wolff, and Douglas J. Ball

7.1 INTRODUCTION

At each phase in the development of a pharmaceutical product, it is important to evaluate the safety profile. An orally inhaled and nasal drug product (OINDP) is generally comprised of a dosage form with its associated packaging and a delivery system, which may be customized or off-the-shelf. Where the delivery system is customized, a parallel development pathway is undertaken for it in conjunction with the development of the dosage form, and the resulting product is classified as a combination product. Combination products are complex from a regulatory perspective in that the safety profile of both the dosage form and the delivery system must be evaluated.

The combination product development process is shown in Figure 7.1 and is characterized by early, mid-, and late phases. Each phase of development includes dosage form development steps and/or device design stages that culminate in achievement of one or more milestones. To support the achievement of these milestones, material testing progresses from qualification to extractables/leachables testing. Throughout the life of a product, a team of individuals with a variety of expertise participates in design, development, or manufacturing; such teams may or

Leachables and Extractables Handbook: Safety Evaluation, Qualification, and Best Practices Applied to Inhalation Drug Products, First Edition. Edited by Douglas J. Ball, Daniel L. Norwood, Cheryl L.M. Stults, Lee M. Nagao.

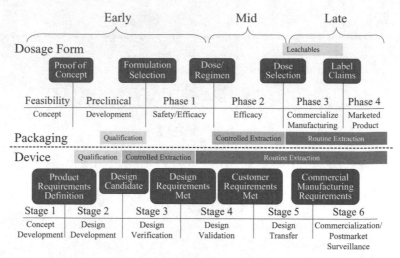

Figure 7.1 Combination product development progression.

may not be formally organized as such depending on the business arrangements for a given product. In any case, there may be a variety of individuals involved in collecting the information relevant to the safety profile of the drug product at various phases and, subsequently, discussing it with the toxicologist(s). Engineers typically select packaging and device materials in the design stage. During development, a variety of activities are undertaken where safety information may be generated: (1) biocompatibility studies may be performed; (2) formulators evaluate product robustness; (3) chemists perform analytical tests for extractables, leachables, product stability, and clinical release; (4) processing engineers and manufacturing experts develop and validate the manufacturing processes; and (5) quality professionals evaluate product returns from the clinic. At commercialization, the quality unit releases the marketed product and evaluates product returns. Whenever information pertaining to the safety profile of the product is to be assessed, it is recommended and appropriate for those individuals collecting the data to discuss it with the toxicologist(s). As a product is developed, there are several points at which a toxicologist may be actively involved. The following sections will discuss the type of materials related, safety assessment that may be needed during the life cycle of the product, and the role of the toxicologist. These will take into consideration the recent emphasis on phase-appropriate testing, risk-based development, and current regulatory expectations. Another important consideration is the fact that each OINDP is unique due to differences in the medical indication being treated, the dosage form, delivery system, and packaging. Therefore, a paradigm of interaction is proposed with specific examples that illustrate the stepwise process by which the experiments are designed, data are analyzed, and results are assessed. Case studies are provided with respect to the setting and use of thresholds not only for the leachables compound evaluation, but also for the development of analytical methods and routine controls.

7.2 INFORMATION EXCHANGE AMONG CHEMISTS AND TOXICOLOGISTS

The safety of the product at each phase of development is of primary importance. The nature of the interactions between team members and the toxicologist takes on different forms depending on the phase of development. A proposed paradigm for these interactions is shown in Figure 7.2. This model assumes that a chemist functions as a point of contact for interactions with the toxicologist. Although this is not required, it does create some efficiency later on in the development process as complex chemical experiments must be designed and performed. The involvement of the chemist at the outset provides a single point of contact for collection of all the background information that relates to material composition and regulatory pedigree (e.g., compliance with food contact regulations, compendial compliance, and transmissible spongiform encephalopathy/bovine spongiform encephalopathy [TSE/BSE] certification). This foundational information can then be readily referenced for the creation of a product-specific leachables and extractables program during development. The chemist is then poised to work with manufacturing personnel and the toxicologist to define adequate controls for both the packaging and delivery system for commercialization. The salient features of the partnership between the chemist and toxicologist are reviewed for the early, mid, and late phases of product development.

7.2.1 Early Phase: General Safety Profile

In the early phases of development (i.e., proof of concept, phase 1 clinical studies), the general safety of the product is ascertained by collecting information on the materials and components during the materials/component selection process. This information is then reviewed with the toxicologist to determine what, if any, studies need to be performed to complete a general safety assessment.

At this point, there are several categories of information to be collected. For packaging and delivery systems, it is expected that the materials/components will be compliant with food contact regulations and meet specific criteria regarding agents or compounds of concern (e.g., TSE/BSE, phthalates). The materials used in pharmaceutical products are expected to meet the regulatory requirements of the countries in which they will be used. For example, in the United States, depending on the route of delivery and duration of contact, there are specific guidelines regarding biocompatibility and physicochemical properties (U.S. Pharmacopeia [USP], general chapters <1031>, <381>, and <661>); plastics are classified based on the sets of requirements that are met. However, in Europe, materials are handled slightly differently in that there are specific compendial requirements for commonly used plastics and elastomers. Recently, the International Organization for Standardization (ISO) 10993 standards have been used preferentially in the United States and the European Union (EU) to aid in establishing the safety profile of materials/components. A common set of requirements for the materials used in OINDP are the following:

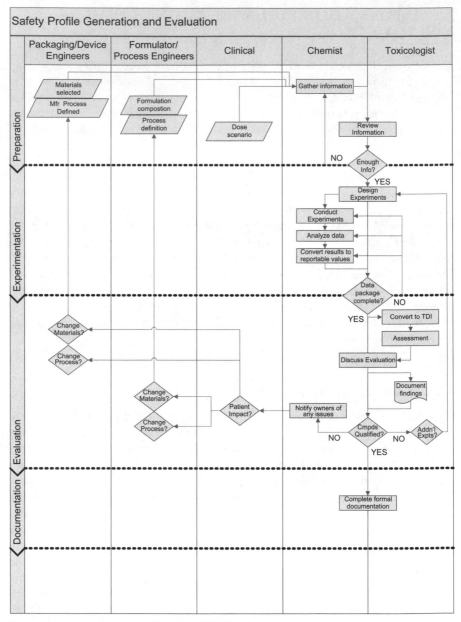

Figure 7.2   Communication flow for OINDP safety assessment.

- mechanically appropriate;
- generally recognized as safe (GRAS), where possible;
- food compliance (21CFR, EU Directives);
- biocompatibility (USP <1031>, <87>, <88>; ISO 10993);
- Physicochemical (USP <661>, <381>; European Pharmacopoeia chapter 3);

- compendial compliance (USP, Japanese Pharmacopoeia, European Pharmacopoeia); and
- suitable for medical use.

Other general requirements can be found in the regulatory guidelines for packaging.[1,2]

Upon receipt of the available information from suppliers, engineers, chemists, and toxicologists can discuss the general safety profile of the packaging materials and delivery system components. At this point, a determination must be made concerning the adequacy of the information. If collected early in the material selection process, this information can be used to aid in material selection. If the information is incomplete for a selected material, it may be necessary to perform specific compendial tests to obtain that information. Although many of these requirements may be viewed as box checking exercises, it is important to note that in cases where custom components are developed, these tests can provide insightful information. For example, a cytotoxicity test on a metal component may reveal use of an inappropriate processing additive. Alternatively, European Pharmacopoeial testing of an elastomer may reveal an undercuring of a component that was not detected by mechanical testing. The general safety profile may be determined from a combination of the materials information available and test results for the materials/components. All of this information taken together provides the information necessary to evaluate the general safety profile of the packaging materials and delivery system.

The risk associated with a general safety profile may be evaluated by considering the information obtained and the intended use of the delivery system. The level of risk can then be used to determine what, if any, additional testing should be conducted at this early phase of product development. Typically, a low-risk profile would include complete information, non-life-threatening treatment (the "treatment" being the drug product) and early clinical phases where the number of patients and duration of dosing is minimal. A medium risk profile would be one where there is incomplete information, a serious but non-life-threatening treatment and early to mid clinical phases. A high-risk profile would involve minimal information, a life-threatening treatment, and late-phase clinical studies where there are large numbers of patients. Additionally, there may be cases where the stage of development of the packaging or delivery system is well ahead of (or behind) that of the dosage form and progression in the clinic. In these cases, the evaluation of the general safety profile can be accomplished by following a risk assessment scheme similar to that used for failure mode and effects analysis (FMEA) or failure mode, effects, and criticality analysis (FMECA).[3] One such scheme based on an FMEA approach is proposed here and may be refined. The material information, treatment type, and clinical phase can be rated in terms of completeness, severity, and numbers of patients, respectively. The completeness of the material information can be thought of as detectability—the more information available, the more that is known. The severity of the treatment type can, in practical terms, be related back to the type of patient contact and duration of use as outlined in ISO 10993. Finally, numbers of patients that are part of a clinical study can be thought of as the potential occurrence. Brief descriptions of these factors and the respective rating numbers are given in Table 7.1. Rating numbers for each of these factors can then be combined by multiplication to obtain a risk priority number.

TABLE 7.1.   General Safety Profile Risk Ranking Scheme

Rating number	Material information	Contact/duration	Clinical phase
1	Meets all food compliance requirements, TSE/ BSE free, confirmed USP class VI plastic, meets compendial requirements	Surface/limited	I
3	Meets several food compliance requirements, unconfirmed USP class VI, not confirmed but likely TSE/BSE free, may meet several compendial requirements	Surface/prolonged	II
5	Meets a few food compliance requirements, USP class unknown, unknown TSE/BSE free, may meet a few compendial requirements	External communicating	III

Thus, with a factor rating range of 1 (low risk) to 5 (high risk), the lowest risk priority number obtainable is 1, and the highest is 125. Based on the risk priority number, a determination can be made regarding the need for any additional testing; the higher the number, the greater the need for mitigation. Typically, in advance of the rating activity, one or more levels would be chosen above which mitigation (e.g., additional testing, material replacement) is recommended or required. These levels vary depending on the application and organizational factors such as risk appetite or budget.

For example, consider a case where a dry powder inhaler (DPI) is used in a surface contact mode for limited duration. This would give a rating of 1 for severity. If the product were in phase 3 clinical studies, there would likely be large numbers of patients involved and the risk rating for potential occurrence would be 5. Additionally, if food compliance statements and TSE/BSE statements were available, but no other device results were available, the rating for the detectability factor would be 5. Taken together, the overall risk priority number would be 25 ($1 \times 5 \times 5$). In this case, the risk could be mitigated by performing biocompatibility testing appropriate to the device type.

Consider another example that involves a nebulizer that is used with a hospitalized patient on a respirator. This external communicating configuration would give a rating of 5 for severity. If the product were in phase 3 clinical studies, there would likely be a relatively large number of patients involved and the risk rating for potential occurrence would be 5. Additionally, if food compliance statements and TSE/BSE statements were available, along with biocompatibility results from a predicate (already marketed) device, the rating for the detectability factor would be 3. Taken together, the overall risk priority number would be 75 ($5 \times 5 \times 3$). In this case, the risk may be best mitigated by performing additional biocompatibility testing to confirm the USP class of the plastic and controlled extraction to determine potential leachables.

7.2.2   Midphase: Detailed Safety Profile

As the product development efforts mature (i.e., phase 2 clinical studies), the safety profile of an OINDP is expected to include some level of leachables and extractables information. There are several specific guidance documents and recommendations in the United States for different types of OINDP.[4–7] In Europe and Canada, there is one guideline for inhalation products.[8,9]

The strategy that is adopted for extractables and leachables testing is dependent on the type of OINDP and general safety profile of the materials/components. For example, for a combination product utilizing a DPI that is a predicate device, there may be minimal risk and minimal testing. Alternatively, for a combination product utilizing a custom nebulizer, there may be higher risk and significant testing.

After consideration of the general safety profile and the regulatory requirements, the chemist and toxicologist must design an appropriate testing strategy. Based on the Product Quality Research Institute (PQRI) recommendations,[7] it is expected that a controlled extraction study will be performed on critical components, for example, those that are in the drug path or contact the mouth or nasal mucosa. At the very outset of designing such experiments, it is desirable and appropriate for the toxicologist to be involved so that the results can be used to provide a relevant and meaningful safety assessment of the chemicals found. Subsequent to the controlled extractables assessment, it may or may not be necessary to perform a leachables assessment or other biocompatibility or physicochemical testing.

The design of the controlled extraction experiment will be based on several factors. The following are considerations:

- dosage form—solid or liquid;
- characteristics of the drug formulation—hydrophilic or hydrophobic;
- nature of contact—mucosal;
- temperature of storage/administration;
- dosage regime—number of doses, time per dose;
- contact time of formulation with packaging/delivery system—continuous or intermittent;
- chemical composition of the material of construction; and
- processing aids used to manufacture packaging/components/delivery system.

Knowledge of these factors will help to determine the following:

- extraction methodology—solvent extraction or volatile analysis;
- extraction solvent—water, saline, isopropanol, hexane;
- extraction temperature—typically not more than 40°C;
- extraction time—minutes to weeks; and
- analytical methodology—high-performance liquid chromatography (HPLC) parameters for nonvolatiles, gas chromatography (GC) parameters for semi-volatiles and volatiles.

Concomitant with the above factors, it is important to discuss the dosing scheme (e.g., doses per day) with the toxicologist so that an appropriate analytical evaluation threshold (AET) can be determined. The AET is based on the component mass, dosing scheme, and threshold—typically the safety concern threshold (SCT). The AET will be used to select an appropriate mode of detection for the analytical methodology. Low-level detection can be accomplished by UV or flame ionization detection (FID), and trace-level detection can be accomplished by mass spectrometry (MS) or other specialized techniques. It is important to establish the limit of quantitation (LOQ) of the particular analytical methodology with representative standards prior to running the samples so that the expected sensitivity is achieved.[10]

After the experiment is designed and performed, the data must be analyzed appropriately. The typical output from the analytical experiment is a chromatogram. The integrated peak areas are converted from units of intensity to concentration with the aid of a standard calibration curve. Unfortunately, it is not likely that a reference standard will exist for every peak in the chromatogram. It is more common that semiquantitative results will be obtained by utilizing the least-squares linear fit for one reference material applied to all or a group of peaks. For compounds that pose safety concerns, it is preferable to use a reference standard for quantification so that accurate estimates of those extractables are obtained. The following calculation is typically used for the conversion of the integrated peak area to concentration:

$$\frac{\text{Peak area}}{\text{Peak area/mg/mL}} - \text{Intercept} = \text{Concentration (mg/mL)}.$$

The next step requires the conversion of the concentration of the measured solution to a concentration associated with a material/component. This is accomplished by considering the final extract volume and the material/component mass. The following equation generally applies:

$$\frac{\text{Ext vol (mL)} \times \text{Concentration (mg/mL)}}{\text{Component mass (g)}} \times \frac{1000\ \mu g}{mg} = \mu g/g.$$

Or, if a diluted aliquot of the final extract is used, the following modified equation generally applies:

$$\frac{\text{Ext vol (mL)} \times \text{concentration (mg/mL)} \times \text{dilution factor}}{\text{Component mass (g)}} \times \frac{1000\ \mu g}{mg} = \mu g/g.$$

After the quantification of all the peaks, those that are above the AET are to be identified. This is most often accomplished with MS through the use of library matching, manual interpretation of the fragmentation patterns, and reference standard matching. Additional tools such as UV-extracted spectra from the chromatographic peak and exact mass measurement can also be useful. It is important for the chemist to provide the toxicologist with as much structural information as possible. Otherwise, it is difficult for the toxicologist to make a proper assessment because literature review requires checking for toxicity of specific compounds. As shown in Figure 7.2, it is at the conclusion of this experimentation process that a complete package must be delivered or additional experiments may be necessary to provide accurate quantification and full structural identification, including the Chemical Abstract Service (CAS) number for each identified compound.

Once the complete controlled extractables package is in the hands of the toxicologist, the safety evaluation process is initiated. The first step in this process entails developing an estimate of the total daily intake (TDI). First, a list of assumptions is created regarding the dosing scheme, duration of exposure, the type of patient interaction with the material/component, the types of compounds that are relevant to the delivery method or packaging system, and rate of appearance of extractables as potential leachables under real-time use conditions. Second, a TDI calculation is performed based on the above considerations and the quantitative levels from the controlled extraction results. Third, a subset of the listed compounds from the controlled extraction studies is generated based on the TDI and categorized as "greater than SCT but below qualification threshold (QT)" or "greater than QT."

There are a few issues that may arise in the preparation of the list of compounds in these two categories. The first is a case where there are compounds that have no library match and no fragmentation information to aid in elucidation of the chemical structure. The amount of necessity is approximated by use of a surrogate standard that may elute in the same vicinity of the chromatogram. Although the amount of the compound may appear to be low, in actuality it could be high because of a difference in detector response to that compound when compared with the detector response to the surrogate. It is also important to keep in mind that the molecule could be highly toxic at low levels. For these reasons, it behooves the chemist to make every effort to obtain at least some information regarding the structural identity of the compounds that appear to be above the SCT.

Another issue that may arise in the preparation of the list of compounds found in the two categories is that of material composition. Where there are several components in one device composed of the same material, it is conceivable that the toxicological evaluation could be done at only the component level. This in the end may be misleading in that individual components may not have levels of a specific compound above the SCT. However, the cumulative total of that compound extracted from all components in a single delivery system may be above the SCT or even the QT. Therefore, it is important to exercise caution in both the design of the extraction experiments and the preparation of the calculated TDIs.

A related issue occurs when materials of the same composition, manufactured by the same supplier but at different sites, are used to construct multiple components used in the same packaging/delivery system. Alternatively, the same material from the same supplier manufactured at the same site might be used at two different component manufacturers. From a macroscopic point of view, the materials may have the same mechanical properties and major ingredients, while at a microscopic level, different additives or amounts of the same additives may be used in the formulation or processing. These components would be expected to have different controlled extraction results and therefore should be extracted individually and assessed separately with respect to TDI.

These issues related to material composition should be known and discussed with the toxicologist prior to preparation of the list of compounds under the two categories. This list is generally the output after iterative conversations between the chemist and toxicologist.

The second step in the evaluation process involves performing an assessment of the compounds in the two categories based on a thorough review of the literature using the general principles described in Chapter 6. Several databases, such as Chemical Carcinogenesis Research Information System (CCRIS), Hazardous

TABLE 7.2. Toxicology Databases

Acronym	Database	Information link
CCRIS	Chemical Carcinogenesis Research Information System	http://toxnet.nlm.nih.gov/ cgi-bin/sis/htmlgen?CCRIS
HSDB	Hazardous Substances Data Bank, National Institutes of Health, National Library of Medicine	http://toxnet.nlm.nih.gov/ cgi-bin/sis/htmlgen?HSDB
IRIS	Integrated Risk Information System EPA	http://cfpub.epa.gov/ncea/iris/ index.cfm
ChemIDplus	Web-based chemical information search system National Library of Medicine	http://chem.sis.nlm.nih.gov/ chemidplus/
HPVIS	High Production Volume Information System EPA	http://www.epa.gov/hpvis/ index.html
TOXNET	Toxicology Data Network National Library of Medicine	http://toxnet.nlm.nih.gov/
PELs	Permissible Exposure Levels Occupational Safety and Health Administration (OSHA)	http://www.osha.gov/SLTC/ pel/index.html
ATSDR	Agency for Toxic Substances and Disease Registry U.S. Department of Health and Human Services	http://www.atsdr.cdc.gov/
PubMed	Public access to biomedical citations National Library of Medicine	http://www.ncbi.nlm.nih.gov/ pubmed/

Substances Data Bank (HSDB), Integrated Risk Information System (IRIS), ChemIDplus, High Production Volume Information System (HPVIS), Agency for Toxic Substances and Disease Registry (ATSDR), and PubMed can be used. Information on these databases is indicated in Table 7.2.

Upon review of the literature, the toxicologist normally looks for the following: data related to mutagenicity or carcinogenicity, single and repeat dose animal toxicity studies, human health data, determination of acceptable daily intakes (ADIs) or reference doses (RfDs) determined by regulatory agencies.

In cases where little or none of this information exists, the following approaches can be taken. The first step is a second look at the literature using a wide range of search engines such as PubMed and perhaps others including Google to determine if there is any information at all related to toxicity of the specific compound. If none is available, then structure–activity relationships (SARs) should be explored from data on compounds with similar structures or functional groups.

For compounds that have a TDI less than the SCT of 0.15 µg/day, no evaluation is necessary unless the compounds are known carcinogens, for example, nitrosamines, in which case a compound-specific assessment is conducted using available data for the given compound.

For compounds that have a TDI greater than the SCT but below the QT of 5 µg/day, the primary evaluation is based on presence or absence of mutagenic or carcinogenic potential and also whether the compound belongs to classes of irritating compounds including isocyanates, short-chain aldehydes, and nitriles.

For compounds that have a TDI greater than QT, the evaluation is similar to that for compounds greater than SCT, but also includes a compound-specific risk assessment using all available information. The latter is essential because existing data may show that a particular compound has an acceptable toxicity profile at a TDI level above the QT of 5 μg/day.

The third step in the evaluation process is to discuss the findings with the chemist and any other affected stakeholders (e.g., packaging/device/process engineers, formulator, clinician). The compounds may range from having no toxicological concerns to having significant toxicological concerns; they may range from being tentatively identified to confidently identified. Consider the following scenarios:

Scenario 1. All compounds pose no toxicological concern. *Recommended action*: Formally document the assessment.

Scenario 2. Some compounds pose a toxicological concern, are confidently identified, and can be qualified based on what is known in the literature. *Recommended action*: Formally document the assessment; consider whether a leachables study or extractables monitoring is warranted.

Scenario 3. Some compounds pose a toxicological concern and are not confidently identified but may be qualified based on their lack of structural alerts. *Recommended action*: Document findings; consider performing a confident identification; consider whether a leachables study or extractables monitoring is warranted.

Scenario 4. Some compounds pose a toxicological concern, are not confidently identified, and may not be qualified based on their structural alerts (if the tentative identification is correct). *Recommended action*: Document findings; consider performing a confident identification; consider performing qualification experiments; consider the trade-offs of the above with selecting a different material.

Scenario 5. Some compounds pose a toxicological concern, are confidently identified, and not qualified based on what is known in the literature. *Recommended action*: Consider the trade-offs between performing qualification experiments and selecting a different material.

As shown by the above scenarios, the decision-making process regarding the results may have a broad impact on the development program for the OINDP. Decisions made regarding compounds that may have toxicological concern become more complex when the compound is not confidently identified. It is preferable to have as much identification information as can be obtained in the time allotted. However, it is recognized that not all compounds are readily identified due to their sheer numbers, chemical nature, and/or low levels. In some cases, it may be more appropriate to either change the material or manufacturing process or perform a toxicological study depending on the phase of development. If the level of toxicological concern is high, it may be most appropriate to replace a material or change a manufacturing process providing the source of the compound(s) can be identified. Such decisions are best made by the key stakeholders (packaging/device/process engineers, chemist, formulator, clinician, and toxicologist) acting as a team after a discussion regarding what information is needed and what testing is possible, processing steps that may be sources, material availability, and patient risk.

Upon completion of the assessment of the controlled extraction results, there may be additional experiments required. These may fall into the category of additional extractables or other structural identification tests. In that case, the experimentation/evaluation cycle shown in Figure 7.2 would be repeated for the new experiments.

However, a leachables study would restart the cycle at the preparation step shown in Figure 7.2. A leachables study may be required as a result of the controlled extraction assessment or because it is expected during the development of a particular type of OINDP.[4-9] Information in addition to that already discussed for controlled extraction studies would be required and includes the following considerations:

- specific details regarding the formulation—chemical composition including degradants, solubility, light and heat sensitivities;
- filling process parameters—time, temperature, relative humidity;
- packaging process parameters—time, temperature, processing aids; and
- composition of any materials that come into direct contact with the formulation during processing.

From the collective information and discussions between the chemist and toxicologist, a list of leachables target compounds is developed. From this list, the chemist begins to develop the analytical methodology that can be used to measure the individual compounds. Analytical methods used for the evaluation of extractables may provide a starting point for leachables analytical methodology. The sample preparation conditions must take into consideration both the properties of the formulation and the compounds to be measured. These considerations will also influence the choice of the analytical technique with respect to selectivity and sensitivity of the analytical technique.

To adequately determine the appropriate sensitivity of the analytical technique, the AET needs to be established. This threshold is based on the SCT and the mass of the formulated dose. The following equation illustrates this calculation:

$$\frac{0.15\,\mu g/day}{\#\ Dose/day} \times \frac{Dose}{Formulation\ (g)} = \frac{\mu g}{g}.$$

The analytical chemist then must convert this value into a number that is meaningful analytically and can be used to select the appropriate analytical detection technique. The calculation for a chromatographic method may be set up as follows:

$$\frac{Concentration\ (\mu g/g) \times mass\ sample \times vol\ injected\ (\mu L)}{Vol\ sample\ (mL)} \times \frac{mL}{1000\,\mu L} = \mu g.$$

As suggested earlier, there are a number of analytical detection techniques that can be used. After the analytical methodology is optimized, the samples must be selected appropriately. Most often samples are taken from stability studies where the packaged product is stored in a controlled environment. Care must be taken when selecting samples that are expected to be representative of the patient experience. For example, if there is a concern about a volatile compound, it would not be appropriate to select samples that have been stored at an elevated temperature since the volatiles may have been evaporated. Similarly, if there is a concern about a nonvolatile migrating from a package into the formulation, taking a low-temperature sample would not be appropriate because of lower migration rates at lower temperatures. It is preferable to take samples that represent the condition that would give the highest realistic levels of the compounds of interest.

After the experiments are performed, the evaluation step in the process should progress just as it does for the controlled extraction experiment. There may be additional activities required to qualify compounds that are found as leachables. These can be found in, for example, Chapters 3, 6, and 8. At the end of the process, it is imperative that the findings and assessment be documented. Information obtained through the process of performing and evaluating results from controlled extractables and leachables studies, combined with that from any toxicological assessment (e.g., SAR information, literature research, genotoxicity, or *in vivo* studies), provides the detailed safety profile for the product and may be useful for future investigations related to the OINDP.

7.2.3 Late Phase: Assurance of Safety Profile

The safety assessments performed on leachables and extractables provide the foundation to determine what manufacturing controls are necessary. A review of the safety assessment by the chemist and toxicologist can be used to prepare a list of compounds to be monitored routinely. Prior to the discussion, it is important for the chemist to determine whether or not there is a qualitative correlation between the leachables observed and the extractables observed in the controlled extraction study. If not, it may be possible to establish a chemical rationale for the presence of the leachable due to degradation of an extractable or chemical reaction between an extractable and either the formulation, a processing aid, or another extractable. Additional experiments may be necessary to determine the source of the leachables that are unaccounted for by the controlled extraction results. After all leachables are accounted for, it is also essential to consider whether there are any extractables identified that, if they were to become leachables, would be a safety concern. Taken together, these considerations can be used to develop a monitoring strategy.

There may be a variety of reasons to monitor specific chemicals. A typical rationale may include

1. compounds that are present as leachables at a level below QT but above SCT, which do not present a safety concern at current levels but would present a safety concern above QT;
2. compounds that are present as leachables below SCT but would present a safety concern at levels above SCT;

3. compounds identified during controlled extraction that would present a concern if present above SCT or QT as a leachable; and

4. compounds identified during controlled extraction that indicate that the material composition is consistent with material used in the clinic where no adverse events were found.

Ideally, it is desirable to establish a routine extraction control to ensure the safety of materials. This control may be either related directly to levels of leachables/ potential leachables or indirectly to clinical experience. To properly establish quantitative controls utilizing an extraction test for the types of compounds in (1) and (2), a quantitative correlation between leachables and extractables levels needs to be established. The chemist may need to perform additional experiments to confirm the quantitative relationship between leachables and extractables. A joint review of the quantitative relationship between leachables levels and extractables levels enables the chemist and toxicologist to agree on limits for the extractables to be monitored routinely.

In the event that the extractables/leachables correlation is not or cannot be made, there are other approaches that can be considered. If the source of the leachable is a result of a processing aid, controls can be put on the processing step to control the levels. If the leachable is a reaction product, the chemist, toxicologist, and other stakeholders may need to discuss alternatives in formulation or materials to effectively control the levels of leachables compounds that present a safety concern. In general, it is not considered an effective routine control to monitor leachables because it may be a matter of weeks or months to make the necessary measurement.

To properly establish controls utilizing an extraction test for the types of compounds in (3) and (4), it is useful to establish the range of levels of these compounds in the materials by performing routine batch testing. For the types of compounds in (4), it is also useful to determine the relationship of the extracted level to the composition level. After the batch history is known, a review with the toxicologist can be utilized to determine realistic limits with due consideration given to a proposed extractable/leachable or extractable/composition relationship.

After the establishment of the limits for routine control, the chemist again must evaluate the analytical methodology that is appropriate for measuring these levels. The routine test will most often include some type of sample preparation followed by chromatography. The typical routine extractables acceptance criteria will include the following:

- Each target compound is present within the specified limits.
- All other peaks in the profile are present.
- No new peaks are present.

There are several occasions on which the chemist and toxicologist may be called to confer. First, if a target compound is above the accepted limit, a safety determination may be required. Second, if a new peak is present, a couple of actions may be necessary. The identity of the peak may need to be determined if it is unknown. The level of the extracted compound may need to be translated into a proposed leachables level and thus evaluated for safety. If there are missing peaks in the profile, it may be an indication that the material has changed and there may be other compounds present that are not detected by the current methodology. This will warrant a discussion between the chemist and the supplier of the material/ component to determine what changes may have been implemented and possibly additional extraction experiments. Subsequently, the change in composition is to be discussed with the toxicologist to determine if there are any safety concerns. Such events may result in the resetting of acceptance criteria.

7.3 CASE STUDIES

7.3.1 Controlled Extractables Safety Evaluation

Controlled extraction studies are expected for OINDPs based on the recent PQRI recommendations. In order to provide data that may be relevant to preliminary risk assessments, at least one of the extraction methods should approximate conditions that do not represent exhaustive extraction but rather are at the upper end of realistic extremes for actual use. This preliminary assessment provides a first step in determining possible worst-case conditions for patient use. A situation that can be frequently encountered is that a particular compound is identified in the controlled extraction studies, and the question posed to the toxicologist is whether this might pose a problem for continued development of the product. Therefore, "an early look" for possible adverse consequences is advisable. The extractables levels derived from gentle extraction methods such as water or saline solvents at physiologically relevant temperatures such as 37 or 40°C should be used as a basis for determining a realistic estimate of possible worst-case elution of such compounds. If the assessment shows that, even for reasonable worst-case conditions, the extractables TDI is less than SCT, then leachables testing for such a compound may not be necessary. Or, if the extractables TDI is less than acceptable levels, such as RfDs that have already been derived by regulatory agencies, then the appearance of such a compound in the extractables studies is not likely to indicate a problem.

The case studies that are described here are taken from OINDP where the general safety assessment risk level was relatively high due to minimal materials information, treatment of serious illnesses, and use in post-phase I clinical trials. These cases demonstrate a variety of extractables levels observed in different OINDP and how such a preliminary assessment may be conducted. In each case, an AET was estimated based on an SCT of 0.15 μg/day, the dose regimen, and component mass as described in the PQRI recommendations. Upon review, it was determined that the AET was exceeded and therefore further evaluation was required. The conclusions presented here are case specific and are tailored to the specific product.

7.3.1.1 Case #1 Component type: Drug path seal, no continuous drug contact
Component material: Elastomer, 0.25 g
AET: 0.3 μg/g

Extractable: Diester bis(2-ethylhexyl)phthalate (DEHP) (CAS #117-81-7; MW 390.56), 170 µg/g

Worst-case TDI estimate: 42.5 µg/day

7.3.1.1.1 Safety Assessment The total amount of DEHP that might be available for patient exposure if all of the DEHP measured by controlled extraction managed to be captured in one dose would be 42.5 µg. It would be extremely unlikely to have all the DEHP captured in one dose since the component is not in direct, continuous contact with the dosage form and would not likely enter the airstream because DEHP is not highly volatile. Thus, a worst-case assumption that all extracted DEHP might be available for exposure results in a large overestimation of possible exposure.

Therefore, two possible scenarios were explored:

1. an extreme case that all of the extract would be available for exposure over the exposure on the first day, or
2. a worst case that all of the extract would be available for exposure over a 10-day exposure period.

Scenario 1. Extreme case—all of the extract delivered during the "puffs" taken on the first day of inhaler use.

This would give rise to a dose of 42.5 µg/day. For a typical 60-kg person, this translates to a dose of 0.71 µg/kg.

This dose is substantially less than the RfD of 20 µg/kg established by the United States Environmental Protection Agency (EPA) for daily dosing to the general population, which should result in no adverse effects with lifetime exposure. It is also less than median environmental exposures of 14 µg/kg/day.[11]

Scenario 2. Worst case—total extracted amount delivered over 10 days.

The dose delivered would be 42.5/10 = 4.25 µg/day. For a typical 60-kg person, this translates to a dose of 0.07 µg/kg/day. This dose is very low compared with the RfD and environmental exposure cited above.

Both of these dose estimates, scenarios 1 and 2, are also considerably lower than doses of 3100 µg/kg/day for long-term hemodialysis patients and 30 µg/kg/day for hemophiliacs undergoing long-term blood transfusions. No adverse effects have been shown in either of these populations that appear to be related to DEHP.[12]

7.3.1.1.2 Conclusion The extractables level of DEHP indicates that there is likely to be minimal health risks in the final product since this case is acceptable and is an overestimate of possible effects. However, since phthalates represent a class of compounds with some safety concerns, a routine test may be warranted.

7.3.1.2 Case #2 Component type: Drug path, continuous drug contact during dosing

Component material: PC/ABS, 4.2 g

AET: 0.5 μg/g

Extractable: Styrene (CAS #100-42-5; MW 104.15), 119 μg/g

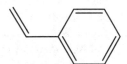

Worst-case TDI estimate: 35.8 μg/day

7.3.1.2.1 Safety Assessment ATSDR, Health Canada, and the Dutch National Institute for Public Health and the Environment (RIVM) have evaluated the carcinogenicity data for styrene, and the consensus was that it is unlikely that styrene is carcinogenic in humans, and therefore, overall assessment is based on noncancer end points.

ATSDR, Health Canada, RIVM, and EPA have evaluated the noncancer inhalation toxicity data for styrene. All four organizations derived risk values, which differ by up to approximately 10-fold. ATSDR derived a chronic minimal risk level (MRL) of 0.2 ppm (0.87 mg/m^3; 1 ppm = 4.33 mg/m^3) based on a lowest observed adverse effect level (LOAEL) of 20 ppm (87 mg/m^3) for biologically significant increases in choice reaction time and decrease in color perception observed in workers exposed to styrene for 8 work-years.[13] ATSDR used an uncertainty factor of 100 (10 each for use of a LOAEL and for human variability). The lowest reference concentration value of 0.09 mg/m^3 was derived by Health Canada by applying a higher safety factor (500) to animal data.

If the lowest negligible effect reference value of 0.09 mg/m^3 is used, this corresponds to 90 μg/m^3 or a daily dose of 1800 μg/day, assuming a ventilation rate of 20 m^3/day, which is typically assumed for the general population with moderate daily activities.

7.3.1.2.2 Conclusion Styrene is unlikely to pose any health risks since the overestimate of possible dose is 1800/35.8 = 50 times less than the RfD that is estimated to have negligible effects on lifetime exposure to the general population.

7.3.1.3 Case #3 Component type: Mucosal contact, drug path, continuous drug contact during dosing

Component material: Polypropylene, 3.0 g

AET: 18 μg/g

Extractable: Caprolactam (CAS #105-60-2; MW 113.16), 203 μg/g

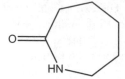

Worst-case TDI estimate: 20.3 µg/day

7.3.1.3.1   Safety Assessment   Caprolactam is extensively used as an intermediate in producing nylon and resins. It has undergone substantial toxicological testing including 2-year carcinogenicity studies in rats and mice, which were negative. An RfD has been proposed as 0.5 mg/kg or 500 µg/kg by the EPA based on noncancer end points. A safety factor of 100 was applied to the no-observed-adverse-effect level (NOAEL) of 50 mg/kg found in a three-generation reproduction study. This study resulted in the most sensitive end point found of reduced pup weights.

7.3.1.3.2   Conclusion   Caprolactam is unlikely to pose any health risks since the overestimate of possible dose is 500/20.3 = 25-fold less than the RfD that is estimated to have negligible effects on lifetime exposure to the general population.

7.3.2   DPI Leachables Evaluation

The study described here involves powder contained in a blister that is dispersed by a DPI. It is significant to note that the blister has a much longer contact time with the powder than the transient limited contact time with the inhaler device. In fact, leachables from powder collected directly from the exit of the inhaler were essentially the same as those collected from the blister packs alone. The leachables were directly traceable to the blister materials. Thus, the evaluation described here is focused only on leachables found in the powder taken directly from the blisters.

The blister is shown schematically in Figure 7.3. The blister material is an aluminum–aluminum (Alu–Alu) type of construction and contains a heat seal coating and polyvinyl chloride (PVC) laminate that are in direct contact with the dosage form. Based on the dosing scheme and mass of powder in the blister, the AET was calculated at 1 ppm based on an SCT of 0.15 µg/day. Several compounds were found to be above this level and are listed in Table 7.3.

Risk assessments of these compounds were performed. The conclusion was that these compounds could be qualified through structure activity relationship analysis and literature review. Risk assessments for those highlighted in gray are provided for illustrative purposes on how risk assessments were conducted for this drug product; 1-butanol and 2-ethyl-1-hexanol required further evaluation because their TDIs were greater than the QT of 5 µg/day, and erucamide required further evaluation because there were very little data available and because there were no mutagenicity studies conducted on this compound.

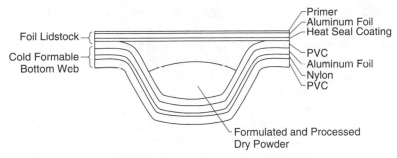

Figure 7.3   Schematic of a blister containing powder for a DPI.

TABLE 7.3.   DPI Leachables Compounds with Calculated TDI (µg/Day)

Ethanol (2.3 µg)	4,6-Dimethyl-2-hepatanone (0.3 µg)	Undecyl dodecanoate (0.2 µg)
Acetone (<0.2 µg)	Diisobutyl ketone (0.5 µg)	Diethylhexyl phthalate (<0.3 µg)
Isopropanol (1.7 µg)	1-Undecene (0.4 µg)	Squalene (0.7 µg)
Butanal (0.3 µg)	Undecane (0.4 µg)	Bis(ethylhexyl)sebacate (1.5 µg)
1-Butanol (6.5 µg)	2,2,4,6,6-Pentamethylheptane (0.3 µg)	2-Methyl-1-propanol (0.2 µg)
Ethyl acetate (<0.2 µg)	2-Ethyl-1-hexyl acetate (0.4 µg)	Methanol (0.6 µg)
Cyclohexanone (0.8 µg)	Di-butyl phthalate (0.5 µg)	Camphene (0.2 µg)
Methyl methacrylate (0.2 µg)	(Z)-9-Octadecenoic acid (0.2 µg)	1-Undecanol (0.3 µg)
Caprolactam (<0.6 µg)	(E)-9-Octadecenoic acid (0.2 µg)	Tetratriacontatetraenoate, methyl ester (0.5 µg)
2-Ethyl-1-hexanol (7.0 µg)	Propanoic acid, 2-methyl-, 1-(1,1-dimethylethyl)-2-methyl-1,3-propanediyl ester (0.4 µg)	2-Ethylhexyl mercaptoacetate (0.2 µg)
1-Methoxy-2-propyl acetate (0.3 µg)	Docosane (0.6 µg)	2-Butanone (<0.2 µg)
Butyl methacrylate (0.3 µg)	Erucamide (<0.6 µg)	–

7.3.2.1   Safety Assessment of 1-Butanol (CAS #71-36-3; MW 74.14)

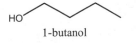

1-butanol

1-Butanol is a primary alcohol with a four-carbon structure and the molecular formula of $C_4H_{10}O$. It belongs to the family of higher and branched-chain alcohols. As a leachable in this drug product, the TDI for 1-butanol is 6.5 µg or ~0.09 µg/kg for a 70-kg person. 1-Butanol was shown to be negative in the Ames and sister chromatid exchange assays[14,15] and has no structural alerts for genotoxicity, carcinogenicity, or general toxicity by Deductive Estimation of Risk through Existing Knowledge (DEREK, Lhasa Ltd., Leeds, UK) software. The acute oral LD50 in rats and mice was determined to be 790 and 2689 mg/kg, respectively,[16,17] and has an acute inhalation LD50 in rats of 8000 ppm (3989 mg/kg).[16]

Additionally, 1-butanol is a regulated substance and is permitted as a direct food additive.[18] The EPA RfD for chronic oral exposure in humans is 100 µg/kg/day.[19] The 8-hour permitted exposure limit (PEL) established in the Unites States is 100 ppm or 0.303 µg/L, which is equivalent to ~19 µg/kg for a 70-kg human.[16,20] Finally, 1-butanol is listed as a class 3 solvent (International Conference on Harmonisation [ICH] Q3C), which allows for acceptable intake as high as 50 mg/day or 714 µg/kg/day intake for a 70-kg human.[21]

Conclusion:

Based on the SAR analysis, toxicity studies reported in literature and the established allowable intakes across several guidance documents the ADI of ~0.09 µg/kg for a 70-kg human presents negligible risk when the drug product is used as recommended.

7.3.2.2 Safety Assessment of 2-ethyl-1-hexanol (CAS #104-76-7; MW 130.23)

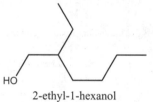

2-ethyl-1-hexanol

The fatty alcohol 2-ethyl-1-hexanol is an organic compound used in the manufacture of a variety of products. It is a branched, eight-carbon alcohol with a molecular formula of $C_8H_{18}O$. 2-Ethyl-1-hexanol can be readily converted into esters that have a variety of uses. The primary use of 2-ethyl-1-hexanol is in the manufacture of the plasticizer, DEHP. As a leachable in this drug product, the TDI for 2-ethyl-1-hexanol is 7.0 µg or ~0.1 µg/kg for a 70-kg person.

Several *in vitro* and *in vivo* genetic toxicology studies have been reported, and 2-ethyl-1-hexanol was negative in the Ames, mammalian cell gene mutation, *in vitro* cytogenetics, *in vivo* cytogenetics, unscheduled DNA synthesis (UDS) (rat hepatocytes), *in vitro* cell transformation, and dominant lethal assay in mice. However, a weak mutagenic response was reported in the 8-azaguanine-resistance assay in *Salmonella*.[22,23]

There have been numerous rodent studies conducted on 2-ethyl-1-hexanol ranging from acute to evaluation of carcinogenicity. Acutely, 2-ethyl-1-hexanol had reported oral LD50s of 3730 and 2500 mg/kg in rats and mice, respectively.[24] In addition, the inhaled LD50 in rats was reported to be >2000 ppm or ~2600 mg/kg over a 6-hour exposure period.[24] In 13-week inhalation rodent studies, the NOAEL was reported in a range of 125–157 mg/kg/day.[25,26] Oral and inhaled reproductive development studies in rodents were negative in doses ranging from 191 to 245 mg/kg.[27,28] However, development effects in rats consisting of hydronephrosis, heart malformations, and tail and limb defects were observed in rats at doses ranging from 800 to 1600 mg/kg.[29] There was no evidence of carcinogenicity in rats at oral doses up to 500 mg/kg/day.[30] In mice, 2-ethyl-1-hexanol was not carcinogenic in males up to a highest dose of 750 mg/kg/day, and in females up to 200 mg/kg/day. A weak or equivocal trend in increased incidence of liver tumors occurred in female mice given the highest dose of 750 mg/kg/day, which may be related to its designation as a weak peroxisomal proliferator.[24,30]

No apparent injury has been reported in humans from its use in industry. The probable oral lethal dose in humans is estimated to be from 500 to 5000 mg/kg.[31]

Conclusion:

Based on available genetic and animal toxicity data and the lack of toxicity of 2-ethyl-1-hexanol in industrial applications, the ADI of 0.1 µg/kg for a 70-kg human presents negligible risk when the drug product is used as recommended.

7.3.2.3   Safety Assessment of Erucamide (CAS# 112-84-5; MW 337.58)

13-Docosenamide (erucamide)

Erucamide is an unsaturated long-chain carboxylic acid amide that is used as a slip agent, antifogging agent, or lubricant for plastic films (polyolefin). It has a molecular formula of $CH_3(CH_2)_7CH=CH(CH_2)_{11}CONH_2$. In this drug product, the TDI for erucamide is 0.6 µg or ~0.01 µg/kg for a 70-kg human. A DEREK SAR analysis was negative for genotoxicity and carcinogenicity. No significant toxicological information is available on erucamide other than the stearoyl derivative of erucamide was negative in the Ames test,[32] has an oral rat LD50 > 5000 mg/kg,[33] and that it is permitted as an indirect food additive in the United States.[34]

Conclusion:

The low TDI and the lack of any indication of mutagenicity or carcinogenicity, as well as its common presence as an indirect food additive, would suggest that there is a negligible risk to humans when the drug product is used as recommended.

7.4   CONCLUSION

Evaluation of the safety profile of an OINDP takes different forms during the product life cycle. The toxicologist interacts with a variety of individuals with different expertise to properly collect and assess material information and test results. A model has been proposed here that positions the analytical chemist as a single point of contact between other stakeholders and the toxicologist. A risk-based, phase-appropriate approach has been outlined. Early in the development process, a general safety profile is established based primarily on material information gathered by the chemist and evaluated by the toxicologist. As development progresses, the chemist and toxicologist collaboratively develop testing strategies based on the level of risk associated with the general safety profile. A detailed, product-specific safety profile may include biocompatibility, physicochemical, and controlled extraction and leachables study results. Any safety concerns are discussed with the appropriate stakeholders. Risk mitigations may include toxicity studies, changes in material or manufacturing processes, or establishment of routine controls for compounds of toxicological concern. In the latter stages of development, the chemist and toxicologist are involved in establishing appropriate control limits for compounds of toxicological concern by performing testing and careful review of the batch history. During the remainder of the life of the product, the chemist and toxicologist are expected to evaluate out of specification results or perform testing and assessments when materials are changed or processes are modified. Through presentation of case studies, the type of information needed and the use of toxicological assessment tools have been demonstrated.

REFERENCES

1   Guidance for industry: Container closure systems for packaging human drugs and biologics. Department of Health and Human Services, Food and Drug Administration, Center for Drug Evaluation and Research (CDER), 1999.

2   Committee for Medicinal Products for Human Use (CHMP) and Committee for Medicinal Products for Veterinary Use (CVMP). Guideline on plastic immediate packaging materials. EMEA, 2005.

3   Tague, N. Mil-Std-1629a. *The Quality Toolbox*, 2nd ed. Procedures for Performing a Failure Mode, Effects and Criticality Analysis. ASQ Quality Press, Milwaukee, WI, 2005; 236–242.

4   Reviewer guidance for nebulizers, metered dose inhalers, spacers and actuators. U.S. Department of Health and Human Services, Food and Drug Administration Center for Devices and Radiological Health (CDRH), 1993.

5   Guidance for industry: Nasal spray and inhalation solution, suspension, and spray drug products—Chemistry, manufacturing, and controls documentation. U.S. Department of Health and Human Services, Food and Drug Administration Center for Drug Evaluation and Research (CDER), 2002.

6   Draft guidance for industry: Metered dose inhaler (MDI) and dry powder inhaler (DPI) drug products. Department of Health and Human Services, Food and Drug Administration Center for Drug Evaluation and Research (CDER), 1998.

7   Norwood, D.L. and Ball, D. Product Quality Research Institute: Safety thresholds and best practices for extractables and leachables in orally inhaled and nasal drug products. Submitted to the PQRI Drug Product Technical Committee, PQRI Steering Committee, and U.S. Food and Drug Administration by the PQRI Leachables and Extractables Working Group, 2006.

8   Guidance for industry: Pharmaceutical quality of inhalation and nasal products. Health Canada, 2006.

9   Committee for Medicinal Products for Human Use. Guideline on the pharmaceutical quality of inhalation and nasal products. EMEA, 2006.

10 Mullis, J.D., Granger, A., Qin, C., and Norwood, D.L. (Paper 4) The analytical evaluation threshold (AET) concept: Sensitivity and analytical uncertainty. In: *Leachables and Extractables 2008*. iSmithers Rapra Publishing, 2008.

11 Koch, H.M., Drexler, H., and Angerer, J. An estimation of the daily intake of di(2-ethylhexyl)pththalate (DEHP) and other phthalates in the general population. *Int J Hyg Environ Health* 2003, *206*, pp. 77–83.

12 Institute for Health and Consumer Protection, Toxicology and Chemical Substance, European Chemicals Bureau, and European Union. Bis (2-ethylhexyl) phthalate (DEHP). Summary Risk Assessment Report, Ispra (VA) Italy, 2008 (EUR 23384 EN, ISSN 1018-5593).

13 Benignus, V.A., Geller, A.M., Boyes, W.K., et al. Human neurobehavioral effects of long-term exposure to styrene: A meta-analysis. *Environ Health Perspect* 2005, *113*, pp. 532–538.

14 Connelly, J., Hasegawa, R., McArdle, J., and Tucker, M. Residual solvents. *Pharmeuropa* 1997, *9*, Suppl., p. 54.

15 Genetox and Chemical Carcinogenesis Research Information System. Available at: http://toxnet.nlm.nih.gov/ (accessed May 2003).

16 Registry of toxic effects of chemical substances, Available at: http://csi.micromedex.com/DATA/RT/RTEO1400000.HTM?Top=Yes (accessed May 2003).

17 Zbinden, G. Acute toxicity. In: Zbinden, G., ed. *Progress in Toxicology*. Springer-Verlag, New York, 1973; 24.

18 Code of Federal Regulations. Synthetic flavoring substances and adjuvants. Part 21, Sec. 172.515.

19 Ollroge, I. Threshold values and recommendations. In: Marquardt, H., Schäfer, S.G., McClellan, R.O., and Welsch, F., eds. *Toxicology*. Academic Press, San Diego, CA, 1999; 1201–1229.

20 National Institute of Occupational Safety and Health Pocket Guide to Chemical Hazards. Available at: http://www.cdc.gov/niosh/npg/npgd0076.html (accessed October 2003).

21 Guidance for industry: Q3C impurities: Residual solvents. U.S. Department of Health and Human Services, Food and Drug Administration Center for Drug Evaluation and Research (CDER), 1997.

22 Woodward, K.N. *Phthalate Esters: Toxicity and Metabolism*. CRC Press, Boca Raton, FL, 1988; 47–75.

23 Seed, J.L. Mutagenic activity of phthalate esters in bacterial liquid suspension assays. *Environ Health Perspect* 1982, *45*, pp. 111–114.

24 Registry of toxic effects of chemical substances. Available at: http://csi.micromedex.com/DATA/RT/RTMP0350000.HTM?Top=Yes (accessed May 2003).

25 Astill, B.D., Deckardt, K., Gembardt, C., Gingell, R., Guest, D., Hodgson, J.R., Mellert, W., Murphy, S.R., and Tyler, T.R. Prechronic toxicity studies on 2-ethylhexanol in F334 rats and B6C3F1 mice. *Fund Appl Toxicol* 1996, *29*, pp. 31–39.

26 Klimisch, H.J., Deckhardt, K., Gembardt, C., and Hildebrand, B. Subchronic inhalation toxicity study of 2-ethylhexanol vapour in rats. *Food Chem Toxicol* 1998, *36*, pp. 165–168.

27 Nelson, B.K., Brightwell, W.S., Khan, A., Kreig, E.F., and Hoberman, A.M. Developmental toxicology evaluation of 1-pentanol, 1-hexanol and 2-ethyl-1-hexanol administered by inhalation to rats. *J Am Coll Toxicol* 1989, *8*, pp. 405–410.

28 Expert panel report on di(2-ethylhexyl)phthalate. Center for the Evaluation of Risks to Human Reproduction, National Toxicology Program, United States Department of Health and Human Services, October, 2000.

29 Ritter, E.J., Scott, W.J., Randall, J.L., and Ritter, J.M. Teratogenicity of di(2-ethylhexyl)phthalate, 2-ethylhexanol, 2-ethylhexanoic acid, and valproic acid and potentiation by caffeine. *Teratology* 1987, *35*, pp. 41–46.

30 Astill, B.D., Gingell, R., Guest, D., Hellwig, J., Hodgson, J.R., Kuettler, K., Mellert, W., Murphy, S.R., Sielke, R.L., and Tyler, T.R. Oncogenicity testing of 2-ethylhexanol in Fischer 344 rats and B6C3F1 mice. *Fund Appl Toxicol* 1996, *31*, pp. 29–41.

31 Hazardous substance database. Available at: http://toxnet.nlm.nih.gov/ (accessed May 2003).

32 Chemical Carcinogenesis Research Information System. Available at: http://toxnet.nlm.nih.gov/ (accessed November, 2005).

33 Thomson Micromedex. LD50 for erucamide. Available at: http://csi.micromedex.com/fraMain.asp?Mnu=0&Restore=Y (accessed September 2003).

34 Code of Federal Regulations. Part 21, Sec. 175.105.

CHAPTER 8

USE OF SAFETY THRESHOLDS IN THE PHARMACEUTICAL DEVELOPMENT PROCESS FOR OINDP: U.S. REGULATORY PERSPECTIVES

Timothy J. McGovern*

8.1 REGULATORY HISTORY RELATED TO CONTROL OF LEACHABLES AND EXTRACTABLES IN ORALLY INHALED AND NASAL DRUG PRODUCTS

Leachables and potential leachables (i.e., extractables) present a potential safety concern for inhalation and intranasal drug formulations, as well as for products intended for parenteral and ophthalmic use. Significantly, their presence can often impact the marketing approval of a drug product.

Federal regulations governing the packaging of drug products, which include reference to extractables and leachables, are extensive and point to the importance of managing extractables and leachables associated with drug product packaging materials and drug products. For example, the Food, Drug and Cosmetic Act (FD&C Act) states that a drug or device shall be deemed to be adulterated "if its container is composed, in whole or in part, of any poisonous or deleterious substance which may render the contents injurious to health."[1,2] Furthermore, the Code of Federal Regulations (CFR) states that "drug product containers and closures shall not be reactive, additive, or absorptive so as to alter the safety, identity, strength, quality, or purity of the drug beyond the official or established requirements."[3] The United

---

* Formerly with the United States Food and Drug Administration, Center for Drug Evaluation and Research (USFDA-CDER).

Leachables and Extractables Handbook: Safety Evaluation, Qualification, and Best Practices Applied to Inhalation Drug Products, First Edition. Edited by Douglas J. Ball, Daniel L. Norwood, Cheryl L.M. Stults, Lee M. Nagao.

TABLE 8.1.   Examples of Packaging Concerns for Common Classes of Drug Products (Taken from the FDA "Packaging Guidance"[2])

Degree of concern associated with the route of administration	Likelihood of packaging component–dosage form interaction		
	High	Medium	Low
Highest	Inhalation aerosols and solutions; injections and injectable suspensions	Sterile powders and powders for injection; inhalation powders	
High	Ophthalmic solutions and suspensions; transdermal ointments and patches; nasal aerosols and sprays		
Low	Topical solutions and suspensions; topical and lingual aerosols; oral solutions and suspensions	Topical powders; oral powders	Oral tablets and oral (hard and soft gelatin) capsules

States Food and Drug Administration (FDA) has translated these high-level regulations into specific and extensive guidance for industry addressing drug product packaging and, in particular, orally inhaled and nasal drug products (OINDPs).

The FDA "Guidance for Industry: Container Closure Systems for Packaging Human Drugs and Biologics" (commonly referred to as the "packaging guidance") provides a tabular overview of the relative importance given to various drug products by the FDA reviewers with respect to container closure systems in contact with drug formulations.[2] Table 8.1, which is reproduced from the packaging guidance,[2] relates the "degree of concern associated with the route of administration" to the "likelihood of packaging component-dosage form interaction." The latter term can be considered as directly related to the probability of container closure system component/drug product formulation interactions resulting in organic leachables appearing in the drug product over its determined shelf life. The inhalation route is of highest concern due, at least in part, to the fact that drug product leachables are delivered directly to a potentially diseased organ system of a compromised patient population. As can be seen in Table 8.1, inhalation aerosols and solutions, as well as nasal aerosols and sprays, are all of very high concern to the regulatory authorities. Also, note that other drug product types, including injections and injectable suspensions, ophthalmic solutions and suspensions, and transdermal ointments and patches are also of relatively high concern.

The packaging guidance also describes, very generally, the types of evaluations that should be performed at a minimum to ensure the safety of a given OINDP. For inhalation solutions and aerosols, and nasal sprays, these include USP biological reactivity test data (United States Pharmocopeia), toxicological evaluation, extractables testing, limits on extractables, and batch-to-batch monitoring of extractables. For inhalation powders (a dosage form of lower concern), this includes a "reference to the indirect food additive regulations for all components except the mouthpiece for which USP biological reactivity test data are provided."[2] The guidance states that a toxicological evaluation of substances, which are extracted in order to determine a safe level of exposure, is appropriate. The approach for toxicological evaluation should be based on good scientific principles and take into account factors such as the specific container closure system, route of administration, and dose regimen. It states that the toxicological aspects for inhalation products are unique in that these products are typically used in respiratory-tract-compromised patients.

Two other FDA guidances—metered dose inhaler (MDI) and dry powder inhaler (DPI) drug products, as well as nasal spray and inhalation solution, suspension, and spray drug products—provide further recommendations regarding the evaluation of leachables and extractables in OINDPs, emphasizing the importance of this issue for these types of drug products.[4,5]

The draft guidance on MDIs and DPIs indicates that toxicological information for extracted materials and residues should be provided in a drug application when appropriate. The toxicological evaluation should include appropriate *in vitro* and *in vivo* tests, and this information should be used in developing a rationale to support acceptance criteria for the various components. It further states that since some extractable components from rubber may be carcinogenic, appropriate risk assessment models may be needed to establish acceptance criteria. The guidance on nasal spray and inhalation solution, suspension, and spray drug products reiterates the points made in the other two documents.

While the above-described guidances provide general information on what types of toxicological evaluations may be required, there is currently no formal FDA guidance on the specific aspects of the safety evaluation of leachables and extractables in OINDPs. The following sections of this chapter provide some general guidance with a particular focus on the use of safety thresholds.

8.2 HISTORICAL PERSPECTIVE OF THE USE OF SAFETY THRESHOLDS FOR OINDP BY THE FDA

The use of safety thresholds is considered acceptable, and the Division of Pulmonary, Allergy and Rheumatology Products (DPARP) in the FDA's Center for Drug Evaluation and Research/Office of New Drugs has incorporated the application of safety thresholds into the qualification process conducted for leachables and extractables in OINDP over the last decade in cases where adequate supporting data are available. It is recognized that the appropriate application of safety thresholds has advantages for overall drug development, including a reduction in the unnecessary expenditure of animals, time, effort, and money. This reduced expenditure may allow greater resources to be applied to drug development areas that present more significant safety concerns.

In the absence of formal guidance, the DPARP developed an internal practice in the late 1990s that included the following general approach:

- identification of the compound and determination of the maximum daily human exposure based on the proposed product specification (e.g., amount of the specific leachable in the drug product, typically in units of parts per million or microgram per canister);

- conduct of a structure–activity relationship (SAR) assessment for genotoxic/ carcinogenic potential through use of published lists[6,7] or software programs such as Deductive Estimation of Risk through Existing Knowledge (DEREK) (https://www.lhasalimited.org/) or Multicase (http://www.multicase.com); and

- review of available toxicology/safety databases or conduct of toxicology studies as deemed necessary (e.g., 14- to 90-day general toxicology, genetic toxicology).

The final safety assessment was based on a consideration of the maximum expected daily human exposure, the intended patient population, and the anticipated duration of use. In general, when adequate safety data are available to support a proposed product specification, conduct of a compound-specific risk assessment rather than a threshold-based approach is recommended. It is anticipated that, in most cases, higher product specifications would be supported from a safety standpoint when data are available to support a compound-specific risk assessment than would be through use of a threshold-based approach.

Specific considerations for the safety assessment of leachables and extractables in OINDP and the use of thresholds include three primary components: systemic toxicity, local toxicity of the respiratory tree, and mutagenic/carcinogenic potential. These three components are discussed in the following sections.

8.3 IDENTIFICATION OF A SAFETY THRESHOLD BASED ON POTENTIAL FOR SYSTEMIC TOXICITY

The DPARP internally identified a safety threshold for systemic toxicity parameters in the late 1990s based on an evaluation of the United States Environmental Protection Agency's (EPA) Integrated Risk Information System (IRIS) and health effects assessment summary tables (HEAST) databases[8] and concluded that there is no significant safety concern for inhaled chemicals with a maximum expected daily dose of 5 µg/day (100 ng/kg) or less. Therefore, no further toxicity data are needed in most cases when the maximum expected daily human exposure is below the stated threshold.

The safety threshold for systemic toxicity was derived from an evaluation of 36 chemicals listed in the EPA database that were administered via the inhalation route and were associated with systemic toxicity. The presumed safe dose derived from the reference concentrations (RfCs) for these compounds were all ≥100 ng/kg with three exceptions; the three exceptions had a "safe" dose of 80 ng/kg. Considering the large safety factors (1000–10,000) that are incorporated into the calculation of RfCs, a safety threshold of 100 ng/kg was considered reasonable.

In cases where the maximum expected human exposure to a leachable is expected to exceed the qualification threshold of 5 µg/day, safety qualification for general toxicity concerns may be provided through evaluation of published toxicity data, relevant regulatory exposure limits such as EPA air quality standards, or through the conduct of inhalation toxicology studies of an appropriate duration (e.g., at least 90 days' duration for chronic indications). In some cases, data for chemicals with well-characterized toxicity profiles that have a high degree of structural similarity to a leachable/extractable for which limited safety data are available may be considered.

When toxicology data are used to support proposed product specifications, the most relevant data should be considered. For example, if data are available from studies using both oral and inhalation administration, the data derived from the inhalation studies are usually considered to be most relevant. Safety margins for anticipated human exposures are calculated based on the no-observed-adverse-effect levels (NOAELs) in animal studies. Generally, a 10-fold safety factor is applied for cross-species extrapolation. In cases where safety data are only available from studies using the oral route of administration, an additional 100-fold safety factor is applied based on an evaluation of a subset of the EPA HEAST database for which data were available following both oral and inhalation administration. For this subset of chemicals, the presumed safe inhalation doses derived from the RfCs were up to 100-fold lower than the identified reference doses (RfDs) after conversion to a milligram per kilogram per day dose. Therefore, a 1000-fold safety factor (10 × 100) is typically applied when using animal data from oral studies to support human inhalation use.

8.4 IDENTIFICATION OF A SAFETY THRESHOLD BASED ON THE POTENTIAL FOR RESPIRATORY TOXICITY

Similar to the approach taken for systemic toxicity, the DPARP identified a safety threshold for local toxicity parameters in the late 1990s by evaluating the EPA's IRIS and HEAST databases[8] and again concluded that there is no significant safety concern for inhaled chemicals with a maximum expected daily dose of 5 μg/day (100 ng/kg) or less. As before, no further toxicity data are needed in most cases when the maximum expected daily human exposure is below the stated threshold.

The safety threshold for respiratory toxicity was derived from an evaluation of 20 chemicals with inhalation data that produced respiratory tract toxicity. All but four of these chemicals had presumed safe daily inhalation exposures greater than 100 ng/kg based on the RfCs, even after incorporation of large safety factors (300–1000). Of note, all of the compounds with a presumed safe dose less than 100 ng/kg had a structural alert associated with respiratory tract irritation. These structural alerts included isocyanates, aldehydes, organic acids, strained heterocyclic rings, and halogenated aromatic rings.

There are some exceptions to the use of the above-described threshold approach for leachables and extractables in inhalation products. These include compounds identified as respiratory irritants and sensitizers. Therefore, chemicals should be evaluated for structural alerts associated with irritation or sensitization. This determination is especially important when considering the indicated population for a given product. Most inhalation products approved to date are indicated for treatment of pulmonary diseases such as asthma. These patients are already considered to have a compromised respiratory function and may be more sensitive to the effects of irritants or sensitizers. If a compound is considered to have an irritant or sensitizing potential, patient risk should be assessed on a case-by-case basis after evaluating the available information for the specific compound. Additionally, the clinical experience with the drug product should be evaluated for evidence of any adverse effects. If no concern is identified for irritancy or sensitization, the safety qualification threshold for systemic and local toxicity of 5 µg/day is appropriate. For anticipated clinical exposures greater than 5 µg/day, safety qualification should be conducted for systemic and local toxicity as described later in this chapter.

8.5 IDENTIFICATION OF A SAFETY THRESHOLD BASED ON THE POTENTIAL FOR MUTAGENIC/ CARCINOGENIC POTENTIAL

The DPARP currently has no formal policy in regard to a safety threshold for leachables or extractables with an identified or suspected genotoxic or carcinogenic potential. Leachables and extractables are considered adequately qualified for genotoxic or carcinogenic potential if they are demonstrated to produce negative results in genotoxicity and/or carcinogenicity assays, or in cases where this type of data is not available if they lack structural alerts for these end points. When no concern for genotoxic or carcinogenic potential is identified, a qualification threshold of 5 µg/day, based on local and systemic toxicity potential, is appropriate in the absence of supporting general toxicology data and an identified potential for respiratory irritation or sensitization.

As stated, the safety thresholds for local and systemic effects were considered applicable as long as no potential for genotoxic or carcinogenic effects was identified with the compound under evaluation. In the past, if a leachable or extractable was a known or suspected genotoxin or carcinogen, the DPARP would generally require that appropriate tests, such as *in vitro* genotoxicity assays, be conducted or a rationale provided to alleviate this concern. Positive findings in these assays would then require follow-up evaluation of the carcinogenic potential.

The Product Quality Research Institute (PQRI) Working Group on Leachables and Extractables proposed a safety concern threshold (SCT) of 0.15 µg/day. This threshold was derived from calculated risk-specific doses of genotoxic (SAL-positive) carcinogens from the carcinogenic potency database (CPDB); the specific derivation of the SCT is discussed in an earlier chapter. The SCT is considered to be a dose below which a leachable would present negligible concern for adverse carcinogenic and noncarcinogenic effects. Although the PQRI recommendations have not yet been formally accepted by the FDA, the DPARP now considers the SCT when conducting a safety evaluation for a compound with an identified concern for genotoxic or carcinogenic effects.

8.6 COMPARISON OF FDA/DPARP PRACTICES WITH THE PQRI RECOMMENDATIONS IN RELATION TO THE SAFETY QUALIFICATION OF LEACHABLES AND EXTRACTABLES

As described in previous chapters, the PQRI Leachable and Extractable Working Group expanded the data analyses initially conducted by the DPARP to include the Agency for Toxic Substances and Disease Registry (ATSDR) and California Environmental Protection Agency (CAL EPA) databases. The evaluation conducted by the Working Group also concluded that a threshold of 5 µg/day presented a negligible safety concern for noncarcinogenic effects and, thus, recommended a qualification threshold of 5 µg/day (100 ng/kg/day, 50-kg person). Therefore, the qualification threshold recommended by the PQRI Working Group for local and systemic toxic effects is in agreement with the DPARP practice.

The proposed SCT for negligible carcinogenic effects expands on the DPARP's previous use of thresholds that focused primarily on general toxicological effects. The PQRI proposal is, however, similar to that described previously by the FDA's Center for Food Safety and Nutrition for the safety assessment of food contact materials,[9] and similar proposals have been made to support safety thresholds for genotoxic impurities.[10–12] The proposed approach is supported by a large database, and the applied cancer risk of 10^{-6} is considered appropriate due to the nature of the chemicals that are commonly encountered as leachables, and the lack of any benefit derived from their presence. Notably, high-potency carcinogens (e.g., N-nitrosamines, polyaromatic hydrocarbons [PAHs]) are excluded from this threshold approach.

8.7 ILLUSTRATIVE CASE EXAMPLES

The following case examples are presented to illustrate the safety qualification process using both a threshold approach and the use of toxicology data.

8.7.1 Bis-(2-Ethylhexyl) Sebacate

Bis-(2-ethylhexyl) sebacate (Chemical Abstracts Service [CAS] #122-62-3)

is an ester of sebacic acid, which is a widely used plasticizer in various plastics. The use of bis-(2-ethylhexyl) sebacate as a plasticizer makes this chemical entity a potential leachable when incorporated into an OINDP container closure system critical component.

The proposed product specification corresponded to a maximum daily human exposure of 9.1 µg/day (182 ng/kg/day for a 50-kg individual). The most relevant toxicity study identified for this compound was a published chronic dietary study in rats in which a no-observed-effect level (NOEL) of 200 mg/kg was observed. This dose corresponds to an acceptable human inhalation exposure of 0.2 mg/kg/day (200,000 ng/kg/day) after application of a 1000-fold safety factor for cross-species extrapolation and for the use of data derived from oral administration to support inhalation use. Even after incorporation of this safety factor, a greater than 1000-fold safety margin for bis-(2-ethyl-hexyl) sebacate was present when comparing the acceptable human daily exposure to the maximum anticipated human exposure associated with the proposed product specification.

8.7.2   4-Toluene sulfonamide

4-Toluene sulfonamide (or *p*-toluene sulfonamide; CAS #70-55-3)

is also a widely used plasticizer in various plastics and resins.

The proposed product specification corresponded to a maximum daily human exposure of 60 µg/day (1200 ng/kg/day for a 50-kg individual). In this case, the sponsor provided no supporting rationale for the proposed specification. However, a review of the literature indicated that only acute toxicity data were available. In this case, the DPARP requested that the sponsor lower the product specification to a level that corresponded to the division's safety qualification threshold of 5 µg/day or provide adequate toxicology data, such as a 3-month inhalation toxicity study, to support their proposed product specification.

8.7.3   Acenaphthene

Acenaphthene (CAS #83-32-9)

is a PAH (or polynuclear aromatic [PNA]), which is one of the group of "special case" leachables/extractables mentioned in Chapters 1 and 11. PAHs are associated with carbon black, which is a widely used filler and reinforcing agent is certain types of rubber, particularly sulfur-cured rubber. Acenaphthene, along with other PAHs, can appear in MDI drug products when carbon-black-containing elastomers are used as seals in the dose metering values.

The proposed product specification corresponded to a maximum daily human exposure of 0.067 µg/day (1.34 ng/kg/day). As in the previous case, only acute toxicity data were available. However, a State of Minnesota drinking water standard for acenaphthene is set at 400 µg/L.[13,14] This standard corresponds to an acceptable daily inhalation exposure of 160 ng/kg assuming a daily intake of 2 L/day, a 50-kg body weight, and incorporation of a 100-fold safety factor for the use of oral data to support inhalation use. The calculated acceptable inhalation exposure for humans provided a greater than 100-fold safety margin when compared with the maximum daily human exposure to acenaphthene through use of the drug product at the sponsor's proposed specification. In addition, the anticipated human exposure was well below DPARP's safety qualification threshold of 5 µg/day.

8.7.4 N-Nitrosamines

N-nitrosamines are also considered to be "special case" leachables (see Chapter 1). Some examples are as follows:

$$H_3C, H_3C > N-N=O$$

N-nitrosodimethylamine (CAS #62-75-9)

N-nitrosomorpholine (CAS #59-89-2)

N-nitrosopiperidine (CAS #100-75-4)

N-nitrosopyrrolidine (CAS #930-55-2)

    N-nitrosamines are formed at relatively trace levels in certain types of rubber during the "vulcanization" (or cross-linking) process. Their formation is most often associated with the use of certain rubber curing agents, such as thiurams (Chapters

1 and 11). As with PAHs, N-nitrosamines can leach into MDI formulations (and other OINDP types) when appropriate elastomeric critical components are included in the container closure system.

In one product, six N-nitrosamines were identified at various levels. The carcinogenic risk assessment was based on total N-nitrosamine exposure using the slope factor calculated for N-nitrosodimethylamine. A maximum human daily exposure up to 0.04 ng/kg, a level associated with a cancer risk estimate of 10^{-5}, was accepted based on the overall risk–benefit analysis and technological considerations, namely, the inability to manufacture rubber components that do not release N-nitrosamine compounds. Although DPARP allowed a level that corresponded to a cancer risk estimate of 10^{-5}, DPARP encourages sponsors to continue to reduce the potential for exposure and to strive to develop methods to eliminate the presence of N-nitrosamines.

8.8   CONCLUSIONS

Issues arising from the need for adequate safety qualification for leachables and extractables can often delay the approval of drug products. Therefore, some consideration should be given toward addressing these issues earlier in product development or by providing more substantive qualification information. Often the DPARP receives submissions containing only the proposed drug product specifications for a given leachable or extractable with no supporting rationale. The use of safety thresholds is considered to be an acceptable approach that can not only address potential safety issues, but also save resources in the conduct of a drug development program.

REFERENCES

1   United States Food, Drug and Cosmetic Act. Section 501(a)(3). U.S. Food and Drug Administration, amended March 2005. December 2004.
2   Guidance for industry: Container closure systems for packaging human drugs and biologics. Department of Health and Human Services, Food and Drug Administration, 1999.
3   Code of Federal Regulations. Sec. 211.94. Drug product containers and closures, April 2011.
4   Draft guidance for industry: Metered dose inhaler (MDI) and dry powder inhaler (DPI) drug products. Department of Health and Human Services, Food and Drug Administration, Center for Drug Evaluation and Research (CDER), 1998.
5   Guidance for industry: Nasal spray and inhalation solution, suspension, and spray drug products— Chemistry, manufacturing, and controls documentation. Department of Health and Human Services, Food and Drug Administration, Center for Drug Evaluation and Research (CDER), 2002.
6   Ashby, J., Tennant, R.W., Zeiger, E., and Stasiewicz, S. Classification according to chemical structure, mutagenicity to *Salmonella* and level of carcinogenicity of a further 42 chemicals tested for carcinogenicity by the U.S. National Toxicology Program. *Mutat Res* 1989, *223*, pp. 73–103.
7   Tennant, R.W. and Ashby, J. Classification according to chemical structure, mutagenicity to *Salmonella* and level of carcinogenicity of a further 39 chemicals tested for carcinogenicity by the U.S. National Toxicology Program. *Mutat Res* 1991, *257*, pp. 209–227.
8   The United States Environmental Protection Agency's Integrated Risk Information System (IRIS). Available at: http://www.epa.gov/IRIS/. Health effects assessment summary tables (HEAST). Risk

assessment information system, chemical specific toxicity values, 1997. Available at: http://cfpub.epa.gov/ncea/cfm/recordisplay.cfm?deid=2877 (accessed April 27, 2009).

9   Guidance for industry: Preparation of food contact notifications for food contact substances: Toxicology recommendations. U.S. Food and Drug Administration, 2002.

10   Müller, L., Mauthe, R.J., Riley, C.M., Andino, M.M., De Antonis, D., Beels, C., DeGeorge, J., De Knaep, A.G.M., Ellison, D., Fagerland, J.A., Frank, R., Fritschel, B., Galloway, S., Harpur, E., Humfrey, C.D.N., Jacks, A.S., Jagota, N., Mackinnon, J., Mohan, G., Ness, D.K., O'Donovan, M.R., Smith, M.D., Vudathala, G., and Yotti, L. A rationale for determining, testing, and controlling specific impurities in pharmaceuticals that possess potential for genotoxicity. *Regul Toxicol Pharmacol* 2006, *44*, pp. 198–211.

11   Committee for Medicinal Products for Human Use (CHMP) of the European Medicines Agency (EMEA). Guideline on the limits of genotoxic impurities, 2006. Available at: http://www.ema.europa.eu/ema/index.jsp?curl=/pages/home/Home_Page.jsp.

12   Draft guidance for industry: Genotoxic and carcinogenic impurities in drug substances and products: Recommended approaches. FDA, December 2008. Available at: http://www.fda.gov/Drugs/GuidanceComplianceRegulatoryInformation/Guidances/ucm065014.htm.

13   U.S. EPA Office of Water, Federal-State Toxicology and Risk Analysis Committee. Summary of state and federal drinking water standards and guidelines, 1993.

14   Hazardous Substances Data Bank (HSDB). Available at: http://toxnet.nlm.nih.gov/cgi-bin/sis/search/f?/temp/~wUSxAF:1.

CHAPTER 9

THE APPLICATION OF THE SAFETY THRESHOLDS TO QUALIFY LEACHABLES FROM PLASTIC CONTAINER CLOSURE SYSTEMS INTENDED FOR PHARMACEUTICAL PRODUCTS: A REGULATORY PERSPECTIVE*

Kumudini Nicholas

9.1 INTRODUCTION

The selection process of a container closure system (CCS) for a pharmaceutical product requires the identification of potential extractables and the qualification of product-specific leachables. The application of this two-tier process is imperative as the presence of impurities originating from packaging materials causes quality concerns due to potential risks to patient safety.[1] To eliminate quality concerns, it is important to utilize a qualification process to determine safe levels of harmful impurities, where applicable, to justify the use of a drug product's specific CCS.

The goal of a qualification process is to obtain sufficient information to enable the identification of potential hazards arising due to chemicals originating from packaging components, followed by the risk assessment of chemicals that leach into

---

* The views expressed in this chapter and the contents are the sole responsibility of the author and do not necessarily reflect those of Health Canada.

Leachables and Extractables Handbook: Safety Evaluation, Qualification, and Best Practices Applied to Inhalation Drug Products, First Edition. Edited by Douglas J. Ball, Daniel L. Norwood, Cheryl L.M. Stults, Lee M. Nagao.
© 2012 John Wiley & Sons, Inc. Published 2012 by John Wiley & Sons, Inc.

the drug. One of the qualification processes to manage the risk is the use of a safety threshold concept, which is considered an acceptable approach when adequate data are available to support this process. It should be noted that for the chemical irritants, sensitizers, and high-potency carcinogens such as nitrosamines and polyaromatic hydrocarbons, the threshold approach is not considered applicable.[2]

With the advancement of analytical technology, the determination of leachables originating from a CCS can be daunting, as many as hundreds if not thousands of individual chemical entities could be detected, identified, and quantified at comparable levels. However, it is well established that there are levels at or below which some organic chemical entities present in a drug product would have no safety concerns for humans. Therefore, the establishment of safety thresholds that can identify the potential safety margins simplifies the analytical practices to determine them.

A qualification process, while taking into account the specific CCS, the drug product formulation, dosage form, route of administration, and dose regime,[3] should also take into consideration the manufacturing formula and the manufacturing process of the plastic material intended to be used to generate a component. A prudent qualification process would ensure that the selected CCS would maintain the quality of drug product throughout its shelf life.

9.2   A TOXICANT AND ITS ASSOCIATED RISK

The word "toxic" can be used to describe fatal poisoning and also be applied to substances that are, to varying degrees, injurious to health. Toxicity can be divided into three categories: acute toxicity, which is displayed by immediate, severe symptoms; chronic toxicity, which is reoccurring or of long continuance; and insidious toxicity, which occurs when the toxicant works "secretly."[4] Insidious toxicity would have unexpected adverse effects as the individual is unprepared for medical action until it is too late. Toxic leachables are considered as a threat to the normal functions of the human body, based on their insidious toxic nature.

There are three distinct classes of insidious toxicants, all of which display what can be termed as "delayed effects." They are mutagens, carcinogens, and teratogens.

A mutagen is a substance or agent that induces heritable change in cells or organisms. A carcinogen is a substance that induces unregulated growth processes in cells or tissues of multicellular animals, leading to cancer. Teratogens are substances which, when ingested by the mother-to-be, can adversely affect her unborn child.[4] Although "mutagen" and "carcinogen" are not synonymous terms, the ability of a substance to induce mutation and its ability to induce cancer are strongly correlated. Such toxicants can be leached into the drug product from various devices or CCS components used to store or administer a drug. It is understood that the duration and frequency of exposure of these impurities determine the depth of effects on humans. For example, chronic exposure, which is known as multiple exposures over more than 3 months, could have detrimental effects to humans compared with an acute exposure, a single exposure, or multiple exposures over 1–48 hours, by the same toxicant.[5]

Toxicants can impact the human body either by affecting the respiratory system, the eyes, and the skin. Such chemicals can be inhaled primarily from the vapors of volatile liquids and solids, could reach the bloodstream if injected as a parenteral drug, or could be swallowed if in a liquid oral drug. Toxic vapors can be irritants, asphyxiants, anaesthetics, and systemic toxicants. Volatile irritants are capable of preferentially affecting specific areas of the respiratory system, such as the respiratory tract, the lung tissues of the terminal air passage, and air sacs. Therefore, the determination of risks associated with potential toxicants originating from a CCS intended for an orally inhaled drug product (OINDP) or any other dosage form is considered imperative to prevent harmful effects on humans.

9.3 THE PLASTIC USE FOR MEDICAL PACKAGING

Plastics or polymers are ubiquitous. The properties of plastics can be tailored to meet specific needs by varying the atomic makeup of the repeat structure, and by varying flexibility as governed by presence of side-chain branching, as well as the length and polarities of the side chains. In addition, by tailoring the crystallinity and the amount of orientation imparted to the plastic during processing, and through copolymerization; by blending with other plastics; and by modifying with an enormous range of additives (fillers, fibers, plasticizers, stabilizers), a variety of plastics can be generated.[6] Based on this variety of availability, a challenge exists for those who select and qualify, and those who regulate the plastics for pharmaceutical products. To meet this challenge, it is the responsibility of the regulator to propose a systematic approach that component and drug product manufacturers can use to generate acceptable data to determine the suitability of a plastic for its intended use.

Plastics can be classified in various ways, but the two major classifications are thermosetting plastic and thermoplastic materials. As the name implies, thermosetting plastics, or "thermosets," are set, cured, or hardened into a permanent shape. The curing, which usually occurs rapidly under heat or UV light, leads to an irreversible cross-linking of the polymer. Thermoplastics differ from thermosetting materials in that they do not set or cure under heat. When heated, thermoplastics merely soften to a mobile flowable state where they can be shaped into useful objects. Upon cooling, thermoplastics harden and hold their shape. Thermoplastics can be repeatedly softened by heat and shaped.

Another important class of polymeric resin is elastomers. Elastomeric materials are rubberlike polymers with a glass transition temperature below room temperature. Below that glass transition temperature, an elastomer will become rigid and lose its rubbery characteristics.

Plastic materials are widely used for medicinal products, such as containers, packaging systems, infusion sets, manufacturing equipment, and various devices. While plastics are frequently available and possess desirable qualities for their intended use, many plastics that have direct contact with the drug are known to leach chemicals into the drug formulation and can therefore deteriorate the quality of a medicine during short-term or long-term storage and transportation. Over several years, quality concerns due to plastic interactions with drugs have been taken into consideration during qualification processes of plastic CCS for medicinal products, and their negative impact on the quality of medicines is well documented.[7] Plastic material used for medical packaging is known as one of the major sources of toxic chemicals that can leach into drug products, based on the monomer/dimer composition, the additive package included during processing to improve physical properties of the plastic, and the process parameters used to manufacture the plastic. Therefore, it is critical to select an appropriate plastic component that will produce minimal leachables.

Based on the diverse nature of the composition of plastics, it is important to understand why a "plastic" should not be selected based on a general name, as one type of plastic (e.g., polypropylene) can be manufactured by more than one plastic manufacturer. The additive package used in the manufacturing formula by each manufacturer of polypropylene can affect the extractables and leachables profile originating from this plastic. Therefore, the diverse nature of the composition of plastic material available in the market today should be well understood in order to implement a qualification process for selecting a suitable plastic intended for a CCS for an identified drug product.

To prevent adverse effects due to leached chemicals from a plastic, the primary step is to gain extensive knowledge of the composition of a plastic material and potential interactions between the plastic and the drug product formulation, which could lead to leaching of harmful chemicals into the drug product. This responsibility should be shared equally between the component manufacturer and the drug product manufacturer to qualify a suitable CCS.

The most commonly used plastic materials in the pharmaceutical industry, specifically for medico-surgical purposes, have been identified in the European Pharmacopeia 3.1 (Ph. Eur. 3.1).[8] They include plasticized polyvinyl chloride (PVC) (human blood and blood components, tubing used in sets for the transfusion of blood and blood components), polyolefins (polyethylene with/without additives, polypropylene for parenteral and ophthalmic drugs), polyethylene–vinyl acetate (containers and tubing for total parenteral nutrition preparations), and materials based on non-plasticized PVC (containers for noninjectable, aqueous solutions). Under this section, the materials described are intended for use in the manufacture of containers for pharmaceuticals. Their use may also be considered for manufacture of part or all of products used for medico-surgical purposes. While the plastics listed under this section are acceptable for the intended purpose, if the listed potential extractables/leachables meet the stipulated limits in conjunction with the requirements discussed in the EMEA Guideline on Plastic Immediate Packaging Material 2005,[9] materials and polymers other than those described in the pharmacopoeia may be used, subject to approval in each case by the competent authority responsible for the approval of the drug product. It should be noted that the type of plastic that is being used to manufacture a CCS for a pharmaceutical product is not limited to the types identified in the Ph. Eur. 3.1.

Table 9.1 lists those plastics that have not been listed under the Ph. Eur. 3.1, but are in general used to manufacture components of a CCS.[10] Based on current practices of the pharmaceutical industry, it is expected that alternative plastics could be used to manufacture CCS. Therefore, it is not feasible to qualify a plastic material

TABLE 9.1.   Plastic Types Used in the Pharmaceutical Industry Not Listed in the Ph. Eur. 3.1

Plastic	Potential use
Cyclic olefin resins	Syringes, vials
Poly(ethylene terephthalate) (PET)	Bottles and implantable mesh for vascular and hernia applications
Polycarbonate	Vials, tubing connectors, housing for dialysis membrane
Polytetrafluoroethylene (PTFE)	Medical stents, fibers
Silicones	Lubricants, surface treatments/coatings for syringes and vials, septum, seals, surgical devices such as catheters
Nylon	Medical fibers
Rubber (SBR, EPDM, isobutylene-isoprene)	Stopper for parenterals (vials and syringes), components for inhalation devices (MDI/DPI)

SBR, styrene-butadiene rubber; EPDM, ethylene propylene diene monomer; MDI, metered dose inhaler; DPI, dry powder inhaler.

not listed under section 3.1 of the Ph. Eur. using the testing protocols identified in the Ph. Eur. as the extractables and leachables originating from such a material are unknown.

9.3.1   Plastic Additives

Plastic additives represent a broad range of chemicals used by resin manufacturers, compounders, and fabricators to improve the properties, processing, and performance of polymers. There are many different plastics that use large volumes of chemical additives, including (in order of total additives consumption) PVC, the polyolefins (polyethylene [PE] and polypropylene [PP]), the styrenes (polystyrene), and engineering resins such as polycarbonates, and each additive category is comprised of an extremely diverse group of chemicals. Given the range of materials used, plastic additives are generally classified by their function rather than chemistry. Over the years, the availability and types of plastic additives have grown with the industry and currently represent over $16 billion in global sales.[6]

Table 9.2 details the additives identified in the Ph. Eur. 3 as commonly used plastic additives.[10] For each additive, the chemical name and the trade name are included, where possible, in addition to the listing reported in the Ph. Eur. However, it should be noted that based on the diverse and broad range of plastic additives available in the current market, one cannot anticipate that a component or resin manufacturer will select one or a combination of additives solely based on this list. Table 9.3 lists additional plastic additives that can be in use in the component manufacturing industry that are not listed under Ph. Eur. 3.1.[6] Given the large variety of additives currently used in plastic CCS, it is often difficult to comply with extractables/leachables requirements for OINDP, even if using "compendia plastic" as referred to in the EMEA Guideline on Plastic Immediate Packaging Materials.

TABLE 9.2. Plastic Additives Listed in the Ph. Eur. 3.1

EU pharmacopoeia number	Chemical name	Trade name	Function
Add01	(2RS)-2-ethylhexyl benzene-1,2-dicarboxylate	DEHP	Plasticizer
Add02	Zinc (2RS)-2-ethylhexanoate	Octoate Z	Vulcanization accelerator
Add03	N,N'-ethylenedialcanamide		Lubricant stabilizer
Add04	Epoxidized soybean oil		Plasticizer and stabilizer for PVC, dispersion agent for pigments
Add05	Epoxidized linseed oil		Plasticizer and stabilizer for PVC, dispersion agent for pigments
Add06	Pigment Blue 29		Pigment
Add07	2,6-Bis(1,1-dimethylethyl)-4-methylphenol	BHT	Antioxidant
Add08	Ethylene bis[3,3-bis[3-(1,1-dimethylethyl)-4-hydroxyphenyl]butanoate]	Hostanox 03	Antioxidant
Add09	Pentaerythritol tetrakis(3-(3,5-di-tert-butyl-4-hydroxyphenyl)propionate)	Irganox 1010	Antioxidant
Add10	4,4′,4″-[(2,4,6-Trimethylbenzene-1,3,5-triyl)tris(methylene)]tris[2,6-bis(1,1-dimethylethyl)phenol]	Irganox 1330	Antioxidant
Add11	Octadecyl 3-[3,5-bis(1,1-dimethylethyl)-4-hydroxyphenyl]propanoate	Irganox 1076	Antioxidant
Add12	Tris(2,4-ditert-butylphenyl)phosphite	Irgafos 168	Melt process stabilizer (secondary antioxidant)
Add13	1,3,5-Tris[3,5-bis(1,1-dimethylethyl)-4-hydroxybenzyl]-1,3,5-triazine-2,4,6(1H,3H,5H)-trione	Irganox 3114	Antioxidant
Add14	3,9-Bis(octadecyloxy)-2,4,8,10-tetraoxa-3,9-diphosphaspiro[5.5]undecane	Weston 618	Melt process stabilizer (secondary antioxidant)
Add15	1′1-Disulphanediyldioctadecane	Hostanox SE10	Antioxidant—costabilizer with phenolic antioxidant for high-temperature applications
Add16	Didodecyl 3,3′-sulphanediyldipropanoate	Irganox PS-800	Antioxidant—costabilizer with phenolic antioxidant for high-temperature applications

TABLE 9.2. (*Continued*)

EU pharmacopoeia number	Chemical name	Trade name	Function
Add17	Dioctadecyl 3,3′-sulphanediyldip ropanoate	Irganox PS-802	Antioxidant—costabilizer with phenolic antioxidant for high-temperature applications
Add18	CAS #119345-01-6	Irgafos P-EPQ	Melt process stabilizer (secondary antioxidant)
Add19	Octadecanoic acid	Stearic acid	Slip agent, lubricant
Add20	(Z)-octadec-9-enamide	Oleamide	Slip agent, mold release agent
Add21	(Z)-docos-13-enamide	Erucamide Atmer such as Atmer SA1753	Slip agent, mold release agent
Add22	Copolymer of dimethyl butanedioate and 1-(2-hydroxyethyl)-2,2,6,6-tetramethylpiperidin-4-ol	Tinuvin 622	Hindered amine heat/light stabilizer

CAS, Chemical Abstracts Service; DEHP, diester bis(2-ethylhexyl)phthalate; BHT, butylated hyroxytoluene.

9.3.2 Plastic Processing

Over the past several years, a number of products that were once metallic are now being manufactured using plastic.[10] Although the processes used to manufacture plastics have similarities to those used for metals, the diversity of processes for manufacturing plastic products exceeds that of metals, largely because plastics are far more versatile. Thus, specific polymer industries may start the manufacturing process using various starting materials, such as liquid resin or solids; the solids may come in sheets, pellets, granules, flakes, and so on. These are usually thermoplastic polymers. In a few cases, the speciality polymer industry works with monomers or partially polymerized resins, which would generally be used to produce thermoset plastics.

Plastic processing involves three key stages: heating, shaping, and cooling. Details of the processing pathway, for example, specific process parameters within these key stages, may vary. Process selection depends on many factors, including quantity and production rate, form and details of the product, nature of material, size of the final product, dimensional accuracy, and surface finish. Most importantly, a final process will be selected based on the type of plastic one wishes to generate, that is, thermoplastic or thermoset plastic. The aim in selecting a processing route is to manufacture the end product to meet all specified objectives at the required production rate and at a reasonable cost.

TABLE 9.3.   Plastic Additives Not Listed in the Ph. Eur. 3.1

Plastic additive	Function
Antiblock slip agent (e.g., diatomeceous earth, talc, etc.)	Roughen the plastic surface to prevent adhesion (blocking) between film layers
Antistatic agent (e.g., quaternary ammonium salts, sulfonium salts, fatty acid esters, ethoxylated alkylamines, etc.)	Prevent buildup of static charge
Biocides (10,10′-omadinen-(trichloro-methylthio)phthalimide, silver-based compounds, etc.)	Protect against mold, mildew, fungi, and bacterial growth
Fillers/plasticizers	Enhance plastic–filler–reinforcement interface Modify the mechanical properties of polymers
Flame retardant (organic or inorganic with bromine, chlorine, phosphorous antimony, or aluminum material)	Prevent fire (environmental toxic concerns)
Light stabilizers (benzophenone, triazines, benzotriazole, etc.)	Prevent polymer degradation from light
Lubricant and mold releasing agent (metallic stearates, fatty amides, etc.)	Improve flow characteristics
Nucleating agent (silicates, highly dispersed silica, sorbitol-based compounds, etc.)	Increase rate of crystallization and decrease cycle times
Organic peroxides (ketone peroxides, peroxyesters, etc.)	Polymerization of thermoplastic resins, curing for unsaturated polyester thermostat resins, cross-linking, and rheology modification in polypropylene
Polyurethane catalysts (polyisocyanates, polyols, etc.)	Change composition and physical properties of plastic
Chemical blowing agents (volatile liquid or compressed gas)	Produce foam structure

The thermoplastic process starts with regular pellets or granules that are then melted to manufacture the desired plastic. The processes used to generate thermoplastics include (but are not limited to) structural foam, rotational molding, injection molding, injection molding (gas assisted), injection blow molding, injection stretch blow molding, and extrusion blow molding, to name a few. Overheating of the original material can lead to degradation of the polymer and loss of physicochemical properties. Such processing mishaps can lead to generation of chemicals that can change the original extractables profile of a given plastic.

It is logical for the plastic manufacturer to select aspects of the polymer formulation based on the required plastic manufacturing process. For example, in reaction injection molding, the polymer melt contains a blowing agent, which may be (1) a dissolved gas that flashes upon a reduction pressure, (2) a volatile liquid that boils upon an increase of temperature, (3) a nonvolatile liquid that thermally decomposes, or (4) a product from a change in the polymer composition.[11] The first two classes are physical blowing agents (PBAs), and the third is a chemical blowing agent (CBA). An example of the fourth class is the release of carbon dioxide from the product of an isocyanate and an acid, which produces an internally foamed polyamide. This example highlights the diverse nature of the plastic formulation based on the selected manufacturing process. The nature of the end product as well as the availability and cost of ingredients can also impact the formulation composition. These examples support the view that potential extractables and leachables from a plastic could be highly diverse.

The diverse nature of plastics used in the manufacture of primary packaging has introduced a challenge for component manufacturers and regulators alike in selecting and qualifying appropriate CCSs for pharmaceutical products. It is understood that the potential extractables/leachables originating from a plastic can vary depending on the starting material used to manufacture the plastic, and the type and amount of additive used to improve properties of the plastic, as well as the method of manufacture used to generate the plastic. Therefore, it is important to understand that a plastic manufactured by two different suppliers can exhibit two different sets of leachables, although the name of two plastic materials is identical. Unfortunately, each leachables profile may or may not match the list of potential leachables identified in the European Pharmacopoeia (EP) monograph, even though the plastic is listed in the Ph. Eur. 3.1. Therefore, it is not surprising that drug product manufacturers have had difficulties meeting the EP requirements stipulated for a particular plastic material with a predetermined list of potential extractables and leachables.

9.4 GENERAL REGULATORY CONSIDERATIONS TO QUALIFY AN IMPURITY

A general concept of impurities qualification is described in International Conference on Harmonisation (ICH) Q3A[12] and Q3B,[13] based on the maximum daily dose, for impurities originating from a new drug substance and a new drug product, respectively. In the case of toxic impurities originating from a CCS, such as leachables, the determination of acceptable levels is generally considered a critical issue in ICH Q8,[14] Pharmaceutical Development. However, a qualification process for these impurities has not been discussed in the existing ICH guidelines. It should be noted that in contrast to the impurities originating from either a drug substance, drug product, or an excipient, leachables originating from a CCS, especially from plastic or rubber components, represent a diverse array of chemical structures and compound classes and may present in widely varying concentrations. Therefore, to qualify a product-specific suitability of a CCS, the drug product manufacturer needs clear direction to follow in order to obtain regulatory acceptance of the CCS to market the drug.

9.4.1 The Regulatory Requirements Proposed by the United States Food and Drug Administration (FDA) to Select a Suitable CCS

Based on the Federal Food, Drug, and Cosmetic Act, and the Code of Federal Regulations, the FDA recommends, in various regulatory guidance, that a CCS should be suitable for its intended purpose. For example, the guidance "Container Closure

Systems for Packaging Human Drugs and Biologics—Chemistry, Manufacturing and Controls Documentation" (May 1999) discusses the submission requirements for the CCS. According to this document, each new drug application should contain enough information to demonstrate that a proposed CCS and its components are suitable for its intended use. The guidance outlines a risk-based approach; packaging suitability is based on four attributes: protection, safety, compatibility, and performance (function and/or drug delivery). The type and extent of information required will depend on the dosage form and route of administration. Inhalation and injection drug products have the highest requirements.

In addition, the guideline "Metered Dose Inhaler and Dry Powder Inhaler Drug Product: Chemistry, Manufacturing, and Controls Documentation" (draft November 1998) requires identity and concentration of leachables in the drug product to be determined through the product's shelf life. Also, toxicological evaluation of extractables/leachables should be performed. Furthermore, according to the guideline "Nasal Spray and Inhalation Solution, Suspension, and Spray Drug Products: Chemistry, Manufacturing and Controls Documentation" (July 2002), the drug product should be evaluated for compounds that leach from elastomeric or plastic components of the CCS. Appropriate analytical procedures should be developed to identify, monitor, and quantify the leached components in the drug product. Although these two guidelines specifically address the safety issues related to nasal and inhalation drugs, a general qualification process of a plastic CCS for any other drug product is based on the qualification requirements stipulated in the U.S. Pharmacopoeia/National Formulary. To meet these requirements, the drug product manufacturer would provide data for a plastic component when tested, according to the USP general chapters <661>, <87>, and <88>.

9.4.1.1   Why the Qualification Process Proposed in the USP General Chapters <87> and <88> Is Inadequate to Qualify a Plastic CCS for a Drug Product

According to the USP general chapters, a plastic material failing to meet requirements stipulated in <87> should be subjected to tests detailed under <88>.[15] Therefore, the test protocol discussed under <87> determines whether the toxicity testing on an extract of a plastic material is necessary. Thus, the analysis of the validity of the tests under <87> is required to determine the validity of the qualification process.

Three tests are described under <87>: the agar diffusion test, the direct contact test, and the elusion test. The decision as to which type of test or the number of tests to be performed depends on the material, the final product, and its intended use. USP <87> also identifies other factors that may also affect the suitability of a sample for a specific use: polymeric composition; processing and cleaning procedures; contacting media; inks; adhesives; absorption, adsorption, and permeability of preservatives; and conditions of storage.

In each of the stated tests under <87>, the testing should be conducted using an extract of the material. The extracting media and conditions under which the extractions should be conducted are described in chapter <87>. Although this test is to be conducted on any plastic that can be used to manufacture a CCS for a pharmaceutical product, it should be noted that only one extracting solvent, sodium chloride injection, is proposed for any test material. Also, this test proposes two alternative extracting media: serum-free mammalian cell culture media or serum-supplemented cell culture media. As tests under <87> determine whether the material should be further tested according to <88>, only one extracting solvent proposed under <87> test is considered as a major limitation in the qualification process of a plastic material using this test.

This limitation could effect the prudent selection of a material of construction for a CCS. For example, the potential chemicals that could leach from a plastic are numerous, based on the manufacturing formula of the material, the manufacturing process selected by the supplier, the additives included prior to molding the material into the component, the conditions used during the priming of the component prior to use, and so on. Therefore, the use of one solvent, such as sodium chloride solution, which has a pH range of 4.5–7.0, may not extract all potential leachables out of the material, under prescribed extracting conditions. Therefore, the decision, based on the test results of <87>, is unreliable to determine the necessity to further test a material using tests under <88>.

The <88> test has been designed to determine the biological response to elastomers, plastics, and other polymeric material with direct or indirect patient contact, or by injection of specific extracts prepared from the material under test. While the testing according to <88> is necessary only if the material would fail requirements in <87>, it should be noted that based on the limitation in the test protocol described in <87> to qualify a material, any test that follows <87> to determine the suitability of a material also has limited value.

In addition, it should be emphasized that there are limitations in the testing protocol described under <88>. For example, the reference standard proposed for this test is "USP high-density polyethylene." This material can be manufactured using varying amounts and kinds of additives, and a variety of manufacturing processes. Therefore, one reference standard, with unknown history of its manufacturing formula and process, is not suitable to select any plastic with respect to potential extractables and leachables. Based on this, it should be noted that the validity of this test based on this reference standard is questionable to qualify a plastic test sample for its intended purpose.

Furthermore, the solvents/solvent systems listed in the protocol for <88> are also considered inadequate to investigate the potential extractability of impurities from a material of construction, as they do not cover the complete range of physicochemical properties of an extracting medium to release all potential extractables. Here, the list includes sodium chloride injection, 1 in 20 solution of alcohol in sodium chloride injection, polyethylene glycol 400, vegetable oil, drug product vehicle, and water for injection. The solvents, other than 1 in 20 solution of alcohol in sodium chloride injection, would extract an impurity based on the solvent properties, including the pH, which would range from 4.5 to 9.5. Although <88> covers adequate pH range in the solvent/solvent system proposed, the usefulness of this test is still questionable due to the major limitation in the prequalification tests listed under <87>.

One key feature of the <88> monograph is the classification of plastics, based on responses to a series of *in vivo* tests for which extracts, materials, and routes of administration are specified. The plastic classes are tabulated from I to VI, where the class VI plastic would be tested for biological reactivity for extracts generated using all listed solvent/solvent systems, while the class I plastic would be qualified based on the response to a single extract using sodium chloride injection. Plastics in all six classes would be qualified based on responses in both mouse and rabbit, and response to implant in rabbit. Although the plastic classifications is only feasible if the material fails to meet the requirements stipulated under USP <87>, the challenge is to make a plastic fail in order to ensure that the plastic can be eventually classified based on the test results of USP<88>. However, under the described test protocol, identifying a plastic that would fail to meet the USP<87> would be a difficult task, since only one medium is proposed to extract all potential impurities originating from the plastic to be used. Therefore, the USP <88> is a test that may never be used either in the plastic industry or in the pharmaceutical industry and may have a limited applicability to classify or qualify a plastic intended for a CCS.

Even if one knows the additives included in a base polymer during plastic manufacturing, the selection and qualification of a product-specific CCS is not a trivial matter. Therefore, if one has to depend entirely on USP <87> followed by <88> test to qualify a plastic and a CCS, the following approaches should be considered: select a diverse array of solvents for extraction studies, take into consideration the additives that can be included into a selected plastic to manufacture the component, the potential prefill priming processes conducted on the CCS, including irradiation sterilization and the potential postfill processes, such as terminal and irradiation sterilization of the drug product. Then the testing of extracts for biological response in rats and rabbits should be conducted. In addition, the list of extracting solvents should be scientifically justified, and the primary extractables studies and leachables studies should be conducted based on product-specific test protocols, regardless of the testing conducted based on USP <87> and <88>.

9.4.2   Regulatory Requirements in the European Union (EU) to Select a Suitable CCS for Food and Pharmaceutical Products

Section 3.2.1.6 of Annex I of the Commission Directive 2003/63/EC of June 25, 2003, identifies the regulatory requirements that are essential for a CCS intended for an active pharmaceutical ingredient (API), and section 3.2.2.2(g) describes the qualification requirements for a CCS intended to store a finished product. However, this section does not detail the test methods to be used for this purpose.

The Commission Directive 2002/72/EC of August 6, 2002, under the Act of the EU is related to plastic materials and articles intended to come in contact with foodstuffs. Directive 2002/72/EC is part of the legislative framework covered by regulation no. 1935/2004 on materials and articles intended to come into contact with food. This directive is applicable to materials, and parts thereof, that may be composed either of plastic material only or of several layers of plastic material or different types of materials. This directive, however, excludes elastomers and natural and synthetic rubber, paper, and paperboard (whether modified or not by the addition of plastics), surface coatings obtained from waxes, ion-exchange resins, silicones, or materials and articles composed of two or more layers, of which at least one does not consist of plastics. This directive describes monomers and other starting substances that may be used in the manufacture of plastic materials and articles (Annex II) and additives, which may be used in the manufacture of plastic materials and articles (Annex III). At the moment, a new additive can always be included in the list after evaluation and authorization by the European Food Safety Authority (EFSA).

9.4.2.1 The Qualification Process Proposed by the EU Although the regulation of plastic use for foodstuffs in the EU is comparable in the United States, the EU regulations for pharmaceuticals do not call for the use of the databases under foodstuffs to select and qualify a plastic material with respect to medicinal products. Instead in 2005, the EMEA published a guideline titled "Guideline on Plastic Immediate Packaging Material" to address the submission requirements in this regard. In addition, the guideline on "Plastic Immediate Packaging Material" is directly linked to EP general chapter 3.1, materials intended for the manufacture of containers. Based on this linkage, the drug product manufacturers are expected to determine the leachables originating from a plastic identified in this chapter. This limits their ability to select a plastic CCS for a specific drug product.

9.4.3 Regulatory Requirements Proposed by Health Canada to Qualify a CCS

Several guidelines and guidance documents have been developed based on the food and drug regulations, and they stipulate various regulatory requirements to select and qualify a CCS for a pharmaceutical product. For example, according to the guidance document "Quality (Chemistry and Manufacturing) Guidance: New Drug Submissions (NDSs) and Abbreviated New Drug Submissions (ANDSs),"[16] a brief description of the CCS, identity, and suitability data based on the USP general chapters <661> and <671> should be submitted. The guideline "Pharmaceutical Quality of Aqueous Solutions" repeats this information. In addition, the guidance document "Pharmaceutical Quality of Aqueous Solutions"[17] suggests that migration studies on the potential leachables from the labels pasted on semipermeable containers should be conducted during the development stage of the product. The guideline "Pharmaceutical Quality of Inhalation and Nasal Products 2006/10/01,"[18] which was developed in collaboration with the EMEA, discusses the requirements for extractables and leachables in the pharmaceutical development section. Analytical data on extractables and leachables should be included in the drug submission and the extractables/leachables profile of the materials that come in direct contact with the formulation should be established. Characterization of compounds that appear as leachables should be attempted, and a safety assessment should be conducted according to established safety thresholds. However, these guidelines do not provide specific methods to determine leachables and extractables originating from a plastic.

In Canada, meeting the requirements stipulated in the general chapters of the U.S. Pharmacopoeia/National Formulary is not considered as a mandatory requirement to qualify a CCS for a pharmaceutical product, as the law does not call for this testing, unlike in the United States. However, the results of the tests in USP <661>, <87>, and <88> are required to be included in the submission to confirm the biocompatibility of the materials of construction of the CCS and the drug product. Although the current drug review process acknowledging these results would determine the product-specific suitability of the CCS, a guidance document is being developed to detail specific processes to follow in order to select and qualify a product-specific CCS.

9.5 THE QUALIFICATION PROCESS PROPOSED BY THE PRODUCT QUALITY RESEARCH INSTITUTE (PQRI) LEACHABLES AND EXTRACTABLES WORKING GROUP IN COMPARISON TO THAT PROPOSED BY THE EMEA

For the first time, "The Safety Threshold and Best Practices for Extractables and Leachables in Orally Inhaled and Nasal Drug Products" document published by PQRI in 2005 introduced the qualification process for chemical compounds that originate from a CCS intended for an OINDP.[2] The qualification process described in this document is based on a threshold concept. The document includes a systematic approach for the safety assessment of potential leachables and explains that the recommended exposure threshold above which the individual organic leachables in an OINDP must be qualified and/or evaluated for safety concerns. In addition, it suggests that one should have extensive knowledge on the quality and the quantity of potential leachables, and the safety threshold must be applied in conjunction with "best practices," which include safety assessment of potential leachables throughout the development program.

The key feature in the qualification process proposed by PQRI is the establishment of a safety concern threshold (SCT) of 0.15 µg/day, based on a one in a million lifetime cancer risk (LCR) of a leachable originating from a plastic component. The SCT is the threshold below which a leachable would present negligible safety concern from carcinogenic and noncarcinogenic toxic effects. Additionally, the PQRI report identifies a 5 µg/day qualification threshold (QT). The QT is the threshold below which a given noncarcinogenic leachable is not considered for safety qualification (toxicological assessment) unless the leachable presents a structure–activity relationship (SAR) concern. Below the SCT, identification of leachables generally would not be necessary. Below the QT, leachables without structural alert for carcinogenicity or irritation would not require compound-specific risk assessment.

The general steps in the proposed qualification process can be summarized. After the identification of the compound of interest, it is important to conduct the SAR assessment to determine the genotoxic or carcinogenic potential of the identified compound. If the compound has the potential to be carcinogenic or genotoxic, by consideration of the potential maximum daily human exposure to this compound, the available toxicology/safety data should be reviewed. If deemed necessary, one should conduct toxicological studies, such as 14- to 90-day toxicology and/or genetic toxicology. The safety assessment of this compound should be conducted based on the maximum expected daily human exposure, the patient population and the duration of the use of the drug.

The threshold concept proposed by PQRI for OINDPs for carcinogens is based on a lifetime risk of 10^{-6} for occurrence of cancer in humans. This concept, however, is only applicable for carcinogen originating from a material of construction of a CCS or a device. In comparison to the process proposed by PQRI for genotoxic or any other impurity originating from a CCS, the toxicological assessment of genotoxic impurities originating from an API is a difficult task, especially in the absence of data, usually needed for the application of any established risk assessment methods. Under such circumstances, the implementation of a generally applicable approach as defined by the threshold of toxicological concern (TTC) is proposed by the EMEA.[19] A TTC value of 1.5 μg/day intake of a genotoxic impurity is considered to be associated with an acceptable risk (excess cancer risk of <1 in 100,000 over a lifetime) for most pharmaceuticals. A cancer risk of 10^{-5} over a lifetime is considered appropriate for a pharmaceutical based on the appreciated benefits of a drug in comparison to the risk associated with a toxicant that is chemically related to the active.

An SCT or TTC value cannot be used in all cases. As emphasized by PQRI and the EMEA, full qualification at detected levels should be performed for known highly potent carcinogenic or genotoxic impurities, such as nitrosamines, to determine their potential toxic effects. The safety estimation of an unknown leachable with respect to lifetime risk factors is discussed in the next section.

9.5.1 Correlation of Thresholds of Concern (TOCs) Values for Carcinogens: A Novel Approach to Correlate Threshold Values at Different Risk Levels (RLs) and to Assess Associated Risks in Relation to the Total Daily Intake (TDI) of a Drug

In this section, a novel framework to correlate TOCs at different RLs is discussed. The framework does not dictate an answer. It facilitates the comparison of approaches for establishing threshold values proposed by EMEA and PQRI. The framework provides a structure for making analysis transparent and assumptions clear. It also allows the assessor of a CCS to understand the facts and reasoning when deciding whether to use a threshold value or a risk assessment for establishing an acceptable limit for a toxic impurity based on an approximate daily dose of a drug product. Such an analysis will enable the assessment of toxic effects of a known or an unknown leachable originating from a plastic material.

9.5.1.1 The Assessment of Toxic Effects Different approaches have been adopted to assess whether a chemical substance is "toxic" depending on whether the critical effect is considered to have or not have a threshold. For many types of toxic effects (i.e., organ specific, neurological/behavioral, immunological, epigenetic, carcinogenic, reproductive, or developmental), it is generally considered that there is a dose or concentration below which adverse effects will not occur (i.e., a threshold). For other types of toxic effects, it is assumed, but not proven, that there is some probability of harm at any level of exposure (i.e., that no threshold exists). At the present time, the latter assumption is generally considered to be appropriate only for mutagenesis and genotoxic carcinogenesis.

As reported in the PQRI recommendations, the FDA threshold of regulation for indirect food additives and an analysis of carcinogenic and noncarcinogenic toxicities, carcinogenic effects typically occur at lower levels of intake than those at which noncarcinogenic toxic effects occur. Therefore, criteria for establishing acceptable cancer RLs would usually correspond to doses being low enough to present negligible safety concerns from noncarcinogenic toxic effects. Thus, a TOC or an SCT based on acceptable carcinogenicity risk has been used as a parameter for protecting human health.

In this section, definitions of key terms and concepts are reviewed. TOC or SCT values for a given substance or group at different RLs are then examined. Finally, a framework to correlate TOC or SCT values at different RLs is developed.

9.5.1.1.1 Terms and Concepts Excess cancer risk is the probability or "risk" (percentage of population affected) that lifetime exposure to a carcinogen at a given dose will result in an excess cancerous effect above the background incidence. A 1 in 100,000 and 1 in 1 million risk for carcinogenicity are two of the most common ratios used to assess cancer risk, and will be described here as the 10^{-5} and 10^{-6} RLs, respectively. Many SCT or threshold values are based on an appropriate "risk-specific dose" (RSD).

The carcinogenicity potency "slope factor" (SF_{CP}) is an estimate of the lifetime risk or probability of a carcinogenic response per unit of exposure. Units are the inverse of dose rate. It is reasonable to assume that the SF_{CP} for a given substance or a defined group will remain constant at low doses or across several orders of magnitude corresponding to low RLs. Also, for nonthreshold carcinogens, it is assumed that a toxic effect may occur no matter how small the dose is.

RSD is the daily dose of a particular carcinogen associated with a specified lifetime excess risk for carcinogenicity, usually chosen as 10^{-5} or 10^{-6}. The daily lifetime dose associated with an excess cancer risk less than 10^{-6} is sometimes referred to as a "virtually safe dose."[2] However, when certain substances, such as APIs, provide certain health benefits, their daily dose levels could be tolerated up to doses associated with an excess cancer risk less than 10^{-5}. In this work, this RL has been termed as a "risk–benefit regulatory safety threshold."

TD_{50} is the chronic dose rate (in milligram per kilogram body weight per day) that would halve the background-corrected percentage of tumor-free animals at the end of a standard experiment time (the *standard lifespan* for the species).[20,21]

9.5.1.2 The Relationship between RSD and the Carcinogenicity Potency "Slope Factor" (SF_{CP})

No single mathematical procedure is recognized as the most appropriate for low-dose extrapolation for carcinogenesis.[22] However, at very low doses, it may be reasonable to assume a linear dose response for carcinogens that do not show a threshold effect (Fig. 9.1). The slope of a line in Figure 9.1 corresponds to the carcinogenicity potency "slope factor" (SF_{CP}) for a particular substance or representative group. Three cases are schematically represented in Figure 9.1, namely, the nongenotoxics (line 1), the genotoxics (line 2), and nitrosamines

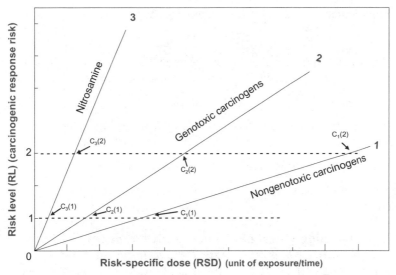

Figure 9.1   Risk level (RL) (carcinogenic response risk) versus risk-specific dose (RSD).

(line 3). The SF_{CP} increases with an increase in the carcinogenic potency. Accordingly,

$$SF_{CP} \text{ (nitrosamine)} > SF_{CP} \text{ (genotoxics)} > SF_{CP} \text{ (nongenotoxics)},$$

where at risk level 1 (RL_1)

$$SF_{CP} \text{ (nongenotoxics)} = RL_1/C_1(1), \qquad (9.1a)$$

$$SF_{CP} \text{ (genotoxics)} = RL_1/C_2(1), \qquad (9.1b)$$

$$SF_{CP} \text{ (nitrosamine)} = RL_1/C_3(1). \qquad (9.1c)$$

$C_1(1)$, $C_2(1)$, and $C_3(1)$ are RSDs at RL_1 in Figure 9.1. Similarly, $C_1(2)$, $C_2(2)$, and $C_3(2)$ are RSDs corresponding to risk level 2 where,

$$C_1(1) > C_2(1) > C_3(1) \text{ and } C_1(2) > C_2(2) > C_3(2).$$

At any given RL X, the magnitude of the RSD (RSD_X)

$$RSD_X \text{ (nitrosamine)} < RSD_X \text{ (genotoxics)} < RSD_X \text{ (nongenotoxics)}.$$

In this work, the SF_{CP} is assumed to be given by Equation 9.2.[2,22] Gold et al. made a more rigorous estimate of slope factors, including an in-depth comparison of low-dose cancer risk assessment methodologies[23]:

$$SF_{CP} = [1/2 \ TD_{50}(\text{rodent})] \times (ASF_{(\text{rodent to human})}), \qquad (9.2)$$

where $ASF_{(\text{rodent to human})}$ is the allometric scaling factor for extrapolating rodent values to humans, and the human equivalent for TD_{50}, termed HTD_{50}, is given by Equation 9.3:

$$HTD_{50} = TD_{50}(\text{rodent})/(ASF_{(\text{rodent to human})}). \qquad (9.3)$$

From Equations 9.2 and 9.3, the value for SF_{CP} is given by

$$SF_{CP} = (2HTD_{50})^{-1}. \qquad (9.4)$$

When the carcinogenicity "slope factor (SF_{CP})" is known, an RSD can be calculated for a given RL using Equation 9.5 obtained from Equation 9.1:

$$RSD = RL/SF_{CP}. \qquad (9.5)$$

RSD can be expressed in terms of HTD_{50} by using Equation 9.4 to substitute for SF_{CP}:

$$RSD\ (at\ RL_X) = (RL_X) \times (2HTD_{50}). \qquad (9.6)$$

Equation 9.6 can be used to estimate an RSD value at any RL if a TD_{50} value is available.

9.5.1.3 Correlating Threshold Values for Different Levels of Cancer Risk and Identification of Isometric Lines for Carcinogenic Potency

Applying Equation 9.6 to RLs 10^{-5} or 10^{-6}, the following Equations 9.7 and 9.8 are obtained:

$$RSD\ (at\ 10^{-5}) = 10^{-5} \times (2HTD_{50}), \qquad (9.7)$$
$$RSD\ (at\ 10^{-6}) = 10^{-6} \times (2HTD_{50}). \qquad (9.8)$$

Dividing Equation 9.7 by Equation 9.8 gives

$$RSD\ (at\ 10^{-5})/RSD\ (at\ 10^{-6}) = 10^{-5}/10^{-6} = 10. \qquad (9.9)$$

The relative magnitude of these RSDs is depicted in Figure 9.2 for the LCR levels of 10^{-5} and 10^{-6}, respectively. As long as the SF_{CP} remains constant, the RSD at the 1 in 100,000 RL is 10 times the RSD at the one in a million RL for any given substance or defined or chosen group.

The logarithmic form of Equation 9.6 can be given by Equation 9.10:

$$Log_{10}\ RSD\ (at\ RL_X) = Log_{10}\ (RL_X) + Log_{10}\ 2 + Log_{10}\ HTD_{50}. \qquad (9.10)$$

For a person (assumed to have a body weight of 50 kg as discussed in Reference 2), the TDI (μg/person/day) for lifetime exposure is equal to 50 times the RSD per kilogram body weight. Therefore,

$$Log_{10}\ TDI\ (at\ RL_X) = Log_{10}\ (RL_X) + Log_{10}\ 2 + Log_{10}\ HTD_{50} + Log_{10}\ 50 \qquad (9.10a)$$

or

$$Log_{10}\ TDI\ (at\ RL_X) = Log_{10}\ (RL_X) + Log_{10}\ HTD_{50} + 2. \qquad (9.10b)$$

Plots of Log_{10} TDI versus Log_{10} (RL_X) are given in Figure 9.2. For a given substance or group with a representative carcinogenic potency, the RL and TDI are linearly related. The linear relationship is represented as an "isometric line" in Figure 9.2. An isometric line corresponds to a locus of constant intrinsic potency (i.e., the line traverses a path that is isometric in potency). The isometric lines are linear because SF_{CP} is assumed to be constant. The gradient of each line is unity with units the inverse of TDI. Therefore, all isometric lines would have the same gradient and hence be parallel to each other.

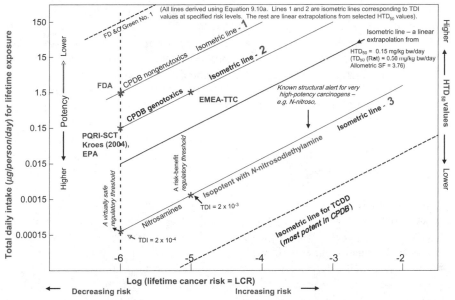

Figure 9.2   Correlating threshold values for different levels of cancer risk and identification of isometric lines for carcinogenic potency. CPDB, carcinogenic potency database; EMEA, European Medicines Agency; EPA, United States Environmental Protection Agency; FDA, United States Food and Drug Administration; PQRI, Product Quality Research Institute; SCT, safety concern threshold; TDI, total daily intake (averaged over a lifetime); TTC, threshold of toxicological concern. References for the respective data values listed in the figure are indicated in parenthesis: EMEA-TTC (Reference 19); EPA and FDA (Reference 24); Kroes (2004) (Reference 25); and PQRI-SCT (Reference 2).

The isometric line for a particular substance or representative group could be obtained by extrapolating downward from the HTD_{50} value after appropriate allometric scaling factors for extrapolation to human are applied to the animal TD_{50} values. The RL corresponding to an HTD_{50} by definition corresponds to LCR = 0.5. Therefore, this RL would be represented as a vertical line in Figure 9.2 (not shown in figure) and would intercept the x-axis at Log_{10} (LCR) = Log_{10} (1/2) = $-Log_{10}$ 2 = -0.3. Lines found lower down on the y-axis would correspond to higher carcinogenic potency (CP) slope factor values. Similarly, an isometric line is found higher up when the CP slope factor is lower. The numbered isometric lines 1, 2, and 3 are in order of increasing CP slope factor or the carcinogenic potency:

nongenotoxics < genotoxics < nitrosamines.

Line #1 in Figure 9.2 depicts a potency representative of carcinogens in the carcinogenic potency database (CPDB)[*] that are neither genotoxic nor have a recognized structural alert or known to have very high potencies. The FDA value of

_____

* The CPDB is a unique and widely used international resource of the results of 6540 chronic, long-term animal cancer tests on 1547 chemicals. The CPDB provides easy access to the bioassay literature, with qualitative and quantitative analyses of both positive and negative experiments that have been published over the past 50 years in the general literature through 2001 and by the National Cancer Institute/National

1.5 μg/person/day was chosen to represent this group.[24] Line #2 depicts a potency representative of carcinogens in the CPDB that is genotoxic. The PQRI value of 0.15 μg/person/day was chosen to represent this group. Line #3 and the other substance-specific isometric lines are linear extrapolations from literature rodent TD_{50} values estimated as described in this chapter.

In Figure 9.2, the 10^{-5} LCR or the 1 in a 100,000 RL is called the *risk–benefit regulatory threshold*, and the 10^{-6} LCR or the one in a million RL is called the *virtually safe regulatory threshold*. In general, SAL-positive (positive for the *Ames Salmonella/microsome mutagenicity assay*) substances would be found below isometric line 1, which is representative of lower potency (or nongenotoxic material in CPDB). For substances with known higher potencies corresponding to the region significantly below line 2 in Figure 9.2, compound-specific risk assessments would have to be conducted as they do not adhere to a common safety threshold. It is expected that substances with known structural alerts related to very high potencies such as N-nitroso, azoxy structures, aflatoxin-like, or polynuclear aromatics (PNAs), would yield isometric lines below line 2 in Figure 9.2. For example, a TDI of 2×10^{-3} and 2×10^{-4} μg/person/day was calculated for nitrosamines at the lifetime risk of 10^{-5} and 10^{-6}, respectively, by linear extrapolation from the Log_{10} (TD_{50}) value of -2.1 reported for N-nitrosodiethylamine.[20,26] The potency locus for N-nitrosodiethylamine is given by line 3 in Figure 9.2.

Many studies indicate that genotoxics are more potent than nongenotoxics, even though there are some significant exceptions.[20] Gold et al.[27] reported over a 10 million-fold range of TD_{50} (rodent) values from the most potent 101 ng for tetrachlorodibenzo-p-dioxin (TCDD) to 5.98 g for the food dye FD&C Green No. 1. The most potent substance in the CPDB (i.e., 2,3,7,8-TCDD) is nongenotoxic.[27] If linear extrapolation is applicable for TCDD, its estimated TDIs with lifetime risk of 10^{-5} and 10^{-6} correspond to 3×10^{-5} and 3×10^{-6} μg/person/day, respectively (see dashed line shown below line 3 in Fig. 9.2).[28] Figure 9.2 schematically displays over a 10 million-fold range of potency values bounded by the dashed isometric lines for TCDD and FD&C Green No. 1. Figure 9.2 also facilitates the comparison of TDI values reported for different RLs.

9.5.1.4 A Schematic Approach to Determine the Safety of a Toxic Substance by Using TOC or Substance-Specific Risk Assessments
The RSD values at the 1 in 100,000 RL and the one in a million RL have been used to define TOCs for carcinogens. Application of Equation 9.9 to TOC values for a given substance or group gives

$$\text{TOC (at } 10^{-5})/\text{TOC (at } 10^{-6}) = 10^{-5}/10^{-6} = 10. \tag{9.11}$$

Toxicology Program through 2004. The CPDB standardizes the diverse literature of cancer bioassays that vary widely in protocol, histopathological examination and nomenclature, and in the published author's choices of what information to provide in their papers. Results are reported in the CPDB for tests in rats, mice, hamsters, dogs, and nonhuman primates. Gold, L.S. Carcinogenic potency database (CPDB). Available at: http://potency.berkeley.edu/cpdb.html, last updated in 2010.

From Equation 9.11, it is evident that the 0.15 µg/person/day TDI value at the one in a million RL is isometric with the 1.5 µg/person/day TDI value at the 1 in 100,000 RL. This equivalency is depicted by the isometric line 2 in Figure 9.2 representing genotoxic carcinogens. In general, SCT values corresponding to a lower risk (e.g., from RSD [at 10^{-6}] to RSD [at 10^{-5}]) can be obtained by moving upward along an isometric line.

PQRI and EMEA have determined TOCs, termed SCT and TTC, respectively, by using data from the same group of genotoxins. These two representative values are isometric in potency as they lie on the same isometric line in Figure 9.2 (i.e., line #2). The principle differences arose because the PQRI and EMEA considered different RLs, namely, the 10^{-6} and 10^{-5} RLs, respectively. Consequently, the PQRI has proposed an SCT value of 0.15 µg/person/day, whereas the EMEA considers 1.5 µg/person/day as the TTC for impurities in APIs necessary to manufacture pharmaceuticals. Application of Equation 9.11 to the two threshold values indicates that they represent equivalent potencies.

As explained in the PQRI report, the PQRI SCT value is based on a 37th percentile potency for SAL-positive carcinogens in the CPDB. Hence, about one-third of the genotoxins considered have lower TD_{50} values than the representative value at the 37th percentile that was extrapolated to yield the threshold value. Rather than lowering the general threshold standard, PQRI considers it better to understand the types of very potent carcinogens that could be leachables (e.g., nitrosamines and PNAs) and set compound-specific limits and specific analytical methods to limit them to acceptable levels. For example, a TDI of 0.15 µg/person/day corresponds to an RL of 5.6×10^{-2} for TCDD, the most potent carcinogen in the CPDB. Hence, the threshold approach is clearly not applicable to TCDD. In general, significantly low TD_{50} values may also signal the need for in-depth risk assessments.

PQRI has noted that a leachable that is a mutagenic carcinogen but not categorized as having special safety concerns may present a scenario where the SCT would not offer sufficient protection to children. Therefore, it may be useful to back-calculate representative TD_{50} values corresponding to threshold values proposed by PQRI and the EMEA. Applying the allometric scaling factors adopted by PQRI for extrapolation from rats to humans (3.76) and mice to humans (6.95), respectively, both the PQRI SCT value of 0.15 µg/person/day and the EMEA TTC value of 1.5 µg/person/day corresponds to an HTD_{50} value of 1.5 mg/kg body weight/day, which is equivalent to the following representative rodent TD_{50} values:

$$TD_{50}(\text{mouse}) = 10 \text{ mg/kg body weight/day and } TD_{50}(\text{rat})$$
$$= 6 \text{ mg/kg body weight/day.}$$

The above values could be used as reference values for comparison. TD_{50}(rodent) values vary much less than these numbers could signal the need for chemical-specific risk assessments to determine safe daily limits for known toxic substances in drugs. A substance-specific TDI, an order of magnitude less than PQRI's representative SCT value, is given by the isometric line corresponding to $HTD_{50} = 0.15$ mg/kg body weight/day shown in Figure 9.2.

In the absence of an experimental determination, a TD_{50} value could be estimated from other toxicity parameters. A correlation between TD_{50} or carcinogenic potency and the maximum tolerated dose (MTD) has been reported.[29] Also, other toxicity parameters such as LTD_{10} have been used to determine TD_{50},[22] where LTD_{10} is the lower 95% confidence limit of TD_{10}. The 10^{-6} and 10^{-5} RL doses have been calculated by dividing $LTD_{10(human)}$ by 100,000 and 10,000, respectively.[20] Parodi et al. recommended that for an unknown, short-term genotoxicity, tests are useful as they tend to detect the fraction of very highly potent carcinogens that are not compatible with the threshold approach.[20] Relatively low MTD or LTD_{10} values obtained from such studies may warrant a substance-specific assessment of the safe dose.

9.5.2 Conclusion

The reader is referred to the PQRI report for discerning the specific conditions when the SCT value could be adopted for leachables.[2] As indicated in the report, even in the absence of structural alerts or other analytical information characteristic of high potency carcinogens, there may be scenarios where a compound-specific risk assessment may be warranted. Such may be the case for compounds with isometric lines significantly lower than the standard line 2 in Figure 9.2 (i.e., TD_{50}(mouse) $\ll$ 10 mg/kg body weight/day and TD_{50}(rat) $\ll$ 6 mg/kg body weight/day). In these cases, substance-specific risk assessments would be recommended. For example, if rodent TD_{50} values of less than 1 mg/kg body weight/day have been reported, such data could be considered as a criterion for conducting a compound-specific TDI calculation at a specified RL such as 10^{-6} and 10^{-5}.

In this work, a framework has been developed for showing TDI values corresponding to chosen LCR levels for potencies spanning a 10 million-fold concentration range from FD&C Green no. 1 to TCDD (Fig. 9.2). The threshold values given by the PQRI SCT and EMEA TTC values have been shown to be representative of the same potency, albeit referring to different RLs. Finally, a TDI or potency estimation using a simple linear extrapolation from TD_{50} values may assist in addressing safety risk, if one is transparent about the assumptions and the limitations of the method. The approach could facilitate the comparison of potencies to determine when in-depth risk assessments to establish safe dose levels of leachables are necessary.

9.6 FINAL THOUGHTS

Patient safety has been jeopardized, and costly product recalls have been noted in the past, when the integrity of a therapeutic product has been compromised because of the release of toxic leachables from pharmaceutical packaging during the shelf life of the drug. Such toxic effects are possible at any stage of the product's life cycle.

The international focus of the use of plastic material in CCSs has increased over the past several years due to the detrimental effects of leaching toxic impurities that could migrate into the drug product during its shelf life. Currently, collaborative international efforts are being focused on risk management approaches for addressing the issue of leachables and extractables through regulation, evolving best practices and guidelines, the work of international bodies, and multistakeholder cooperation for finding solutions to related technical challenges. Although this field of science is still evolving and is being continuously discussed at international meetings, the safety threshold concept proposed by PQRI is considered suitable to qualify a CCS intended for a pharmaceutical product.

ACKNOWLEDGMENTS

The author thanks Martin Nicholas (Head, Public Awareness, Surveillance and National Compliance Coordination, National Office of WHMIS, Health Canada) for his contribution to Section 9.5 of this chapter; Caroline Vanneste (Project Manager, Good Review Practices, Office of Business Transformation, Therapeutic Products Directorate, Health Canada) for her assessment of the contents and editorial review; and Andrew Adams (Director, The Bureau of Pharmaceutical Sciences, Therapeutic Products Directorate, Health Canada) for his valuable comments and careful editing.

It is with great pleasure that the author acknowledges fruitful communications with Dr. Michael Ruberto (CEO, Material Needs Consulting, LLC, 110 Chestnut Ridge Road, #311, Montvale, NJ 07645) and his assistance in creating the list of additives related to plastic manufacturing.

REFERENCES

1   Nicholas, K. Extractables and leachables determination: A systematic approach to select and qualify a container closure system for a pharmaceutical product. *Am Pharm Rev* 2006, *9*(3), pp. 21–27.
2   Norwood, D.L. and Ball, D. Product Quality Research Institute: Safety thresholds and best practices for extractables and leachables in orally inhaled and nasal drug products. Submitted to the PQRI Drug Product Technical Committee, PQRI Steering Committee, and U.S. Food and Drug Administration by the PQRI Leachables and Extractables Working Group, 2006.
3   Guidance for industry: Container closure systems for packaging human drugs and biologics. U.S. Department of Health and Human Services, Food and Drug Administration, Center for Drug Evaluation and Research (CDER), Center for Biologics Evaluation and Research (CBER), 1999.
4   Annual report of the committees of toxicity, mutagenicity, carcinogenicity of chemicals in food. Consumer Products and the Environment, Department of Health, United Kingdom, 2002.
5   Klaassen, C.D. *Casarett & Doull's Toxicology: The Basic Science of Poisons*, 6th ed. McGraw-Hill, New York, 2001.
6   Harper, C.A. *Modern Plastics Handbook*. McGraw-Hill, New York, 2000.
7   Jenke, D. Evaluation of the chemical compatibility of plastic contact materials and pharmaceutical products; safety considerations related to extractables and leachables. *J Pharm Sci* 2007, *96*(10), pp. 2566–2581.
8   European Pharmacopoeia. 2008, Vol. 1.
9   European Medicines Agency Inspections. Guideline on plastic immediate packaging materials. Committee for Medicinal Products for Human Use (CHMP), Committee for Medicinal Products for Veterinary Use. London (CVMP), 2005.
10   Ruberto, M. Material needs consulting, LLC, 110 Chestnut Ridge Road, #311, Montvale, NJ. Personal communication, 2009.
11   Kutz, M. *Handbook of Material Selection*. John Wiley & Sons, New York, 2002; article 11, 335–355. Peters, E.N. *Plastics: Thermoplastics, Thermosets and Elastomers*.

12   ICH harmonised tripartite guideline: Q3A(R2) impurities in new drug substances. International Conference on Harmonisation of Technical Requirements for Registration of Pharmaceuticals for Human Use, 2006.

13   ICH harmonised tripartite guideline: Q3B(R2) impurities in new drug products. International Conference on Harmonisation of Technical Requirements for Registration of Pharmaceuticals for Human Use, 2006.

14   International Conference on Harmonization. Guidance for industry: Q8: Pharmaceutical development. U.S. Department of Health and Human Services, Food and Drug Administration, Center for Drug Evaluation and Research (CDER), Center for Biologics Evaluation and Research (CBER), 2006.

15   USP NF: The official compendia of standards, Vol. 1, 2009.

16   Draft quality (chemistry and manufacturing) guidance: New drug submissions (NDSs) and abbreviated new drug submissions (ANDSs). Health Canada, 2001.

17   Guidance for industry: Pharmaceutical quality of aqueous solutions. Health Canada, 2005.

18   Guidance for industry: Pharmaceutical quality of inhalation and nasal products. Health Canada, 2006.

19   European Medicines Agency (EMEA). Guideline on the limits of genotoxic impurities. Committee for Medicinal Products for Human Use (CHMP). London, 2004.

20   Parodi, S., Malacarne, D., Romano, P., and Taningher, M. Are genotoxic carcinogens more potent than nongenotoxic carcinogens? *Environ Health Perspect* 1991, *95*, pp. 199–204.

21   Peto, R., Pike, M.C., Bernstein, L., Gold, L.S., and Ames, B.N. The TD50: A proposed general convention for the numerical description of the carcinogenic potency of chemicals in the chronic-exposure animal experiments? *Environ Health Perspect* 1984, *58*, pp. 1–8.

22   Krewski, D., Szyszkowicz, M., and Rosenkranz, H. Quantitative factors in chemical carcinogenesis: Variation in carcinogenic potency. *Regul Toxicol Pharmacol* 1990, *12*, pp. 13–29.

23   Gold, L.S., Gaylor, D.W., and Slone, T.H. Comparison of cancer risk estimates based on a variety of risk assessment methodologies. *Regul Toxicol Pharmacol* 2003, *37*, pp. 45–53.

24   Guidance for industry: Genotoxic carcinogenic impurities in drug substances and products: Recommended approach. U.S. Food and Drug Administration, Center for Drug Evaluation and Research (CDER), Center for Biologics Evaluation and Research (CBER), 2008.

25   Kroes, R., Renwick, A.G., Cheeseman, M., Kleiner, J., Mangelsdorf, I., Piersma, A., Schilter, B., Schlatter, J., van Schothorst, F., Vos, J.G., and Würtzen, G. Structure-based thresholds of toxicological concern (TTC): Guidance for application to substances present at low levels in the diet. *Food Chem Toxicol* 2004, *42*, pp. 65–83.

26   Gold, L.S., Slone, T.H., and Bernstein, L. Summary of carcinogenic potency and positivity for 492 rodent carcinogens in the carcinogenic potency database. *Environ Health Perspect* 1989, *79*, pp. 259–272.

27   Gold, L.S., Sawyer, C.B., Magaw, R., Backman, G.M., Veciana, M., de Levinson, R., Hooper, N.K., Havender, W.R., Bernstein, L., Peto, R., Pike, M.C., and Ames, B.M. A carcinogenic potency database of the standardized results of animal bioassays. *Environ Health Perspect* 1984, *58*, pp. 9–319.

28   Jones, T.D. Toxicological potency of 2,3,7,8-tetrachlorodibenzo-p-dioxin relative to 100 other compounds: A relative potency analysis of in vitro and in vivo test data. *Arch Environ Contam Toxicol* 1995, *29*, pp. 77–85.

29   Krewski, D., Gaylor, D.W., Soms, A.P., and Szyszkowicz, M. An overview of the report "correlation between carcinogenic potency and the maximum tolerated dose: Implication for risk assessment." *Risk Anal* 1993, *13*, pp. 383–398.

BEST PRACTICES FOR EVALUATION AND MANAGEMENT OF EXTRACTABLES AND LEACHABLES IN ORALLY INHALED AND NASAL DRUG PRODUCTS

ANALYTICAL BEST PRACTICES FOR THE EVALUATION AND MANAGEMENT OF EXTRACTABLES AND LEACHABLES IN ORALLY INHALED AND NASAL DRUG PRODUCTS

Daniel L. Norwood, Cheryl L.M. Stults, and Lee M. Nagao

10.1 INTRODUCTION

The impact to orally inhaled and nasal drug product (OINDP) pharmaceutical development programs of a defined leachables safety qualification process, including the development of the safety thresholds (qualification threshold [QT] and safety concern threshold [SCT]), is clearly significant. However, leachables safety qualification is but one part of the larger OINDP pharmaceutical development process related to extractables and leachables (see Fig. 10.1). Additional aspects include

- selection of container closure/delivery system components and materials of construction (raw materials);
- accomplishment of controlled extraction studies on all "critical components";
- accomplishment of definitive drug product leachables studies, including the development and validation of leachables analytical methods (as guided by the safety thresholds and analytical evaluation threshold [AET]);

Leachables and Extractables Handbook: Safety Evaluation, Qualification, and Best Practices Applied to Inhalation Drug Products, First Edition. Edited by Douglas J. Ball, Daniel L. Norwood, Cheryl L.M. Stults, Lee M. Nagao.

155

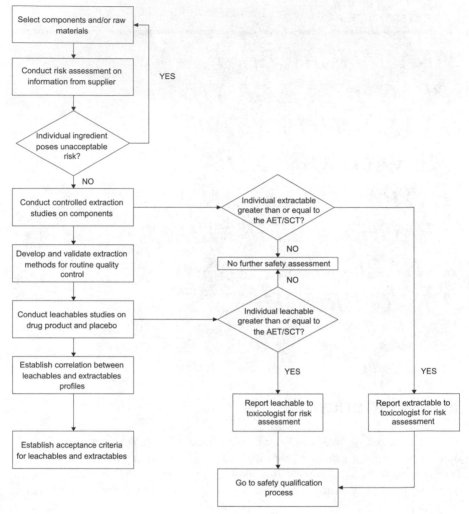

Figure 10.1   Example OINDP pharmaceutical development process related to extractables and leachables.

- development and validation of analytical methods for routine extractables testing of appropriate critical components;
- development of a qualitative and quantitative leachables/extractables correlation; and
- development of appropriate leachables (and extractables) specifications.

The successful completion of a pharmaceutical development process with this level of complexity requires both a significant amount of time, typically measured in years, and a diversity of scientific and engineering expertise. Selection of container closure/delivery system components requires knowledge of materials of construction, which includes elastomers, plastics, metals, and glass, along with the engineering processes that are involved in the manufacture of these materials and the fabrication of finished components. Also required is a detailed understanding of the total supply chain, from raw materials (e.g., plastic resins, raw rubber, additives) to finished components, to complete container closure/delivery systems. Accomplishment of controlled extraction studies requires not only detailed knowledge of materials and components, but also expertise in analytical chemistry and in particular trace organic analysis. Development and validation of drug product leachables methods and routine extractables control methods, conduct of definitive leachables studies, development of a leachables/extractables correlation, and development of appropriate leachables/extractables specifications all require analytical chemistry expertise as well. Therefore, in most pharmaceutical development programs for OINDP, it falls to the analytical chemist to drive the process to a successful conclusion.

As a result of this reality, Part II of this book focuses on analytical chemistry, and in particular those best demonstrated laboratory practices and their accompanying required expertise that are necessary for a successful OINDP pharmaceutical development program. These "best practices" are based on the work of the Product Quality Research Institute (PQRI) and the International Pharmaceutical Aerosol Consortium for Regulation and Science (IPAC-RS), along with the knowledge and work of the individual chapter authors. What follows in this introductory chapter to Part II are a detailed discussion of the current regulatory environment for extractables and leachables in pharmaceutical development, a "glossary" of key terminology and definitions, a brief summary of the various chapters and their relationship to the pharmaceutical development process (Fig. 10.1), and a brief look at how current initiatives in the industry may affect future developments in the area of leachables and extractables.

10.2 REGULATORY ENVIRONMENT FOR EXTRACTABLES AND LEACHABLES

The current regulatory environment for OINDP pharmaceutical development is complex. Since OINDPs are in fact pharmaceutical products, the general regulations for such apply to their development and manufacture, for example, 21CFR210, 211, and 820. At the most fundamental level, one must consider pharmaceutical packaging requirements. Both the European Medicines Agency (EMEA) "Guideline on Plastic Immediate Packaging Materials" May 19, 2005 (CPMP/QWP/4359/03),[1] and the United States Food and Drug Administration (FDA) "Guidance for Industry: Container Closure Systems for Packaging Human Drugs and Biologics,"[2] May 1999, require compliance to food additive regulations[3-6] and the respective pharmacopoeias. Since many OINDPs rely on medical devices for proper delivery, the medical device regulations[7,8] must also be considered. These regulations set forth requirements for container closure systems and device components based on the level of risk associated with the route of administration, likelihood of dosage form interaction, or type of patient contact/duration of use, respectively. For example, as is discussed in Chapter 18 ("Development and Optimization of Methods for Routine

Testing") the FDA packaging guidance[2] categorizes inhalation products as highest risk for route of administration and high-to-medium risk for type of patient contact. This translates to more rigorous requirements for the packaging and associated materials of construction used in these products. These requirements are particularly noteworthy when it comes to compounds that may be extracted or leach out of the container closure system. According to both USP <1031>[9] and the International Organization for Standardization (ISO) standard for biological evaluation of medical devices (ISO 10993-1),[10] there are specific tests required that are described in either USP <87> and <88>[11,12] and/or ISO 10993, for example, subparts 5, 6, 10, and 11.[13–16] Plastics may be classified as described in USP <88> based on their compliance to these tests, for example, USP class VI is common for medical applications. These tests are often referred to as biocompatibility tests and involve the extraction of a finished component with a solvent. Subsequently, the extract is dosed in some form *in vitro* using cell cultures or *in vivo* using a representative animal model.

The U.S. and European Union (EU) pharmaceutical packaging guidelines have physicochemical requirements for component materials. These requirements can be met either by compliance with the food additive regulations as detailed in 21CFR172-189,[17] or by performance of compendial tests, for example, European Pharmacopoeia chapter 3,[18] USP <660> Containers-Glass,[19] USP <661> Containers-Plastic,[20] or USP <381> Elastomeric Closures for Injections.[21] Several of these tests involve extracting the finished component with a solvent with subsequent wet chemistry limit tests against specified criteria. However, neither the biocompatibility nor the physicochemical tests represent the type of trace-level chemical analysis recommended by PQRI for OINDP.

It should be noted that, in the case of injection, inhalation, ophthalmic, or transdermal drug products, the FDA packaging guidance[2] does specify a comprehensive extraction study to determine amounts and toxicologically safe levels of extractables. In contrast, according to the EU guideline on plastic immediate packaging,[1] an extraction study is only necessary for inhalation, parenteral, or ophthalmic drug products if a noncompendial material is used and the dosage form is nonsolid. Thus, there is a conflict between these regulations regarding the applicability of extractables testing. This position on extraction testing is further born out in the guidance for inhalation and nasal products that has been adopted by both Canada[22] and Europe.[23]

Leachables testing is treated slightly differently. The U.S. and EU packaging guidelines do require that no harmful or undesirable amounts of substances leach out of the packaging.[1,2] Interestingly, for medical devices, leachables studies are not a requirement unless the device is implantable and therefore not considered relevant for inhalation devices. On the other hand, leachables studies are to be considered for medicinal or combination products.

In addition to the general pharmaceutical regulations and guidelines, there are several in the United States and EU that are specific to inhalation products. These address product-specific performance requirements as well as leachables and extractables requirements, which are summarized in Table 10.1. The first FDA document to detail requirements regarding nebulizers and metered dose inhalers (MDIs) was issued in 1993 as a reviewer guidance[24] and does not address leachables or

TABLE 10.1.   Regulatory Guidelines and Requirements for Various OINDPs

	Nasal spray, inhalation solution	Metered dose inhaler	Dry powder inhaler	Nebulizer
FDA nebulizer reviewer guidance[24]	N/A	None	N/A	None
FDA draft MDI/DPI[25]	A, B, C	A, B, C	A, B, C	A, B, C
FDA nasal spray and inhalation solutions[26]	A, B, C	N/A	N/A	N/A
Canada[22]				
EU[23]	B, D, E	B, D, E	None	B, D, E

A, controlled extraction study for critical components; B, leachables study; C, routine extractables testing on critical components; D, extraction profile only for noncompendial materials; E, routine extractables testing to monitor for leachables of safety concern; N/A, not applicable.

extractables. The FDA MDI/dry powder inhaler (DPI) draft guidance[25] that origi-nated in 1998 mandates a comprehensive evaluation of leachables and extractables for critical components and presumes that routine extraction testing will be per-formed for purposes of control. The FDA nasal spray and inhalation solution guid-ance[26] that was released in 2002 has similar requirements. The PQRI recommendations[27] for OINDP were submitted in 2006 and provide extensive detail and clarification regarding these requirements for leachables and extractables evaluation. In contrast, the jointly crafted European[23] and Canadian[22] guidelines for inhalation products that were also published in 2006 exclude DPIs from the extractables and leachables requirement. Additionally, according to these guidelines, routine extraction testing is only required if there is a safety concern with the levels and types of leachables found. The U.S. and EU regulations have the following expectations in common: extractables profiles for noncompendial materials, leachables for liquid-based prod-ucts, and routine extraction can be used as a surrogate to monitor leachables where a correlation between leachables and extractables has been established.

The safety evaluation and qualification of leachables have been discussed at length in Part I of this book. Regarding leachables in pharmaceutical products, the regulations that are often cited and incorrectly applied to OINDP are those for geno-toxic impurities as detailed in International Conference on Harmonisation (ICH) Q3B.[28] Leachables that arise from the container closure systems are specifically excluded from applicability in ICH Q3B. The PQRI recommendations have been informally adopted as the governing guidance for OINDP safety evaluation. In addi-tion to the assessment of leachables, there are specific regulations for certain entities that are considered to represent safety concerns in pharmaceutical products. These include transmissible spongiform encephalopathy/bovine spongiform encephalopa-thy (TSE/BSE),[29–32] diethylhexyl phthalate (DEHP),[33] bisphenol A (BPA),[34] and phthalates.[35] As can be seen from this discussion, there is a variety of regulations that are operative at different levels and thus, particularly in the general area of extractables and leachables, require constant vigilance for OINDP manufacturers and their suppliers.

10.3   KEY TERMINOLOGY AND DEFINITIONS

This section contains a list of significant terms, their definitions ("official" definitions with appropriate citations where available), and accompanying discussions that elaborate on those definitions. Understanding these important terms will facilitate the reader's appreciation and understanding of the chapters in Part II of this book.

10.3.1   OINDPs

According to the IPAC-RS,[36] OINDPs "are intended to deliver therapeutic benefit by delivery of a pharmaceutical substance to the lungs (Orally Inhaled) or nasal cavity (Nasal). Both of these routes of administration by OINDP have some common characteristics:

- Delivery of the drug as a specific range of particle sizes, which may be the drug particle alone, or bound to a carrier particle (dry powder), or dissolved or suspended in a liquid droplet.
- Targeted deposition to specific membranes (for example specific point of pulmonary tract, specific mucous membrane in the nasal cavity)."

Often, OINDPs are referred to simply as "inhalation drug products." There are several general types of OINDP, which are discussed in the following sections.

10.3.2   MDIs

MDIs are defined by the FDA as "drug products that contain active ingredient(s) dissolved or suspended in a propellant, a mixture of propellants, or a mixture of solvent(s), propellant(s), and/or other excipients in compact pressurized aerosol dispensers. An MDI product may discharge up to several hundred metered doses of drug substance(s)."[25] MDIs are also sometimes referred to as "pressurized" or pMDIs.

10.3.3   DPIs and Inhalation Powders

DPIs and inhalation powders are defined by the FDA as "drug products designed to dispense powders for inhalation. DPIs contain active ingredient(s) alone or with suitable excipient(s). A DPI product may discharge up to several hundred metered doses of drug substance(s). Current designs include **pre-metered** and **device-metered DPIs**, both of which can be driven by patient inspiration alone or with power-assistance of some type. Pre-metered DPIs contain previously measured doses or dose fractions in some type of units (e.g., single or multiple presentations in blisters, capsules, or other cavities) that are subsequently inserted into the device during manufacture or by the patient before use. Device-metered DPIs typically have an internal reservoir containing sufficient formulation for multiple doses which are metered by the device itself during actuation by the patient."[25]

10.3.4 Inhalation Solutions, Suspensions, and Sprays

Inhalation solutions, suspensions, and sprays are defined by the FDA as "drug products that contain active ingredients dissolved or suspended in a formulation, typically aqueous-based, which can contain other excipients and are intended for use by oral inhalation. Aqueous-based drug products for oral inhalation must be sterile (21 CFR 200.51). Inhalation solutions and suspensions are intended to be used with a specified nebulizer. Inhalation sprays are *combination products* where the components responsible for metering, atomization, and delivery of the formulation to the patient are a part of the container closure system."[26]

10.3.5 Nasal Sprays

Nasal sprays are defined by the FDA as "drug products that contain active ingredients dissolved or suspended in a formulation, typically aqueous-based, which can contain other excipients and are intended for use by nasal inhalation. Container closure systems for nasal sprays include the container and all components that are responsible for metering, atomization, and delivery of the formulation to the patient."[26]

10.3.6 Combination Product

Many inhalation products are combination products, which have a unique regulatory submission process in the United States. According to the FDA's definition, the term combination product includes:

- "a product comprised of two or more regulated components, that is, drug/device, biologic/device, drug/biologic, or drug/device/biologic, that are physically, chemically, or otherwise combined or mixed and produced as a single entity;
- two or more separate products packaged together in a single package or as a unit and comprised of drug and device products, device and biological products, or biological and drug products;
- a drug, device, or biological product packaged separately that according to its investigational plan or proposed labeling is intended for use only with an approved individually specified drug, device, or biological product where both are required to achieve the intended use, indication, or effect and where upon approval of the proposed product the labeling of the approved product would need to be changed, for example, to reflect a change in intended use, dosage form, strength, route of administration, or significant change in dose; or
- any investigational drug, device, or biological product packaged separately that according to its proposed labeling is for use only with another individually specified investigational drug, device, or biological product where both are required to achieve the intended use, indication, or effect."[37]

10.3.7   Container Closure System

Defined by the FDA as "the sum of packaging components that together contain, protect, and deliver the dosage form. This includes *primary packaging components* and *secondary packaging components* if the latter are intended to provide additional protection to the drug product (e.g., foil overwrap). The container closure system also includes the pump for nasal and inhalation sprays. For nasal spray and inhalation solution, suspension, and spray drug products, the critical components of the container closure system are those that contact either the patient or the formulation, components that affect the mechanics of the overall performance of the device, or any protective packaging."[26] "A *primary packaging* component means a packaging component that is or may be in direct contact with the dosage form. A *secondary packaging* component means a packaging component that is not or will not be in direct contact with the dosage form."[2] Container closure systems can also act as components of *delivery systems* for the drug product formulation (e.g., in the case of MDIs).

10.3.8   Medical Device

Although the terms delivery system and medical device are often used interchangeably, the term medical device has a specific regulatory meaning while delivery system does not. A drug delivery system has been described as "a formulation or device that delivers the API(s) in site-directed applications or provides timely (i.e., immediate, delayed, or sustained) release of the API(s). The system, on its own, is not pharmaceutically active, but improves the efficacy and/or safety of the API(s) that it carries."[38] The term medical device carries with it several regulatory obligations that must be met. The United States and EU each have their own definitions of medical device and accompanying requirements:

- *United States.* If a product is labeled, promoted, or used in a manner that meets the following definition in section 201(h) of the Federal Food Drug & Cosmetic (FD&C) Act,[39] it will be regulated by the FDA as a medical device and is subject to premarketing and postmarketing regulatory controls. A device is "an instrument, apparatus, implement, machine, contrivance, implant, in vitro reagent, or other similar or related article, including a component part, or accessory which is:
    - recognized in the official National Formulary, or the United States Pharmacopeia, or any supplement to them,
    - intended for use in the diagnosis of disease or other conditions, or in the cure, mitigation, treatment, or prevention of disease, in man or other animals, or
    - intended to affect the structure or any function of the body of man or other animals, and which does not achieve its primary intended purposes through chemical action within or on the body of man or other animals and which is not dependent upon being metabolized for the achievement of any of its primary intended purposes."

The FDA has established classifications for approximately 1700 different generic types of devices and grouped them into 16 medical specialties that are referred to as "panels." Inhalation devices are typically found in either the anesthesiology (21CFR868) or the ear, nose, and throat (21CFR874) panels.

Each of the generic types of devices is assigned to one of three regulatory classes based on the level of control necessary to assure the safety and effectiveness of the device. The three classes and the requirements that apply to them are

- class I—general controls, with or without exemptions;
- class II—general controls and special controls, with or without exemptions;
- class III—general controls and premarket approval.

The class to which a device is assigned along with its exemption status determines the type of premarketing submission/application required for FDA clearance to market. If a device is classified as class I or II, and it is not exempt, a 510k is required for marketing. Inhalation devices such as nebulizers and spacers are typically either class I or II and may or may not be marketed as stand-alone devices. Other devices such as inhalers and nasal sprays are, by intercenter agreement, treated as drug products.

- *EU*. The medical device directive is applicable to medical devices entering the EU and requires that medical devices bear the *conformité européenne* (CE) mark, which indicates conformity to the applicable provisions of the directive. The directive applies to medical devices or accessories that are considered devices for the purpose of the directive, and the following definitions apply:

"(a) 'medical device' means any instrument, apparatus, appliance, software, material or other article, whether used alone or in combination, including the software intended by its manufacturer to be used specifically for diagnostic and/or therapeutic purposes and necessary for its proper application, intended by the manufacturer to be used for human beings for the purpose of:

- diagnosis, prevention, monitoring, treatment or alleviation of disease,
- diagnosis, monitoring, treatment, alleviation of or compensation for an injury or handicap,
- investigation, replacement or modification of the anatomy or of a physiological process,
- control of conception, and which does not achieve its principal intended action in or on the human body by pharmacological, immunological or metabolic means, but which may be assisted in its function by such means;

(b) 'accessory' means an article which whilst not being a device is intended specifically by its manufacturer to be used together with a device to enable it to be used in accordance with the use of the device intended by the manufacturer of the device."[40]

Based on this directive, medical devices are classified as class I, IIa, IIb, or III, based on risk associated with duration of use and level of invasiveness, where class I has the lowest risk. Various levels of engagement of a notified body are required depending on the class. These may range from certification of a device manufacturer's compliance to ISO 13485 to conformity assessment of a specific device. In any case, it is expected that a technical file is maintained by the manufacturer to support the claims made regarding conformity to the directive's provisions.

10.3.9 Packaging Component

Per the FDA, packaging component "means any single part of a container closure system. Typical components are containers (e.g., ampules, vials, bottles), container liners (e.g., tube liners), closures (e.g., screw caps, stoppers), closure liners, stopper overseals, container inner seals, administration ports (e.g., on large-volume parenterals (LVPs)), overwraps, administration accessories, and container labels. A *primary packaging component* means a packaging component that is or may be in direct contact with the dosage form. A *secondary packaging component* means a packaging component that is not and will not be in direct contact with the dosage form."[2]

10.3.10 Critical Component

"The critical components of the container/closure system are defined as those that contact either the patient, that is, the mouthpiece, or the formulation, components that affect the mechanics of the overall performance of the device, or any necessary secondary protective packaging."[27]

10.3.11 Materials of Construction

Materials of construction are defined by the FDA as referring "to the substances (e.g., glass, high density polyethylene (HDPE) resin, metal) used to manufacture a packaging component. This term is used in a general sense for the basic material, which should be defined in the application in terms of its specific chemical composition for a given drug application (e.g., the specific polymer and any additives used to make the material)."[2] Materials of construction are also often referred to as *raw materials*.

10.3.12 Supplier

In this context, a "supplier" can be defined as an organization or company that "supplies" either raw materials (e.g., raw rubber, plastic resin pellets, rolls of aluminum, or stainless steel), processed materials (e.g., compounded rubber), finished components (e.g., rubber gaskets and seals, dose metering valve plastic components, metal canisters), or assembled delivery system components (e.g., dose metering valves, DPIs) to a pharmaceutical manufacturer. These suppliers are often referred to (in reverse order from above) as n-1, n-2, and so on, to designate their level away from the pharmaceutical manufacturer, which is designated n, in the overall supply chain.

10.3.13   Elastomer

"An elastomer is a polymer with the property of viscoelasticity (colloquially 'elasticity'), generally having notably low Young's modulus and high yield strain compared with other materials. The term, which is derived from *elastic polymer*, is often used interchangeably with the term **rubber**, although the latter is preferred when referring to vulcanisates. Each of the monomers which link to form the polymer is usually made of carbon, hydrogen, oxygen and/or silicon. Elastomers are amorphous polymers existing above their glass transition temperature, so that considerable segmental motion is possible. At ambient temperatures rubbers are thus relatively soft (E ~ 3 MPa) and deformable. Their primary uses are for seals, adhesives and molded flexible parts."[41] Elastomers are used in the fabrication of many types of critical (and other) components for OINDP, such as MDI gaskets and seals. The reader interested in further technical information regarding elastomers and rubber is referred to the comprehensive treatise by Morton.[42]

10.3.14   Plastic

The word "plastic" comes from the Greek word "plastikos" meaning "to form."[43] In more technical terms, a plastic is a material that can be heated and formed (i.e., molded, extruded, pressed) so that it keeps its desired shape after it cools.[43] Plastics are commonly polymeric materials (e.g., polyethylene, polybutylene terephthalate) and are used to fabricate many types of critical (and other) components for OINDP, such as mouthpieces, MDI valve components or DPI powder dispersing units.

10.3.15   Extractables

Extractables are compounds that can be extracted from OINDP device components or surfaces of the OINDP container/closure system in the presence of an appropriate solvent(s) and/or condition(s). Thus, extractables are individual chemical entities that can be extracted from individual component types, for example, rubber seals and plastic valve parts, of an OINDP container/closure system under relatively vigorous laboratory conditions using appropriate solvents or solvent systems. Extractables can, therefore, be considered as potential leachables in OINDP.[27]

10.3.16   Leachables

Leachables in OINDP are compounds that are present in the drug product due to leaching (i.e., migrating) from container/closure system components. Leaching can be promoted by the formulation, or components of the formulation, for example, chlorofluorocarbon (CFC) or hydrofluoroalkane (HFA) propellants in MDIs. Leachables are often a subset of, or are derived directly or indirectly from, extractables.

Due to the time-dependent nature of the leaching process, leachables appear in an OINDP formulation over the shelf life of the product as determined during appropriate stability and accelerated stability studies.[27] Chemical entities that appear in drug product as a result of contact between drug substance and/or the formulation (or formulation ingredients) with processing equipment are also usually referred to as leachables, even though these chemical entities do not appear as extractables from container closure system critical components. In such cases, achieving extractables/leachables correlations requires additional investigative laboratory studies.

10.3.17 Extractables/Leachables Profile

According to the FDA "an *extraction profile* refers to the analysis (usually by chromatographic means) of extracts obtained from a packaging component. A *quantitative extraction profile* is one in which the amount of each detected substance is determined."[2] The terms *extractables* and *leachables profiles* are used in this book to refer to various chromatographic representations of the extractables content in a particular extracting medium, or the leachables content of a drug product. However, other representations can also be termed extractables/leachables profiles, such as Fourier transform infrared (FTIR) spectra and ultraviolet (UV) spectra.

10.3.18 QT

The QT for OINDP is a threshold below which a given noncarcinogenic leachable is not considered for safety qualification (toxicological assessments) unless the leachable presents structure–activity relationship (SAR) concerns.[27] The QT for an individual organic leachable in any OINDP is defined as 5 μg/day total daily intake (TDI).[27]

10.3.19 SCT

The SCT for OINDP is the threshold below which a leachable would have a dose so low as to present negligible safety concerns from carcinogenic and noncarcinogenic toxic effects.[27] The SCT for an individual organic leachable in any OINDP is defined as 0.15 μg/day TDI.[27]

10.3.20 AET

The AET is defined as the threshold at or above which an OINDP pharmaceutical development team should identify and quantify a particular extractable and/or leachable and report it for potential toxicological assessment[27]:

- *Estimated AET.* The AET can be estimated from the SCT by converting the SCT from units of daily exposure (microgram per day) to units of amount per product unit or dose, for example, microgram per canister, microgram per dose, and microgram per blister. This value is then converted into amount per gram of component, for example, microgram per gram, using the weight and amount of component used per drug product. This resulting value is the estimated AET. The required sensitivity of the analytical method(s) (the limit of quantitation [LOQ]) can then be determined from the estimated AET.[27]

- *Final AET.* The uncertainty of each analytical method used for definitive extractables/leachables profiling should be estimated. One way to accomplish this is to develop a response factor (RF) database of extractables using authentic standards (where available). The estimated uncertainty, for the given method, should be applied to the estimated AET to calculate the final AET. This determination allows the analytical chemist to refine the original estimated AET, and if necessary, to identify any extractables/leachables that were not assessed previously.[27]

10.3.21   RF Database

An RF is defined as

$$RF = A_a/C_a,$$

where

A_a = response of an individual analyte, for example, chromatographic peak area; and

C_a = concentration (or mass) of the individual analyte.

For a gas chromatography/mass spectrometry (GC/MS) method, for example, the chromatographic peak areas for individual analytes, that is, leachables or extractables, as determined from either the total ion chromatogram (TIC) or individual mass chromatograms (extracted ion current profiles), are divided by individual analyte concentrations in a known sample of authentic reference compounds. The concentration levels of the authentic reference compounds chosen for RF determination must be within the linear dynamic range of the analytical system. For GC/MS, this means not overloading the GC column or saturating the mass spectrometer's detector.

A somewhat more precise uncertainty evaluation can be obtained through the use of relative response factors (RRFs), which are defined as follows:

$$RRF = C_{is}A_a/A_{is}C_a,$$

where

C_{is} = concentration (or mass) of an internal standard,

A_{is} = response of the internal standard,

A_a = response of an individual analyte,

C_a = concentration of the individual analyte.

The RRF normalizes individual RFs to the RF of an internal standard. The use of internal standards is a well-established procedure for improving the accuracy and precision of trace organic analytical methods.

An RF database is therefore a compilation of RFs (and/or RRFs) for authentic reference standards representing identified (or known) extractables/leachables from a particular extractables/leachables profile. Statistical analysis of an RF database can be used to estimate analytical uncertainty and convert an estimated AET into a final AET.

10.3.22   Internal Standard

An internal standard is used in trace organic analysis (i.e., extractables/leachables profiling) to correct for minor variations in detector response due to, for example, suppression or enhancement of ionization efficiency[44] and analytical method recovery. An internal standard appropriate to the particular analytical technique/method used to generate the extractables/leachables profile should be chosen. Some characteristics of a good internal standard are the following[27]:

- It should be compatible with the particular analytical technique.
- It should be "well behaved" in the particular analytical method. A "well-behaved" internal standard in a GC method, for instance, will not have a significant tailing factor, will not irreversibly adsorb onto the column, and so on.
- It should be stable in the analytical matrix.
- It should not be interfered with by other analytes or components in the analytical matrix.
- It should possess a response similar to those of other analytes in the particular analytical technique/method.

10.3.23   Controlled Extraction Study

A controlled extraction study is a laboratory investigation into the qualitative and quantitative nature of extractables profiles of critical components of an OINDP (or other) container/closure system.[27] The purpose of a controlled extraction study is to systematically and rationally identify and quantify potential leachables, that is, extractables, to the extent practicable, and within certain defined analytical threshold parameters. Controlled extraction studies typically involve vigorous extractions of representative lots of components using multiple solvents of varying polarity, with both qualitative and quantitative evaluation of the resulting extractables profiles. Multiple analytical techniques/methods with compound specific detection, for example, MS, are usually employed to establish extractables profiles. It is often the case that the analytical techniques/methods used in controlled extraction studies, along with the qualitative and quantitative results of these studies, are used to

1. establish a basis for the development and validation of routine quality control methods and acceptance criteria for critical component extractables profiles;
2. establish a basis for the development and validation of leachables methods suitable for use in drug product leachables studies as well as for potential use as routine quality control methods for drug product leachables (should such be required by regulatory authorities).
3. allow for the "correlation" of extractables and leachables.

10.3.24 Asymptotic Levels

An optimized extraction method is defined as one that yields a high number and concentration of extractables, and achieves steady-state levels, that is, "asymptotic levels." Optimization of the extraction technique(s)/method(s) prior to conducting quantitative controlled extraction studies ensures that the extractables profile represents at least a "worst-case" scenario of potential leachables and their levels.

10.3.25 Leachables Studies/Leachables Stability Study

A leachables study is a laboratory investigation into the qualitative and quantitative nature of a particular OINDP (or other) leachables profile(s) over the proposed shelf life of the product.[27] The purpose of a leachables study is to systematically and rationally identify and quantify drug product leachables to the extent practicable, and within certain defined analytical threshold parameters. Leachables studies typically involve the development and validation of analytical methods capable of detecting and quantifying all potential leachables characterized in the controlled extraction studies, as well as identifying "unspecified" leachables, which may have escaped prior to characterization or formed via chemical reaction in the drug product formulation matrix. Leachables studies are most often accomplished as part of a larger drug product stability program on multiple batches of drug product, using multiple component batches, stored under a variety of conditions through the intended shelf life of the product, designed to support registration activities. Since these large drug product stability studies involve analysis of samples at multiple time points, it is possible to discern trends in drug product leachables profiles over time and storage condition. Like the controlled extraction study, the leachables study can be framed as a trace organic analysis problem, with the sample matrix being the drug product formulation. Analytical methods for leachables analysis must quantitatively recover leachables from the drug product matrix and separate and individually detect them with appropriate sensitivity. Analytical techniques most often employed for leachables studies are the same as those used in controlled extraction studies, namely GC/MS, liquid chromatography (LC)/MS, and LC/UV detection. Leachables studies provide information in support of developing an extractables/leachables correlation, and for the establishment of drug product leachables specifications and acceptance criteria.

10.3.26 Validation

Validation is a process of demonstrating that an analytical method (e.g., quantitative leachables profiling method and routine quantitative extractables control method) is capable of meeting defined acceptance criteria for its key performance characteristics.[45] These key performance characteristics (also termed "validation parameters") include accuracy, precision (repeatability, intermediate precision), specificity, limit of detection/LOQ, linearity, range, system suitability, and robustness.[46]

10.3.27   Leachables/Extractables Correlation

The significance of a *correlation* between extractables and leachables profiles cannot be overstated. A correlation should be both qualitative and quantitative, and should be demonstrable over multiple batches of drug product to end of shelf life, and multiple batches of container closure system critical components[27]:

- *Qualitative correlation.* A *qualitative correlation* can be established if all compounds detected in validated leachables studies can be linked qualitatively either directly or indirectly to an extractable identified in comprehensive controlled extraction studies or during routine extractables testing.[27] A direct qualitative correlation is relatively simple, for example:

  I.   Stearic acid is a known ingredient in a particular MDI dose metering valve critical component, that is, as technical grade calcium stearate.

  II. Stearic acid is *confirmed* by GC/MS in methylene chloride Soxhlet extracts of the critical component in question during controlled extractions studies. Stearic acid is also confirmed in 30 batches of the critical component during routine extractables testing with a validated GC/flame ionization detection (FID) method.

  III. Stearic acid is *confirmed* by a validated GC/MS method to be present in definitive registration batches of drug product, at various time points over the proposed shelf life of the product, under different storage conditions, and different product orientations.

  An indirect qualitative correlation is only slightly more challenging:

  I.   Stearic acid is a known ingredient in a particular MDI dose metering valve critical component, that is, as technical grade calcium stearate.

  II. Stearic acid is *confirmed* by GC/MS in methylene chloride Soxhlet extracts of the critical component in question during controlled extractions studies. Stearic acid is also confirmed in 30 batches of the critical component during routine extractables testing with a validated GC/FID method.

  III. Ethyl stearate is *confirmed* by a validated GC/MS method to be present in definitive registration batches of drug product, at various time points over the proposed shelf life of the product, under different storage conditions, and different product orientations.

  IV. The MDI drug product formulation is known to contain 10% ethanol, which can react with stearic acid to form ethyl stearate.

  Qualitative correlations obviously require some knowledge and understanding of the chemistry and reactivity of extractables and chemical additives to rubber and plastic. It is important to be aware that many of these chemical additives, such as polymerization agents, accelerators, antioxidants, and stabilizers, are by their very nature reactive species. Note that one does not need to have *confirmed* identifications of particular leachables and extractables in order to establish a qualitative correlation. Information available from analytical techniques such as GC/MS and LC/MS allows for leachables/extractables qualitative correlations of chemical entities with *confident* and *tentative* identifications. *Confirmed* and *confident* levels of identification are generally required for toxicological evaluation of leachables (see Chapter 13).

• *Quantitative correlation.* A *quantitative correlation* between a leachable and an extractable can be made if the level of the leachable is demonstrated to be consistently less than that of the extractable(s) to which it is qualitatively correlated. For an individual batch of OINDP, this quantitative correlation should be valid through the proposed end of shelf life, and across all accelerated storage conditions and product orientations. Quantitative correlations are best accomplished using data from a significant number of critical component batches, acquired using validated routine extractables testing analytical methods. For example:

I.   Stearic acid is shown to have a qualitative leachables/extractables correlation (as defined above) in an MDI drug product.

II. Comprehensive leachables studies show stearic acid to have a maximum level in drug product of 50 µg/canister, across all definitive registration batches of drug product, stability storage conditions, drug product orientations, and stability time points to the proposed end of shelf life.

III. A database of 50 critical component batches analyzed by a validated routine extractables testing analytical method quantitates stearic acid at 800 µg/g ± 100 (standard deviation, i.e., 12.5% relative standard deviation).

IV. Given that there is one 150-mg critical component per MDI valve, the anticipated maximum level of stearic acid as a drug product leachable would be 120 ± 15 µg/canister. This result represents a positive quantitative correlation.

   In establishing both qualitative and quantitative leachables/extractables correlations, it is highly recommended that the analytical chemist compare the following:

– Leachables profiles from multiple (at least three) drug product definitive registration batches using specific batches of critical components, with qualitative and quantitative extractables profiles of those *specific component batches*. For example, the leachables profiles from MDI registration batches should be compared with the extractables profiles of the components that make up the valves used in those registration batches.

– Leachables profiles from multiple drug product registration batches with extractables profiles from multiple batches of critical components (which may not have been used in the drug product registration batches). This comparison is intended to check the consistency of correlations between extractables profiles from multiple component batches and leachables profiles from multiple drug product batches.

– If a qualitative and quantitative correlation cannot be established, the source of the problem should be determined and corrected. Potential sources include excessive variability in component composition and/or manufacturing processes, changes in drug product formulation, inadequate controlled extraction studies, and inappropriate or poorly validated leachables and extractables methods.

10.3.28 Routine Extractables Testing

Routine extractables testing is the process by which OINDP (or other) container closure system critical components are qualitatively and quantitatively profiled for extractables, for purposes of either establishing extractables acceptance criteria or release according to already established acceptance criteria. Like the analytical methods used in leachables studies, those used for routine extractables testing must be capable of detecting and quantifying all relevant extractables characterized in the controlled extraction studies, as well as identifying "unspecified" extractables, which could result from unanticipated changes in critical component ingredients or some external contamination. However, routine extractables testing analytical methods must also be highly rugged and robust, making them easily transferable and useful in quality control and manufacturing environments. As a result of these requirements, it is common practice to employ analytical techniques that lend themselves to methods with the desired characteristics. For example, when GC/MS was used for controlled extraction studies, a routine extractables testing method could be based on the more rugged and robust GC/FID. LC/MS controlled extraction study methods could be converted to LC/UV routine extractables testing methods, as long as the UV detector is sufficiently sensitive for the extractables under consideration. Again, like leachables study analytical methods, analytical methods used for routine extractables testing must be validated according to accepted industry practice. Early in the OINDP development process, routine extractables testing is used to create a qualitative/quantitative extractables database, which can be used to help establish an extractables/leachables correlation and to develop critical component extractables specifications. Later in the development process and post-approval, routine extractables testing is used to release critical components for drug product manufacture according to previously established acceptance criteria.

10.3.29 Special Case Compounds

Special case compounds are individual (or classes) of chemical entities that have special safety or historical concerns as drug product leachables in OINDP and, as a result, must be evaluated and controlled (either as extractables, leachables, or both) by specific analytical techniques and technology-defined thresholds.[27] Special case compounds are therefore not to be evaluated relative to either the QT or SCT for OINDP leachables. Special case compounds include polyaromatic hydrocarbons (PAHs or polynuclear aromatics [PNAs]), *N*-nitrosamines, and 2-mercaptobenzothiazole (2-MBT).

10.3.30 Leachables/Extractables Specification/Acceptance Criteria

A pharmaceutical "specification" for OINDP is defined by the FDA as "a list of tests, references to analytical methods, and appropriate acceptance criteria that are numerical limits, ranges or other criteria for the tests described. Specifications establish a set of criteria to which a drug substance or drug product should conform using the approved analytical procedure to be considered acceptable for its intended use. *Acceptance criteria* are numerical limits, ranges, or other criteria for the tests described."[25]

Leachables specifications should therefore include a fully validated analytical test method. The acceptance criteria for leachables should apply over the proposed shelf life of the drug product, and should include

1. quantitative limits for known drug product leachables monitored during product registration stability studies, and

2. a quantitative limit for "new" or "unspecified" leachables not detected or monitored during product registration stability studies.

Quantitative acceptance criteria should be based on leachables levels, and trends in leachables levels, observed over time and across various storage conditions and drug produce orientations during product registration stability studies, with the application of appropriate statistical analysis (or other equally valid approaches). A comprehensive extractables/leachables correlation, as defined and elaborated above, may obviate the need for routine implementation of drug product leachables specifications. This, of course, further assumes the following:

1. adequate information from critical component suppliers, with an adequate evaluation of this information;

2. complete understanding and control of critical component fabrication and manufacturing processes;

3. adequate and comprehensive controlled extraction studies on all critical components;

4. validated leachables analytical methods and a comprehensive leachables study;

5. validated routine extractables testing analytical methods and an adequate database of critical component extractables profiles;

6. appropriate specifications for extractables from critical components.

Note that the requirement for establishment and implementation of leachables specifications for any particular OINDP is a regulatory policy matter.

Routine extractables testing should be performed on OINDP critical components prior to drug product manufacture for the purposes of marketing. Critical components should be released to drug product manufacture based on carefully defined specifications established through

1. a complete understanding of critical component composition(s), ingredients, and compounding/fabrication processes;

2. comprehensive controlled extraction studies;

3. a significant database of extractables profiles obtained with fully optimized and validated routine extractables testing analytical methods; and

4. a complete leachables/extractables correlation. The actual form and statement of specifications will depend on factors such as the type of OINDP, the type of critical component (such as contact or noncontact with the drug product formulation), and adequacy of leachables/extractables correlation.

Acceptance criteria for OINDP critical component extractables can include the following:

1. confirmation of extractables in the profile, for example, identified in controlled extraction studies or routine extractables batch testing;
2. quantitative limits for extractables, for example, identified in controlled extraction studies or found as correlated leachables;
3. a quantitative limit for "new" or "unspecified" extractables, for example, not detected during controlled extraction studies or routine extractables batch testing.

It is recognized that there are many possible ways for setting acceptance criteria, and the authors do not recommend any particular approach to establishing such criteria. The point in the development process at which these criteria are established may determine the nature of the criteria depending on the studies that have been performed and material/processing information available. For example, quantitative limits need not necessarily be established for all extractables identified in controlled extraction studies, but could be established for major extractables representative of major chemical additives in the component formulation. Such criteria may be refined as experience with the critical components is gained. Failure of a particular batch of critical component to meet established acceptance criteria suggests either an unapproved change in critical component ingredients, or an unapproved change (or problem) with critical component compounding/fabrication processes. To prevent critical component extractables profile failures, and to ensure that critical component quality is maintained, it is important that the pharmaceutical development organizations work closely with component suppliers to control critical component compounding/fabrication processes. The analytical chemist should also clarify to the supplier any expectations regarding changes to component ingredients, compounding, fabrication, or other manufacturing processes, including prior notification of such changes. It is recommended that procedures be developed for investigating routine extractables testing acceptance criteria failures, that is, out of specification, or OOS, procedures. Furthermore, it is recommended that all critical component extractables profiles be monitored for qualitative or quantitative changes that are within established acceptance criteria, and procedures be developed for investigating and understanding the root causes of such variations. Careful monitoring of critical component extractables profiles will likely result in fewer failures and OOS investigations.

10.4   BEST PRACTICE RECOMMENDATIONS

The chapters in Part II of this book are primarily concerned with various aspects of the pharmaceutical development process for OINDP related to extractables and leachables (Fig. 10.1). These chapters were significantly influenced by a series of "best practice" recommendations issued by a working group of the PQRI related to critical component selection, controlled extraction studies, leachables studies, and routine extractables testing.[27,47] The recommendations represent a consensus of the fundamental scientific principles appropriate to OINDP pharmaceutical development programs, and with due consideration to the pharmaceutical development programs of other drug product types relative to extractables and leachables. The recommendations are summarized as follows:

A. Component Selection

- The pharmaceutical development team should obtain all available information on the composition and manufacturing/fabrication processes for each component type to the extent possible, and determine which components are "critical."
- Component formulation should inform component selection.
- Risk assessment should be performed during the selection of components and materials.
- Extractables testing, including controlled extraction studies and the development and validation of routine extractables testing methods, should be accomplished for all critical OINDP components.

B. Controlled Extraction Studies

- Controlled extraction studies should employ vigorous extraction with multiple solvents of varying polarity.
- Controlled extraction studies should incorporate multiple extraction techniques.
- Controlled extraction studies should include careful sample preparation based on knowledge of analytical techniques to be used.
- Controlled extraction studies should employ multiple analytical techniques.
- Controlled extraction studies should include a defined and systematic process for identification of individual extractables.
- Controlled extraction study "definitive" extraction techniques/methods should be optimized.
- During the controlled extraction study process, sponsors should revisit supplier information describing component formulation.
- Controlled extraction studies should be guided by an AET that is based on an accepted SCT.
- PAHs (or PNAs), N-nitrosamines, and 2-MBT are considered to be "special case" compounds, requiring evaluation by specific analytical techniques and technology-defined threshold.
- Qualitative and quantitative extractables profiles should be discussed with and reviewed by pharmaceutical development team toxicologists so that any potential safety concerns regarding individual extractables, that is, potential leachables, are identified early in the pharmaceutical development process.

C. Leachables Studies and Routine Extractables Testing

- Analytical methods for the qualitative and quantitative evaluation of leachables should be based on analytical technique(s)/method(s) used in the controlled extraction studies.

- Leachables studies should be guided by an AET that is based on an accepted SCT.

- A comprehensive correlation between extractables and leachables profiles should be established.

- Specifications and acceptance criteria should be established for leachables profiles in OINDP as required.

- Analytical methods for routine extractables testing should be based on the analytical technique(s)/method(s) used in the controlled extraction studies.

- Routine extractables testing should be performed on critical components using appropriate specifications and acceptance criteria.

- Analytical methods for leachables studies and routine extractables testing should be fully validated according to accepted parameters and criteria.

- PAHs (or PNAs), N-nitrosamines, and 2-MBT are considered to be "special case" compounds, requiring evaluation by specific analytical techniques and technology defined thresholds for leachables studies and routine extractables testing.

- Qualitative and quantitative leachables profiles should be discussed with and reviewed by pharmaceutical development team toxicologists so that any potential safety concerns regarding individual leachables are identified as early as possible in the pharmaceutical development process.

Part II also includes a chapter on foreign particulate matter (FPM). Although FPMs are not considered leachables (and the techniques to evaluate them are very different), their successful management in OINDP benefit also from early knowledge and understanding of sources, characterization, and quantification.

10.5 THE FUTURE: RESPONDING TO CURRENT TRENDS

Leachables and extractables management is a topic of relatively recent concern and is associated with the broader topic of product quality. In recent years, several initiatives have been launched to directly or indirectly address product quality both with respect to the pharmaceutical industry and manufacturing in general. There is continued emphasis on producing high-quality products with fewer resources and in shorter time periods. Many industries rely on "lean" initiatives or "six sigma" programs, which are designed to utilize systematic approaches to improve manufacturing process efficiency and maintain product quality.[48] The constant emphasis on measurement of process capability and use of control charts in the pharmaceutical industry is accompanied by process analytical technology (PAT)[49,50] and a strong movement toward reducing end-product testing. The introduction of ICH Q8[51], Q9,[52] and Q10[53] has prompted much discussion on the topics of critical quality attributes, design space, risk management, and continual improvement. The FDA outlined a similar set of underlying principles implicit in the pharmaceutical current good manufacturing practices (cGMPs) for the 21st century as follows:

- risk-based orientation,
- science-based policies and standards,
- integrated quality systems orientation,
- international cooperation,
- strong public health protection.[54]

At the heart of these initiatives is the concept of taking an informed, risk-based approach. Taken together, these were embodied in the FDA's pilot program that was launched in 2005 to allow participating firms an opportunity to submit chemistry, manufacturing, and controls (CMC) information demonstrating application of quality by design (QbD).[55,56] Although the concept first originated with Juran,[57] many articles and several books have been written on the topic in an attempt to address how QbD can be implemented in the pharmaceutical industry. A recent article by Snee pointed out that "while the benefits of the approach are generally acknowledged, its implementation has been slower than expected."[58] There have been several discussions, presentations, and articles written regarding the implementation of QbD to various aspects of pharmaceutical development. The IPAC-RS consortium has formed the Leachables and Extractables (L&E) Development Paradigm Working Group that is investigating how QbD concepts can be integrated with more traditional approaches to create a development paradigm for extractables and leachables in OINDP. This effort builds on what has been accomplished through the PQRI recommendations for best practices as described above. Specific goals of the L&E Development Paradigm Working Group are to

- develop an understanding of manufacturing process factors that affect L&E;
- determine if quantitative models can be developed that predict the effect of key factors on L&E;
- develop a systematic approach that uses quantitative models in the initial development and subsequent continuous improvement of a product.

The first consideration for application of QbD concepts to leachables and extractables evaluation is that of linkage to critical quality attributes. This involves the consideration of the product profile (some may refer to this as product definition). As an example, consider the MDI profile diagrammed in Figure 10.2, where the specifics related to extractables and leachables are detailed as a subset of the critical attribute of safety and efficacy.

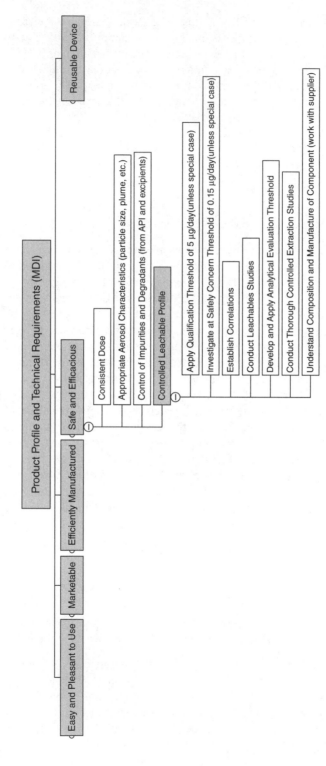

Figure 10.2   Example product profile for a metered dose inhaler (MDI) that incorporates leachables and extractables. API, active pharmaceutical ingredient.

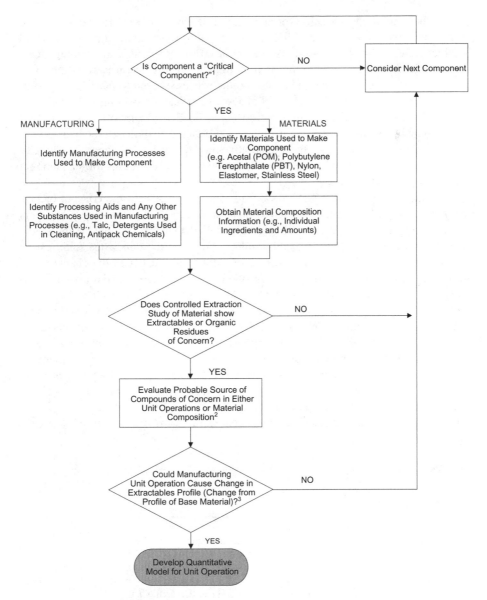

1. For the purposes of this exercise, "Critical Components" are those components of the container/closure system or device that contact the patient, the formulation, that affect the mechanics of the overall performance of the device, or any necessary secondary protective packaging. (PQRI Recommendations, 2006)
2. This evaluation is a risk assessment investigating the possible sources of compounds.
3. This decision could in some cases be based on results of controlled extraction studies on manufactured component.

Figure 10.3 A risk-based approach to critical component determination.

The second consideration for application of QbD concepts to leachables and extractables evaluation is that of linkage to critical process parameters. A risk-based evaluation as outlined in Figure 10.3 can be used to determine critical components of the delivery system that should be investigated. This decision is based not only on the materials of construction but also on the processes used to fabricate the component. Effective evaluation of the linkage of critical processing parameters to leachables from a critical component so determined could involve experimental assessment of specific processing parameters and their effect on extractables (potential leachables). It is proposed that quantitative models can be developed that directly correlate critical process parameters with extractable levels and thus form an indirect linkage with product safety.

Although the term "QbD" has been reduced by some to the concept of performing design of experiments (DOEs) on a small subset of processing parameters or analytical methodologies, it should be recalled that the original intent of the concept is to design quality into the product. This is often contrasted with testing quality into the product. As will become evident to the reader in the chapters that follow, certain types of testing (e.g., controlled extraction studies and routine extractables) when performed at early stages of development can be well utilized in a risk-based product development paradigm. To properly design quality into OINDP requires a fundamental knowledge of the materials of construction, establishing effective relationships with suppliers, setting appropriate quality standards, implementation of change management procedures, and holistic life cycle management. It is anticipated that as new approaches to leachables and extractables are formalized that incorporate the concepts of science-based and risk-based decision making, these development paradigms will provide a firm foundation to address future concerns regarding sustainability, counterfeiting, and manufacturing in emerging markets.

REFERENCES

1   European Medicines Agency. Guideline on plastic immediate packaging materials, 2005. Available at: www.emea.europa.eu/pdfs/human/qwp/435903en.pdf (accessed February 3, 2011).
2   Guidance for industry: Container closure systems for packaging human drugs and biologics. Chemistry, manufacturing, and controls documentation. Department of Health and Human Services, Food and Drug Administration Center for Drug Evaluation and Research (CDER), 1999.
3   European Parliament and Council Directive: 2002/72/EC as amended—Plastic materials and articles intended to come into contact with foodstuffs, 2002. Available at: http://eurlex.europa.eu/LexUriServ/LexUriServ.do?uri=CONSLEG:2002L0072:20080327:EN:PDF.
4   Code of Federal Regulations. Parts 172–189, Title 21. Available at: http://www.accessdata.fda.gov/scripts/cdrh/cfdocs/cfcfr/cfrsearch.cfm (accessed February 3, 2011).
5   European Parliament and Council Directive: 1935/2004 materials and articles intended to come into contact with food, 2004. Available at: http://ec.europa.eu/food/food/chemicalsafety/foodcontact/legisl_list_en.htm (accessed February 3, 2011).
6   European Parliament and Council Directive: 94/62/EC—Packaging and packaging waste, 2005. Available at: http://eur-lex.europa.eu/LexUriServ/site/en/consleg/1994/L/01994L0062-20050405-en.pdf (accessed February 3, 2011).

7   Code of Federal Regulations. Establishment registration and device listing for manufacturers and initial importers of devices. Part 807, Title 21.

8   Council of European Committees Directive: 93/42/EEC as amended, Medical Devices Directive, 1993.

9   U.S. Pharmacopeia (USP). Chapter 1031. The biocompatibility of materials used in drug containers, medical devices, and implants.

10   International Organization for Standardization (ISO). ISO 10993-1: Biological evaluation of medical devices. TC 194 Biological Evaluation of Medical Devices, 2010. Available at: http://www.iso.org/iso/search.htm?qt=10993&sort=rel&type=simple&published=on (accessed January 28, 2009).

11   U.S. Pharmacopeia (USP). Chapter 87. Biological Reactivity Tests, In Vitro.

12   U.S. Pharmacopeia (USP). Chapter 88. Biological Reactivity Tests, In Vivo.

13   International Organization for Standardization (ISO). ISO 10993-5: Tests for in vitro cytotoxicity. TC 194: Biological Evaluation of Medical Devices, 2009. Available at: http://www.iso.org/iso/iso_catalogue/catalogue_tc/catalogue_detail.htm?csnumber=36406 (accessed January 21, 2011).

14   International Organization for Standardization (ISO). ISO 10993-6: Tests for local effects after implantation. TC 194: Biological Evaluation of Medical Devices, 2007. Available at: http://www.iso.org/iso/iso_catalogue/catalogue_tc/catalogue_detail.htm?csnumber=44789 (accessed January 21, 2011).

15   International Organization for Standardization (ISO). ISO 10993-10: Tests for irritation and skin sensitization. TC 194: Biological Evaluation of Medical Devices, 2010. Available at: http://www.iso.org/iso/iso_catalogue/catalogue_tc/catalogue_detail.htm?csnumber=40884 (accessed January 21, 2011).

16   International Organization for Standardization (ISO). ISO 10993-11: Tests for systemic toxicity. TC 194: Biological Evaluation of Medical Devices, 2006. Available at: http://www.iso.org/iso/iso_catalogue/catalogue_tc/catalogue_detail.htm?csnumber=35977 (accessed January 21, 2011).

17   Code of Federal Regulations. Food and drugs. Parts 174–182, 184, and 189, Title 21.

18   European Pharmacopoeia. Chapter 3. Materials and containers.

19   U.S. Pharmacopeia (USP). Chapter 660. Containers-glass.

20   U.S. Pharmacopeia (USP). Chapter 661. Containers-plastic.

21   U.S. Pharmacopeia (USP). Chapter 381. Elastomeric closures for injections.

22   Guidance for industry: Pharmaceutical quality of inhalation and nasal products. Health Canada, 2006. Available at: http://www.hc-sc.gc.ca/dhp-mps/prodpharma/applic-demande/guide-ld/chem/inhalationnas-eng.php.

23   European Medicines Agency, Committee for Medicinal Products for Human Use (CHMP). Guideline on the pharmaceutical quality of inhalation and nasal products, 2006. Available at: http://www.ema.europa.eu/docs/en_GB/document_library/Scientific_guideline/2009/09/WC500003568.pdf (accessed February 3, 2011).

24   Center for Devices and Radiological Health. Reviewer guidance for nebulizers, metered dose inhalers, spacers and actuators, 1993. Available at: http://www.fda.gov/downloads/MedicalDevices/DeviceRegulationandGuidance/GuidanceDocuments/ucm081293.pdf (accessed February 3, 2011).

25   Draft guidance for industry: Metered dose inhaler (MDI) and dry powder inhaler (DPI) drug products. Department of Health and Human Services, Food and Drug Administration Center for Drug Evaluation and Research (CDER), 1998.

26   Guidance for industry: Nasal spray and inhalation solution, suspension, and spray drug products—Chemistry, manufacturing, and controls documentation. Department of Health and Human Services, Food and Drug Administration Center for Drug Evaluation and Research (CDER), 2002.

27   Norwood, D.L. and Ball, D. Product Quality Research Institute: Safety thresholds and best practices for extractables and leachables in orally inhaled and nasal drug products. Submitted to the PQRI Drug Product Technical Committee, PQRI Steering Committee, and U.S. Food and Drug Administration by the PQRI Leachables and Extractables Working Group, 2006.

28   ICH harmonised tripartite guideline: Q3B(R2) impurities in new drug products. International Conference on Harmonisation of Technical Requirements for Registration of Pharmaceuticals for Human Use, 2006.

29   European Parliament and Council Directive: 2003/32/EC. *Official J EU L* 105, 26.4.2003, pp. 18–23.

30   International Organization for Standardization (ISO). ISO 22442-1: Application of risk management. TC194/SC1: Medical devices utilizing animal tissues and their derivatives, 2007. Available at: http://www.iso.org/iso/iso_technical_committee.html?commid=363330 (accessed February 3, 2011).

31   European Commission. CPMP/EMEA 410/01. Note for guidance on minimizing the risk of transmitting animal spongiform encephalopathy agents via human and veterinary medicinal products. *Official J EU C* 73, 5.3.2011, pp. 1–18.

32   European Commission. MEDDEV 2.11/1 guideline on medical devices, 2008.

33   Health Canada. Notice to manufacturers of licensed class II, III, and IV medical devices: DEHP content, 2008. Available at: http://www.hc-sc.gc.ca/dhp-mps/alt_formats/hpfb-dgpsa/pdf/md-im/md_notice_im_avis_dehp_bpa-eng.pdf (accessed February 3, 2011).

34   Health Canada. Notice to manufacturers of licensed class II, III, and IV medical devices: BPA content, 2008. Available at: http://www.hc-sc.gc.ca/dhp-mps/alt_formats/hpfb-dgpsa/pdf/md-im/md_notice_im_avis_dehp_bpa-eng.pdf (accessed February 3, 2011).

35   European Parliament and Council Directive 2007/47/EC Medical Devices Directive, 2007.

36   International Pharmaceutical Aerosol Consortium on Regulation and Science. IPAC-RS Regulatory Requirement Toolkit for OINDP, 2010. Available at: http://www.ipacrs.com/PDFs/Regulatory%20Requirements%20Toolkit.pdf (accessed November 12, 2010).

37   Code of Federal Regulations. Definitions. Part 3.2(e), Title 21.

38   Definition of a Drug Delivery System. *AAPS Newsmagazine*, March 2007, p. 38.

39   Federal Food, Drug and Cosmetic Act. Chapter II—Definitions. Section 201 [21 U.S.C. 321], 1(h), 2009.

40   European Parliament and Council Directive: 93/42/EEC as amended, medical devices directive.

41   Elastomer, 2010. Available at: http://en.wikipedia.org/wiki/Elastomer (accessed January 21, 2011).

42   Morton, M., ed. *Rubber Technology*, 3rd ed. Kluwer Academic Publishers, Dordrecht, 1999.

43   Plastic, 2011. Available at: http://en.wikipedia.org/wiki/Plastic (accessed January 21, 2011).

44   Boyd, R.K., Basic, C., and Bethem, R.A. *Basic Quantitative Analysis by Mass Spectrometry*. John Wiley & Sons, Chichester, England, 2008.

45   Jenke, D. *Compatibility of Pharmaceutical Products and Contact Materials—Safety Considerations Associated with Extractables and Leachables*. John Wiley & Sons, Hoboken, NJ, 2009.

46   Swartz, M.E. and Krull, I.S. *Analytical Method Development and Validation*. Marcel Dekker, New York, 1997.

47   Norwood, D.L., Paskiet, D., Ruberto, M., Feinberg, T., Schroeder, A., Poochikian, G., Wang, Q., Deng, T.J., DeGrazio, F., Munos, M.K., and Nagao, L.M. Best practices for extractables and leachables in orally inhaled and nasal drug products: An overview of the PQRI recommendations. *Pharm Res* 2008, 25(4), pp. 727–739.

48   Green, R. Lean: What it is, where it started, and where it might be going. Bare Bones Production, *Quality Digest*, February 2002.

49   Guidance for industry: PAT—A framework for innovative pharmaceutical development, manufacturing and quality assurance. U.S. Department of Health and Human Services, Food and Drug Administration Center for Drug Evaluation and Research (CDER), September 2004.

50   Scott, B. Process analytical technology in the pharmaceutical industry: A toolkit for continuous improvement. *PDA J Pharm Sci Technol* 2006, 60(1), pp. 17–53.

51   ICH harmonised tripartite guideline: Q8(R2)pharmaceutical development. International Conference on Harmonisation of Technical Requirements for Registration of Pharmaceuticals for Human Use, 2009.

52   ICH harmonised tripartite guideline: Q9 quality risk management. International Conference on Harmonisation of Technical Requirements for Registration of Pharmaceuticals for Human Use, 2005.

53   ICH harmonised tripartite guideline: Q10 pharmaceutical quality system. International Conference on Harmonisation of Technical Requirements for Registration of Pharmaceuticals for Human Use, 2008.

54   Pharamaceutical cGMPs for the 21st century: A risk based approach. Final Report, September 2004.

55 Winkler, H. Evolution of the global regulatory environment: A practical approach to change. Presented at *PDA/FDA Joint Regulatory Conference*, 2007. Available at: http://www.fda.gov/downloads/AboutFDA/CentersOffices/CDER/ucm103453.pdf (accessed November 10, 2008).

56 Van Arnum, P.A. FDA perspective on quality by design, pharmaceutical technology sourcing and management, December 5, 2007.

57 Juran, J.M. *Juran on Quality by Design: The New Steps for planning quality into goods and services.* Simon and Schuster, New York, 1992.

58 Snee, R.D. Implementing quality by design. *Pharm Pro Magazine*, March 23, 2010.

CHAPTER 11

CHEMICAL AND PHYSICAL ATTRIBUTES OF PLASTICS AND ELASTOMERS: IMPACT ON THE EXTRACTABLES PROFILE OF CONTAINER CLOSURE SYSTEMS

Michael A. Ruberto, Diane Paskiet, and Kimberly Miller

11.1 INTRODUCTION

Components of pharmaceutical container closure systems and medical devices can be fabricated with a wide variety of materials. Polymers, such as plastics or elastomers (rubber), are most commonly used in these applications and represent a very broad class of materials with many different properties that can provide functionality and security, as well as aesthetics to primary packaging systems. Metals, such as stainless steel or aluminum, and glass are also used in container closure systems but to a lesser degree than polymers. Some packaging structures include one or more layers of polymer combined with a layer of metal foil, or, coated onto a metal or glass container. In any case, a container closure system or delivery device may be the source of leachable or extractable compounds. The types and origins of these compounds in polymeric materials are the focus of this chapter, while those from inorganic materials are dealt with in Chapter 20.

Polymer components are lightweight, flexible, and often more durable than traditional metal or glass and can be modified by the addition of polymer additives to exhibit many of the desirable properties of these conventional materials, such as strength and clarity. The fairly recent direct consumer marketing and advertising campaigns for drug products can also be enhanced by the use of plastics and elastomers since these materials can be easily colored by the use of pigments or dyes to create a unique package for the drug product with a subsequent brand image. Typical

Leachables and Extractables Handbook: Safety Evaluation, Qualification, and Best Practices Applied to Inhalation Drug Products, First Edition. Edited by Douglas J. Ball, Daniel L. Norwood, Cheryl L.M. Stults, Lee M. Nagao.
© 2012 John Wiley & Sons, Inc. Published 2012 by John Wiley & Sons, Inc.

pharmaceutical polymeric components are bottles, vials, caps, connectors, bags, blister packs, syringes, stoppers, and seals. Additionally, inhalation products often incorporate molded plastic components into the delivery system as mouthpieces, holding chambers, or mechanical assemblies used in the dosing mechanism.

Although polymers are often the materials of choice for use in container closure systems and have many advantages over metal and glass, they are not without shortcomings of their own. In the presence of light, heat, oxygen, and other environmental factors, polymers can degrade. The degree of polymer degradation depends on the nature of its chemical composition, the manner in which it is processed or molded, and the end use of the finished products. For example, the inherent stability of a polymer substrate will be influenced by its molecular structure, polymerization process, presence of residual catalysts, and finishing steps used in its production. The processing conditions during extrusion such as temperature, shear, and residence time in the extruder can dramatically impact polymer degradation. End-use conditions that expose the polymer to excessive heat or light, such as outdoor applications, or sterilization techniques used in medical applications can foster premature failure of polymer products as well. This degradation can manifest itself visually as discoloration, followed by cracking and flaking, and physically as loss of flexibility or strength. If left unchecked, the result can often be total failure of the polymer component, with subsequent impact on the pharmaceutical container closure system or device, and drug product performance.

Polymer degradation can be controlled by the addition of chemical additives into the plastic or elastomer system. Additives are specialty chemicals that provide a desired effect to a polymer. This effect can be one of stabilization that allows a polymer to maintain its strength and flexibility, or it can be a performance improver that adds color or special characteristics to a substrate such as antistatic or antimicrobial properties. There are typically three classes of stabilizers: (1) melt processing aids such as phosphites and hindered phenols that are antioxidants and protect the polymer during extrusion and molding; (2) long-term thermal stabilizers that provide defense against heat encountered in end-use applications, and include hindered phenols and hindered amines; and (3) light stabilizers that provide UV protection through mechanisms such as radical trapping, UV absorption, or excited-state quenching.

Another limitation of polymers is that they do not possess the barrier properties of glass and metal. Some substrates are very permeable to small molecules such as residual solvents found in secondary packaging components or inks, adhesives, and photoinitiators used on labels. This can often make the drug that is in a plastic container closure system susceptible to leachables from secondary packaging components or labeling on the primary package. Flexible foil laminates can be used to isolate container closure systems from the chemicals present in secondary packaging, especially for products that have been historically problematic such as polyethylene-packaged inhalation drug products. However, the foil laminates themselves may contribute residual solvents or compounds that result from sealing, since they are either heat sealed or joined together by an adhesive. Judicious selection of labeling materials can minimize the potential leachables from labels.[1] In some cases, the primary packaging can be composed of multilaminate systems that consist of several polymer substrates held together by an adhesive, for example, a blister pack.

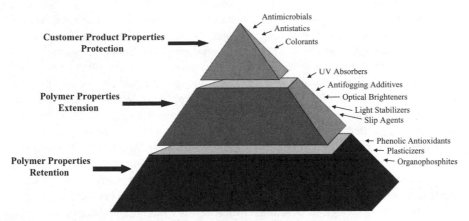

Figure 11.1   Roles and types of polymer additives.

This combination of polymer substrates, especially when a layer of foil is part of the laminate, often improves the performance of the container closure system and subsequently provides protection to the drug product. The trade-off is that care must be taken in the selection of the laminates to minimize potential leachables. When utilizing polymers in packaging or device applications, one must be mindful of three aspects, which are graphically represented in Figure 11.1: (1) retaining the properties of the polymer; (2) extending the properties of the polymer; and (3) protecting the contents of the packaging or branding the pharmaceutical product. Retaining the properties of polymers is accomplished by the addition of stabilizers as described above. Other additives can also be included to extend the properties of a polymer to make it especially useful for packaging, such as light stabilizers, slip agents, clarifiers, or antifogging agents. Specialty additives are often used to protect the contents of polymer packaging or facilitate proper functioning of the device. This can include providing packaging with antimicrobial properties to maintain sanitary conditions or using antistatic agents in the mouthpiece of a dry powder inhaler (DPI) to ensure that the drug product powder flows evenly through the component and no clinging or clumping occurs. In mechanical assemblies, it is sometimes desirable to extend the properties of the polymer by adding other materials such as glass fibers to extend strength or polytetrafluoroethylene (PTFE) to improve lubricity. Each of these materials in turn may contain their own additive package, thus adding to the total number of additives in the final material.

To protect the contents of the packaging or facilitate branding, color is often used. Color can be used to absorb certain wavelengths of light to protect the contents of a vial or bottle from photodegradation or to standardize and even personalize packaging like amber pill bottles with a colored band to indicate a prescription medication that belongs to a specific individual in a household. Color is also widely used for branding purposes. This may occur in the form of a colorant added to a plastic or ink applied to a surface. For example, a different color plastic may be used for the same delivery system to differentiate either the dosage strength or the drug that it contains. Great care is often taken to ensure that a certain shade of color is consistent throughout all of the components of a packaging system. If each component is fabricated from a different polymer substrate, obtaining the same color throughout is not a trivial task. There are several commercial color matching computer programs that will help to ensure a harmonized color scheme throughout all the components of a packaging system. The solution often utilizes several different pigments either alone or in blends to capture the same color palette in each substrate. Color may be applied to the surface of a pharmaceutical product by utilizing laser marking or inks. In the case of laser marking, different additives may be incorporated into the polymer substrate to obtain different colors. Inks may be used on molded components or polymer laminates not only for labeling but also for purposes of brand recognition. In each case where color is added to a pharmaceutical product, there is an introduction of additives that may later be found as leachables.

An often neglected or overlooked source of leachables and extractables is the fabrication process itself. Many polymeric components are molded, blown, or cut into the desired shape. Processing aids such as mold release, talc, or cleaning solvents are commonly used. Postprocessing steps may include heat treatment, washing, autoclaving, or gamma irradiation. Assembly of certain components may require the incorporation of adhesives or lubricating oils to ensure proper functioning. If the fabrication process includes external or added chemicals, these may contribute to the trace-level chemicals identified as part of an extractables or leachables profile.

Polymer degradants, additives, colorants, and processing aids are typical categories of compounds associated with polymers and may be extracted under various conditions. These chemicals or their breakdown products may be found in an extractables profile of the polymeric component or in the leachables profile of the drug product. The physical state of the polymer may also play a role in determining the nature of the extractables or leachables profile. The origin of extractables and leachables from polymeric components can best be understood by taking a more in-depth analysis of the materials used in container closure systems, including their chemistry, processing, and uses. Each of these aspects will be discussed in the sections that follow. It is important to develop an understanding of the polymer-associated factors that impact the extractables or leachables profile so that the material selection process is effective and the interpretation of profiles is efficient.

11.2 POLYMER CHEMISTRY

The behavior of plastics and elastomers is inherent to the polymerization process, chemical composition, morphology, and rheological characteristics. Many textbooks have been written on these subjects, both individually and collectively, and the reader is referred to recommended texts at the end of this chapter.[2] The purpose of this section is to describe the fundamentals of polymer chemistry, its influence on physical properties, and the contributions that these might make to extractables or leachables profiles. The nature of typical polymers used for pharmaceutical packaging and medical devices will be described in terms of their polymerization process, structure, and properties that are related to the potential for extractables and leachables.

11.2.1   Basic Polymer Chemistry

A polymer is a large macromolecule composed of many smaller repeating units. This "buildup" of monomers can occur in a linear fashion, like a chain, or take on a three-dimensional framework having a branched configuration. Regardless of the configuration, each repeating unit of a polymer is usually composed of the starting material or monomer from which the polymer is formed. Polymeric macromolecules can exist as natural products as well as synthetic analogs having a variety of properties with many applications in the pharmaceutical and medical device industry, for example, delivery system components, container closures, flexible films, adhesives, and coatings.

Most polymers are made up of carbon, hydrogen, and oxygen; nitrogen, silicon, chlorine, fluorine, bromine, phosphorus, and sulfurs are also used. Molecular weight is customarily used to represent the molecular size. Since polymeric materials can be mixtures of different size molecules with varying degrees of polymerization, the chain lengths can vary by thousands of monomer units, therefore polymer molecular weights (MWs) are given as averages. An example calculation of polymer molecular weight is[4]:

MW ethylene $(28) \times (1000)$ repeat units = 28,000 MW polyethylene.

A number average MW, M_n, is calculated from the mole fraction distribution of different sized molecules; the weight average MW, M_w, is calculated from the weight fraction distribution. Weight average is skewed to higher values for larger molecules and it is always greater than the number average. M_w impacts the fabrication process, while M_n attributes to the mechanical properties such as tensile strength or brittleness. The ratio of M_w/M_n is a measure of the polydispersity of the system. Typically, polymers with a wide distribution are more stiff/brittle than those with a narrow distribution. Most polymers used in plastics or elastomers have MWs in the range of 10,000 to several million.

There are two mechanisms for polymerization to occur—addition (chain growth) or condensation (step growth). Addition polymerization is the combination of monomers by exothermic reaction of carbon–carbon double bonds (C=C) with no loss or gain of material. The double bond will open when exposed to an initiator and the reaction propagates until it is terminated. A common type of an initiator is a free radical, a compound with an unpaired electron, such as a peroxide or azo compound. Termination of a growing chain can occur by reaction with another free radical or growing chain. Trace amounts of contamination can also terminate the polymer chains, resulting in low MWs. Ionic chain growth polymerization can occur using carbocations or carbanion as well as catalytic polymerization using transition metal complexes. Residual chain initiators or chain terminators, catalysts, monomers, or oligomers may be subsequently extracted.

Condensation polymers are formed by a series of chemical reactions often requiring heat. These reactions involve two active sites that join to form a chemical bond, and typically, by-products are formed. For example, as shown in Figure 11.2, polyethylene terephthalate is formed by the reaction of terephthalic acid with ethylene glycol to form water and ethylene terephthalate monomers, which then react with each other. Terminal functional groups on the chain remain active, and the shorter chains combine into longer chains in the late stages of polymerization. The

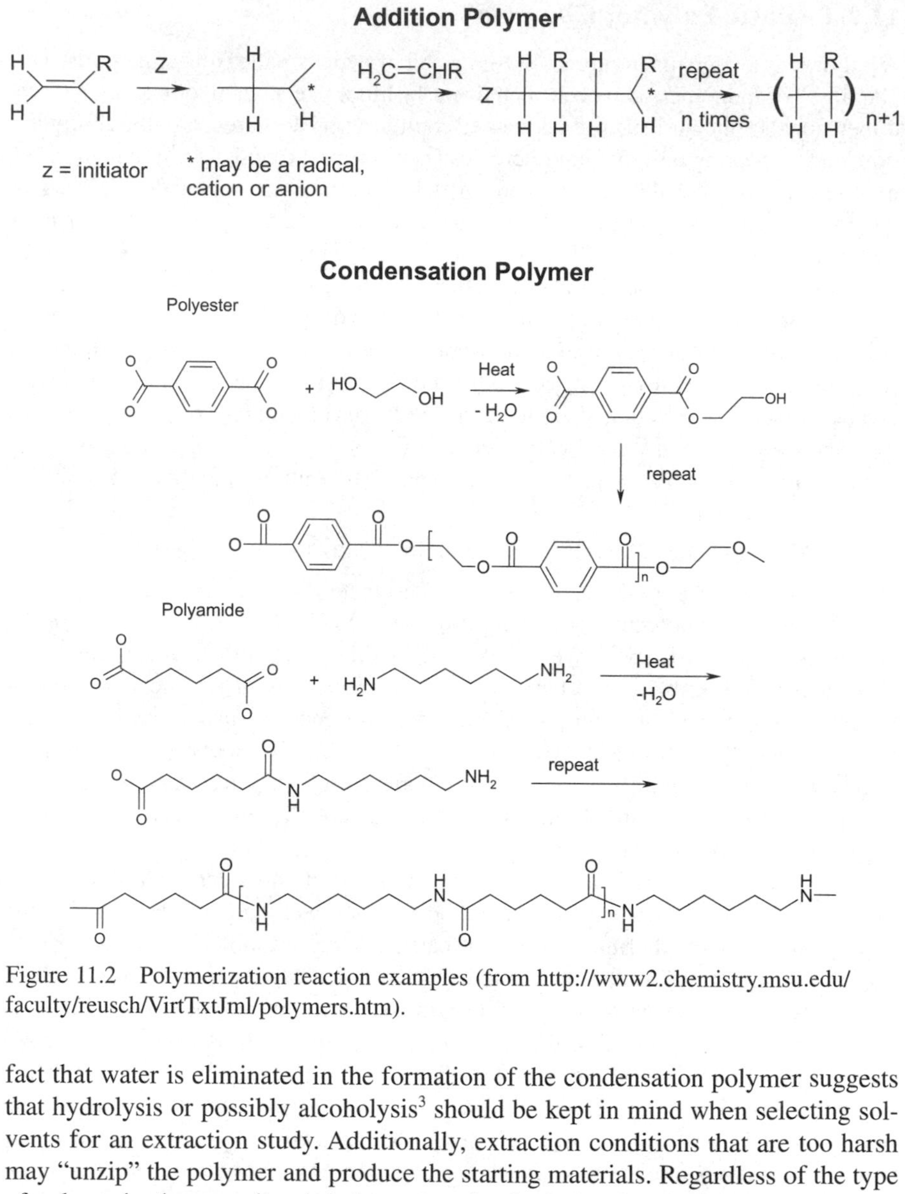

Addition Polymer

z = initiator   * may be a radical,
cation or anion

Condensation Polymer

Polyester

Polyamide

Figure 11.2   Polymerization reaction examples (from http://www2.chemistry.msu.edu/faculty/reusch/VirtTxtJml/polymers.htm).

fact that water is eliminated in the formation of the condensation polymer suggests that hydrolysis or possibly alcoholysis[3] should be kept in mind when selecting solvents for an extraction study. Additionally, extraction conditions that are too harsh may "unzip" the polymer and produce the starting materials. Regardless of the type of polymerization reaction, it is important that both the chemical starting materials and polymerization by-products be considered as potential leachables.

11.2.2   Polymer Structure

The polymerization process produces long chains that are subsequently organized in a three-dimensional space. The configuration of polymer chains will vary based on the location of the substituent groups. The carbon chain symmetry defined by the relative position of each substituent group results in three configurational isomers: isotactic (all substituents on one side of the chain); syndiotactic (substituents

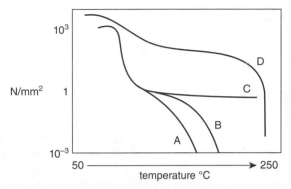

Figure 11.3   Elastic modulus variation with temperature for different polystyrenes. (A) Atactic polystyrene (PS), $M_w = 140,000$; (B) atactic PS, $M_w = 217,000$; (C) semicrystalline isotactic PS; (D) elastomeric lightly cross-linked atactic PS (adapted from http://gertrude-old.case.edu/276/materials/04.htm).

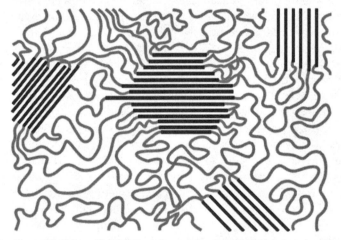

Figure 11.4   Crystalline (parallel lines) and amorphous (coiled lines) regions in a semicrystalline material (based on http://www2.chemistry.msu.edu/faculty/reusch/VirtTxtJml/polymers.htm).

alternate from side to side); and atactic (random). Several properties of a polymer will vary greatly depending on its tacticity. For example, as shown in Figure 11.3, the tacticity of different polystyrene species significantly affects the elastic modulus behavior with temperature. Curve D for the isotactic polymer shows a notably different pattern than curves A and B for the atactic polymers. This is because the packing of the chains is different in each species and results in varied degrees of crystallinity (i.e., ordered, symmetrical packing of chains).

The terms crystalline and amorphous are used to describe the ordered and disordered regions in a polymer system, respectively.[4] Figure 11.4 illustrates the microenvironment of a semicrystalline polymer where there is a mixture of highly ordered, lamellar, densely packed chains and disordered, loosely packed random coils. Alternatively, depending on the polymer, it may adopt an entirely crystalline or amorphous form. Factors that can influence the degree of crystallinity include

- chain length,
- chain branching, and
- interchain bonding.[5]

If a polymer chain has structural regularity and is long with little branching, it will have some degree of flexibility, and will tend to pack in such a way as to produce an ordered, crystalline morphology. The degree of secondary intermolecular forces, such as those from polar side groups or hydrogen bonding, will influence the stability of the chains and their ability to assume a more crystalline morphology. In contrast, if a polymer chain is highly branched or has random substitutions, it will be more likely to take on a disordered, amorphous morphology.

In addition to the spatial considerations that play a role in determining the morphology of a polymer, the time domain must be considered. As a molten polymer is cooled, its vibrational, rotational, and translational energies are reduced to a point where it is possible for the regions of a polymer that have a favored symmetry to pack into an organized lattice framework and crystallize. The temperature at this point is the melt temperature, T_m, and, for a purely crystalline polymer, it would also be considered the temperature of crystallization. Alternatively, if a polymer is cooled very rapidly and there is insufficient time or symmetry for long-range order to be established, a glassy state may be created. The temperature at this point is the glass transition temperature, T_g, which is dependent on the cooling rate and the moisture content of the polymer. When a polymer has both crystalline and amorphous regions, it is possible that the amorphous regions may be converted to crystalline regions by heating the material above the T_g to allow rotational motion sufficient to create long-range order. The temperature at this point is the crystallization temperature, T_c, and will be below the T_m. An example thermogram depicting these transitions is shown in Figure 11.5. The majority of plastics and elastomers will have both a T_m and T_g. Knowing the T_m and T_g for a specific polymer can help to provide some insight as to the physical characteristics (such as flexibility or rigidity) of the substrate at a given external temperature. Table 11.1 provides a list of T_m and T_g values for various

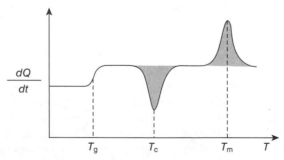

Figure 11.5 Thermal transitions in a semicrystalline polymer. Data are plotted as heat flow (dQ/dt) against temperature (based on http://faculty.uscupstate.edu/llever/Polymer%20 Resources/DSC.htm).

TABLE 11.1.   Some Reported Glass Transition and Melt Temperatures of Selected Polymers Used in Pharmaceutical Applications

Polymer name (abbreviation)	T_g (°C)	T_m (°C)	Typical morphology[f]
Low-density polyethylene (LDPE)[a]	−110	110	Amorphous
High-density polyethylene (HDPE)[a]	−100	130	Semicrystalline
Polypropylene (PP)[a]	5	150	Semicrystalline
	−10	175	Crystalline
	−20 (atactic)		
	0 (isotactic)		
	−8 (syndiotactic)		
Polytetrafluoroethylene (PTFE)[a]	−115	327	Semicrystalline
	−110	330	
Nylon 6[a]	50	215	Semicrystalline
	53	223	
Polyethlyene terephthalate (PET)[a]	67	256	Semicrystalline
	70	265	
Polycarbonate (PC)[a]	150	267	Amorphous
Polyurethane (PU)[a]	52	–	–
Polybutylene terephthalate (PBT)[c−e]	52–66	224	Semicrystalline
Polyvinyl chloride (PVC)[a]	80	180	Amorphous
Polystyrene (PS)[a]	90	175	Amorphous
Polymethylmethacrylate (PMMA)[a]	105	180	Amorphous
	100 (atactic)		Crystalline
	130 (isotactic)		
	120 (syndiotactic)		
Butyl rubber (IIR)[b]	−63	N/A	Amorphous
Ethylene-propylene-diene (EPDM)[b]	−63	N/A	Amorphous
Natural rubber (NR)[b]	−58	N/A	Amorphous
Isoprene rubber (IR)[b]	−60	N/A	Amorphous
Styrene-butadiene rubber (SBR)[b]	−47	N/A	Amorphous
Chloroprene rubber (CR)[b]	−36	N/A	Amorphous

[a] http://www2.chemistry.msu.edu/faculty/reusch/VirtTxtJml/polymers.htm.

[b] Sircar A.K. and Chartoff, R.P. Measurement of the glass transition temperature of elastomer systems. In: R.J. Seyler, ed. *Assignment of the Glass Transition, ASTM STP 1249.* American Society for Testing and Material, Philadelphia, 1994; 226–238.

[c] Fakirov, S., Balta Calleja, F.J., and Krumova, M. *J Polym Sci B Polym Phys* 1999, *37*, pp. 1413–1419.

[d] http://www.polymerprocessing.com/polymers/PBT.html.

[e] Brunelle, D.J. *J Polym Sci A Polym Chem* 2008, *46*, pp. 1151–1164.

[f] Rodriguez, E.L. The glass transition temperature of glassy polymers using dynamic mechanical analysis. In: R.J. Seyler, ed. *Assignment of the Glass Transition, ASTM STP 1249.* American Society for Testing and Material, Philadelphia, 1994; 255–268.

N/A, not applicable.

types of polymers. A crystalline polymer looses its strength above the T_m, and a similar loss occurs for an amorphous entity above its T_g. Additional strength may be imparted to a semicrystalline polymer by annealing it at a temperature above the T_g (but below the T_m) so that the rigid amorphous region can be reduced.[6]

The degree of crystallinity can vary by polymer type and will govern many of its properties. The properties of a polymer are also influenced by the way crystallization has taken place—processing operations are critical to the nucleation and growth rate and nucleating agents may be necessary. Crystalline polymers are opaque and rigid with low permeation to gases and moisture, while amorphous polymers have good clarity and toughness but are less inert. Amorphous polymers have good impact resistance and are flexible when the temperature of use is above T_g. The hardness of an amorphous entity is directly proportional to the T_g.[7] When crystalline domains alternate with amorphous segments, the resulting material is strong and stiff yet retains a degree of flexibility. Migration is more likely to occur in lightly cross-linked amorphous polymers and in semicrystalline polymers at temperatures above their glass transition temperatures, following Fick's law[8] of diffusion for a simple system.

In more complex systems, the polymer structure is not only a matter of physical orientation of the polymer chains, but also chemical bonding. Two or more polymer chains can join together by chemical reactions of unsaturated atoms or functional groups containing heteroatoms in the side chains to form a network structure. Alternatively, cross-linked polymers can often form from side reactions during polymerization. Two different types of cross-linked structures are shown in Figure 11.6. In some cases, particularly elastomers, sites of unsaturation in the polymer backbone can react with a curing agent to actively introduce cross-linking. Historically, sulfur-based curing agents were used for cross-linking (note: "vulcanization" is a term that was originally used to describe sulfur-based curing processes, but is commonly applied to any curing process today), and more recently, peroxide- or platinum-based curing agents have been employed. The polymer structure is significant when consideration is given to developing appropriate extraction

(a)   (b)

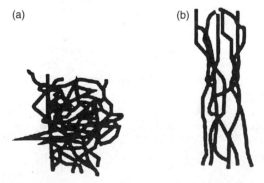

Figure 11.6   Cross-linked structures in polymers. (a) Short-chain cross-link network. (b) Long-chain cross-link network.

conditions. Amorphous entities are traditionally more readily solubilized than crystalline entities, and this should be considered when choosing extraction solvents. Polymers that are more prone to migration will be easier to extract. This along with the hardness of a polymer has implications for sample preparation—the greater the hardness, the more likely the sample will need to undergo particle size reduction, extended extraction times, or higher temperatures. The T_g and T_m must be taken into consideration to maximize mobility of chemical constituents out of the polymer but minimize polymer degradation when identifying appropriate extraction temperatures. When evaluating cross-linked polymers for potential leachables, it is important to be aware of curing agents or reaction by-products that may be present.

11.3 MATERIAL PROPERTIES

A suitable container closure system must protect the drug product, function properly, and be safe for use in the intended application. When selecting polymer materials for use in pharmaceutical packaging or delivery devices, many characteristics must be considered such as resistance to chemicals or biocompatibility; however, the main attribute that will dictate the utility of a polymer material for any application is its mechanical properties over the temperature range of use. Remember that a container closure system acts as a receptacle for the drug formulation and often is an integral part of a drug delivery device. Protection of the drug product is paramount, so that mechanical properties such as impact resistance are important qualities for a polymer to possess. However, polymer components of container closure systems often require a desired flexibility (or rigidity) if they are expected to accurately deliver exact quantities of drug over the lifetime of the packaging or delivery system. The mechanical behavior of a polymer is evaluated by its deformation or flow characteristics under a given stress or strain at a specified temperature.

There are four attributes that characterize this stress–strain relationship as depicted in Figure 11.7:

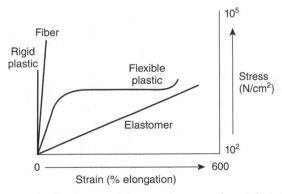

Figure 11.7   Stress–strain plot for material properties comparison (adapted from Odian, G. *Principles of Polymerization*, 3rd ed. John Wiley, New York, 1991).

- modulus—the resistance to deformation (slope of the curve),
- tensile strength—the stress required to rupture a polymer (maximum stress),
- ultimate elongation—the extent of elongation at the point where the sample ruptures (maximum strain), and
- elastic elongation—the elasticity as measured by the extent of reversible elongation (linear portion of the curve).

Properties such as T_m, T_g, degree of crystallinity, and degree of polymer cross-linking all impact the mechanical behavior of the material. Varying these combinations will determine the utility of the polymer—it may be a fiber, rigid plastic, flexible plastic, or elastomer. For example, synthesizing a polymer with a high degree of crystallinity or cross-linking will result in a substrate having good strength, but low extensibility (ability of a material to stretch without tearing). This type of material might be used as a rigid plastic. Synthesizing an amorphous polymer having a low T_g and sufficient cross-linking to make it elastic would result in a flexible material.

Rigid plastics, like polystyrene and polycarbonate, can withstand a good deal of stress, but will not deform or elongate to a great extent before breaking. Flexible plastics, such as polyethylene, do not resist deformation as well as rigid plastics, but they will not break as easily. Their ability to deform allows the flexible plastics to absorb more energy than rigid plastics prior to breaking, making them tougher. The other two polymer types in the plot above are elastomers and fibers. Elastomers, like a rubber band, have a very low moduli and deform quite easily. The mark of a good elastomer is not a high degree of elongation alone—it should not remain in the deformed state after the stress is removed, but must, instead, revert back to its original form. Therefore, elastomers have highly reversible elongation or elastic elongation. Fibers are very similar to rigid plastics in that they have a high moduli. A highly oriented polymer, in most cases, will not have the toughness of flexible plastics, but it is superior for strength. Fibers, such as Kevlar™*, a highly oriented aromatic polyamide, can have better tensile strength than steel.

Most polymers require additives in addition to those used for polymerization or cross-linking to create the desired performance characteristics of the material. Plastic and elastomers are the terms commonly associated with pharmaceutical and medical polymers. The term plastic infers plasticity, the ability to be formed by a thermomechanical process, and elastomers are rubberlike polymers and generally extensible, compliant, and return to the original shape after deformation.

The chemical and physical nature of a polymer material will indicate suitability for use based on certain properties such as ability to flow, resistance to heat, crystalline/transparent nature, mechanical strength, and resistance to solvents. Physical and chemical properties can be used to distinguish three types of polymer material types: thermoplastic, thermoset, and elastomer (see Table 11.2). A fourth type of structure, the thermoplastic elastomer (TPE), has been developed to incorporate the properties of a thermoplastic and an elastomer.

---

* Kevlar™ is a trademark of E. I. du Pont de Nemours and Company.

TABLE 11.2. Polymer Material Types

Structure	Melt	Cross-linking	Deformation
Thermoplastic	Reversible	None	Resistant
Thermoset	One time then decomposition	Not reversible	Resistant
Elastomer	One time then decomposition	Not reversible	Not resistant
Thermoplastic elastomer	Reversible	Reversible	Not resistant

11.3.1 Plastic Materials

Plastics can be thermoplastic or remeltable, while thermosetting plastics chemically react during processing and subsequently are generally nonmelting. Plastic performance characteristics are a function of the polymer structure. Stiffness, toughness, and strength are a measure of its mechanical properties, and the ability of the polymer to bend and deform without breaking is a measure of its elasticity. Typically, mechanical properties such as ductility, tensile strength, and hardness increase and will then level off as the chain grows. Linear polymers have long chains with few side chains and are soft and tough with high creep (i.e., dimensional change with time under a load). Polyethylene and PTFE (also called Teflon®*) are examples of linear polymers. Branched polymers can have several side chains making the polymer material hard and tough with medium creep, such as polypropylene and polymethylmethacrylate. Side rings like that of polystyrene make the polymer material hard and brittle with low creep; polycarbonate and polysulfone have rings incorporated into the polymer backbone that make the polymer material hard and tough with low creep.

Linear and branched polymeric materials are usually thermoplastic in nature and can be melted and molded into many different shapes and forms with the application of heat and pressure. Saturated polymers are more resistant to UV radiation than unsaturated polymers—the double bonds are subject to oxidation or cross-linking. Thermosets involve polymer cross-linking that is initiated by heat or light and may take place during fabrication of the final article. Polymers that are highly cross-linked will be more stable to physical stress and heat. Cross-linked polymers, or thermosets, cannot be solubilized and can be so resistant to heat that they do not melt at elevated temperatures, but decompose under very extreme conditions. Cross-linked plastics are usually brittle at room temperature and will maintain the same properties at cryogenic conditions; this is typical behavior for amorphous materials at temperatures below the T_g. Brittleness is observed when the material looses its strength at yield indicating low-impact strength. Brittleness can be an advantage in certain applications.

Properties of plastics can be improved by the addition of various fillers, additives, processing aids, and reinforcements. Hundreds of materials or combinations thereof can be used in any given formulation. Strength and stiffness are critical to the performance of a container closure system, and as temperature increases, the strength decreases, which can be improved by using silicate type fillers or fibers as reinforcements. Adhesion promoters or coupling agents are typically used to create

* Teflon® is a registered trademark of E. I. du Pont de Nemours and Company.

a high filler–resin bond. Plasticizers are liquid or low MW solids added to a plastic formulation to increase flexibility and processability. Consider polyvinyl chloride (PVC) that is used in the home for water pipes. This material is very rigid by nature. However, with the addition of plasticizers, PVC becomes very flexible and can be used as IV bags or inflatable devices. The surface of the polymeric components can be enhanced to prevent sticking to machinery or reduce static charge or wettability. The degree of crystalline morphology can be increased with the use of a nucleating agent. Colorants, biocides, cross-linking agents, blowing agents, and lubricants are other additives that are commonly used to modify plastics. It is the compounding ingredients that may have some solubility in the drug product matrix and have the potential to diffuse out of the polymer system and partition into the drug product or react with a drug product constituent. Knowledge of the identity of the compounding ingredients is a key factor in conducting an extraction study.

Polymers can also be combined by either copolymerization, blending, or making composite materials to obtain more desirable qualities. For example, TPEs exhibit the end-use properties of a cured elastomer and the processing properties of a thermoplastic. Unlike thermosets, TPEs form reversible, physical cross-links that are formed as an elastomeric moiety interspersed in a neighboring plastic domain immobilized at the onset of a thermal phase change. For example, when monomers from a thermoplastic, polystyrene, and an elastomer, butadiene, are copolymerized, a poly(styrene-butadiene-styrene) (SBS) block copolymer can be formed as shown in Figure 11.8. Although polystyrene is a rigid polymer, when polybutadiene particles are physically mixed with it, the result is high-impact polystyrene or HIPS that has both the strength of polystyrene and the toughness of polybutadiene. When polymers are combined, it is important to keep in mind that additives may be required to facilitate miscibility or copolymerization.

Moisture may impact the performance of polymeric materials. Strength and stiffness of plastic materials are often reduced with high moisture absorption. Moisture uptake is directly related to the makeup of the polymer—carbon, hydrogen, and fluorine atoms have very low moisture adsorption properties; oxygen and chlorine slightly increase adsorption levels, while nitrogen increases adsorption significantly.

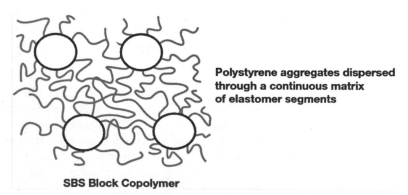

Polystyrene aggregates dispersed through a continuous matrix of elastomer segments

SBS Block Copolymer

Figure 11.8   Cross-linked TPE illustrates the chain network for a block copolymer.

Moisture transmission is affected by the order of the polymer chains, and dimensional stability is directly affected by moisture adsorption. Chemical resistance or permeability to water and other solvents depends on both the chemical composition and amount of cross-linking. A good understanding of the material permeability and adsorption characteristics is important for proper material selection and identification of leachables targets.

Plastic materials vary substantially due to different monomer chemistries, additives, reinforcements, MW, and many other variables. This results in hundreds of families of plastic materials altogether yielding thousands of different grades. This can be an asset for supplier differentiation but a detriment to standardization of pharmaceutical materials that results in a lack of common extractable profiles. Some of the benefits of plastics are ease of fabrication, low cost of tooling, and ability to reinforce for high tensile strength. However, plastics generally have low compression and shear, can fatigue under stress, and performance is limited by temperature extremes. Table 11.3 lists typical plastic materials used in the pharmaceutical industry.

11.3.2 Stabilization of Plastic

As stated above, chemicals are routinely added to polymers to provide a variety of effects, including stabilization. Some of the common stabilizers for plastics are listed in Table 11.4.[9,10] When a plastic resin is processed, it is often introduced into an extruder where it is melted at high temperatures and mixed with a series of screws into a homogenous molten mixture. Additional heat and shear are encountered when the extruded resin is then molded or shaped into its final product form such as a gasket, mouthpiece, or dose metering valve component (for a metered dose inhaler drug product). Stabilizers are often introduced during extrusion and molding to avoid degradation of the polymer under these extreme conditions. Figure 11.9 depicts a schematic of the autoxidation cycle.[11] In this figure, the solid lines represent pathways of degradation and the dashed lines are the paths to stabilization. It is clear from this diagram that the origins of degradation in polymers are radical species such as

R •   alkyl radical,

RO •   alkoxy radical,

ROO •   peroxy radical, and

ROOH   hydroperoxide.

Good stabilizers should, therefore, be efficient radical scavengers. For plastics, a two-tiered approach is often used to protect these polymers from the heat and shear that they encounter during processing. Primary antioxidants, for example, hindered phenols, such as butylated hydroxytoluene (BHT), Irganox®* 1010 or Irganox 1076, are added to the polymer to provide protection during processing as well as the

---

* Irganox is a registered trademark of Ciba Specialty Chemicals Corporation.

TABLE 11.3. Common Pharmaceutical Plastics

Name(s)	Abbreviations	Formula	Uses
Addition polymers			
Polyethylene (low density)	LDPE	$-(CH_2-CH_2)_n-$	Films, overwraps, bags
Polyethylene (high density)	HDPE	$-(CH_2-CH_2)_n-$	Bottles/containers caps/closures
Polypropylene	PP	$-[CH_2-CH(CH_3)]_n-$	Bottles/containers caps/closures
Polyvinyl chloride	PVC	$-(CH_2-CHCl)_n-$	Tubing, bags
Cyclic olefin copolymers	COCs		Containers, syringe barrels well plates
			Films
Polyoxymethylene	POM		Rigid device components
Polytetrafluoroethylene	PTFE	$-(CF_2-CF_2)_n-$	Lubricous coatings/films, containers
Copolymers			
Acrylonitrile butadiene styrene terpolymer	ABS	$(C_8H_8)_x \cdot (C_4H_6)_y \cdot (C_3H_3N)_z$	Device components
High-impact polystyrene	HIPS	$(C_8H_8)_x \cdot (C_4H_6)_y$	Device components
Polymethyl methacrylate	PMMA		Films, clear components

Condensation polymers

			Applications
Polybutylene terephthlate	PBT		Device components
Polyethylene terephthlate	PET		Bottles/containers Films, blister
Polycarbonate	PC		Bottles/containers Films
Polyamide	PA		Films Rigid components
Polyurethane	PU		Films, rigid components

TABLE 11.4. Antioxidants Commonly Used in Plastics

Chemical name	Ciba trade name[a]
Polyolefins	
Pentaerythritol tetrakis(3-(3,5-di-*tert*-butyl-4-hydroxyphenyl)propionate)	Irganox 1010
Octadecyl-3-(3,5-di-*tert*-butyl-4-hydroxyphenyl)-propionate	Irganox 1076
Thiodiethylene bis[3-(3,5-di-*tert*-butyl-4-hydroxyphenyl)propionate]	Irganox 1035
Tris(3,5-di-*tert*-butyl-4-hydroxybenzyl)isocyanurate	Irganox 3114
Tris(2,4-di-*tert*-butylphenyl)phosphite	Irgafos 168
Bis(2,4-di-*tert*-butylphenyl)pentaerythritol diphosphite	Irgafos 126
Product by process CAS #119345-01-6	Irgafos PEP-Q
Engineering plastics (e.g., POM, PA, PC, PET, PMMA)	
Pentaerythritol tetrakis(3-(3,5-di-*tert*-butyl-4-hydroxyphenyl)propionate)	Irganox 1010
Octadecyl-3-(3,5-di-*tert*-butyl-4-hydroxyphenyl)-propionate	Irganox 1076
Ethylenebis(oxyethylene)bis-(3-(5-*tert*-butyl-4-hydroxy-*m*-tolyl)-propionate)	Irganox 245
3,3'Bis(3,5-di-*tert*-butyl-4-hydroxyphenyl)-*N*-*N*'hexamethylenedipropion amide	Irganox 1098
3,5-bis(1,1-dimethylethyl)-4-hydroxy-benzene propanoic acid, isoocty lester	Irganox 1135
Tris(2,4-di-*tert*-butylphenyl)phosphite	Irgafos 168
Bis(2,4-di-*tert*-butylphenyl)pentaerythritol diphosphite	Irgafos 126
Product by process CAS #119345-01-6	Irgafos PEP-Q
Polyurethane	
Pentaerythritol tetrakis(3-(3,5-di-*tert*-butyl-4-hydroxyphenyl)propionate)	Irganox 1010
Octadecyl-3-(3,5-di-*tert*-butyl-4-hydroxyphenyl)-propionate	Irganox 1076
Ethylenebis(oxyethylene)bis-(3-(5-tert-butyl-4-hydroxy-*m*-tolyl)-propionate)	Irganox 245
3,5-bis(1,1-dimethylethyl)-4-hydroxy-benzene propanoic acid, isoocty lester	Irganox 1135
Bis(2,4-di-*tert*-butylphenyl)pentaerythritol diphosphite	Irgafos 126
Polyvinyl chloride	
Pentaerythritol tetrakis(3-(3,5-di-*tert*-butyl-4-hydroxyphenyl)propionate)	Irganox 1010
Octadecyl-3-(3,5-di-*tert*-butyl-4-hydroxyphenyl)-propionate	Irganox 1076
Ethylenebis(oxyethylene)bis-(3-(5-*tert*-butyl-4-hydroxy-*m*-tolyl)-propionate)	Irganox 245

[a] Irganox and Irgafos are registered trademarks of Ciba Specialty Chemicals Corporation.

CAS, Chemical Abstracts Service; POM, polyoxymethylene; PA, polyamide; PC, polycarbonate; PET, polyethylene terephthlate; PMMA, Polymethyl methacrylate.

long-term heat stability that the polymer will require. Secondary antioxidants, such as organic phosphites like Irgafos®* 168, are also added to the plastic as process stabilizers. These compounds are typically hydroperoxide decomposers and protect the polymer during extrusion and molding while also acting as "sacrificial lambs" by defending the primary antioxidants against decomposition (Fig. 11.10).

* Irgafos is a registered trademark of Ciba Specialty Chemicals Corporation.

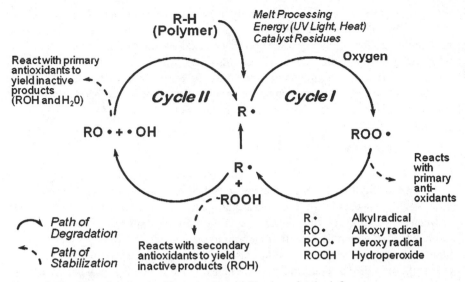

Figure 11.9   The role of antioxidants in the stabilization of polyolefins.

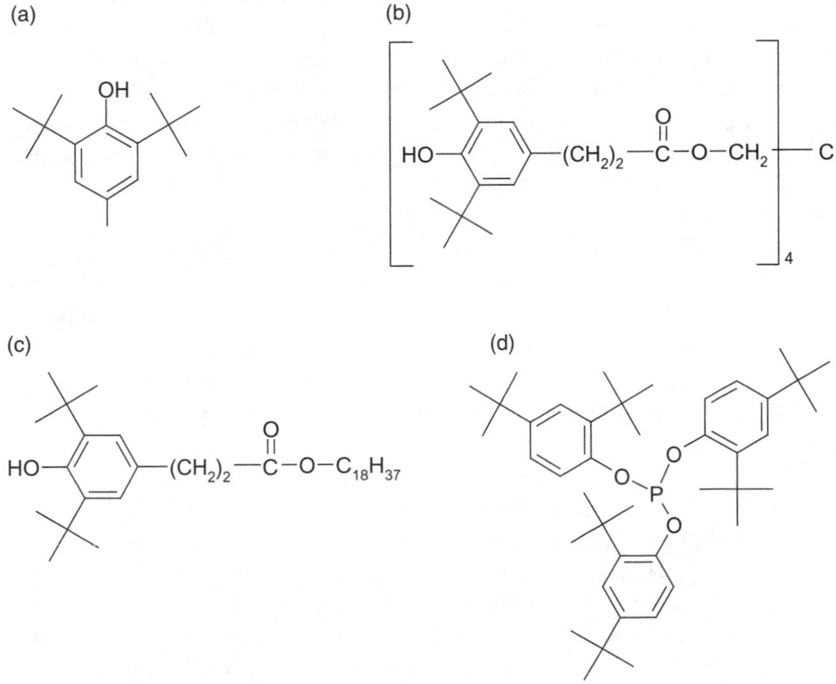

(a)

(b)

(c)

(d)

Figure 11.10   (a) BHT, (b) Irganox 1010, (c) Irganox 1076, and (d) Irgafos 168.

Light stabilizers are also commonly used to protect polymers from sunlight as well as artificial lighting; however, these additives are not commonly used in medical packaging applications. It is still important to be knowledgeable of these light stabilizers since they could unintentionally find their way into a polymer formulation through incidental contamination or the use of masterbatches that contain these compounds. Many of these chemistries include benzophenones, benzotriazoles, or triazines. This topic will be discussed in greater detail below.

11.3.3   Elastomers

Elastomers typically are thermosets and are members of the polymer family having the unique ability to recover from being stretched or deformed beyond the extent of their original state. It is from this elastic property that this type of polymer derives its name, "elastomer." Elastomers are composed of molecular segments called monomers that generally consist of carbon, hydrogen, silicon, or oxygen chemically bound to form a backbone. Thermosets are produced by cross-linking elastomer chains through the use of curing agents. The irreversible bonds formed or cross-links produce memory within the thermoset rubber assuring recovery from deformations at applied stress.

Curing or vulcanizing rubber increases its viscosity, hardness, modulus, tensile strength, abrasion resistance, and rebound while decreasing elongation at break, compression set, hysteresis, and solubility in solvents. All of these properties depend on the degree of cross-linking. The strength of the cross-links determines the degree of chain stiffness or modulus (λ) of the rubber. Subsequently, the elasticity of rubber is due to its ability to redistribute a load (stress) through realignment of its chains (*strain*). Based on this intrinsic ability to rearrange, the elongation of a specific elastomer can range from less than 100% to more than 1000%. The elastic property of rubber is temperature and humidity dependent.

These materials, either amorphous or semicrystalline in structure, exist above their glass transition temperature and possess sufficient chain motion within their molecular segments to render them soft and rubbery. At lower temperatures, rubber goes through strain-induced transition eventually becoming a rigid amorphous glass at a specific temperature known as the glass transition temperature (T_g). This property is used as one of several indicators for service temperature limits.

Thermoset elastomers can possess varied degrees of unsaturation within their backbone. Unsaturation permits the reactivity required for polymerization and cross-linking. Subsequently, as unsaturation increases, resistance to degradation decreases. The addition of functional side groups to an elastomer can attribute properties not inherent to the parent polymers. For example, elastomers that have oxygen incorporated into their backbone tend to have better low-temperature properties than those that contain only carbon–hydrogen bonds. These heteroatom side chains may also provide reactive cure sites on the backbone of an otherwise saturated polymer.

Saturated rubbers, for example, ethylene propylene copolymer, tend to be less reactive and more chemically resistant than polymers having higher diene content or those bearing functional side groups, for example, ethylene propylene diene ter-polymer. Because of their innate impermeability to gases, superior heat resistance, and good chemical resistance, formula compositions containing saturated elastomers are the best choice for closures used in parenteral applications.

Elastomers are selected for pharmaceutical packaging based primarily on the physical, chemical, and environmental demands of the application. Natural rubber, known for its superior resilience, resistance to abrasion, and compression set, is best suited for packaging articles intended for dynamic use (i.e., plunger tip, needle puncture). Because of its low permeability to moisture vapor and gases, butyl rubber is a good choice for stopper formulations intended for lyophilized drug packaging. A list of elastomers commonly used in the pharmaceutical and medical device industry and their respective American Society of Testing and Materials (ASTM) designations are listed in Table 11.5 along with their attributes and limitations.

Elastomer compounds can be designed to meet specific requirements. Manu-facturers typically offer several grades of the gum rubbers that are then optimized through compounding and fabrication operations. There are immeasurable combina-tions of elastomer grades and formulation ingredients that influence the physico-chemical and processing behavior of a compound. An understanding of the elastomer, compounding, and fabrication processes provides useful knowledge for the develop-ment of extraction methods and identification of potential leachables.

By itself, rubber has poor properties and must have other raw materials added to it to make it useful for the intended application. Rubber compounding is the science of formulating rubber with various ingredients to yield an effective product that can be efficiently manufactured. Compounders must have a keen knowledge of the following:

- elastomer properties,
- raw material costs,
- cure systems,
- antidegradants,
- mix methods, and
- test methods for determining processability and cure-related properties.[12]

In the pharmaceutical packaging industry, stringent regulatory requirements compel developers to perform due diligence when selecting polymers and chemical additives for elastomer formulations. While compliance with food additive regula-tions is expected, it is not required for compounding considerations.[13,14] Concur-rently, compounds developed for use in parenteral packing applications should have an extractables profile, be nontoxic, zinc free, noncarcinogenic, radiation or steam resistant, and meet the requirements of global compendial testing.

Elastomer formulations are multicomponent, and each ingredient has a specific role as described below. For these formulations, the mixing process (time, tempera-ture, number of cycles, equipment) is equally as important to the proper functioning of the finished component as composition of the formulation.

TABLE 11.5. Common Elastomers in the Pharmaceutical Industry

Common name	ASTM designation	Composition/structure	Attributes	Limitations
Butyl (*also* halo-butyl)	IIR *also* CIIR, BIIR	Isobutylene-isoprene (chlorinated or brominated)	Extremely low permeability to gases and moisture; good ozone, oxidation, steam, UV, and water resistance	Radiation resistance; max service temp. ~100°C
EPDM	EPDM, EPM	Ethylene-propylene-diene; ethylene-propylene	Very good resistance to ozone, oxidation, UV, chemical, aging, steam, water and compression set	Max service temp. ~125°C
Silicone	MQ, VMQ	Polysiloxane	Excellent high- and low-temperature properties; compression set and resilience; ozone, oxidation, UV, and water resistance	Gas permeability; radiation resistance; max service temp. ~225°C
Urethane	AU, EU	Polyethylene-apdate, poly(oxy-1,4-butylene)ether	Low gas permeability; ozone, radiation, UV, and oxidation resistant; excellent physical properties, abrasion, tear, and solvent resistance	Compression set, resilience; steam and water resistance; max service temp. ~120°C

206

Name	Abbr.	Monomer/Structure	Properties	Temperature/Resistance
Fluoroelastomer	FPM	Hexaflouropropylene-vinylidene fluoride $*-[CF_2-CF_2]_n-*$	Low gas permeability; excellent ozone, oxidation, steam, UV, heat, and moisture resistance; very good chemical resistance	Max service temp. ~250°C
Natural/ synthetic polyisoprene	NR/IR	Isoprene, natural and synthetic	Excellent physical properties including abrasion, resilience, compression set, tear properties, and water resistance	UV and ozone resistance; max service temp. ~70°C
Styrene butadiene	SBR	Styrene-butadiene	Good resistance to oxidation, radiation, steam and water; good compression set and abrasion resistance	Tear, UV and ozone resistance; max service temp. ~70°C
Chloroprene (Neoprene)	CR	Chloroprene	Good abrasion and tear resistance; resistant to ozone, oxidation and UV	Compression set resistance; max service temp. ~120°C
Nitrile (Buna N)	NBR	Nitrile-butadiene	Good abrasion, compression set and tear resistance; low gas permeability and resistance to water	Ozone and UV resistance; max service temp. ~1000°C
Butadiene	BR	Polybutadiene	Oxidation and water resistance; excellent physical properties, abrasion resistance, and low-temperature properties; high resilience	Similar to SBR; max service temp. ~700°C

Elastomers	Materials provide inherently unique properties required for the intended product use.
Fillers	Materials also referred to as extenders that provide structure, processing, and barrier properties. These can affect the pH of a solution wherein intimate contact exists and affects cure rate and state.
Pigments	Materials used to impart hue to a compound.
Process aids	Materials that function as internal lubricants for ease of miscibility in mixing and flow in preforming and molding operations.
Antidegradants	Materials employed to improve and stabilize the resistance of pre- and postmolded compound to thermal, oxidative, and chemical degradation.
Plasticizers	Materials used to improve compound flexibility and processability. Can impart low-temperature properties to an elastomer system. Also impacts rate and state of cure.
Curatives	Materials employed to produce irreversibly cross-linked networks throughout a polymer matrix to provide permanent elastic properties.

Cross-linking additives (or curatives) for pharmaceutical and medical device elastomers include the following:

- sulfur/sulfur-bearing accelerators
- resins
  - halogenated alkyl phenolic heat reactive resins
- dimaleimides
  - M-phenylenedimaleimide
- metallic oxides
  - magnesium oxide
- peroxides
  - 2,5-Dimethyl-2,5-di(t)butylperoxyhexane
  - Dicumyl peroxide
  - α,α'-Bis(t)butylperoxydiisopropylbenzene
  - 2,5-Dimethyl-2,5-dibenzoylperoxyhexane
- methacrylate monomers.

11.3.4 Stabilization of Elastomers

Elastomers (i.e., rubber) differ from other polymers in that their molecular structure imparts a high degree of elasticity to these materials. Elastomers, like rubber bands, have very low moduli (ratio of stress to strain) and therefore deform quite easily. The mark of a good elastomer is not a high degree of elongation alone—the elastomer should not remain in the deformed state after the stress is removed, but must, instead, revert back to its original form. Elastomers, therefore, also have high reversible elongation, or elastic elongation.[15]

TABLE 11.6.   Antioxidants Commonly Used in Elastomers

Chemical name	Ciba trade name[a]
Ethylenebis(oxyethylene)bis-(3-(5-*tert*-butyl-4-hydroxy-*m*-tolyl)-propionate)	Irganox 245
4,6-Bis(octylthiomethyl)-*o*-cresol	Irganox 1520
Thiodiethylene bis[3-(3,5-di-*tert*-butyl-4-hydroxyphenyl)propionate]	Irganox 1035
Pentaerythritol tetrakis(3-(3,5-di-*tert*-butyl-4-hydroxyphenyl)propionate)	Irganox 1010
2,6-Di-*tert*-butyl-4-(4,6-bis(octylthio)-1,3,5-triazin-2-ylamino)phenol	Irganox 565
Octadecyl-3-(3,5-di-*tert*-butyl-4-hydroxyphenyl)-propionate	Irganox 1076
Didodecyl 3,3′-thiodipropionate	Irganox PS 800
Benzenamine, *N*-phenyl-, reaction products with 2,4,4-trimethylpentene	Irganox 5057
Tris(2,4-di-*tert*-butylphenyl)phosphite	Irgafos 168

[a] Irganox and Irgafos are registered trademarks of Ciba Specialty Chemicals Corporation.

Rubber articles are exposed to oxidation, flex, fatigue, ozone, and light during their shelf life. Additives are often introduced to provide durability to the rubber. Protection via additives is especially important when rubber goods are used in dynamic and outdoor applications. The oxidation of rubber can be considerably slowed by the addition of antioxidants to the substrate, which scavenge oxyradicals before they have the opportunity to react with rubber chains. Table 11.6 contains a list of antioxidants commonly used in elastomers.[9]

In the rubber industry, acid scavengers are used to neutralize traces of halogen anions formed during aging of halogen-containing rubbers (e.g., chloroprene and halobutyl rubbers; for properties, see Table 11.5). If not neutralized, the anions would cause premature aging of rubber, and performance of rubber articles would significantly deteriorate with time. Very effective acid scavengers include lead oxides and lead salts, but because of environmental pressure, these are being phased out from rubber formulations. Magnesium oxide could be another option; however, it is sometimes difficult to mix in the elastomer compounding process and can induce high water swell of vulcanizates. Hycite®* 713 is an environmentally friendly product that allows effective acid neutralization without mixing disadvantages and without compromising water resistance. The structure of this compound is proprietary.

Ions of copper (Cu), iron (Fe), cobalt (Co), nickel (Ni), and other transition metals that have different oxidation states with comparable stability, and which are easily oxidized or reduced by one-electron transfer, are called "rubber poisons." They are very active catalysts for hydroperoxide decomposition. Hydroperoxides are species within the "autoxidation cycle" that contribute to the degradation of rubber vulcanizates. Rubber poisons present in vulcanizates, even in trace amounts (below 5 ppm), increase the decomposition rate of hydroperoxides and thus acceler-

* Hycite is a registered trademark of Ciba Specialty Chemicals Corporation.

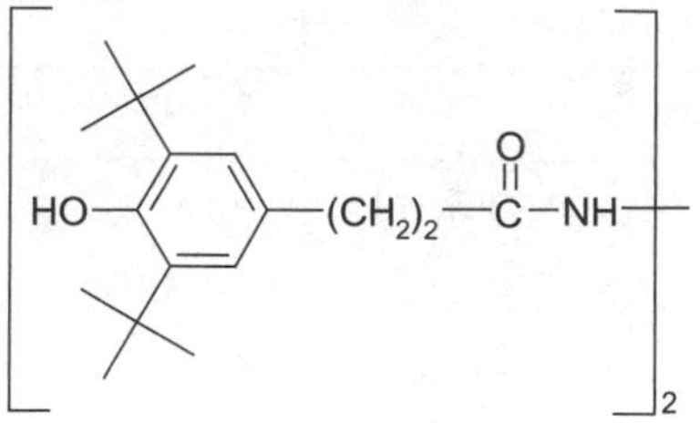

Figure 11.11   Irganox MD 1024.

ate oxidation and aging of rubber goods. Therefore, rubber that contains, or could be in contact with, rubber poisons requires a specific stabilizer, a so-called metal deactivator. The metal deactivator, 2′,3-bis[[3-[3,5-di-*tert*-butyl-4-hydroxyphenyl]propionyl]]propionohydrazide also known as Irganox MD 1024, binds ions into stable complexes and deactivates them. Rubber articles most affected by rubber poisons are cables and hoses, but the problem can occur in other applications as well (Fig. 11.11).

Many polymer materials and end products are irradiated prior to being used in certain applications, especially in the pharmaceutical and medical device industries. Gamma irradiation is the second most widely used sterilization technique, following the ethylene oxide process. Gamma irradiation is characterized by deep penetration at low dose rates. In such applications, radiation-induced degradation in the form of postirradiation effects is very important in the determination of the shelf life of the polymer article.[16,17] The impact of the high-energy irradiation leads to the generation of carbon-centered radicals, which react with oxygen and accelerate autoxidation and the subsequent degradation of the elastomer. These radical species are long-lived and may cause degradation well after the irradiation process. The best way to counteract the effects of irradiation is to use antioxidants and radical scavengers.[17] Since the antioxidants are consumed during this sterilization process, it is important to use sufficient concentrations to ensure the long-term protection of the rubber substrate.

Microbial growth on the rubber surface can lead to discoloration, staining, odor development, or the formation of biofilms that ultimately result in deterioration of mechanical properties of the product. The patient exposed to a contaminated rubber article of this sort can also be compromised. As a result, antimicrobials and fungicides can be used by resin manufacturers and molders to maintain the freshness, durability, and aesthetics of the elastomer.

11.3.5   Material Processing

Different processing techniques must be coordinated with the desired properties, nature of the material, and shape of the product. The processing conditions will contribute to the performance characteristics of the material in its finished form. It is beyond the scope of this section to describe all the different processes but an attempt is made to bring about an awareness of the manufacture and fabrication processes in relation to the chemical and physical nature of the materials of construction. In general, plastics or elastomers are heated above the T_g or T_m during compounding with additives. The intermediate material may be cooled and formed into pellets by extrusion or sheets by calendering. Depending on the desired shape of the finished product and the properties of the material to be used, a variety of different forming techniques may be used. Heat and pressure are common to each of the forming processes. Heating temperature and time will impact the levels of polymerization, cross-linking, and levels of remaining processing initiators, stabilizers, or transformation products. Cooling temperatures and time will impact the level of crystallinity or amorphous content.

Thermosets can be formed by compression molding or injection molding. Compression molding involves a set of matched die molds that form the component by use of heat and pressure. Cure takes place in the mold by applying further heat and pressure. Injection molding is accomplished by feeding materials through a heated barrel then injecting them into a hot mold to form the component. Cross-linking may be completed during a postmolding heating of the molded components.

Thermoplastics are commonly injection molded or blown. Resin pellets are dried and added with any other necessary ingredients, for example, colorant, to the hopper of the molding machine. After heating the ingredients to melting, the molten mixture is processed through a single or twin screw processor to obtain a homogenous mixture that is injected into the mold. The two halves of the mold are held together in a press and cooled slightly prior to ejecting the finished part. Rod shapes or sheets are formed by a process that is similar to injection molding—extrusion through an appropriately shaped orifice. Hollow products can be made by blow molding from a preform.

In all cases, temperature and time are the variables that can have an impact on the properties of resins, compounding ingredients and surface enhancements. Furthermore, the mixing of ingredients to ensure uniform blending and dispersing will also impact the characteristics of the materials. It is important to note that processing aids, such as mold release or slip agents, or residue from postmolding cleaning agents or surface treatments may contribute potential leachables to the component.

11.4 EXTRACTABLES PROFILE FOR PLASTIC AND RUBBER COMPONENTS OF CONTAINER CLOSURE SYSTEMS

Given the information presented above, it would seem that the typical extractables profile for a polymer component of a pharmaceutical container closure system would include antioxidants, acid scavengers, metal deactivators, colorants, and possibly antimicrobials and light stabilizers. However, there are many other chemical species that could be present. As depicted in Figure 11.9, polymer stabilization is a dynamic process. The stabilizers transform into other chemical species as they scavenge the radical moieties in the polymer. These transformation products are also typically a part of the extractables profile. Many stabilizers, such as pentae-rythritol tetrakis(3-(3,5-di-*tert*-butyl-4-hydroxyphenyl)propionate)—also known as Irganox 1010—have more than one active sight and could potentially have multiple transformation products. Stabilizers could also react in unwanted pathways that do not result in stabilization of the polymer. For example, an antioxidant could undergo a hydrolysis reaction, which would render the stabilizer inactive. These degradation products could also show up in the extractables profile. Other species could include residual starting materials or solvents used in the synthesis of the elastomers, or monomers and high MW oligomers resulting from an incomplete polymerization reaction.

As briefly mentioned above, unexpected additives can sometimes find their way into a polymer formulation. Additives can be included into a polymer as either a neat material or as part of a masterbatch. A masterbatch, sometimes referred to as a concentrate, is a mixture of an additive at a high concentration (5%–30% or more) in a polymer matrix. The polymer that is used in the additive concentrate can be the same chemical species, for example, polypropylene, as the polymer, for example, polypropylene, that it is used to stabilize, but may be a different grade or manufac-tured by a different vendor, and as a result can contain a different base stabilizer system. Alternatively, the polymer that is used in the masterbatch, for example, nylon, may be a different chemical species than the polymer, for example, polypro-pylene, it is used to stabilize so that not only a different base stabilizer system, but also different polymer oligomers/degradants are added to the polymer formulation. The masterbatch could also be earmarked for other applications than medical and contain specialty additives, for example, fire retardant, that provide no value in medical and pharmaceutical applications. A masterbatch is often the source of unan-ticipated light stabilizers or several primary antioxidants in container closure system or device components.

Elastomers can also have some so-called "special case" compounds associated with them that require particular attention due to special safety and historical con-cerns.[18] These compounds include volatile *N*-nitrosamines, mercaptobenzothiazole, and polyaromatic hydrocarbons (PAHs).[19] Rubbers are often cured using curing agents that contain either sulfur or peroxide. These compounds, along with *N*-nitrosamines, mercaptobenzothiazole, and other benzothiazole compounds that result from the curing process or vulcanization, could also be detected in an extract-ables study.[18] Carbon black is routinely used as a filler or colorant in elastomers. The color of these rubber substrates could range from light gray to black depending on the concentration of carbon black. PAHs also referred to as polynuclear aromatics (PNAs) are impurities in carbon black and must be analyzed in all extractables studies whenever this filler is present or suspected to be present.

So, in summary, the extractables profile of a polymer could be quite complex and may include many different classes of additives, their corresponding transforma-tion and degradation products, residual starting materials and solvents, monomers and high MW oligomers from incomplete polymerization reactions, and trace levels of special case compounds such as PAHs, *N*-nitrosamines, and mercaptobenzothia-zole. An example of a complex extractables profile of an elastomer is shown in Chapter 15.

11.5   ADDITIVES ANALYSIS IN POLYMERS

The polymer industry often performs quantitative in-polymer additives analyses to determine the concentrations of stabilizers in a polymer, to elucidate the composition of an unknown formulation, to determine why a polymer did not perform as expected, or to obtain some insights as to the mechanism of an additive's stabilization process.[11] Typically, the analyses involve extracting the additives from the plastic or elastomer followed by chromatographic analysis of the extract. There are also some techniques by which the additive concentration can be determined directly in the polymer without extraction. This is particularly advantageous in that it eliminates the efficiency of the extraction process as a variable and will also quantitate stabilizers that are grafted to the polymer. Both techniques will be discussed below and are similar for plastics and elastomers. The general term "polymer" is used to refer to both plastics and elastomers. It should be noted that the intent of these analyses is deformulation, which is different from the intent of the controlled extraction study as described in Chapter 14.

The first step in performing an additives analysis in a polymer material is to obtain as much information as possible about both the stabilizers and substrates. For example,

- chemical structure of additives and polymer;
- MW of additives;
- pKa values of additives;
- solubility of additives and polymer;
- concentration range of additives; and
- degree of cross-linking in the polymer.

Based on this information, extraction and chromatography techniques as well as associated solvents can be selected. Typically, the extraction process utilized in the polymer industry involves refluxing the plastic or elastomer in a suitable solvent that will ideally dissolve or, at the very least, swell the polymer. Next, a co-solvent is added to precipitate any dissolved polymer while leaving the additives in solution. Other, more automated, extraction techniques can be performed, such as microwave-assisted extraction or supercritical fluid extraction if desired. The mixture is then filtered and analyzed chromatographically by either high-performance liquid chromatography (HPLC), gas chromatography (GC), or gel permeation chromatography (GPC). HPLC is the workhorse for the analysis of most stabilizers including primary and secondary antioxidants and some light stabilizers. UV detection is used in combination with either evaporative light-scattering detector (ELSD) or mass spectrometry (MS) detection since some stabilizers do not contain a suitable chromophore. GC is routinely used for some of the smaller and more volatile additives, while GPC is required for quantitative analysis of polymeric additives such as some of the hindered amine light stabilizers.

Other techniques are available for analyzing the additives in polymers that do not require an extraction method. In some cases, extraction is not a viable technique, for example, if the polymer is highly cross-linked and cannot be adequately swelled. Extractions are not always done efficiently. Some polymers have a tendency to re-encapsulate the extracted additive when precipitation occurs by addition of the co-solvent. Fillers, such as carbon black, can adsorb a significant amount of an additive lowering the efficiency of an extraction. Typically, nonextraction techniques are not as selective as chromatography, but instead focus on either a specific chromophore or element. The advantages of these techniques are that there is minimal amount of sample preparation and they can be used to screen for whole classes of compounds rather than just one entity. They can also be used to quantitatively determine the concentration of insoluble additives such as pigments and some fillers. Total nitrogen analysis, using either a chemiluminescence nitrogen analyzer or the Kjeldahl method, is a very accurate way to determine the concentration of a hindered amine stabilizer or nitrogen-containing pigment in a plastic or elastomer. Ultraviolet/visible spectrophotometry is often utilized to analyze antioxidants or light stabilizers in films by measuring the absorption at a specific wavelength. For measuring the concentration of organic phosphites (secondary antioxidants) or sulfur-containing additives in a wide variety of substrates, X-ray fluorescence is an industry proven tool. And finally, the total concentration of active antioxidants can be determined by differential scanning calorimetry by measuring the oxidative induction times of the samples. This method is particularly useful in determining the amount of antioxidant in exposed samples.

11.6 SUMMARY

Polymers, such as plastic and elastomers, are used in all areas of the medical industry, especially in container closure systems and delivery devices. In addition to patient safety and formulation compatibility, the mechanical properties of a material are a major factor in its selection for use in pharmaceutical products. Additives, copolymerization, and specific processes are used to modulate strength, hardness, stiffness, toughness, crystallinity, cross-linking, thermal stability, and extensibility. These properties, in addition to the polymer chemistry and structure, may impact the extractables or leachables profiles.

Like all polymers, rubber and plastic must be stabilized with additives to protect from the stresses of extrusion, molding, and the environment. Rubber is also susceptible to degradation from residual metal catalysts and halogen ions left behind from its fabrication. Specialty additives can be added to provide color or antimicrobial properties to polymers. The extraction profile of polymers can be quite complex and be composed of the above-mentioned additives, as well as their respective transformation and degradation products, residual starting materials, monomers, and high MW oligomers. Unexpected additives can also be present in plastic and elastomers due to incidental contamination or issues associated with the polymer supply chain. Additive analysis in polymers is typically done by extraction followed by chromatography. The specific techniques utilized are dependent on the additives and polymers. Several analytical techniques that do not require extraction can also be employed for additive analysis.

REFERENCES

1   Paskiet, D.M. *Principles of Quality by Design Applied to the Selection of Labels for Inhalation Drug Products*. RDD Europe, Lisbon, 2009.

2   Harper, C.A., ed. *Handbook of Plastics and Elastomers*. McGraw-Hill Book Company, New York, 1975.

3   Yang, J., Huang, J., Chyu, M.K., Wang, Q.M., Xiong, D., and Zhu, Z. Degradation of poly(butylene terephthalate) in different supercritical alcohol solvents. *J Appl Polym Sci* 2010, *116*, pp. 2269–2274.

4   Driver, W.E. *Plastics Chemistry and Technology*. Van Nostrand Reinhold Company, New York, 1979.

5   Polymers, in Reusch, W., Virtual Text of Organic Chemistry, Michigan State University, 1999. Available at: http://www2.chemistry.msu.edu/faculty/reusch/VirtTxtJml/polymers.htm (accessed January 21, 2011).

6   Wunderlich, B. The nature of the glass transition and its determination by thermal analysis. In: Seyler, R.J., ed. *Assignment of the Glass Transition, ASTM STP 1249*. American Society for Testing and Material, Philadelphia, 1994; 17–31.

7   Fakirov, S., Balta Calleja, F.J., and Krumova, M. On the relationship between microhardness and glass transition temperature of some amorphous polymers. *J Polym Sci Phys Ed* 1999, *37*, pp. 1413–1419.

8   Bart, J.C.J. Chapter 2: Physical chemistry of additives in polymers. In: *Polymer Additive Analytics: Industrial Practice and Case Studies*. Firenze University Press, Firenze, 2006; 67–177.

9   Ciba Technical Brochure: Effects for specialty polymers and products, Publication Number: 07-06-09.01. Ciba Specialty Chemicals, 2007.

10 Ciba Technical Brochure: Effects for polyolefins, Publication Number: 07-07-09.01. Ciba Specialty Chemicals, 2007.

11 Zweifel, H. *Plastics Additives Handbook*, 5th ed. Hanser, Munich, 2001.

12 Layer, R.W. Introduction to rubber compounding. In: Ohm, R.P., ed. *The Vanderbilt Rubber Handbook*, 13th ed. RT Vanderbilt Company, Norwalk, CT, 1990; 11–21.

13 Guidance for industry: Container closure systems for packaging human drugs and biologics. U.S. Department of Health and Human Services, Food and Drug Administration, Center for Drug Evaluation and Research (CDER), Center for Biologics Evaluation and Research (CBER), 1999.

14 European Medicines Agency Inspections. Guideline on plastic immediate packaging materials. Committee for Medicinal Products for Human Use (CHMP), Committee for Medicinal Products for Veterinary Use(CVMP), London, 2005.

15 Odian, G. *Principles of Polymerization*, 2nd ed. Wiley Interscience, New York, 1981.

16 Clough, R.L., Gillen, K.T., and Dole, M. *Irradiation Effects on Polymers*. Elsevier Applied Science, London, 1991.

17 Plastic Design Library Staff. *Effect of Sterilization Methods on Plastics and Elastomers*. William Andrew Publishing, 1994.

18 Norwood, D.L. and Ball, D. Product Quality Research Institute: Safety thresholds and best practices for extractables and leachables in orally inhaled and nasal drug products. Submitted to the PQRI Drug Product Technical Committee, PQRI Steering Committee, and U.S. Food and Drug Administration by the PQRI Leachables and Extractables Working Group, 2006.

19 Norwood, D.L., Prime, D., Downey, B.P., Creasey, J., Sethi, S.K., and Haywood, P. Analysis of polycyclic aromatic hydrocarbons in metered dose inhaler drug formulations by isotope dilution gas chromatography/mass spectrometry. *J Pharm Biomed Anal* 1995, *13*(3), pp. 293–304.

CHAPTER 12

PHARMACEUTICAL CONTAINER CLOSURE SYSTEMS: SELECTION AND QUALIFICATION OF MATERIALS

Douglas J. Ball, William P. Beierschmitt, and Arthur J. Shaw

12.1 SELECTION, CHARACTERIZATION, AND APPROPRIATE ANALYTICAL TESTING OF MATERIALS FOR THE CONTAINER CLOSURE SYSTEM (CCS): THE DEVELOPMENT CHEMISTRY AND ENGINEERING PERSPECTIVE

A packaging system for a drug product is often one of the last items on a checklist for companies involved with development of a new drug product; however, if speed to market is an important concern, packaging selection cannot be left to the last minute. Historically, medicines were packaged using materials already in use by food manufacturers. Glass containers were usually a good choice because of the availability and inert nature of the material. As modern-day food packaging materials became available, the selection choices for drug manufacturers also greatly expanded.

The needs of food products and drug products are similar in many ways, but two important differences need to be kept in mind. First, food products for the most part are taken orally while drug products have many routes of delivery (oral, nasal, inhalation, injectable, etc.). Each delivery route has specific toxicological concerns and design needs that can affect the packaging choice. Second, drug products often demand longer shelf-life requirements than food products. This dictates that drug product packaging needs to be more robust compared with short shelf-life food products.

Shelf-life requirements and leachables testing requirements of drug products drive the timing needed to select packaging systems, forcing selection to occur

Leachables and Extractables Handbook: Safety Evaluation, Qualification, and Best Practices Applied to Inhalation Drug Products, First Edition. Edited by Douglas J. Ball, Daniel L. Norwood, Cheryl L.M. Stults, Lee M. Nagao.

several years ahead of the projected commercial marketing date. Most drug product manufacturers desire a shelf life of 2 or more years. Regulatory requirements necessitate proof that the packaging systems can protect and preserve the drug product over the stated shelf life. This often requires manufacturers to store the drug product in the commercial packaging system over the stated shelf life and retest the product to demonstrate the continued efficacy and safety of the drug. If either efficacy or safety cannot be demonstrated over a commercially viable shelf life, then it is unlikely the product will be marketed if the packaging is deficient.

If the cause of failing a shelf-life test was the packaging, the manufacturer can choose another packaging system, but the delays to market can be very costly to patients in need of the drug product. A motivational expression sometimes posted in pharmaceutical development areas reads—"The patient is waiting." The manufacturer will also be faced with significant losses. These include potential patent lifetime loss and costs associated with retesting the new packaging (manufacturing more drug product, repackaging, and repeated shelf-life testing).

12.1.1   When to Begin the Selection Process

While the selection of a packaging system can rarely begin too early in a new drug development program, it can begin too late. Time must be allotted for testing the packaging components to avoid approval delays due to compliance with regulations. Figure 12.1, recommended time of extractables and leachables process steps, relates stages in a new drug development program to the process of selection and extractables and leachables qualification of a packaging system/delivery device. Depending on the function and degree of complication of a packaging system/device, the timing of each step in the process will vary. Figure 12.1 compares the timing of a novel CCS or delivery device system to that of a standard CCS. When dealing with new delivery systems, selecting materials of construction for the device must begin early in the drug development process. If the packaging system is not complex or has been used in other similar drug products, then the extractables and leachables testing process can be delayed until approximately 2 years prior to the start of registration lot manufacture. Traditionally, three registration lots of drug product are manufactured and packaged on a large scale, and samples are set up for stability testing according to the International Conference on Harmonisation (ICH) recommended conditions to demonstrate the shelf life of the product. Leachables testing on ICH drug product stability samples is typically used to demonstrate the commercial manufacturing and packaging system compliance to regulations.

It is usually convenient (to save both time and additional costs) to initiate leachables testing on samples from the ICH batches. Many companies target the availability of leachables analytical methods to meet this goal. Working back against the timeline (see Fig. 12.1), this dictates when controlled extraction studies must be completed, which in turn determines when selection of the packaging system components/materials must be completed.

The logic of the extractables and leachables process (Fig. 12.1) shows two decision points early in the process where unacceptable materials can be screened out (rejected). The first decision point is usually based on available data obtained

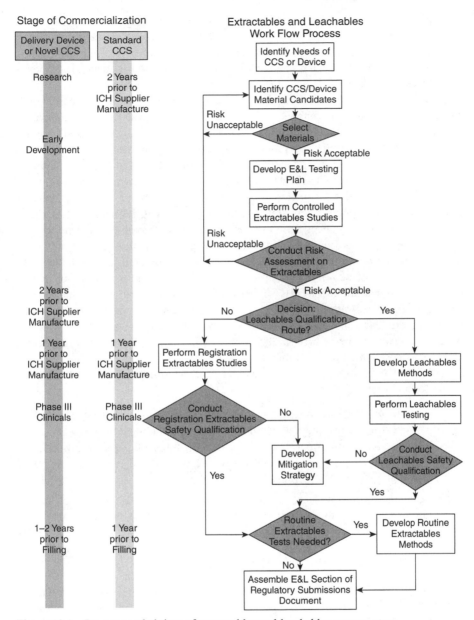

Figure 12.1   Recommended time of extractables and leachables process steps.

from the supplier or other literature sources. The second decision point is based on information (controlled extractables study data) that is obtained with significant expense of time and money. Also shown on this diagram is the "leachables safety qualification" decision point that can lead to rejection of a packaging material as part of a mitigation strategy. After leachables testing and a failed safety qualification of the leachables in the drug product, the next step in the extractables and leachables work process is "develop mitigation strategy." Getting to this step may occur because something changed in the packaging or risks were taken that did not result as projected. It is important to recognize that correcting a packaging choice failure at this stage will result in delaying a project timeline by 1–2 years plus associated additional unbudgeted costs, many times that of a controlled extraction study.

The third decision point in the process, "decision: leachables qualification route," leads to either leachables testing or qualification based on just controlled extractables data. A positive result from the safety qualification of a packaging system via extractables testing requires that toxicology analysis of "worst-case" levels of all observed controlled extractables from all materials in direct and indirect contact with the drug product is deemed safe. "Worst case" means assuming complete migration of all extractables present in the materials into the drug product immediately upon contact. This situation also requires that the analytical methods used to determine the concentration of the controlled extractables were performed to good manufacturing practice (GMP) or ICH quality standards. This often means repeating all the controlled extractables studies using validated methods. Needless to say, this route of qualification can only be used when the packaging or device materials demonstrate low levels of extractables and can be safety qualified at "worst-case" concentrations. In practice, few drug products can meet these demands. Following the leachables testing route and testing the actual leachables concentrations in the drug product, which rarely ever reach the concentrations of "worst case," is the practice adopted by most manufacturers.

Risk analysis on the extractables and leachables work flow process should lead to the following conclusions:

- Inadequate screening of materials or components will lead to an increased risk for repeating controlled extractables studies with a new material.
- The project timeline increase associated with failed material realized at the controlled extractables stage will be greater for a device material compared with a simple packaging component.*
- Inadequate controlled extractables studies will increase the risk of failure during leachables qualification. Having to cycle back to select and qualify a new material can delay a product from the market by 1–3 years.

12.1.2 Selection of Packaging/Device Materials

The materials selected to make a CCS or a device typically go through a series of selection filters.† The filters are driven by the needs of the product. The selection process begins by making a list of the product needs. This is a required step if a device will be used to deliver the drug product and is recommended when following

---

* Controlled extraction studies for a single component take between 2 and 4 months to complete; however, qualifying a new material for a critical device component may take over 1 year.
† An example of a selection filter is the moisture permeability of a packaging component. Many drug products will decompose quickly if moisture levels exceed a critical threshold during the shelf life of the drug product. The amount of moisture (water) that migrates through a plastic packaging film is often published by the supplier. Using this information, calculations can project the moisture absorption of the drug product as a function of time, Comparing results of different potential packaging materials will lead to a subset of materials that meet the needs of this drug product.

a quality by design development program.[1–4] While building the needs list, it is important to review the current regulations around packaging and device components for the drug product.[5–7] Once the needs list is complete, the process of selecting a packaging or device component/material can begin. Unfortunately, extractables is one of the last selection filters usually applied. This can be due to the lack of extractables data supplied by material/component manufacturers. While many manufacturers commonly supply important design properties of their materials like oxygen and moisture permeation data and/or strength and durability data, extractables data are rarely supplied. If extractables data suitable for a toxicological evaluation are not available, data generation falls to the drug product manufacturer.

Understanding the many reasons for the lack of extractables data supplied by material/component manufacturers can help in devising solutions to obtain the needed information. Some of the reasons suppliers invoke include cost of testing, uncertainty in how to test, the fear of what testing might uncover, lack of complete understanding of their incoming ingredients, skepticism regarding the relevance of extractables profiles to leachables profiles, and the lack of control over the extractables that may have been added to the material prior to their purchase. The lack of control in the plastic supply chain can be understood by examination of Figure 12.2, plastics—a segmented industry,[8] one example of a plastic supply chain. The pharmaceutical manufacturer is considered the "N" user of a finished plastic component. The finished component was molded by the "n-1" supplier who obtained the plastic resin from the "n-2" supplier called a masterbatcher. At each stage in the supply chain, the supplier adds chemicals to the polymer material to enhance the properties or facilitate the processing of the plastic. Also added to thermally formed polymers at each stage is additional thermal history that can change the decomposition products. Each supplier also protects the

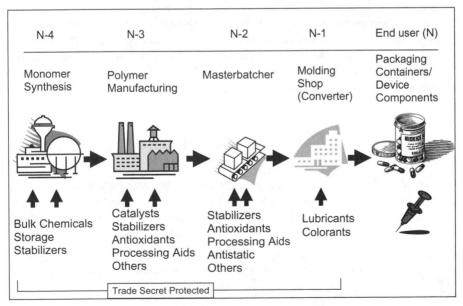

Figure 12.2 Plastics—a segmented industry.

confidential information of its process by taking trade secret precautions. This greatly limits the knowledge available to each supplier in the chain regarding the potential extractables that may exist in the polymer. To further complicate the extractables picture, many of the additives can be purchased from multiple secondary suppliers (antioxidants, plasticizers, processing aids, etc.). Each additive carries impurities that are often specific to the secondary suppliers' manufacturing process. Without detailed knowledge and control over all the ingredients, being able to rely on a consistent extractables profile of the final component becomes difficult.

12.1.3 Material Selection Using Quality by Design Principles

Selection implies that a choice is to be made. In packaging materials, this usually means that components of the CCS/device will be purchased from specialized suppliers. The selection of a supplier is as important as the selection of a good material. As mentioned, plastic suppliers often have some information about the key physical design parameters of a component/material but little useful information about the extractables that may be present.* On the other hand, obtaining detailed controlled extractables data on several potential candidate materials/components using traditional solvent-controlled extraction/analytical techniques can be expensive and time-consuming. This leaves the drug product manufacturer in a difficult selection situation—without sound extractables knowledge, a low risk, good decision cannot be made.

Quality by design principles infer that all the product packaging components must be selected and risk assessed to consistently meet the critical quality attributes (CQAs) of the components.[4,9] With regard to extractables and leachables issues, the key CQA is the absence or minimal level of extractables that become leachables.[10] Furthermore, common sense implies that selecting among components or materials based on their extractables profile is difficult to impossible without comparative data. The process used by many companies is to obtain as much readily available information as possible (vendor information, literature information, prior experience information, quick analytical screening data) prior to making a decision. The choice is then validated via controlled extractables testing and toxicology assessment of the observed extractables.

12.1.4 Predicting Potential Leachables from Extractables Data

Caution should be taken when using controlled extractables data to assess which potential chemicals may migrate into a drug product. When using controlled extractables data, the extraction potential of the drug product must be considered. It has

---

* Starting in 2007, some progress has been made by a few innovative pharmaceutical packaging suppliers. These innovators have started to test their product for extractables using techniques recommended by the Product Quality Research Institute (PQRI) Leachables and Extractables Working Group. These techniques may allow for a preliminary toxicological assessment without additional testing.

been found that the polarity of a solvent system can be related to the extractables that migrate from the target materials. Solvent systems that bracket the drug product's polarity should be used to estimate what chemicals may leach into the drug product. As an example, consider a parenteral single-use syringe for a biological drug product. The drug product is formulated in an aqueous system buffered to pH 8.2 with a surfactant to keep the active ingredient in solution. The choice of an elastomeric tip on the plunger of the syringe is under consideration. Controlled extractables data are available on two elastomeric tips. The data consist of controlled extractables information from seven solvent systems:

- water pH = 2.5,
- water pH = 7.0 (neutral no buffer),
- water pH = 9.5,
- isopropyl alcohol : water (1:1 mix),
- isopropyl alcohol,
- dichloromethane, and
- hexane.

To bracket the drug product's extraction potential, three extractables data sets from the above system were used to evaluate potential leachables: water pH 7 and water pH 9 (to bracket ionized leachables) and IPA : water mix (to bracket the organic extraction potential due to the surfactant used in the formulation).

Attempting to use all seven solvents to evaluate the elastomeric materials is not recommended. Many extractables observed in solvent systems differing significantly in polarity from that of the drug product will not be observed as leachables and will complicate leachables testing if targeted by the method. In this example, hexane or dichloromethane were not selected because many nonpolar chemicals extracted by these solvents would never reach concentrations in the aqueous drug product that would pose a safety concern.

12.1.5 Vendor and Literature Information

Vendor and literature information can give clues to potential extractables that might be present in a component/material. While vendors usually do not have extensive information on chemicals that migrate out of a component or material, they often have knowledge of some of the additives that have been mixed in with the polymer. This information can be helpful when attempting to narrow down potentially suitable components/materials. Through the literature, it may be possible to learn of potential thermal and oxidative degradation pathways of both the plastic and known additives.*[11–14]

When approaching component/material suppliers for information about ingredients/extractables, the requester should be prepared to sign a confidential

---

* Dennis Jenke has several publications on degradation pathways of common additives. Ciba Specialty Chemicals has published information on reaction pathways of some common antioxidants used in plastic and rubber materials.

disclosure agreement (CDA). It should not be surprising, if, even with a CDA, key information about additives (exact chemical names, amounts added, vendor names that supply the ingredients, etc.) is not disclosed.

12.1.6 Drug Master File (DMF)

To meet United States Food and Drug Administration (FDA) regulatory requirements for full disclosure of materials and processes used in drug products, the supplier may have established a DMF. A DMF is a file provided confidentially to the FDA. Information in the file is not disclosed to the drug product manufacturer, but through a letter of authorization, reference is made to the DMF in regulatory submissions. The FDA's DMFs are currently limited to four types[15]:

- *type I*: manufacturing site, facilities, operating procedures, and personnel;
- *type II*: drug substance, drug substance intermediate, and material used in their preparation, or drug product;
- *type III*: packaging material;
- *type IV*: excipient, colorant, flavor, essence, or material used in their preparation;
- *type V*: FDA-accepted reference information.

Sometimes the supplier will allow a drug product manufacturer to view this information after the signing of a CDA. This is perhaps the best way to get vendor information because both the drug manufacturer and the FDA get detailed information about the packaging material ingredients. Obtaining information this way can be very helpful when defending a new drug application (NDA). While the drug manufacturer (sponsor) cannot place confidential information obtained from a material supplier into an NDA, the sponsor can refer to the DMF. Since the FDA reviewer also has access to the DMF, the sponsor can now support statements in the NDA based on information from the supplier's DMF. Often, logical arguments about the absence or maximum concentration of a given chemical can be justified when information like this is available, without the need of experimental data.

12.1.7 Thermal Desorption Gas Chromatography Mass Spectrometry (TD/GC/MS)

Another method for obtaining timely and useful extractables information for a toxicological assessment of the materials is by TD/GC/MS testing.[16] The method allows for quick analysis of volatile and semivolatile compounds within the component/material. The extractables are released from the material by application of heat in an inert atmosphere. Vapor pressure differences between matrix and extractables in the material are used to separate the volatile components for identification and quantification. The method is orthogonal to solvent extraction and is not dependent on polarity of analytes to be detected. One drawback is that nonvolatiles cannot be detected by this methodology; however, volatiles and semivolatiles comprise most of the observed extractables. The analysis time is quick (1–2 hours including sample preparation time) in comparison to traditional multisolvent extraction and analysis techniques. TD/GC/MS has been successfully used to identify packaging components/ materials that contain high safety risk chemicals. An example of a material selection/ screening using TD/GC/MS results coupled with a toxicological assessment is given in the case study provided later in this chapter.

The component/material selection process should also include thoughts about how quality control will be maintained over the commercial life of the drug product or device. Routine testing of incoming components/materials (see Chapter 18) may be required to assure patient safety and/or component functional quality. Costs associated with testing potentially thousands of batches of components/materials over a drug product's lifetime can be significant. Selection of a less expensive component/material that would force an increase in testing surveillance can negate any calculated saving based on cost differences. This would be especially important if the material supply chain is under constant flux due to suppliers continuously seeking the lowest priced additives or base polymer resin. As changes are detected, GMP compliance would force placing the offending batches of components/materials under quarantine until suitable testing and additional safety qualifications could be completed. If out-of-stock issues arise due to quarantined supplies, additional costs could be attributed to the selected components/materials.

12.1.8 Level of Detection Requirements

When screening materials for extractables, it is equally important to obtain an estimate of the concentration as well as an identification of the extracted chemical. Advances made by the PQRI Orally Inhaled and Nasal Drug Product (OINDP) Leachables and Extractables Working Group in establishing a safety concern threshold (SCT) for leachables have been directly transferred, with a few assumptions, to extractables.[17] This allows the analyst to ignore extractables below a calculated threshold concentration. This section will deal with approaches to the establishment of an analytical evaluation threshold (AET) for extractables encountered in material screening.

Information that needs to be available to calculate the threshold includes the SCT that will be used to assess leachables in the final drug product, the weight of the CCS or device component that contacts the drug product, and the maximum number of doses per day a patient will be taking when using the drug product.

A worst-case extractables threshold can be calculated as follows:

$$\text{AET } (\mu g/g) = \left[\left(\frac{\text{SCT } (\mu g)}{\text{day}} \right) \div \left(\frac{\text{max \# of doses}}{\text{day}} \right) \div \left(\frac{\text{wt of component } (g)}{\text{dose}} \right) \right] \times (\text{UF})$$

AET	= extractables analytical evaluation threshold for a component of a CCS or device ($\mu g/g$).
SCT	= safety concern threshold for the drug product ($\mu g/day$).
Max \# of doses	= maximum number of doses a patient would take per day.
Wt of component	= weight of the component per dose (g/unit dose).
UF	= uncertainly factor used to correct for sources of uncertainty with the methodology. The authors recommend starting with

a default value of 0.5 and adjusting this factor downward to account for known uncertainty in the design.

12.1.8.1 Example: Application of an Extractables AET

A design team wants to select a plastic material for use in a nasal drug product. The bottle will be made from a pliable material to allow the patient to discharge a dose by squeezing the bottle. Several potential plastic materials were identified, and now the team must narrow the choice down to one material that will be advanced for controlled extractables testing/confirmation and into full device development. Running through a checklist of material information needed prior to making a final selection, it was realized that an evaluation of potential leachables was missing. The regulatory chemistry, manufacturing, and control (CMC) team member indicated that an SCT of 0.15 µg/day will be required for this drug product. The engineering team member stated that the delivery bottle is projected to hold sixty 0.5-mL doses of drug product solution and the bottle weight will be approximately 20 g. A maximum of five doses per day was supplied by the clinical team's representative.

The project team calculated the AET for the bottle as

$$AET_{bottle} = [(0.15)/(5)/(\sim 20/60)] \times (0.35) = 0.032 \text{ µg/g of plastic bottle material.}$$

In this calculation, the weight of the component per dose is ~20 g divided by 60 doses or ~0.333 g/dose. An uncertainly factor of 0.35 versus the default value of 0.5 was chosen to account for the uncertainty in design parameters at this stage (i.e., the approximate weight of the plastic bottle material). The engineering group indicated that the bottle weight would not vary by more than ±30%; therefore, the UF was calculated as $0.5 \times (1 - 0.3) = 0.35$ uncertainty factor.

A key and necessary assumption in using this technique is the concept of "worst-case" consideration. Since many parameters may not be exactly known at this stage in the development process, one must always consider the worst case within the boundaries of the uncertainty. Failure to follow this practice can result in unpleasant surprise leachables in the drug product discovered during ICH lot testing.

The AET calculation allows the team to draw a cutoff line in terms of concentration below which potential extractables can be ignored. Exceptions to this are special case compounds.[17] In the above example, the team is now alerted that additives and impurities in the plastic material above a level of 0.032 µg/g may show up as chemicals requiring a safety evaluation and monitoring by analytical methods. Knowing the extractables AET for the component material is very useful if chemical composition data are available on the material (formulation data, TD/GC/MS data, literature data, etc.). Potential extractables in the material can be assessed for risk based on chemical identification and concentration level.

12.2 ANALYTICAL DATA REQUIREMENTS FOR A TOXICOLOGICAL ASSESSMENT

Extractables analytical data requirements for assessments submitted to a toxicologist are listed in Table 12.1. A toxicological assessment is a preliminary review of the toxicological data for the chemical in question. This should be distinguished from

TABLE 12.1.   Analytical Data Requirements

Information supplied	Description	Example
Chemical Abstracts Service (CAS) #	A unique identification number assigned by the CAS to substances that have been reported in the scientific literature	128-37-0
Chemical name	The common chemical name is often preferred if a CAS # is available. Otherwise, the full chemical name should be supplied.	Butylated hydroxytoluene (BHT) [IUPAC name: 2,6-bis(1,1-dimethylethyl)-4-methylphenol]
Chemical structure	An electronic structure of the chemical is often included to facilitate searching for toxicological information in specialized databases. Must have structure if no CAS # is provided.	H3C / CH3 / C—CH3 / CH3 / H3C / C / OH / CH3 / H3C
Max. TDI (µg/day)	Estimate of maximum patient daily exposure from taking the maximum label claim dose of the drug product.	130 µg/day

IUPAC, International Union of Pure and Applied Chemistry.

a toxicological safety qualification of a leachable. The subject of toxicological risk assessment is detailed in the next section of this chapter. Briefly, the difference in level of assessment can be described as follows:

- *Toxicological assessment of an extractable (or potential extractable)*—based on an estimated maximum dose a patient might receive by taking the drug product using only a preliminary review of the toxicological data.

- *Toxicological qualification of a leachable (or an extractable determined by validated method)*—based on the actual projected dose a patient would receive by taking the maximum recommended label claim of the drug product. Must be determined by an analytically validated method. The safety qualification will include an extensive search for all safety data associated with the chemical and may include animal testing specially conducted for the qualification report.

12.3   THE RISK ASSESSMENT OF EXTRACTABLES AND LEACHABLES: THE TOXICOLOGIST'S PERSPECTIVE

There are currently no universally accepted regulatory guidelines established addressing the appropriate risk assessment of extractables and leachables. Therefore, the toxicologist should approach this task with the same degree of due diligence that should be employed when risk assessing other impurity-related issues or, for that matter, any other chemical that a human will or possibly be exposed to. In doing so, the toxicologist is assured that the risk from such chemicals has been adequately assessed, and therefore, the safety of the product that contains them is determined to the greatest degree of certainty. In addition, an effective coordinated strategy integrating the examination of safety data for extractables versus leachables is warranted, even though only the latter chemicals are actually found in the drug. In this regard, the risk assessment of extractables by the toxicologist is as critical to the successful and timely development of a potential drug as determining the safety of the leachables. Toward this end, in this section, we outline an integrated approach to the risk assessment of both extractables and leachables, in addition to describing appropriate research processes needed for both types of chemicals to ensure a successful resolution to this critical drug development issue.

12.3.1 Level of Detection Requirements

The inclusion of the toxicologist in the process of selecting materials needed for the manufacture of drug delivery devices and CCSs for parenteral products is often overlooked, but represents the first critical step where such input is needed. In this regard, the toxicologist brings to such discussions knowledge of important safety issues that need to be considered when dealing with the purchase of material components from suppliers. For example, the selection of a black rubber stopper for a vial to hold an intravenously administered drug may seem innocuous, but such stoppers may possess carbon black, a filler that contains the known human carcinogen benzo[a]pyrene. In addition, some rubber products used to manufacture stoppers may contain latex, a material that can elicit a serious allergic reaction if it is leached into a product that is subsequently administered to a sensitive individual. Therefore, including the toxicologist in discussions dealing with material selection may help to identify a potential safety issue at its earliest stage, thereby avoiding the unfortunate scenario of first discovering it during the conduct of extractables or leachables studies.

It is important to note that some suppliers are reticent to offer a great deal of information about their materials for a number of reasons including propriety issues, but many will make some or all relevant data available if these are requested. Not only can the toxicologist help to ascertain from the manufacturer's information any possible safety issues from a potential component as described previously, but they may also help facilitate the selection process by determining what important compendial testing may or may not have been performed on the materials being considered. One must remember that a manufacturer may not be producing materials that were meant or perceived to be used in a CCS for a drug, and as such would not have subjected them to compendial testing required of materials meant for such use. For example, the appropriate application of container regulatory requirements in USP general chapters <87> (Biological Reactivity Tests; *In Vitro*), <88> (Biological Reactivity Tests; *In Vivo*), <381> (Elastomeric Closures for Injection), and <661> (Containers; Plastic) are very important to consider during the material selection process. If such testing has not been performed, or if a potential safety issue is apparent in a particular constituent, the toxicologist can help assess the resources and effort needed to rectify the situation, thereby aiding in determining if further consideration of the materials in question is warranted. This is the risk determination that is made at the first decision point illustrated in Figure 12.1.

Collectively, as a member of the material selection team, the toxicologist helps to ensure that appropriate safety and compendial testing information is obtained from the manufacturer on all of the components being considered. Upon examining such data, the toxicologist may help identify a potential issue that otherwise might not have become apparent until much later in the development process.

12.3.2 The Risk Assessment of Extractables

As discussed earlier in this chapter, extractables data may be generated during material selection. Once the material selection process is completed, the individual components are subjected to controlled extraction studies. These data are generated prior to determining the leachables profile. It is important to note at this point that the extractables and leachables studies, and the ensuing risk assessment procedures we describe for each, are best performed on packaging components that are as close to, if not the exact ones intended for the market. In this regard, alterations in packaging components can change the impurity profiles that are generated in both tests, which could compromise the risk assessments that were previously performed.

From a toxicology perspective, formal reportable risk assessments are typically not performed on extractables. In this regard, extractables are not representative of "real-life" situations, since aggressive manipulation of the components (e.g., with various solvents, elevated temperatures) is used to determine what chemicals might possibly leach into the drug (i.e., the extractables), not what actually makes it into the product under more realistic circumstances (i.e., leachables). Nevertheless, since the ensuing leachables will typically be a subset of the extractables, the latter warrant examination from a safety perspective, albeit not to the extent of the leachables. Thus, from a practical scientific and resource standpoint, more thorough and exhaustive safety assessments should be reserved for the leachables that are identified. For extractables, we recommend a more targeted and resource sparing procedure to assess their potential safety impact. In this regard, limiting the research to assess the potential for an extractable to elicit genotoxic, carcinogenic, sensitization, and adverse reproductive effects addresses the most critical concerns at this stage. In addition, *in silico* analysis of chemical structures by systems such as Deductive Estimation of Risk through Existing Knowledge (DEREK) is important, especially if there are little safety data available for a particular extractable. Using this research scenario, the toxicologist can assess potential risk from a purely qualitative perspective, and if no concerns arise, the risk is considered acceptable (see Fig. 12.1) and it is justified to continue development with a fair degree of certainty that a problematic issue will not arise in the subsequent leachables studies. If a particular extractable does prove to have potential liability, then the toxicologist may move to a quantitative analysis and assume, however unlikely, that the entire amount of extractable identified will occur in the leachables study. In assuming this unlikely scenario, if the risk can be mitigated, no further action is required and development should continue. Lastly, if the risk in the quantitative analysis cannot be mitigated, the potential benefit of switching materials to obtain a more favorable extractables profile should be considered.

Thus, in performing this targeted analysis, the toxicologist concentrates on the most problematic of toxicology issues without sparing an inordinate amount of resources, and in doing so can help mitigate a potential problem that might otherwise not be identified until the leachables work. In addition, knowing the number of extractables helps the toxicologist plan for proper resource deployment that will be needed for the more extensive leachables risk assessment work. As part of an examination of potential resource use, the toxicologist should consider at this stage the amount of safety information available for a particular extractable, and whether it is sufficient to support a subsequent safety assessment if the chemical were to later be identified as a leachable. This is a critical step, in that in some instances, the lack of appropriate safety data may lead to the consideration of performing additional animal safety studies to account for the gaps that are identified if the chemical is later found to leach into the drug.

12.3.3 The Risk Assessment of Leachables

The toxicologist should use all resources available to obtain safety data to perform an exhaustive risk assessment for leachables. In addition to performing a comprehensive search of the published literature, there are various databases available (e.g., Micromedex®, TOXNET®, and the National Toxicology Program) that offer a wide range of data to aid in performing the assessment. In addition, organizations such as the Agency for Toxic Substances and Disease Registry (ATSDR), the World Health Organization (WHO), the Occupational Safety and Health Administration (OSHA), the United States Environmental Protection Agency (EPA), the National Institute of Occupational Safety and Health (NIOSH), and the American Conference of Governmental Industrial Hygienists (ACGIH®) often have valuable toxicology data on a wide range of chemicals, as well as recommended or legally enforceable acceptable daily intakes for many of them. Regarding the latter, while they may not deal specifically with exposure to a chemical as a leachable, they remain valuable as part of a risk assessment since they are based on sound scientific and toxicological principles. An *in silico* structural analysis of the chemical by DEREK or a similar platform can also be performed if deemed necessary in the light of the literature information, or lack thereof, obtained. Once the research has been completed, the art of performing the risk assessment begins, based on the data at hand. The toxicologist must consider a wide range of different factors that will ultimately determine the safety of a particular leachable in a specific drug. Thus, the risk assessment is tailor made for a particular situation, and what is acceptable in one product may not be for another. We outline some of these critical factors below and define their relevance to the process.

12.3.3.1 Dose Paracelsus, the "father" of toxicology (see Fig. 3.1) wrote, "Alle Ding' sind Gift und nichts ohn' gift: allein die Dosis macht, dass ein Ding kein Gift ist"—"All things are poison and nothing is without poison, only the dose permits something not to be poisonous." Thus, he established the mantra of the science of toxicology, which is often shortened to "the dose makes the poison." To the toxicologist, there is no such thing as a nontoxic substance, but there is such a thing as a nontoxic dose of it. Not surprisingly therefore, the most critical factor to consider when risk assessing a leachable is to determine under conditions of maximum daily use of the drug, how much of a total daily dose of each individual leachable will be coadministered to humans. The analytical chemist typically reports leachables concentrations in the drug using parts per million, a nomenclature that has little use to the toxicologist. In this regard, the parts per million of a leachable in a drug can often be misleading from a toxicology standpoint until the dose calculation is performed. For example, if the same leachable is present in drug A and drug B at 1 and 100 ppm, respectively, one might think that the larger amount of the impurity in the latter drug would impart more potential risk. If, however, the total daily doses of drug A and drug B happen to be 1 g/day and 10 mg/day, respectively, the corresponding dose of the leachable in both cases would be the same (i.e., 1 μg). Thus, establishing the actual daily dose of the leachable is the first critical step in the risk assessment process.

12.3.3.2 Duration
How long the patient will be exposed to a leachable must be considered. Chronic, daily exposure to a potential toxin carries a higher theoretical risk of eliciting an adverse effect than short-term administration. Thus, greater scrutiny of safety data supporting a leachable should be paid to those in chronic use drugs. If there are chronic animal and/or human safety data available for a leachable that will be administered in a short-term use drug, the toxicologist can support safety from the perspective of duration of exposure as long as appropriate margins of safety are present (i.e., differences between no effect levels in the research data vs. the dose humans will receive), and the intrinsic toxicological profile of the leachable is not deemed unreasonably hazardous. Alternatively, if only short-term animal and/or human safety data are found for a leachable that will be administered chronically, the toxicologist may have to consider performing more animal toxicity testing with the chemical to support the assessment, or apply additional safety factors (to account for the likelihood that no effect levels would become lower with longer treatment) as deemed appropriate. Regarding the latter, if there are sufficient data available characterizing the dose and time relationship of a leachable's toxicity in shorter duration studies, the toxicologist can use them to extrapolate what would likely occur following longer-term administration.

12.3.3.3 Route of Administration
The toxicological profile of any chemical can be affected by the route of administration, and is therefore a critical point to consider when performing a risk assessment. A poorly absorbed leachable may not produce toxicity by the oral route, but intravenous administration of that same chemical could elicit an adverse event. Thus, ideally, an assessment should be based on animal and human toxicity data via the expected route of administration. Realistically, however, such data may not be available. In such cases, it is appropriate to support the assessment by toxicity data from other routes of administration. Such assessments should take into account differences in bioavailability via different

routes, and appropriate correction factors applied. For this reason, animal and human intravenous toxicity data (representing 100% bioavailability) tend to easily support an assessment for a leachable that will be administered orally, but the converse may require adjusting the no effect doses in such studies to appropriately apply them to the actual route of human exposure. One special circumstance deserves mention regarding route-related adverse toxicity. Secondary, tertiary, and quaternary amines, when administered by the oral route in the presence of nitrites (common in food) and the acidic environment of the stomach, can transform into cancer causing nitrosamines. There are data available in many cases to estimate what percent conversion may occur to aid in risk assessing such cases, but in the absence of such information, the toxicologist must assume a stoichiometric 1:1 conversion.

12.3.3.4 *Patient Population and Indication*   The overall health status of the patient, and the indication that they are being treated for, must be considered when assessing the potential risk of a leachable. Thus, if the target organ toxicity of a leachable is determined to be the kidney, and the drug containing it is to treat diabetes where renal impairment is not uncommon, additional scrutiny in assessing risk would be warranted. Alternatively, if a drug is indicated to treat a life-threatening condition such as cancer, a greater degree of latitude regarding the potential safety issues for any given leachable is acceptable. It is difficult to justify withholding such a drug from a patient whose life could benefit from receiving it, but follow-up work on improving the packaging to help diminish the potential exposure to such a leachable, in concert with continued development of the drug, should be considered if possible.

12.3.3.5 *"Special" Populations*   Factors such as potential administration of a drug to a pediatric population, or to the elderly, need to be understood to better assess the potential risk of a leachable. In this regard, age-associated increases to toxicity of a wide range of different chemicals are well documented in animals and humans. If such animal or human toxicity data are not available for a leachable in this type of situation, the risk assessment should include appropriate leverage of additional safety factors to ensure the risk is minimized.

12.3.3.6 *Women of Childbearing Potential*   Reproductive toxicology data for the leachable would be needed to support safety if women of childbearing potential are to be administered the drug. Special attention should be paid to data addressing teratogenicity, and if such data are not available, additional safety factors should be applied to a calculated safe dose to support the assessment.

12.3.3.7 *Standard Layout of Data for a Risk Assessment*   We utilize a standard layout of presenting the data, where available, to support the assessment of a leachable. We recommend a bulleted presentation of each critical study or relevant data, as it allows for more succinct presentation of the information to allow easier review by a regulator. The major areas for data search and presentation we utilize

are presented below. It is recommended that where such data are not available, that it be clearly stated in the assessment:

- *DEREK structural analysis (if needed).* The use of *in silico* systems to estimate the potential liabilities of a leachable is at the discretion of the toxicologist. Usually, such assessments are not necessary if there are adequate published safety data for a particular leachable, but the use of such technology is an important tool if a paucity of information exists. While many such systems can help to predict potential systemic toxicity issues, they are typically stronger at estimating genotoxic and carcinogenic potential since the structure toxicity relationship data for these effects are more comprehensive and understood. Importantly, however, a "hit" for genotoxicity and/or carcinogenicity warrants follow-up analysis of the structure by the toxicologist to gain the proper perspective, since the alert triggered by a particular moiety of the molecule may not be relevant in the light of the additional chemical constituents surrounding it.

- *Genetic toxicology.* These data are critical to understanding the potential for the most serious of adverse events. If sufficient genetic toxicology data are not available, *in silico* analysis of the chemical structure is surely warranted.

- *Acute toxicity.* While acute animal toxicity data tend to only report mortality (e.g., LD50 values) and not morbidity, they are still important to consider since they aid in determining the overall potency of a toxicant. They can be particularly useful in supporting the assessment of a leachable in a short-term use product, but should be included in all cases to help establish the overall safety profile of the chemical.

- *Subchronic toxicity.* Studies up to 6 months' duration in animals.

- *Chronic toxicity.* Studies up to 1 year duration in animals. These data, as well as from subchronic toxicity studies listed previously, are appropriate for inclusion in multidose risk assessments.

- *Carcinogenicity.* Of obvious importance is the inclusion of carcinogenicity data when available, and there is no conceivable circumstance not to do so. It is important to note that a leachable that is determined to be a nongenotoxic carcinogen can be risk assessed, since the concept of a threshold dose can be applied.

- *Reproductive toxicology.* Of obvious importance if women of childbearing potential could be exposed to the leachable.

- *Irritation and sensitization.* For in intravenous or ocular drug containing a leachable, determining if it is an irritant can be of great importance. Sensitizers are a potential major issue to consider, especially for a drug to be administered via the pulmonary route.

- *Human toxicity data.* The most critical and relevant of all data, but often the most difficult to find. Nevertheless, acute poisoning data, as well as any available clinical toxicity data greatly strengthen the assessment, and diminish the use of applying safety factors in cases when only animal toxicity data are available. Often, if a particular leachable is found to be a normal constituent of the human body, risk can often be assessed by calculating the additional body burden that will result from the administration of the drug and comparing the results (e.g., anticipated blood levels) to known human toxicity data.

- *Special populations (e.g., pediatrics).* It is important to know the target population. A pediatric indication will require consideration of applying additional safety factors to the data if only adult animal and/or adult human safety information are available.

- *Regulatory guidance information (e.g., OSHA, EPA, or WHO).* While not meant to address the safety of a chemical as a leachable, regulatory guideline information (e.g., OSHA permissible exposure limits) is extremely useful to an assessment, since it is calculated by considering a wide range of animal and human safety information.

- *Conducting the assessment.* Once all of the information has been gathered for the assessment, and the toxicologist understands the potential liabilities inherent in the leachable, the risk can then be determined based on the various factors previously described. The toxicologist brings to bear knowledge of the leachable's end-organ toxicity, as well as dose and time relationship information, and applies it to the clinical situation at hand. No effect levels for animal and human toxicity data are compared with the anticipated human dose, and the toxicologist determines if appropriate safety margins exist to support safety. If there are a great deal of data available, this can be readily determined. On the other hand, if some data are lacking, the toxicologist should employ appropriate safety factors to account for unknowns. By convention, factors of 10 are used for each relevant unknown. For instance, if there are no human oral safety data available, but corresponding animal toxicity data are plentiful, the toxicologist may take the lowest no effect level in the most relevant animal study and apply safety factors to account for extrapolation to humans (10×), and to account for individual human sensitivity that may occur (10×). In this instance, if the extrapolated lower no effect dose still provides coverage for the anticipated human dose, the assessment is completed. Having human toxicity data available may prevent the need to apply such safety factors, but the toxicologist needs to be assured of the quality of the data at hand to make such a judgment. Moreover, the quality of the animal toxicity studies also needs to be considered when performing an assessment.

- *The threshold concept.* The concept of thresholds is becoming a very important tool in the science of toxicology. In this regard, it is conceivable that for just about any given chemical, there is a dose below which the risk to humans would be universally acceptable. This concept has recently been employed by the PQRI for establishing safe thresholds for leachables present in OINDPs. For compounds delivered by this route, the PQRI recommendations should be followed, as they are based on extensive retrospective examination of animal and human safety data for a wide range of toxic compounds. Thus, the threshold concept developed by PQRI, described in previous chapters, would ease the burden of establishing safe doses of leachables delivered by the pulmonary route.

12.4   CASE STUDY: SELECTION OF A SUITABLE DRUG PRODUCT CONTAINER MATERIAL FOR A DRY POWDER INHALER

12.4.1   Situation

A dry powder inhaler development team was in the process of selecting a plastic film that could be laminated to other materials to form a blister package for holding a single dose of drug product. Moisture penetration through the packaging material was being addressed by use of an aluminum foil to form the outer laminate of the packaging blister. For the inner layer of the laminate, which would be in direct contact with the drug surface, several materials were under consideration. The engineering development team needed to control the stiffness of the laminate plastic film and minimize its thickness. Polyvinyl chloride (PVC) was a material of choice for this application due to the large range of film stiffness control achievable by variation of plasticizer concentration incorporated into the PVC plastic. The team was also concerned with potential leachables that might migrate into the drug product powder over a planned 2-year shelf life for the drug product.

The team narrowed down the choice to two types of PVC films:

- PVC-S is made from a suspension polymerization process. In this process, vinyl chloride monomer (VCM) is suspended in water and allowed to polymerize to PVC. Proprietary additives are added to help maintain the suspension. After the polymerization step, the PVC is recovered and dried. The PVC is then formulated with additional proprietary additives including plasticizers to give the desired properties prior to being rolled into a thin film.

- PVC-B is made using a waterless process. Polymerization of the VCM occurs in a special reactor that removes the excess heat by condensation of VCM on the jacketed reactor walls via an external coolant. After completion of the polymerization process, the PVC is treated to remove residual VCM and proprietary additives are mixed into the polymer. A potential disadvantage of the PVC-B process is possible higher trace concentrations of the VCM than those in the PVC-S material.

The engineering team was interested in learning about potential leachables that might migrate from the PVC films. The team attempted to get a list of additives used in making both films. This was refused by the PVC manufacturers stating trade secret concerns. Offers to sign confidentiality agreements for the information were explored. After 2 months of legal team discussions, no mutually acceptable agreement could be reached. At this point, the analytical team was contacted and asked to evaluate both types of PVC film.

12.4.2   Analytical Evaluation of the PVC Films

The analyst calculated an AET for each film thickness based on the amount of film surface area required for a blister:

TABLE 12.2. Analysis of PVC Films

RT (minutes)	Name	CAS #	PVC-B 15 μm thickness	PVC-B 30 μm thickness	PVC-S 60 μm thickness
3.837	Methyl methacrylate	80-62-6	58.0	55.5	16.4
5.683	Maleic anhydride	108-31-6	160.9	140.0	ND
7.544	Glycerin	56-81-5	ND	ND	60.3
8.382	2-Ethyl-1-hexanol	104-76-7	ND	ND	1208.3
10.201	Acetic acid, 2-ethylhexyl ester	103-09-3	ND	ND	45.4
14.178	2-Ethylhexyl mercaptoacetate	7659-86-1	ND	ND	2429
15.169	Butylated hydroxytoluene	128-37-0	6.9	23.5	ND
16.857	Octane, 1,1′-oxybis-	629-82-3	69.5	65.4	240.6

RT, retention time; CAS, Chemical Abstracts Service; ND, not detected.

$$AET\ (\mu g/g) = \left[\left(\frac{SCT\ (\mu g)}{day}\right) \div \left(\frac{max\ \#\ of\ doses}{day}\right) \div \left(\frac{wt\ of\ component\ (g)}{dose}\right)\right] \times (UF),$$

$$AET\ 15\ \mu m\ PVC = \left[\left(\frac{0.15\ \mu g}{day}\right) \div \left(\frac{10}{day}\right) \div \left(\frac{0.005\ g}{dose}\right)\right] \times (0.5) \rightarrow 1.5\ \mu g/g\ PVC,$$

$$AET\ 30\ \mu m\ PVC = \left[\left(\frac{0.15\ \mu g}{day}\right) \div \left(\frac{10}{day}\right) \div \left(\frac{0.01\ g}{dose}\right)\right] \times (0.5) \rightarrow 0.75\ \mu g/g\ PVC,$$

$$AET\ 60\ \mu m\ PVC = \left[\left(\frac{0.15\ \mu g}{day}\right) \div \left(\frac{10}{day}\right) \div \left(\frac{0.03\ g}{dose}\right)\right] \times (0.5) \rightarrow 0.25\ \mu g/g\ PVC.$$

The analytical team proposed to screen both types of PVC films using TD/GC/MS analysis. Three samples of film were received for analysis: PVC-B samples of 15- and 30-μm-thickness film and a PVC-S 60-μm-thick film. Approximately 5 mg of each film was analyzed using a Gerstel TDU/CIS4 autosampler (Gerstel Inc., Linthicum, MD) and Agilent 6890GC/5975MSD (Agilent Technologies, Santa Clara, CA) with an HP-5MS 30 m × 0.25 mm 0.25 μm film column. The maximum desorption temperature was set to 180°C.

Results of the analysis are presented in Table 12.2. Estimated concentrations reported were calculated using an external standard of butylated hydroxytoluene (BHT). Tentative identification of extractables was made by a fragmentation pattern match to a 2006 NIST/Wiley library database.

Initial inspection of the concentration of thermal extractables from the PVC films indicated that the PVC-B films have significantly lower levels of additives than the PVC-S film. Maleic anhydride seen in the PVC-B films was confirmed as a thermal stabilizer for PVC by literature.[18] The 2-ethyl-1-hexanol, 2-ethyl-1-hexanol acetic acid ester, and the 2-ethylhexyl-mercaptoacetate are likely degradation products of dioctyltin bis(2-ethylhexyl mercaptoacetate), a known thermal stabilizer for PVC.[19] The octane, 1,1′-oxybis is commonly used as an antistatic agent and/or lubricant.[20] Based on this information, the PVC-B films seemed favorable and the data were provided to the toxicologist for evaluation.

Calculations of the amount of polymer film in direct contact with the drug product powder on a per dose basis enabled the determination of an estimated maximum total daily intake (TDI).

12.4.3 Toxicological Assessment of the Extraction Profile of the Drug Product Container Material for a Dry Powder Inhaler

Referring back to Table 12.2, and focusing on the PVC-B films, eight chemicals were identified through TD extraction studies. Employing the methods just described for toxicological assessment, only one chemical, maleic anhydride, was cause for concern. A summary of the risk assessment and the concern for potential human risk follows.

Short et al. chronically exposed rats, hamsters, and rhesus monkeys to maleic anhydride by inhalation.[21] Four groups of each species were exposed to concentrations of 0, 1.1, 3.3, or 9.8 mg/m^3 maleic anhydride for 6 months in stainless steel and glass inhalation chambers. All species exposed to any level of maleic anhydride showed signs of irritation of the nose and eyes, with nasal discharge, dyspnea, and sneezing reported frequently. The severity of symptoms was reported to increase with increased dose. No effects on pulmonary function in monkeys were observed. Dose-related increases in the incidence of hyperplastic change in the nasal epithelium occurred in rats in all exposed groups, and in hamsters in the mid- and high-dose groups. Neutrophilic infiltration of the epithelium of the nasal tissue was observed in all species examined at all exposure levels. All changes in the nasal tissues were judged to be reversible.

In many occupational situations, workers are exposed to mixtures of acid anhydrides, including maleic anhydride, phthalic anhydride, and trimellitic anhydride. For example, Barker et al. studied a cohort of 506 workers exposed to these anhydrides.[22] In one factory, workers were exposed only to trimellitic anhydride, which has the lowest acceptable occupational exposure limit (40 $\mu g/m^3$) of the three anhydrides. In that factory, there was an increased prevalence of sensitization to acid anhydride and work-related respiratory symptoms with increasing full-shift exposure even extending down to levels below the current occupational standard. However, none of the workplaces had exposure only to maleic anhydride, and a dose–response relationship was not seen with mixed exposures.

There are several case reports describing asthmatic responses possibly resulting from exposure to maleic anhydride. An individual showed an acute asthmatic reaction after exposure to dust containing maleic anhydride.[23] Workplace concentrations of maleic anhydride were 0.83 mg/m^3 in the inspirable particulate

BOX *12.1*

HOW TO CALCULATE A PULMONARY DOSE OF A MATERIAL BASED ON A PARTS PER MILLION CONCENTRATION

Exposure to materials in the workplace and in animal inhalation toxicology studies are often reported as a concentration, typically in parts per million. To calculate a dose, parts per million (a generic term denoting the number of units of one substance relative to 1 million units of another substance) must first be converted to a compound-specific volume-related concentration (e.g., milligram per liter). Thus, in the example given, 1.5 ppm of maleic anhydride at standard temperature and pressure equates to a compound-specific volume-related concentration of 6 mg/m^3, which is subsequently converted to units of milligram per liter by dividing by 1000, the latter being the number of liters in a cubic meter of air. The conversion of a given parts per million of a substance to a compound-specific volume-related concentration is dependent on the molecular weight of the gas or vapor in question. There are various published sources[27] as well as online computational tools available to perform this calculation. In the given example, the concentration of 0.006 mg/L is multiplied by the time of exposure (i.e., 1 minute) and the average minute ventilation of a human (i.e., 9 L/min) to obtain a value of 0.05 mg (i.e., 50 µg), which when divided by the weight of the subject (i.e., a 50 kg human) results in a final dose of 1 µg/kg.

mass and 0.17 mg/m^3 in the respirable particulate mass. Bronchial provocation testing was performed with phthalic anhydride, lactose, and maleic anhydride. Exposure of this individual to maleic anhydride (by bronchial provocation testing) at 0.83 and 0.09 mg/m^3 in inspirable and respirable particulate mass, respectively, showed a response of cough, rhinitis, and tearing within 2 minutes. Within 30 minutes, rales developed in both lungs and peak flow rate decreased 55%. An individual occupationally exposed to maleic anhydride developed wheezing and dyspnea upon exposure.[24] Another case report described occupational asthma due to exposure to maleic anhydride.[25]

Based on concentration information obtained from the extraction study, the maximum TDI was estimated to be 50 µg/day. The National Toxicology Program[26] reports that human nasal irritation occurs within 1 minute following exposure to 1.5 ppm maleic anhydride. This concentration equates to a dose of 1 µg/kg (1.5 ppm [6 mg/m^3]/1000 = 0.006 mg/L × 1 minute × 9 L/min/50 kg = 1 µg/kg). This dose of 1 µg/kg/min for nasal irritation is the same dose from exposure to maleic anhydride from the blisters (50 µg/50 kg human).

There was no safety margin based on these estimates, and humans would be exposed to repeated daily doses of the inhaled dry powder possibly increasing the sensitization and/or irritation potential to maleic anhydride. Therefore, the likelihood for maleic anhydride to leach into the drug substance was considered a significant safety risk. The team was informed of this assessment with the recommendation that the team should not use the proposed blister configuration as the a primary container closure for the drug substance. Although the drug product had already been prepared, the team acknowledged that there was significant risk and decided to develop an alternative blister packaging and not use the prepared drug product.

12.5 CONCLUSION

This chapter highlighted the criticality of the chemical composition of the drug product CCS. A significant expenditure of time and investment needs to be allotted early in the drug product life cycle to ensure that there are no significant risks to patients from chemicals that may leach from the CCS. The disadvantage and potential clinical risk one could face by selecting materials late in the drug product life cycle was also shown. As was demonstrated by the case example, the primary drug product container closure for a dry powder inhaler could have leached a potentially toxic chemical from the materials used to form a multilaminate blister. Fortunately, the potential hazard was identified before the drug product was used in clinical trials and the potential for a significant adverse event was prevented.

REFERENCES

1 Code of Federal Regulations. Medical devices, quality system regulation, design controls. Title 21, Volume 8, Section 820.30, revised April 1, 2011.

2 Council of the European Communities. Council directive 93/42/EEC of 14 June 1993 concerning medical devices. *OJ no. L 169*, 1993.

3 International Organization for Standardization. ISO13485: Quality management systems for medical devices, 2003.

4 ICH harmonised tripartite guideline: Q8(R2) pharmaceutical development. International Conference on Harmonisation of Technical Requirements for Registration of Pharmaceuticals for Human Use, 2005.

5 Guidance for industry: Container closure systems for packaging human drugs and biologics chemistry, manufacturing, and controls documentation. US Food and Drug Administration, May 1999.

6 Guidance for industry: Nasal spray and inhalation solutions, suspension, and spray products—Chemistry, manufacturing, and controls documentation. US Food and Drug Administration, July 2002.

7 Guideline on plastic immediate packaging materials, CPMP/QWP/4359/03 and EMEA/CUMP/205/04, EMEA, 2005.

8 Shaw, A.J. Pharmaceutical manufacturer's perspective on extractables and leachables. *PIRA Conference*, October 16, 2007.

9 ICH harmonised tripartite guideline: Q9 quality risk management. International Conference on Harmonisation of Technical Requirements for Registration of Pharmaceuticals for Human Use, 2005.

10 Chen, C.W. Quality by design approach to extractables and leachables—An FDA perspective. PDA Extractables /Leachables Forum, Bethesda, MD, November 6–8, 2007.

11 Shlyapnikov, Y.A., Kiryushkin, S.G., and Mar'in, A.P. *Antioxidative Stabilization of Polymers*. Taylor & Francis, London, 1996.

12 Zweifel, H. *Stabilization of Polymeric Materials (Macromolecular Systems—Materials Approach)*. Springer-Verlag, New York, 1997.

13 Hamid, S.H., ed. *Handbook of Polymer Degradation*, 2nd ed. Marcel Dekker, New York, 2000.

14 Crompton, T.R. *Thermo-Oxidative Degradation of Polymers*. iSmithers Rapra Publishing, Shrewsbury, 2010.

15 FDA website. Drug master files (DMFs), August 9, 2011. Available at: http://www.fda.gov/Drugs/DevelopmentApprovalProcess/FormsSubmissionRequirements/DrugMasterFilesDMFs/default.htm#guidance (accessed September 5, 2011).

16 Zweiben, C. and Shaw, A.J. Use of thermal desorption GC-MS to characterize packaging materials for potential extractables PDA. *J Pharm Sci Technol* 2009, *63*(4), pp. 353–359.

17 Norwood, D.L. and Ball, D. Product Quality Research Institute: Safety thresholds and best practices for extractables and leachables in orally inhaled and nasal drug products. Submitted to the PQRI

Drug Product Technical Committee, PQRI Steering Committee, and U.S. Food and Drug Administration by the PQRI Leachables and Extractables Working Group, 2006.

18 Kelen, T., Ivan, B., Nagy, T.T., et al. Reversible crosslinking during thermal degradation of PVC. *Polym Bull* 1978, *1*, pp. 79–84.

19 Parnell, S. and Min, K. Reaction kinetics of thermoplastic polyurethane polymerization in situ with poly(vinyl chloride). *Polymer* 2005, *46*, pp. 3649–3660.

20 The Physical and Theoretical Chemistry Laboratory Oxford University. Safety data for dioctyl ether. Available at: http://msds.chem.ox.ac.uk/DI/dioctyl_ether.html (accessed September 5, 2011).

21 Short, R.D., Johannsen, F.R., and Ulrich, C.E. A 6-month multispecies inhalation study with maleic anhydride. *Fundam Appl Toxicol* 1998, *10*, pp. 517–524.

22 Barker, R.D., van Tongeren, M.J., Harris, J.M., et al. Risk factors for sensitization and respiratory symptoms among workers exposed to acid anhydrides: A cohort study. *Occup Environ Med* 1998, *55*(10), pp. 684–691.

23 Lee, H.S., Wang, Y.T., Cheong, T.H., et al. Occupational asthma due to maleic anhydride. *Br J Ind Med* 1991, *48*, pp. 283–285.

24 Gannon, P.F.G., Burge, P.S., Hewlett, C., and Tee, R.D. Haemolytic anaemia in a case of occupational asthma due to maleic anhydride. *Br J Ind Med* 1992, *49*, pp. 142–143.

25 Guerin, J.C., Deschamps, O., Guillot, T.L., Chavallion, J.M., and Kalb, J.C.A. Propos d'un Casd'asthme a L'anhydride Maleique [A case of asthma due to maleic anhydride]. *Poumon Coeur* 1980, *36*, pp. 393–395.

26 National Toxicology Program. Maleic anhydride. CAS registry number: 108-31-6 Toxicity effects. Available at: http://ntp.niehs.nih.gov/index.cfm?objectid=E87C0F18-BDB5-82F8-F9C-5C1E3F6AE01ED (accessed September 2011).

27 Derelanko, M.J. *Toxicologist's Pocket Handbook*. CRC Press, New York, 2000; 57–60.

CHAPTER 13

ANALYTICAL TECHNIQUES FOR IDENTIFICATION AND QUANTITATION OF EXTRACTABLES AND LEACHABLES

Daniel L. Norwood, Thomas N. Feinberg, James O. Mullis, and Scott J. Pennino

13.1 INTRODUCTION

As detailed in earlier chapters of this volume, organic extractables and leachables can represent a wide variety of chemical types and classes and be present in both container closure system critical component extracts and drug products at widely varying concentrations. These realities suggest that the problems of qualitative and quantitative analysis of extractables/leachables are more akin to environmental trace analysis problems than to the typical pharmaceutical impurity problem, where one has the known structure and synthetic scheme for the drug substance along with well-established analytical thresholds as starting points. In fact, many of the chemical entities that commonly appear as extractables and leachables are also common environmental target analytes, including such compounds as phthalates, phenols, polyaromatic hydrocarbons (PAHs), and *N*-nitrosamines. Typical organic pharmaceutical impurities can be present in both drug substance[1-3] (e.g., starting materials, by-products, intermediates, degradation products, reagents, ligands, catalysts, residual solvents) and drug product[4] (e.g., degradation products of drug substance or reaction products of drug substance with an excipient and/or the immediate container closure system). Processes for identification/quantitation and control of these typical pharmaceutical impurities are described and referenced in the scientific literature[5,6] and various regulatory guidance documents.[1-4] However, it is important to note that the most widely applied regulatory guidance for drug product impurities, International Conference on Harmonisation (ICH) Q3B(R2),[4] specifically excludes from

Leachables and Extractables Handbook: Safety Evaluation, Qualification, and Best Practices Applied to Inhalation Drug Products, First Edition. Edited by Douglas J. Ball, Daniel L. Norwood, Cheryl L.M. Stults, Lee M. Nagao.

consideration "impurities . . . extracted or leached from the container closure system,"[4] recognizing the uniqueness of this particular issue. Proposals and guidance related to extractables/leachables for orally inhaled and nasal drug products (OINDPs) have been developed and published by the Product Quality Research Institute (PQRI),[7–9] again as detailed in earlier chapters.

Environmental analysis problems, and by inference extractables/leachables problems, are included within the general analytical chemistry classification termed "trace organic analysis" (TOA),[10,11] which also includes organic geochemical analysis, metabolite/drug profiling in biological matrices, and other similar problems. TOA can be defined as "the qualitative and/or quantitative analysis of a complex mixture of relatively trace level organic compounds contained in a complex matrix."[12] For environmental analysis, complex matrices include soil, natural waters, sludge, and ambient air. For extractables/leachables, the complex matrices are the container closure system critical component materials of construction (i.e., rubber and plastic) and the drug product formulation, respectively. This chapter outlines and describes the processes and technical considerations for the qualitative, quantitative, and structural analysis of extractables and leachables as guided by TOA principles. It is not practical within the bounds of a single chapter to describe in detail each of the analytical techniques that can be applied to the extractables/leachables problem. Neither is it feasible to present a significant level of detail regarding interpretation of data from these analytical techniques (e.g., volumes have been written about the interpretation of fragmentation reactions in mass spectrometry [MS]). Instead, analytical techniques and data interpretation will be presented within the context of actual extractables/leachables examples and case studies, with the goal of providing an appreciation and understanding of the philosophy and process of analytical chemistry as applied to the extractables/leachables issue.

13.2 TOA

In the comprehensive treatise entitled *Principles of Analytical Chemistry—A Textbook*,[13] Valcarcel presented and described a classification scheme for analytical chemistry according to

1. the purpose and type of information required (i.e., qualitative, quantitative, structural);
2. the analytical technique (i.e., classical, instrumental);
3. the nature of the sample analyzed and analytes to be determined (i.e., inorganic, organic, biochemical);
4. the initial size of the sample (i.e., macro, semi-micro, micro, ultra-micro); and
5. the relative proportions of analyte mass in the sample (i.e., macro, micro, trace).

As a "specially adapted analytical process,"[13] TOA is classified based primarily on the type of analyte (i.e., organic) and the analyte level (i.e., trace). In this context, the term "organic" refers to organic carbon-containing molecules generally below

1000 Da, and excludes from consideration traces of synthetic polymer substances[11] and biopolymers (e.g., DNA, proteins),[11] which both rely on highly specialized analytical techniques and strategies. The term "trace" refers to an analyte present in minor concentration relative to the matrix.[11] Valcarcel defined trace as 0.01% (100 ppm) or less of analyte relative to matrix.[13] Based on the PQRI safety concern threshold (SCT)[7,8] and analytical evaluation threshold (AET)[7,9] concepts presented and discussed in earlier chapters, the entire issue of extractables/leachables clearly falls within the bounds of TOA. Typical estimated AET values for individual organic leachables in metered dose inhaler (MDI) drug products are in the single microgram per canister range (i.e., approximately <0.5 ppm),[7] and the recommended threshold for the so-called noncontact critical component extractables evaluation is 20 ppm.[7]

Based on Valcarcel's classification scheme,[13] TOA problems can be further classified as *qualitative*, *quantitative*, or *structural*, considering the purpose and type of information required. *Qualitative* TOA problems are concerned with the identification of trace analytes present in a sample matrix, and ask the question "Are the target analytes present above a specified detection limit, yes or no?" *Quantitative* TOA problems are concerned with determination of the levels, or relative levels, of trace analytes present in a matrix as defined by predetermined parameters of sensitivity, precision, and accuracy. *Structural* TOA problems are concerned with elucidating the molecular structure of unknown trace-level analytes to a predetermined level of confidence. The sensitivity of detection required for TOA further classifies it based on analytical technique as *instrumental* rather than *classical* (i.e., weighing and titrations to visible end points have little application for TOA problems). TOA cannot be classified based on initial sample size, which distinguishes TOA from "microanalysis,"[13] since these can range from hundreds of liters of river water or kilograms of soil, to microliter samples of blood or plasma.

Any problem in TOA, including extractables/leachables, can be broken down into four stages:

1. *Information.* Acquiring knowledge and understanding of the chemical nature of the potential extractables/leachables and the matrix in which they are contained.

2. *Extraction.* Removal/extraction of the complex mixture of extractables/leachables from the matrix, and preparing/concentrating the extract in a manner consistent with the requirements of subsequently applied analytical techniques and the information required from the analysis (e.g., partitioning behavior, thermal lability, detection limit, quantitation limit, and accuracy).

3. *Separation.* Separation of the complex mixture of extractables/leachables into individual chemical entities in a manner suitable for sequential introduction into an appropriate detector.

4. *Detection.* Detection of the individual chemical entities within the complex mixture, generating qualitative, quantitative, and/or structural information for each individual analyte.

Consider for example, the MDI drug product with its container closure system critical components shown in Figure 13.1. For each critical component, the

Figure 13.1   Metered dose inhaler drug products (bottom) with "critical components," including canisters, valves, elastomeric and plastic valve components, and mouthpiece/actuators.

following information (step 1 above) should be available from the component supplier/manufacturer and the scientific/technical literature[7,14]:

- the elastomeric/polymeric or base material of construction of the component;
- the additive composition of the component, including the detailed chemical composition and reaction/degradation chemistry of each individual additive;
- the polymerization process and associated polymerization/curing agents;
- the fabrication process, including any additives designed to assist in fabrication or processes that could result in chemical modification of any additives or the polymer, for example, temperature;
- any cleaning washing processes for finished components, including knowledge of cleaning agents; and
- the storage/shipping environment for both components and drug product.

Obtaining complete information of this type from component suppliers is often a difficult task as multiple suppliers will likely be involved for any individual component (termed n-1, n-2, n-3, etc., suppliers); however, it is essential that due diligence be employed in the effort, and as much appropriate information as possible be obtained.

Consider, for example, the composition information available for the sulfur-cured elastomer (Table 13.1) whose controlled extraction study is described in detail in Chapter 15. How can this limited information be used to facilitate the design and conduct of the controlled extraction study, assuming that this elastomer is used to fabricate an MDI valve component?[7]

TABLE 13.1.   Ingredients in Sulfur-Cured Elastomer Test Article[7]

Ingredient	Registry #(s)	Percent (w/w)
Calcined clay	308063-94-7	8.96
Blanc fixe (barium sulfate)	7727-43-7	25.80
Crepe	9006-04-6	38.22
Brown sub MB (ingredients below)	Not available (NA)	16.84
Brown sub loose	NA	33.30
Crepe	9006-04-6	66.70
1722 MB (ingredients below)	NA	2.11
Standard Malaysian rubber (SMR)	NA	60.00
FEF carbon black (low PNA)	1333-86-4	40.00
Zinc oxide	1314-13-2	4.04
2,2'-Methylene-bis(6-*tert*-butyl-4-ethylphenol)	88-24-4	0.56
Coumarone-indene resin	164325-24-0	1.12
	140413-58-7	
	140413-55-4	
	68956-53-6	
	68955-30-6	
Paraffin	8002-74-2	1.12
	308069-08-1	
Tetramethylthiuram monosulfide	97-74-5	0.11
Zinc 2-mercaptobenzothiazole	149-30-4	0.29
	155-04-4	
Sulfur	7704-34-9	0.84

Consider the following:

- The presence of carbon black indicates the likely presence of PAHs in the elastomer (also termed polynuclear aromatics or PNAs). Note that PAHs are the so-called special case compounds (see Chapter 1) requiring specific consideration.[7,14,15]
- Sulfur-curing agents, such as tetramethylthiuram monosulfide, indicate the likely presence of *N*-nitrosamines, which is a second class of "special case" compound.[15]
- 2-Mercaptobenzothiazole is itself a "special case" compound.[15]
- Paraffin and coumarone-indene resin are natural product materials and therefore are likely to produce complex extractables profiles containing many structurally related chemical entities.
- Some of the individually named additives, such as 2,2'-methylene-bis(6-*tert*-butyl-4-ethylphenol), are known to be amenable to analysis by gas chromatography (GC) methods. Hence, GC and GC/MS will be required for separation and analysis of extractables mixtures.

So, it is known that all three types of "special case" compounds will require attention (i.e., highly specific analytical methods for their qualitative and quantitative analysis), and that GC-based analytical techniques will be required for more general extractables profiling. In addition, the fact that MDI drug products contain organic propellants and co-solvents indicates that organic solvents with similar solvating properties should be used in the extraction studies of corresponding container closure system critical components. Elastomers of this type are known to swell in organic solvents, which will serve to release larger quantities of extractables for analysis.

Fortunately, information on the drug product formulation, which is the matrix containing any leachables, is readily available to the pharmaceutical development scientist. For an MDI drug product, it is essential to know

- the identity of the propellant or drug product vehicle (e.g., chlorofluorocarbon, hydrofluorocarbon, aqueous);
- the identity and quantity of each active ingredient; and
- the identities and quantities of excipients and co-solvents (e.g., soy lecithin, ethanol).

Obviously, it is important to know whether the drug product is a solution or suspension, or perhaps both as is the case for certain combination products.

Once a thorough understanding of potential analytes, sample matrix, and required detection limits (e.g., the information desired, the appropriate AET) is acquired, laboratory procedures can be developed to remove/extract potential analytes from the sample matrix and prepare the resulting extract(s) for analysis. For example, an elastomeric critical component could be Soxhlet extracted with an organic solvent to remove potential leachables from the rubber matrix for analysis. An MDI suspension drug product could be "cold-filtered" to remove active pharmaceutical ingredient particles from the organic propellant, which contains the leachables mixture. The details of extractions will be discussed more extensively in subsequent chapters of this book. Once extraction of a complex mixture of analytes from the sample matrix has been accomplished, further sample preparation steps, such as extract concentration, solvent switching, or cleanup, can be applied to prepare an appropriate sample extract for analysis. Although there have been many attempts to accomplish qualitative and quantitative TOA without prior separation of complex mixtures into individual chemical analytes, the most generally and widely successful techniques are "hyphenated," that is, those which combine the separation power of gas or liquid chromatography (GC or LC) with various highly sensitive and specific detection technologies.

Hyphenated analytical techniques typically applied to the qualitative and structural analysis of extractables/leachables include GC/MS and LC/MS (or high-performance liquid chromatography [HPLC]/MS). Other hyphenated analytical techniques that include detectors with somewhat lower selectivity than the mass spectrometer can also be applied to this problem. The most widely applied of these is LC combined with diode array detection (LC/DAD), which because of the non-destructive nature of the DAD is often used in combination with LC/MS. GC/MS,

LC/MS, and LC/DAD can also be used for quantitative analysis of extractables/leachables. However, for applications such as routine quality control of critical component extractables and routine monitoring of drug product leachables, detection technologies considered to be more rugged, robust, and economical in pharmaceutical quality control environments are usually employed. These include GC combined with flame ionization detection (GC/FID), and LC combined with variable wavelength ultraviolet (UV) detection (LC/UV). Compound class-specific detectors are also used in certain particular applications, most notable of these is the GC/thermal energy analysis® (TEA) system historically used to identify and quantify target N-nitrosamines.

13.3 IDENTIFICATION OF EXTRACTABLES/LEACHABLES

Based on Valcarcel's classification system[13] for analytical chemistry, the term "identification" can be clarified as either "qualitative analysis," which generates a yes/no response (i.e., Is a target analyte present in a sample at or above a specified limit of detection?), or "structural analysis" (i.e., What is the molecular structure of an unknown analyte?).

13.3.1 Qualitative Analysis of Extractables/Leachables

As an example of a leachables qualitative analysis, consider the problem of N-nitrosamines in an MDI drug product. The structure of N-nitrosodimethylamine is shown below:

$$H_3C \diagdown \atop H_3C \diagup N{-}N{=}O$$

As mentioned above, N-nitrosamines in rubber, certain foods, and other matrices have been historically analyzed by GC/TEA (Fig. 13.2).[14,15] GC/TEA combines the analyte-separating capability of GC with a detection system based on the phenomenon of chemiluminescence. As described in reviews by Norwood et al.,[14,15] target N-nitrosamines extracted from a sample matrix are separated by either a packed or capillary column GC and eluted sequentially into a pyrolyzer where nitrosyl radicals are released and subsequently oxidized with ozone to form electronically excited nitrogen dioxide. This decays to ground-state and emits a photon at a characteristic wavelength, which is detected by a spectrophotometer:

$$H_3C \diagdown \atop H_3C \diagup N{-}N{=}O \xrightarrow{\text{Heat}} NO\cdot + O_3 \longrightarrow {}^{*}NO_2 \longrightarrow NO_2 + h\nu$$

The detector is highly selective and, being based on a spectrophotometer that monitors only a single wavelength, highly sensitive. Controlling the pyrolysis temperature to 500°C provides an additional level of selectivity by allowing only nitroso and not nitro compounds to form nitrosyl radicals. Figure 13.3 shows two

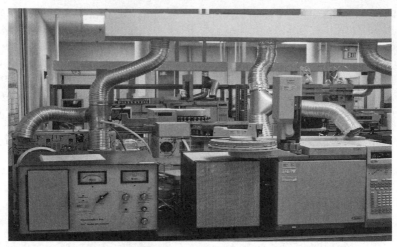

Figure 13.2   A gas chromatography/thermal energy analysis (GC/TEA) system for N-nitrosamine analysis (Agilent Technologies, Santa Clara, CA, 6890 GC with a Thermedics Detection Model 543 TEA). Note the gas chromatograph (far right), pyrolysis unit (center), and detector (left).

representative GC/TEA chromatograms from extracts of an MDI drug product.[16] In the top trace, each MDI canister was spiked with 3 ng/canister of six target N-nitrosamines prior to extraction and extract concentration. The bottom trace was derived from an unspiked MDI canister sample extract. Clearly, no target N-nitrosamines are present in this particular MDI drug product at the 3 ng/canister level. As discussed in other chapters, there are no accepted threshold levels for "special case" compounds, so this analysis represents the limits of currently applied analytical technology. For a complete description of the development of an analytical method for N-nitrosamines in an MDI drug product, the reader is referred to Norwood et al.[16]

Qualitative analysis of extractables/leachables can be accomplished with many different chromatographic techniques and detection systems, including GC/MS and LC/MS. For GC, both general/relatively nonselective and selective detectors (like the TEA) can be applied. The most commonly employed general detectors include thermal conductivity (GC/TCD) and flame ionization (GC/FID). Selective detectors include electron capture (GC/ECD), nitrogen-phosphorus (GC/NPD), Hall electrochemical, flame photometric (GC/FPD), and others. For hyphenated HPLC analytical instruments, there is no truly general detection system. The systems closest to realizing the dream of a "universal" detector for HPLC include the evaporative light-scattering detector (LC/ELSD) and the corona assisted discharge detector (LC/CAD). However, by far the most commonly used detectors for HPLC are fixed and variable wavelength UV/visible absorbance detectors (LC/UV), especially including the aforementioned diode array detector (LC/DAD), which is capable of generating an absorbance spectrum for a given analyte, thus providing an additional degree of selectivity.

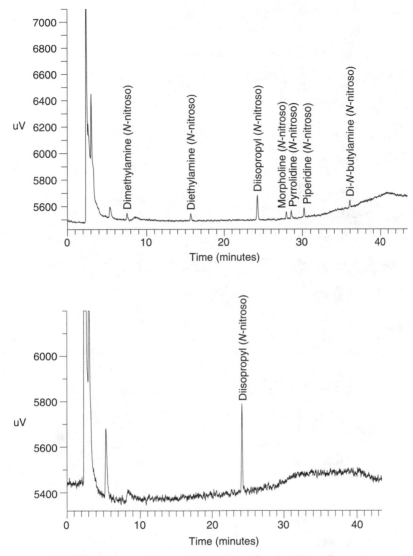

Figure 13.3   GC/TEA chromatograms of spiked (top; 3 ng/canister of each analyte) and unspiked (bottom) metered dose inhaler drug product extracts. Note that *N*-nitrosodiisopropylamine is an internal standard.

13.3.2   Structural Analysis of Extractables/Leachables

A mass spectrometer is an analytical instrument that can generate compound-specific information for trace-level organic analytes through ionization of the neutral species, mass separation of the resulting ions in the gas phase, and detection of these ions with signal intensity directly proportional to ion current. The basic components of any organic mass spectrometer include (1) the sample inlet system; (2) the ionization

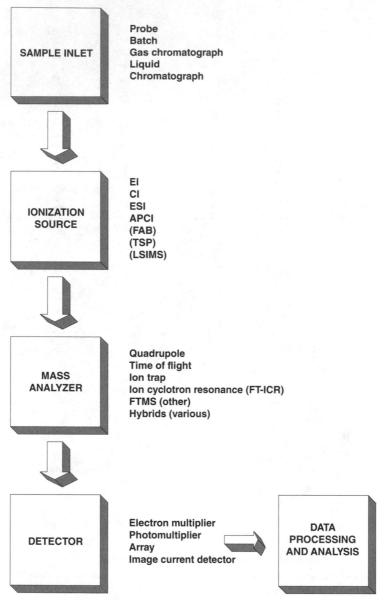

Figure 13.4 Block diagram of the basic components of a mass spectrometer.

source; (3) the mass analyzer; (4) the ion detection system; and perhaps most importantly, (5) the computerized data acquisition/processing system. Figure 13.4 highlights these basic components and can be compared with a typical GC/MS system such as the one shown in Figure 13.5. For extractables/leachables analysis, we will consider only the GC and HPLC as inlet systems with the corresponding ionization processes of electron ionization (EI) and chemical ionization (CI) for GC/MS, and

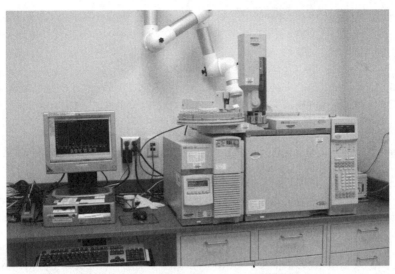

Figure 13.5 A gas chromatography/mass spectrometry (GC/MS) system (Agilent Technologies 6890 GC with 5973 MS). Note the gas chromatograph (far right), mass spectrometer (center), and data system (left).

electrospray ionization (ESI) and atmospheric pressure chemical ionization (APCI) for LC/MS. Mass analyzers will be considered only within the context of the mass spectral data elements that each is capable of producing. Additional details regarding organic MS, including theory, chemistry of gas-phase ions, instrumentation and ionization processes, structure elucidation, and analytical method development, are available in the comprehensive treatises by Gross,[17] de Hoffmann and Stroobant,[18] and Boyd et al.[19]

For structural analysis, the mass spectrometer is capable of producing the following data elements:

1. *Molecular weight.* The nominal monoisotopic molecular weight of an organic compound is the single most important piece of information available in any mass spectrum. The isotopic composition of the molecular ion as visualized by the molecular ion's isotopic "pattern" in a mass spectrum can also reveal the presence and numbers of certain heteroatoms.

2. *Molecular formula.* Mass analyzers with sufficient mass accuracy can allow elemental composition information to be determined for both molecular and fragment ions. In addition, application software is available for certain instruments, which can predict molecular formulas based on the isotopic pattern of the molecular ion.

3. *Fragmentation behavior.* Ionization of neutral molecules initially produces either positive or negative "molecular ions" that can undergo fragmentation, either in the ion source, within a "collision chamber" filled with inert collision gas, inside the cyclotron cell of a Fourier transform ion cyclotron resonance

(FT-ICR), or in other ways according to defined processes. Structure elucidation is facilitated by an understanding of a molecular ion's fragmentation behavior.

In addition to these basic data elements, certain types of mass spectra, most notably EI spectra, can be compared with compiled databases or libraries of spectra from known compounds, facilitating structure elucidation through computerized and/or manual pattern recognition. The hyphenated techniques of GC/MS and LC/MS also yield a compound's chromatographic retention behavior, which can be compared with known structures, as in the qualitative analysis example described above. As also mentioned above, LC/MS includes nondestructive detection systems such as DAD, which can acquire qualitatively useful information (i.e., UV spectra). Table 13.2 lists the instrumental configurations most commonly used for extractables/leachables identifications, along with the data elements and unique features that each can provide.

It is important to understand what is meant by the term "identification" as it is applied to the structural analysis of extractables and leachables. There is a school of thought that suggests that an unknown organic compound is truly "identified" only when its spectral and/or chromatographic properties are matched with those of an authentic reference compound, either synthesized or obtained commercially, as with the N-nitrosamine qualitative analysis described above. In many cases of extractables/leachables profiling, this criterion is neither practical nor necessary. Consider, for example, the GC/MS extractables profile of the sulfur-cured elastomer (Table 13.1) shown in Figure 13.6. The unambiguous identification of each extractable organic compound in this profile (as represented by the individual peaks in the chromatogram) would be a daunting task measured in years for a team of experts, especially considering that many of the peaks are derived from natural product materials that will likely require many custom synthesized reference compounds. The recommendations established by PQRI for this task are based on proposed criteria for the identification of environmental trace organics by GC/MS.[7,20] Mass spectral data elements are assigned identification categories that are used to designate individual extractables/leachables identifications as *confirmed*, *confident*, or *tentative*, as shown in Table 13.3. The level of identification that is sufficient for an individual extractable or leachable can only be determined with appropriate input from safety experts and perhaps regulatory authorities. There are cases in which either *confident* or even *tentative* identifications could be sufficient. For example, if a leachable slightly above an established AET is determined by fragmentation behavior to be an *n*-alkane, this *tentative* identification would likely be enough to evaluate safety. On the other hand, one would not want to recommend the termination of a pharmaceutical development program based on anything short of a *confirmed* identification.

13.3.3   Structural Analysis by GC/MS

GC, being a gas-phase separation process as implied by its name, interfaces well with the gas-phase ionization processes EI and CI. In most GC/MS instruments, ion sources are available that can be switched between these two ionization processes.

TABLE 13.2. GC/MS and LC/MS Instrumental Configurations Commonly Used for Extractables/Leachables Identifications

Instrumental configuration	Data elements available	Comments
GC (LC)—single quadrupole (Q)	Molecular weight In-source fragmentation Library searchable spectra (EI)	• Molecular weight confirmation can be achieved with a combination of EI and CI (GC/MS) and APCI and ESI (LC/MS).
GC (LC)—triple quadrupole (QQQ)	Molecular weight In-source fragmentation Library searchable spectra (EI) MS/MS	• MS/MS is also known as tandem mass spectrometry. • Outstanding instrumental configuration for quantitative analysis due to selectivity provided by multiple reaction monitoring (MRM). • Note that instruments with hexapole or octapole mass analyzers are also referred to as "triple quadrupoles."
GC (LC)—time of flight (TOF)	Molecular weight In-source fragmentation Library searchable spectra (EI) Accurate mass measurements	• Various hybrid configurations are available, including quadrupole (QTOF) and ion trap. These hybrid configurations are capable of MS/MS.
GC (LC)—quadrupole ion trap	Molecular weight In-source fragmentation Library searchable spectra (EI) MS/MS (MS^n)	• Certain ion traps produce EI spectra that are difficult to compare with those acquired on other instrumental configurations, or with available mass spectral libraries. • Hybrid configurations are available with TOF and FT-ICR. • MS^n designates multiple levels of MS/MS.
GC (LC)—Fourier transform mass spectrometer (FTMS)	Molecular weight In-source fragmentation Library searchable spectra (EI) Accurate mass measurements MS/MS (MS^n)	• Instrumental types available include ion cyclotron resonance (FT-ICR) and Orbitrap™. • Various hybrid configurations are available, including quadrupole and ion trap. These are capable of various MS/MS experiments and induced fragmentation processes.

A schematic of a typical EI/CI source is shown in Figure 13.7. Compounds eluting as peaks from the GC column interact in the gas phase with a beam of relatively energetic (70 eV) electrons, which remove a single electron from the neutral analyte molecule forming a radical cation, termed the molecular ion (M^+). Excess energy from the interaction is distributed throughout the chemical bonds of the molecular

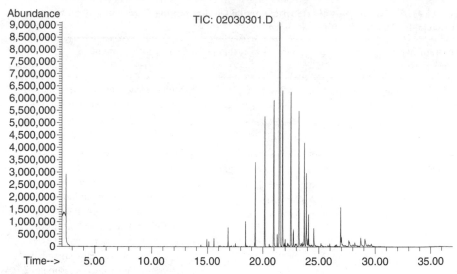

Figure 13.6   A GC/MS extractables "profile" in the form of a total ion chromatogram (TIC). This profile was produced by GC/MS analysis of a solvent extract of the sulfur-cured elastomer whose composition is given in Table 13.1.

TABLE 13.3.   Identification Categories for the Identification of Extractables and Leachables by GC/MS and LC/MS

Identification category	Mass spectral data element
A	Mass spectral fragmentation behavior
B	Confirmation of molecular weight
C	Confirmation of molecular formula
D	Mass spectrum matches automated library/database or literature reference
E	Mass spectrum and chromatographic retention index match authentic reference compound

- A *confirmed* identification means that identification categories A, B (or C), and D (or E) have been fulfilled.
- A *confident* identification means that sufficient data to preclude all but the most closely related structures have been obtained.
- A *tentative* identification means that data have been obtained that are consistent with a class of molecule only.

ion in a "quasiequilibrium," resulting in bond cleavages that produce fragment ions, either even-electron or odd-electron depending on the nature of the fragmentation process. It is important to note that since (1) there are no atoms or molecules that cannot be ionized at 70 eV, (2) ionization efficiency plateaus at around 70 eV, and (3) the pressures in the EI source (10^{-4}–10^{-5} Pa) preclude ion–ion and ion–molecule interactions inside the source, EI spectra are very reproducible, which allows comparison of spectra from different instruments/different days/different analysts to each other and to mass spectral databases (i.e., libraries) (note that for certain types of

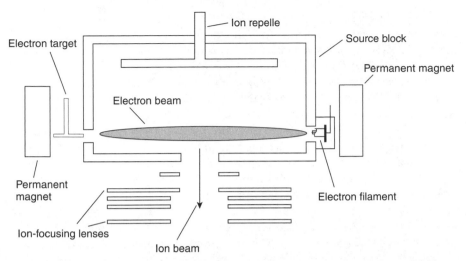

Figure 13.7   Schematic diagram of an electron ionization (EI) ion source for an organic mass spectrometer. Note that a capillary GC column would interface with this source perpendicular to this two-dimensional representation, eluting separated compounds directly into the electron beam.

ion trap mass spectrometer, EI spectral comparisons with other instrument types and/ or mass spectral libraries can be problematic).

CI, unlike EI, relies on ion–molecule reactions inside the ion source. A CI source is basically an EI source that has been "tightened up" to reduce gas leaks and allow for equilibrium pressurization by a "reagent gas." The reagent (or reactant) gas interacts with the electron beam to produce a steady-state plasma of reagent gas ions within the source. Consider methane, which is a commonly used reagent gas:

$$CH_4 + e^- \rightarrow CH_4^{+\cdot}, CH_3^+, CH_2^{+\cdot}, CH^+, C^{+\cdot}, H_2^{+\cdot}, H^+$$
$$CH_4^{+\cdot} + CH_4 \rightarrow CH_5^+ \, (m/z \; 17) + CH_3^{\cdot}$$
$$CH_3^+ + CH_4 \rightarrow C_2H_7^+ \rightarrow C_2H_5^+ \, (m/z \; 29) + H_2$$
$$CH_2^{+\cdot} + CH_4 \rightarrow C_2H_3^+ + H_2 + H^{\cdot}$$
$$C_2H_3^+ + CH_4 \rightarrow C_3H_5 + (m/z \; 41) + H_2$$

The reagent gas plasma is composed primarily of CH_5^+ (m/z 17), $C_2H_5^+$ (m/z 29), and $C_3H_5^+$ (m/z 41), along with various relatively minor species. At typical CI source pressures (i.e., 10^2 Pa), ion–molecule collision frequencies are quite high, so many collisions can occur between analyte molecules and reagent gas ions. These collisions can result in four fundamental types of gas-phase reactions,[17] including (1) proton transfer, (2) electrophilic addition, (3) anion abstraction, and (4) charge exchange. Using methane reagent gas as an example (where M is an analyte neutral molecule):

1. $M + CH_5^+ \rightarrow [M + H]^+ + CH_4$,
2. $M + C_2H_5^+ \rightarrow [M + C_2H_5]^+$,
3. $M + CH_3^+ \rightarrow [M–H]^+ + CH_4$ (in this case *hydride ion abstraction*),
4. $M + CH_4^{+\cdot} \rightarrow M^{+\cdot} + CH_4$.

The ions observed in any particular CI mass spectrum as well as their relative abundances depend on the thermodynamic properties of M (e.g., proton affinity) and the reagent gas. For most small molecules such as extractables/leachables, one would expect to see $[M+H]^+$, referred to as the "protonated molecule" or more colloquially as the "protonated molecular ion," as an important spectral feature. A very important property of CI is that collisions of analyte ions with reagent gas ions and molecules within the ion source also serve to stabilize the analyte ions by removing excess internal energy. This is why CI is often referred to as a "soft" ionization process in which molecular ions dominate and fragmentation is lessened. This makes CI a good complement to EI for molecular weight confirmation.

The above discussion of CI is limited to positive ions; however, it is also possible to form negative ions from analyte molecules by CI processes. The most widely applied of these processes for GC/MS, and therefore of potential utility for extractables/leachables analysis, is resonance electron capture to form a collisionally stabilized molecular anion (M^-). The energetics of resonance electron capture is such that it is a rare event in EI, where electron energies are relatively high (70 eV). In CI, however, within the reagent gas plasma is an equilibrium population of thermal energy electrons (0–2 eV), which can be captured by analyte molecules with relatively high electron affinity. Extractables/leachables particularly amenable to negative ion CI GC/MS analysis include (predictably) chlorine-, bromine-, and fluorine-containing molecules. This technique is noted for extremely high sensitivity and selectivity for appropriate analytes. Negative ion formation mechanisms other than electron capture are possible, and these will be mentioned as part of the presentation on LC/MS. For additional information on CI, the comprehensive treatise by Harrison[21] is recommended as well as the volumes previously cited.

Structural analysis by GC/MS involves interpretation of EI spectra, often assisted by interpretation of corresponding CI spectra. A general process recommended for extractables/leachables structural analysis is as follows:

1. Search the unknown EI spectrum through any available mass spectral libraries. Consider the quality of the search results both visually and by noting appropriate numerical pattern recognition quality of "fit" factors. Is there a definite positive hit? If not is there any information that the search results contain, such as compound class information?

2. Consider the general features of the EI spectrum:

    a. Can the molecular ion be identified (or confirmed by CI)?

    b. Based on stable isotope patterns, are there obvious heteroatoms present?

    c. Considering the nitrogen rule,[22,23] is the number of nitrogens in the molecule even or odd?

    d. Are the higher mass ions of greater relative abundance suggesting greater stability (possibly aromatic)?

    e. Is there an identifiable low mass ion series suggesting compound class?

    f. If accurate mass data are available, can the molecular formula be determined?

3. Consider fragment ions and their relationship to the molecular ion:
   a. Can reasonable and stable structures be determined for the most abundant fragment ions?
   b. Can the higher mass fragment ions be mechanistically linked to the molecular ion?
   c. If accurate mass data are available, can hypotheses from (a) and (b) be confirmed?
4. Finally, can authentic reference materials or literature references be obtained to confirm/support proposed structures?

The process is best illustrated by consideration of Case Study 13.1.

CASE STUDY *13.1*

STRUCTURAL ANALYSIS OF AN UNKNOWN EXTRACTABLE BY GC/MS

Problem

Consider the partial GC/MS extractables profile shown in Figure CS 13.1.1 and note the relatively minor peak indicated by the asterisks (this partial chromatogram is a portion of a total ion chromatogram [TIC] similar to, and from the same elastomeric test article as, that shown in Fig. 13.6). The problem is to elucidate the molecular structure of (i.e., to identify) this unknown extractable.

Procedure

1. Consider the EI mass spectrum of the unknown and immediately search this unknown mass spectrum through any available computerized mass spectral libraries.

   Figure CS 13.1.2 suggests that the unknown extractable is tetramethylthiourea (molecular weight 132), which is known to be related to rubber vulcanization with

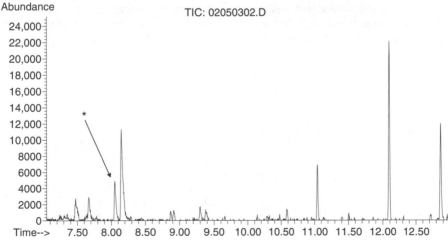

Figure CS 13.1.1   Partial total ion chromatogram (TIC) from the GC/MS analysis of an elastomer solvent extract (i.e., an extractables "profile"). Note the extractable of interest indicated (*). Data acquired on an Agilent 5973 GC/MSD.

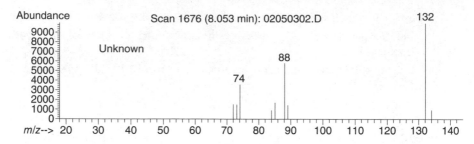

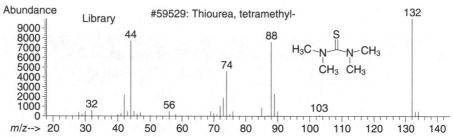

Figure CS 13.1.2   Electron ionization (EI) mass spectrum of the unknown extractable (*) from Figure CS 13.1.1 (top), along with the best fit result from a computerized mass spectral library search (bottom).

certain sulfur-containing cross-linking agents and accelerators. Note from Table 13.1 that this elastomer was vulcanized with tetramethylthiuram monosulfide and 2-mercaptobenzothiazole, which are both sulfur containing. Tetramethylthiuram monosulfide also contains the basic structural unit of tetramethylthiourea:

Tetramethylthiuram monosulfide

Tetramethylthiourea

This analysis confirms identification category D from Table 13.3.

2. Consider the EI fragmentation behavior (note that the favorable library search result obviously also considers fragmentation behavior):

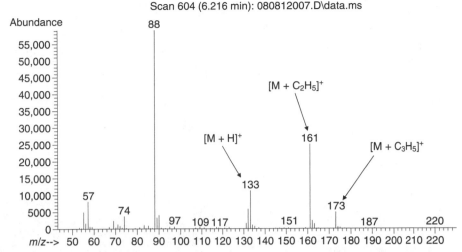

Figure CS 13.1.3   Methane chemical ionization (CI) mass spectrum of the unknown extractable (*) from Figure CS 13.1.1.

This analysis confirms identification category A from Table 13.3.

3. Consider the methane CI mass spectrum in Figure CS 13.1.3. The $[M+H]^+$, $[M+C_2H_5]^+$, and $[M+C_3H_5]^+$ ions confirm the molecular weight as 132.

This analysis confirms identification category B from Table 13.3.

4. Consider the accurate mass measured EI molecular ion, which confirms the molecular formula

$$132.0723 - C_5H_{12}N_2S \ (1.5 \text{ ppm error}).$$

(Data acquired on a VG Micromass Autospec (Waters Corporation, Milford, MA) double-focusing sector GC/MS system.)

This analysis confirms identification category C from Table 13.3.

5. Note the EI mass spectrum of authentic tetramethylthiourea in Figure CS 13.1.4. (Also note that chromatography retention behavior of the authentic material matched that of the unknown [data not shown]).

Based on the criteria in Table 13.3, this structural analysis (identification) is confirmed.

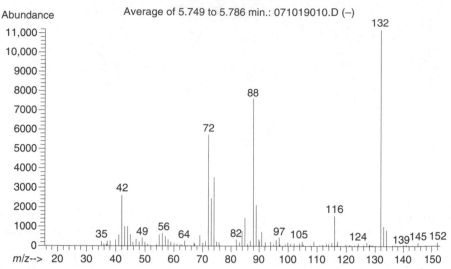

Figure CS 13.1.4 Electron ionization (EI) mass spectrum from authentic tetramethylthiourea.

13.3.4 Structural Analysis by LC/MS

LC is obviously not a gas-phase separation technique and therefore does not interface well with gas-phase ionization processes, especially in high vacuum (although this was attempted in the early days of LC/MS with somewhat mixed results). The history of LC/MS ionization processes and instrumentation development can be traced back to the late 1970s, and the interested reader is referred to the comprehensive text by Willoughby et al.[24] as well as previous citations. The following discussion will focus on the two LC/MS ionization processes and their variants that are most applicable to extractables/leachables structural analysis, ESI, and APCI. Because of their almost universal application over approximately the past 15 years, these two processes and their variants are likely to be central in LC/MS analysis for many years to come.

Like EI and CI for GC/MS, APCI and ESI are available on most LC/MS instruments in some combination form. This combination is feasible because, as the name of the former states, both ionization processes are accomplished at atmospheric pressure, that is, outside the high vacuum region of the mass spectrometer. ESI, APCI, their variants, and other related ionization processes are therefore collectively referred to as atmospheric pressure ionization (API). The development of API processes has greatly facilitated the wide application of LC/MS in many scientific fields due to the relative ease of interfacing liquid chromatography systems to API ion sources, and the relative simplicity of sending analyte ions formed by API processes into the high vacuum mass analyzer/detector regions of mass spectrometers via differential pumping systems.[18] The term "relative" in this context refers to earlier LC/MS ionization processes and interfaces, particularly including thermospray (TSP) and continuous-flow fast atom bombardment (CF-FAB), in which analyte ionization occurred inside the mass spectrometer's high vacuum region. However, it should be noted that during the 1980s and early 1990s, both TSP and CF-FAB established LC/MS as a viable and significant tool for analytical sciences and, in the authors' direct experience, facilitated the solution of numerous problems in qualitative, structural, and quantitative TOA.

APCI, as the term implies, accomplishes analyte ionization through CI processes (i.e., ion–molecule reactions in the gas phase). In the APCI process (see Fig. 13.8B), eluent from the HPLC column at flow rates of about 0.2–2 mL/min[18] passes through a heated tube (e.g., quartz at 500°C) and, assisted by a relatively high flow

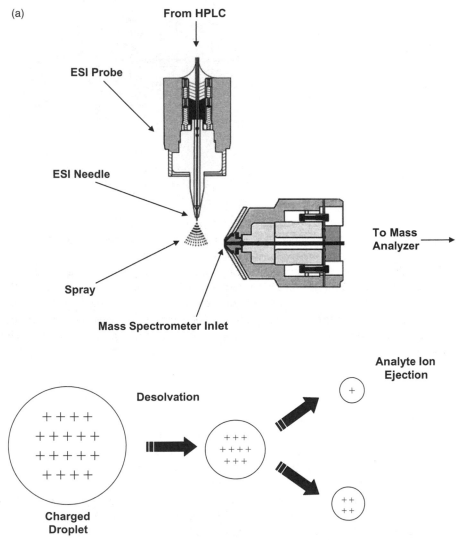

Figure 13.8   (a) Schematic diagram of a typical electrospray ionization (ESI) ion source for LC/MS, along with a pictorial representation of the basic ESI process. (b) Schematic diagram of a typical atmospheric pressure chemical ionization (APCI) ion source for LC/MS.

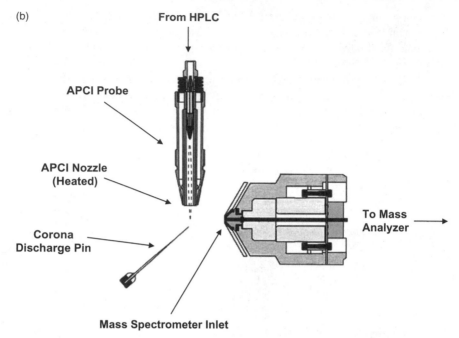

Figure 13.8   (*Continued*)

of nitrogen gas, is flash vaporized into the region of a corona discharge. The corona discharge forms ions such as N_2^+ and O_2^+ by EI from the corresponding atmospheric species, which react with HPLC mobile-phase molecules in the gas phase (e.g., water, acetonitrile, methanol) to form a steady-state reagent gas plasma in the source. Analyte molecules in the gas phase are then ionized through reactive collisions with these ions. Positive molecular ions are formed by typical CI processes described earlier, such as proton transfer and addition of electrophilic species such as sodium and potassium. Negative molecular ions are also formed in the gas phase through processes such as (1) proton abstraction and (2) negative ion "attachment":

1. $M+OH^- \rightarrow [M-H]^- + H_2O$
2. $M+Cl^- \rightarrow [M+Cl]^-$.

The relative sensitivities of analyte species in APCI are related to their gas-phase thermochemical properties such as proton affinity and negative ion stability. Therefore, relatively high proton affinity analytes such as amines show good positive ion sensitivity, and analytes that can form relatively stable negative ions such as carboxylic and sulfonic acids show good negative ion sensitivity.

Relative to APCI, ESI is mechanically simpler but significantly more complex theoretically. In the ESI process (see Fig. 13.8A), eluent from the HPLC column at flow rates of about 1–10 $\mu L/min^{18}$ passes through a metal capillary tube whose tip is at a high electrical potential (3–6 kV)[18] relative to a counter electrode about 0.3–2 cm away.[18] The resulting electric field (10^6 V/m)[18] assisted by a flow of nitrogen "nebulizing gas" causes charged droplets to be emitted from the apex of the

Figure 13.9   Picture of a Waters Micromass combination ESI/APCI orthogonal ion source for LC/MS. Note that in this particular picture, the ion source is being operated in ESI mode (see the ESI probe needle).

liquid at the capillary tip. This liquid apex is referred to as the Taylor cone.[17,18] As the charged droplets move toward the counter electrode, they break into ever smaller droplets and desolvate, increasing the surface charge on the droplets. Eventually, analyte ions (either positive or negative) are either emitted from the charged droplet surfaces or totally desolvated into the gas phase. Note that, unlike APCI, gas-phase ion–molecule reactions are not required for the ESI process. Analyte molecular ions exist in the liquid/solution phase and are emitted/desolvated into the gas phase for mass spectral analysis. Molecular ion species, such as $[M+H]^+$ and $[M-H]^-$ are stabilized through collisions with gas molecules. The reader is advised that the above discussion of the ESI process is simplified, and those readers interested in further detailed treatments are referred to de Hoffmann and Stroobant.[18]

Analyte molecular and fragment ions from APCI and ESI are drawn into the analyzer region of the mass spectrometer via electrical potentials and differential pumping systems. Note from Figure 13.9 that ESI and APCI probes are typically orthogonal to the actual inlet to the mass spectrometer's analyzer/detector region, thus allowing mostly ions, and not potentially contaminating HPLC mobile-phase molecules, to enter the mass spectrometer. This orthogonal geometry results in significant robustness advantages for LC/MS systems, and has facilitated the wide use of quantitative LC/MS for pharmaceutical applications. Figure 13.10 shows a variety of API probes.

Functionally for the analytical scientist, APCI and ESI have both common and unique features. Along with the most obvious common feature that both ionization processes occur at atmospheric pressure is the fact that both are "soft" ionization processes, resulting in mostly molecular ion species with little in-source fragmentation. However, if desired, fragmentation can be induced by varying certain electrical potentials in the ion source resulting in increased collision-induced dissociation (CID) processes. Some typical APCI and ESI spectra from selected extractables/leachables are shown in Figures 13.11 and 13.12.

Figure 13.10 Picture of various Thermo-Finnigan LC/MS probes (Thermo-Finnigan, San Jose, CA); APCI (left), ESI (center), heated ESI (right).

As stated above, APCI and ESI have both common and unique features:[25]

1. *Both ESI and APCI occur at atmospheric pressure.* This is one of the most important factors in the success of these two ionization processes for LC/MS, as well as the exponential increase in LC/MS application. The fact that the "dirty" part of the LC/MS process occurs outside the high vacuum region of the mass spectrometer, which contains the contamination-sensitive mass analyzer and detector systems, enables LC/MS to be both rugged and robust. With current instrumentation, it is a rare event indeed when a mass analyzer (e.g., a quadrupole mass filter) requires disassembly and cleaning. This was not the case several decades ago.

2. *Both ESI and APCI are "soft" ionization processes.* As previously discussed for CI GC/MS, this means that molecular ion species are stabilized, which preserves molecular weight information for an unknown. With ESI, it is also possible to observe multiply charged molecular ion species for molecules that have multiple sites for protonation in solution. Although not particularly useful for extractables/leachables structural analysis, this feature of ESI is extremely important for the analysis of biological molecules and macromolecules such as peptides and proteins (see Fig. 13.13 for an example of a multiply charged ion). As the mass spectrometer measures mass-to-charge ratio, increasing numbers of charges on a molecular ion has the effect of increasing the mass range of the mass spectrometer.

3. *ESI mass spectra tend to reflect solution chemistry while APCI mass spectra reflect gas-phase chemistry.* This suggests that ESI spectral appearance and sensitivity can be affected by altering solution chemistry, and APCI spectra by altering gas-phase chemistry. In fact, sensitivity in positive ion ESI can be improved for certain analytes by lowering the pH of the HPLC mobile phase.

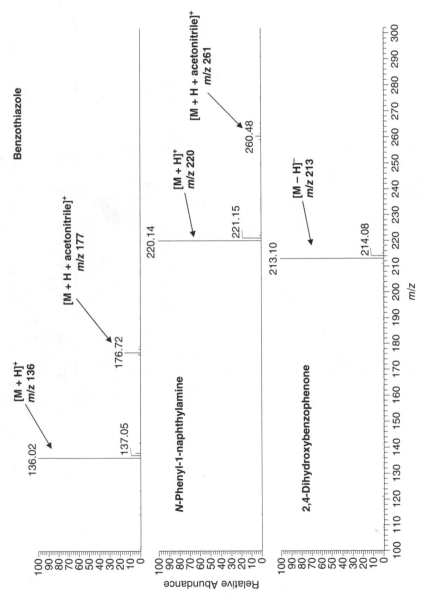

Figure 13.11 Some representative electrospray ionization (ESI) mass spectra: positive ion benzothiazole (top), positive ion *N*-phenyl-1-naphthylamine (middle), and negative ion 2,4-dihydroxybenzophenone (bottom).

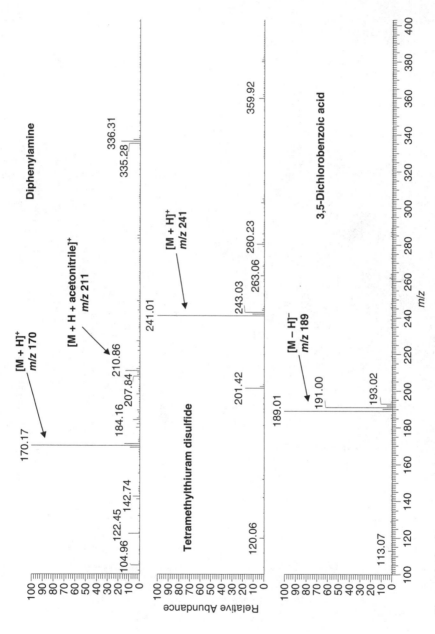

Figure 13.12 Some representative atmospheric pressure chemical ionization (APCI) mass spectra: positive ion diphenylamine (top), positive ion tetramethylthiuram disulfide (middle), and negative ion 3,5-dichlorobenzoic acid (bottom).

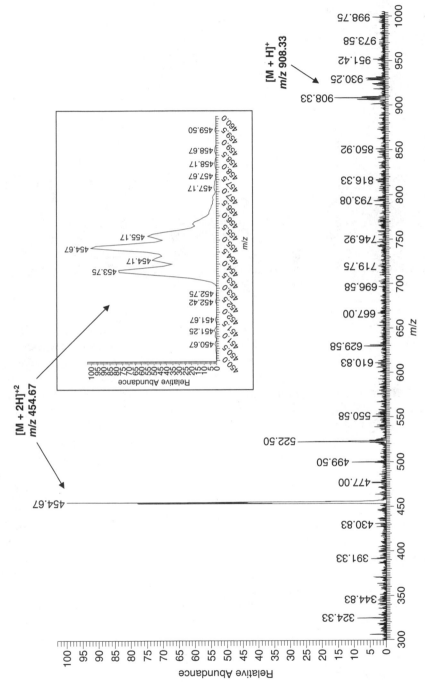

Figure 13.13 A positive ion ESI spectrum showing a doubly charged molecular ion. Note the [M+H]⁺ at m/z 908.33 and the [M+2H]⁺² at m/z 454.67. The doubly charged molecular ion can be recognized by its nonintegral mass and relatively poor resolution (see insert).

267

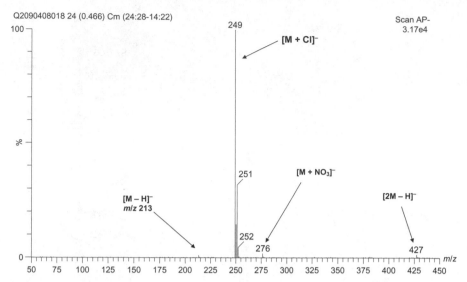

Figure 13.14   A negative ion chloride ion attachment APCI mass spectrum of 2,4-dihydroxybenzophenone. Note also the $[M+NO_3]^-$ at m/z 276 and $[2M-H]^-$ at m/z 427. There is also a small $[M-H]^-$ at m/z 213.

On the other hand, negative ion sensitivity can be severely reduced by including strongly acidic buffers such as trifluoroacetic acid (TFAA) in the mobile phase. TFAA can suppress the ionization of analyte species through mechanisms such as ion pairing.[19] For APCI, mobile-phase additives that have no effect on analyte solution ionization can affect gas-phase ionization. For example, the addition of small amounts of chloroacetonitrile to a reversed-phase HPLC system can result in chloride ion attachment (see Fig. 13.14). The corona discharge produces a steady-state plasma of Cl^- in the ion source.

4. *APCI tends to be more amenable to higher HPLC flow rates, while ESI in general prefers lower flow rates.* ESI has some of the characteristics of a concentration-dependent detector in that it has been demonstrated through a controlled experiment that ESI analyte peak height response was independent of mass flow rate for a given analyte concentration.[19] It has been further demonstrated that ESI accomplished at very low flow rates (nanoliter per minute) can significantly reduce or eliminate interferences and suppression effects from salts and other mobile-phase species.[19] Therefore, there is no clear advantage in ESI to higher HPLC flow rates (milliliter per minute) from the analyte sensitivity point of view. For APCI, on the other hand, there is no clear advantage to lower flow rates and APCI probes, and ion sources are physically designed to work well at approximately 0.2–2 mL/min, which makes APCI a good match for the so-called analytical scale HPLC.

5. *Both ESI and APCI prefer "reversed-phase" HPLC platforms.* Reversed-phase chromatography platforms and methods dominate pharmaceutical applications of HPLC, including those involving extractables/leachables. Aqueous-based

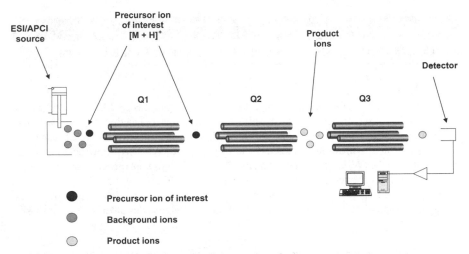

Figure 13.15   Schematic diagram of a triple quadrupole mass spectrometer.

mobile phases interface better with both ESI and APCI for many reasons, both chemical and practical. For ESI, surface tension and desolvation rates of charged droplets are issues, as is the ability to control solution pH. For APCI, gas-phase ions that can easily give up or accept a proton are important.

The fact that ESI and APCI are soft ionization processes can limit the utility of the resulting mass spectra for structural analysis, in that in most cases, the ion current is concentrated in molecular ion species with little fragmentation. In most instruments, fragmentation can be induced in the ion source by altering various source potentials, which results in increased CID of analyte molecular ions in the source. CID processes can also occur in the so-called collision cells (as mentioned earlier) or collision regions. For example, in a triple quadrupole mass spectrometer (see Table 13.2 and Fig. 13.15), the first quadrupole mass analyzer can be set to pass only the $[M+H]^+$ of an unknown analyte into the second quadrupole, which is filled with an inert gas (e.g., argon) and acts as a collision cell for CID. The third quadrupole is then scanned to separate and pass to a detector all of the resulting fragment ions, which are referred to as "product ions" to distinguish them from ions produced by in-source fragmentations. This and related processes on other instrumental configurations (see Table 13.2) are collectively referred to as MS/MS or "tandem mass spectrometry." The advantage of MS/MS over in-source fragmentation is that all product ions are linked directly to a precursor ion. Also, note that multiple levels of MS/MS can be accomplished on several instrumental configurations (see Table 13.2 and Case Study 13.2).

As ESI and APCI LC/MS have become mature analytical techniques, additional ionization processes have been developed, which can be (or have the potential to be) interfaced with HPLC. These include[18] atmospheric pressure photoionization (APPI), desorption electrospray ionization (DESI), and matrix-assisted laser desorption (MALDI). The reader is advised that LC/MS remains an advancing area, and it is certain that ionization processes and instrumentation will continually improve.

CASE STUDY *13.2*

STRUCTURAL ANALYSIS OF AN UNKNOWN EXTRACTABLE BY LC/MS

Problem

Consider the partial LC/MS extractables profile shown in Figure CS 13.2.1, and note the peak indicated by the asterisk (designated 12). This particular sample is a methylene chloride extract of a polypropylene test article (see Chapter 16). This LC/MS analysis was accomplished in negative ion APCI mode with in-line UV detection at 280 nm. Note that the unknown peak of interest appears in both the UV and TICs, indicating that it has a UV chromophore and significant negative ion sensitivity. Remember that in APCI, compounds with easily abstractable protons (e.g., carboxylic acids, phenols, sulfonic acids) usually show significant negative ion sensitivity. The problem is to elucidate the molecular structure of (i.e., to identify) this unknown extractable.

Procedure

1. Consider the expanded negative ion APCI mass spectrum of the unknown shown in Figure CS 13.2.2. Note that the spectrum has only one significant ion at m/z 1175. A reasonable hypothesis is that this ion is an $[M-H]^-$ suggesting a molecular weight of 1176 for the unknown extractable. Closer examination of the spectrum reveals a very small ion at m/z 1211, which is a possible confirming $[M+Cl]^-$ adduct ion.

2. Consider the positive ion APCI mass spectrum shown in Figure CS 13.2.3. Note what appears to be a very weak molecular ion region (expanded in Fig. CS 13.2.4), with significant fragmentation. Given the hypothesis of molecular weight 1176, the most significant ion in the positive ion molecular ion region corresponds to $[M+NH_4]^+$ (m/z 1194), which is a commonly observed adduct ion. Also visible is a possible $[M+K]^+$ (m/z 1215)

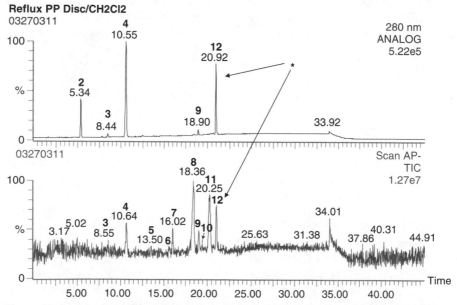

Figure CS 13.2.1 Total ion chromatogram (TIC; bottom) and UV at 280 nm (top) from the negative ion APCI LC/MS analysis of a polypropylene solvent extract (i.e., an extractables "profile"). Note the extractable of interest indicated (*). Data acquired on a Micromass Platform II.

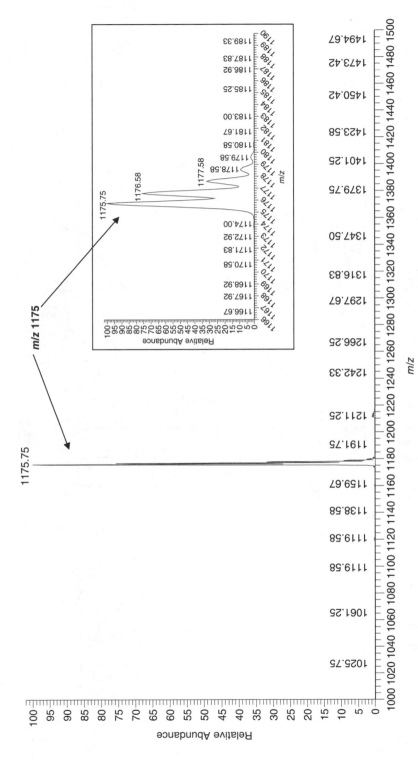

Figure CS 13.2.2   Negative ion APCI mass spectrum of the unknown extractable (*) from Figure CS 13.2.1. Data acquired on a Thermo-Finnigan LTQ-FT Ultra.

271

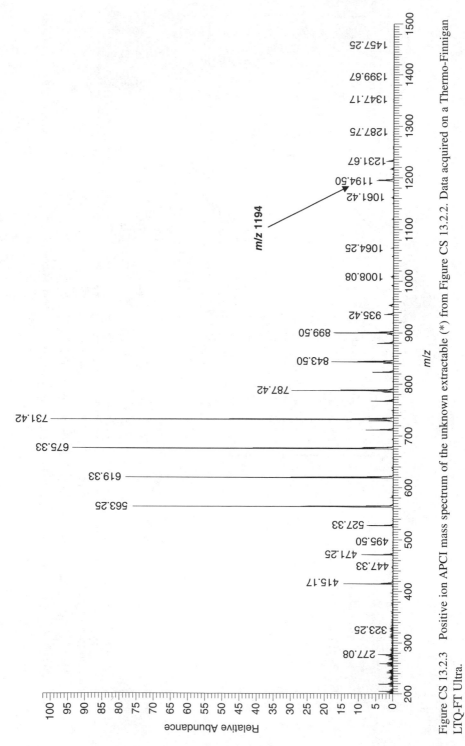

Figure CS 13.2.3   Positive ion APCI mass spectrum of the unknown extractable (*) from Figure CS 13.2.2. Data acquired on a Thermo-Finnigan LTQ-FT Ultra.

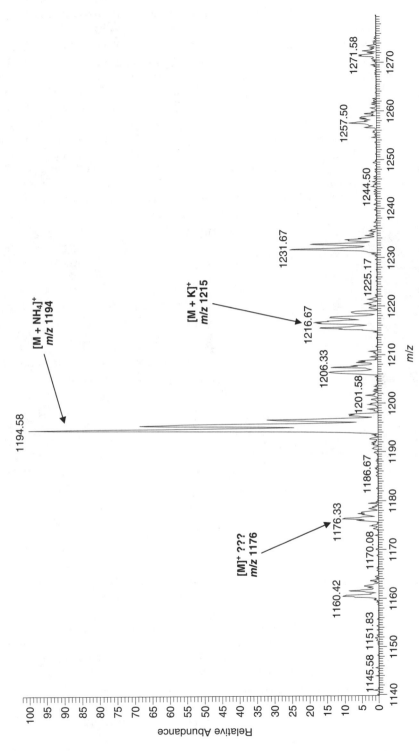

Figure CS 13.2.4 Expanded (molecular ion region) positive ion APCI mass spectrum of the unknown extractable (*) from Figure CS 13.2.1.

and a possible $M^{+\cdot}$ (m/z 1176) formed by charge exchange. The evidence considered in total can be taken to confirm the molecular weight of the unknown to be 1176.

This analysis confirms identification category B from Table 13.3.

As noted above, the significant fragmentation is an interesting feature of the positive ion spectrum. Observe that many of these fragment ions are separated by 56 mass units (i.e., $899 \rightarrow 843$, $843 \rightarrow 787$, $787 \rightarrow 731$, $731 \rightarrow 675$, $675 \rightarrow 619$, $619 \rightarrow 563$), which suggests the presence of multiple *tert*-butyl groups in the molecule.

The ionization behavior along with the presence of multiple *tert*-butyl groups strongly suggests that the unknown extractable is a phenolic antioxidant or stabilizer. These compounds are commonly added to polypropylenes to react with singlet oxygen and prevent polymer degradation. An examination of the formulation of this particular polypropylene indicates that it contains the phenolic antioxidant Irganox 1010:

Registry number: 6683-19-8

Chemical Abstracts (CA) index name: Benzenepropanoic acid, 3,5-bis(1,1-dimethylethyl)-4-hydroxy-, 1,1'-[2,2-bis[[3-[3,5-bis(1,1-dimethylethyl)-4-hydroxyphenyl]-1-oxopropoxy]methyl]-1,3-propanediyl] ester

3. Consider the accurate mass measured negative ion APCI mass spectrum shown in Figure CS 13.2.5 (molecular ion region only) with best fit elemental composition for the $[M-H]^-$ of $C_{73}H_{107}O_{12}$ (0.627 ppm mass error). The elemental composition of the $[M-H]^-$ confirms the known molecular formula of Irganox 1010 ($C_{73}H_{108}O_{12}$).

This analysis confirms identification category C from Table 13.3.

4. Consider reasonable proposed structures for the two principal fragment ions (MS^2 and MS^3) in Figure CS 13.2.6, which are consistent with the structure of Irganox 1010 and confirmed by accurate mass measurement:

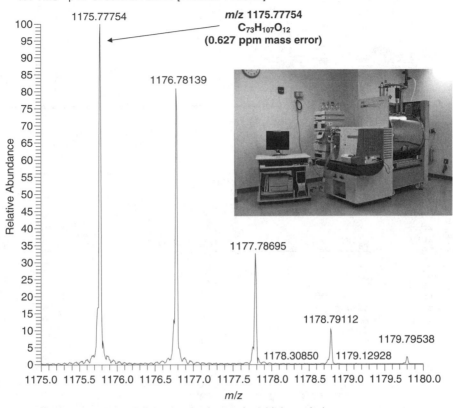

SP071016006 #174   RT: 4.00 AV: 1   NL: 2.97E5
T: FTMS - p APCI corona Full ms [1170.00-1185.00]

m/z **1175.77754**
$C_{73}H_{107}O_{12}$
(0.627 ppm mass error)

1175.77754

1176.78139

1177.78695

1178.79112

1179.79538

1178.30850

1179.12928

Relative Abundance

m/z

Figure CS 13.2.5   Expanded (molecular ion region) high resolution accurate mass measured negative ion APCI mass spectrum of the unknown extractable (*) from Figure CS 13.2.1. Data acquired on a Thermo-Finnigan LTQ-FT Ultra (see insert).

m/z 957 (957.61089; $C_{58}H_{85}O_{11}$; 1.203 ppm error)

275

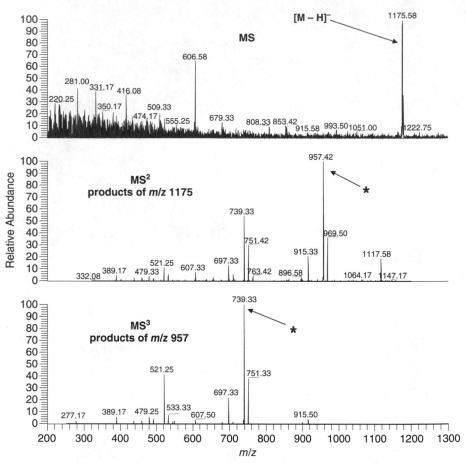

Figure CS 13.2.6 Negative ion APCI mass spectrum (top), along with MS² (middle) and MS³ (bottom) of the unknown extractable (*) from Figure CS 13.2.1. Data acquired on a Thermo-Finnigan LTQ-FT Ultra.

m/z 739 (739.44241; $C_{43}H_{63}O_{10}$; 0.354 ppm error)

This analysis confirms identification category A from Table 13.3.

Based on the criteria in Table 13.3, data presented here would qualify this structural analysis (identification) as confident.

Note that authentic reference material is readily available to compare with this unknown extractable and enable the promotion of the identification from *confident* to *confirmed*.

Structural analysis by LC/MS involves the acquisition and interpretation of APCI and ESI spectra. Given a reversed-phase HPLC platform separation of an extractables/leachables mixture, a general process recommended for structural analysis is as follows:

1. Accomplish APCI LC/MS analysis and consider the general features of the APCI spectrum:
   a. Can the molecular ion ($[M+H]^+$, $[M-H]^-$, etc.) be identified (or confirmed by ESI)?
   b. Based on stable isotope patterns, are there obvious heteroatoms present?
   c. Considering the nitrogen rule, is the number of nitrogens in the molecule even or odd?
   d. If accurate mass data are available, can the molecular formula be determined?

2. Consider fragment ions and their relationship to the molecular ion (note that fragmentation can be source induced, collision induced, or FT-ICR induced by radiation absorption or electron capture):
   a. Can reasonable and stable structures be determined for the most abundant fragment ions?
   b. Can the higher mass fragment ions be mechanistically linked to the molecular ion?
   c. If accurate mass data are available, can hypotheses from (a) and (b) be confirmed?

3. Finally, can authentic reference materials or literature references be obtained to confirm/support proposed structures?

The process is best illustrated by consideration of Case Study 13.2.

For a comprehensive review of HPLC and LC/MS analysis of extractables and leachables, the interested reader is referred to Norwood et al.[26]

13.4 QUANTITATIVE ANALYSIS OF EXTRACTABLES/LEACHABLES

Quantitation of extractables/leachables is required for the accomplishment of controlled extraction studies (see Chapter 14) and for the routine monitoring and control of critical component extractables and drug product leachables. To be useful for this task, an analytical system must be capable of producing a detector response that is directly proportional to the amount or concentration of individually separated analytes, over a relatively wide dynamic range. Many analytical techniques have these capabilities with varying degrees of specificity, selectivity, and ruggedness. For routine monitoring of extractables/leachables and quality control applications, techniques such as HPLC/UV and GC/FID are generally considered most appropriate (these will be discussed in other chapters). Selective detectors are also used for specific applications, such as GC/TEA for N-nitrosamine quantitation discussed earlier. However, for pharmaceutical development applications such as controlled extraction studies, the hyphenated techniques used for qualitative and structural analysis of extractables/leachables (i.e., GC/MS, LC/MS) can also be used for quantitative analysis, and the following discussion will therefore focus on these.

13.4.1 Quantitative MS: General

Mass spectrometers, along with providing compound-specific structural information, are capable of providing highly selective and sensitive quantitative information on trace-level organic analytes separated by either GC or HPLC. There are three general acquisition modes for quantitative GC/MS and LC/MS:

1. *"Full scan" with mass chromatograms.* This is the acquisition mode most often used for structural analysis, in which the mass spectrometer is repeatedly scanned over a specified mass range (e.g., m/z 50–650 for GC/MS; m/z 100–1000 for LC/MS) at a rate adequate to collect at least 5–10 mass spectra across any eluting chromatographic peak (Note: While the term "scanned" can be applied directly to mass analyzers such as quadrupoles, it is not strictly accurate when applied to time of flight, ion cyclotron resonance, or certain other instrument types. However, the concept is the same.). Quantitation can be accomplished using computer-generated chromatograms of single ions, for example, the molecular ion of an analyte of interest, and integrated peak areas. These chromatograms are referred to as "mass chromatograms" or "extracted ion current profiles." Figure 13.16 shows an example.

2. *Selected ion monitoring (SIM).* In this acquisition mode, the mass spectrometer is programmed to acquire data only for target analyte ions of interest, such as molecular ions. SIM chromatograms are generated for each acquired ion and appropriate peak areas obtained by integration. Since far less structural information is available from this acquisition mode, it is typically used for target compound analyses where higher sensitivity is required. This higher sensitivity is available because SIM increases the *duty cycle* of the mass spectrometer relative to full scan mode.[19] The duty cycle is the fraction of total acquisition time spent measuring the ions of interest.[19] Modern computerized data acquisition systems allow many ion channels to be acquired either simultaneously or over chromatographic retention time windows during a GC/MS or LC/MS analysis (see Figs. 13.18 and 13.19 along with the discussion below). It should be noted that certain quadrupole ion trap mass spectrometers

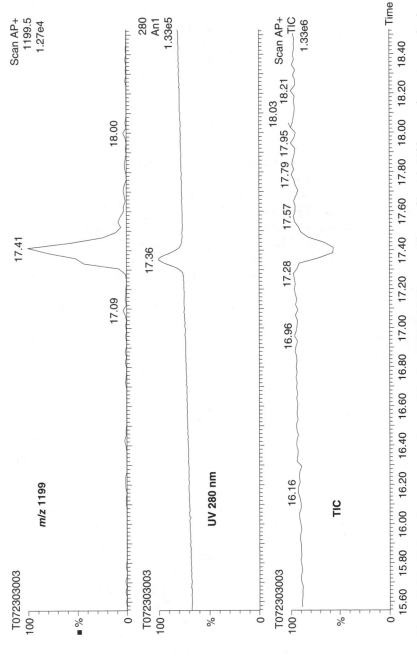

Figure 13.16 Various chromatograms for a positive ion APCI LC/MS extractables profile: *m/z* 1199 mass chromatogram (top), UV at 280 nm (middle), and total ion chromatogram (bottom).

279

show sensitivities in full scan mode, which are roughly equal to quadrupole mass filters in SIM mode.

3. *Selected reaction monitoring (SRM).* SRM provides an increased level of selectivity relative to SIM but, since it relies on CID processes, requires MS/MS capability. The triple quadrupole mass spectrometer is the most commonly employed instrument for SRM acquisition. Although SRM can be used with either GC/MS or LC/MS, it is historically most often used with LC/MS. This is because (1) ESI and APCI produce mostly molecular ions (as described above) while EI does not; (2) adequate selectivity for GC/MS is supplied by the powerful separating capability of capillary columns; and (3) chemical noise levels are relatively high with LC/MS as compared with GC/MS.

SRM on a triple quadrupole mass spectrometer can be accomplished with either of three scanning modes: *product ion scan, precursor ion scan,* or *constant neutral loss scan.* With reference to Figure 13.15 and its associated discussion above, a product ion scan requires Q1 be set to transmit a selected precursor ion (the $[M+H]^+$ of a target analyte) to Q2, which acts as a collision cell for CID thus producing product ions that are transmitted to Q3. For structural analysis, Q3 would be scanned to create a product ion spectrum (see Fig. 13.17), but for SRM, Q3 would be set to pass a selected product ion for quantitation, again increasing the duty cycle with the resulting sensitivity increase. A product ion scan for 2-mercaptobenzothiazole is given in Figure 13.17. As with SIM, modern computerized data acquisition systems allow many reaction channels to be acquired either simultaneously or over chromatographic retention time windows during an LC/MS analysis.

The reader interested in greater detail in SRM, or any aspect of quantitative MS, is referred to the comprehensive text of Boyd et al.[19]

13.4.2   Quantitative MS: Method Development and Validation

For GC/MS, few constraints are placed on the chromatography system. Fused-silica capillary GC columns can be inserted directly into the mass spectrometer's EI/CI source, with carrier gas flows of 1–1.5 mL/min easily accepted by the high vacuum system. LC/MS analysis, however, does potentially constrain the chromatography system in certain ways.[25] As mentioned above, reversed-phase separation platforms are preferred over normal-phase platforms. This is usually not a problem since reversed-phase platforms dominate pharmaceutical applications of HPLC. Analytical scale HPLC platforms must also consider that APCI typically works best at 0.2–2 mL/min flow rates, but ESI tends to show better performance at much lower flows (microliter per minute) which requires postcolumn flow splitting. ESI can operate at higher flows on most instruments, but performance can be compromised.

The most significant constraint placed on HPLC platforms by LC/MS regards mobile-phase additives, particularly including buffers. In general, involatile buffers and ion-pairing reagents, such as potassium phosphate, borate, and sodium dodecyl sulfate, are not compatible with LC/MS as they contaminate and foul the ion source.

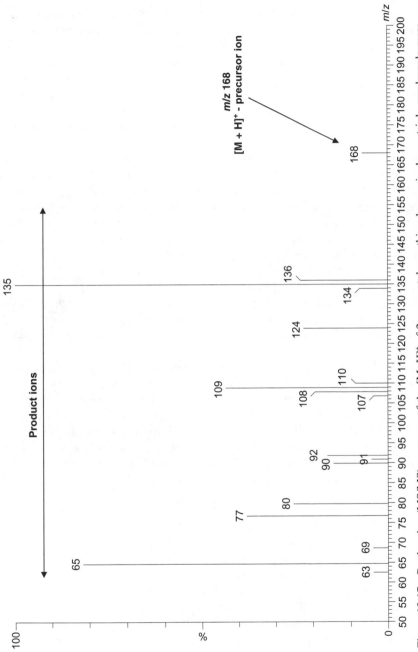

Figure 13.17 Product ion (MS/MS) spectrum of the $[M+H]^+$ of 2-mercaptobenzothiazole acquired on a triple quadrupole mass spectrometer.

HPLC platforms that include such additives require redevelopment to incorporate volatile additives, such as ammonium formate, ammonium acetate, TFAA, and heptafluorobutyric acid.[25] Mobile-phase composition and additives also affect the resulting ESI and APCI spectra by altering solution chemistry with ESI, or gas-phase ion chemistry with APCI. As mentioned above, chloride ion attachment is accomplished in negative ion APCI by adding a chlorine-containing reagent such as chloroacetonitrile in small quantities to the mobile phase.

Perhaps the most significant advantage for quantitative TOA gained by using a mass spectrometer as the detector is the ability to use stable isotope-labeled analogs (^{13}C, ^{2}H, ^{15}N) of target analytes as internal standards. This is because the mass spectrometer is capable of mass separating the ions from natural abundance and isotope-labeled analytes even if they co-elute chromatographically. Stable isotope-labeled internal standards can be used with GC/MS and LC/MS, and with any acquisition mode (full scan, SIM, SRM). It has been repeatedly demonstrated that trace organic analytical method precision and accuracy are directly related to the molecular structure of the internal standard. The closer are the two structures, the greater are the precision and accuracy. Norwood et al.[27] described a method for the quantitative analysis of PAHs as leachables in MDI drug products based on GC/MS and SIM, which incorporated deuterium-labeled analogs of each target PAH as internal standards. Figures 13.18 and 13.19 show example SIM chromatograms from a quantitative LC/MS method for target fatty acids in an MDI drug product that incorporates deuterium-labeled internal standards for each target analyte. Note that for each target analyte, the corresponding perdeuterated internal standard has a slightly shorter chromatographic retention time due to kinetic isotope effects.

GC/MS and LC/MS methods for extractables/leachables are typically validated as ICH quantitative impurity methods, with the appropriate validation parameters assessed relative to defined acceptance criteria. Table 13.4 presents a validation summary for the quantitative LC/MS method for target fatty acids in an MDI drug product shown in Figures 13.18 and 13.19. When appropriate robustness studies are added to the overall validation protocol, this would represent a typical validation process (note that additional examples of quantitative extractables/leachables methods can be found in other chapters of this volume).

13.5 CONCLUDING SUMMARY

The identification and quantitation of extractables and leachables is a TOA problem, and the analytical techniques most often used to solve such problems are directly applicable to extractables and leachables. Hyphenated techniques involving GC and HPLC interfaced with MS provide both compound-specific qualitative and high sensitivity quantitative information suitable for the qualitative, structural, and quantitative analysis of extractables/leachables during pharmaceutical development studies. Other analytical techniques, including those with selective (although not compound specific) detectors, are usually employed for monitoring leachables on stability and routine quality control applications.

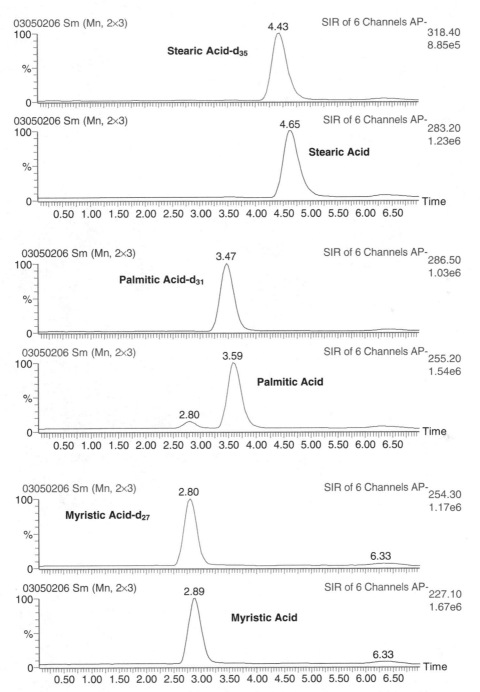

Figure 13.18   Example SIM chromatograms ([M–H]⁻ for each analyte and internal standard) from a standard mixture of target medium-chain fatty acids and deuterium-labeled internal standards acquired by negative ion APCI LC/MS.

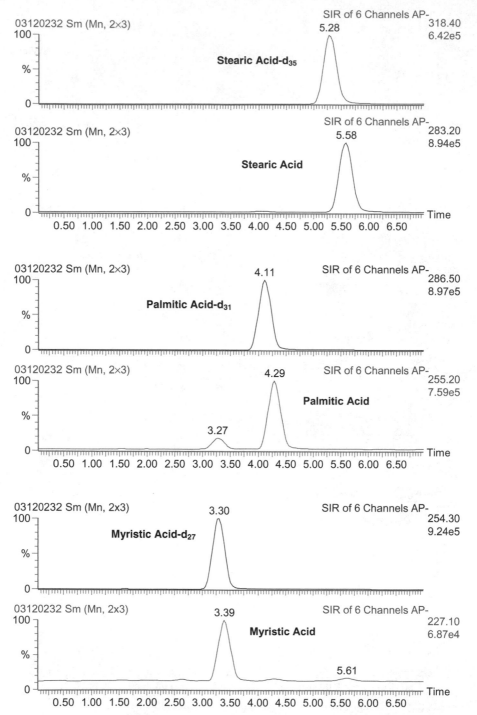

Figure 13.19   Example SIM chromatograms ([M–H]⁻ for each analyte and internal standard) from a metered dose inhaler drug product extract of leached medium-chain fatty acids (spiked with deuterium-labeled internal standards) acquired by negative ion APCI LC/MS.

TABLE 13.4. Validation Summary for Target Extractables Quantitation in a Metered Dose Inhaler Drug Product by LC/MS with Selected Ion Recording

Validation parameters	Acceptance criteria	Validation results		
		Myristic acid	Palmitic acid	Stearic acid
Instrument precision ($N = 10$; 1 µg/canister)	%RSDs $\leq 10\%$	2.49%	4.75%	6.63%
Chromatographic resolution (R) tailing factor ($N = 10$)	$R \geq 2$ (critical peak pair), $T \leq 2$	$R = 2$ (myristic acid-d$_{27}$ and palmitic acid), $T = 1.19$	$T = 1.23$	$T = 1.42$
Linearity (0–2000 µg/canister)	Coefficients of determination (r^2) ≥ 0.99	$y = 1.801x + 0.0635$ $r^2 = 0.9995$	$y = 1.2452 + 0.0345$ $r^2 = 0.9999$	$y = 1.2512x + 0.0275$ $r^2 = 1.0000$
Accuracy ($N = 3$; 100, 200, 300 µg/canister)	$75\% \leq$ average level recovery $\leq 125\%$	102%, 92%, 89%	98%, 92%, 82%	103%, 96%, 106%
Repeatability ($N = 6$; endogenous level)	%RSDs (of the determined amounts) $\leq 20\%$	2.55%	3.70%	1.13%
Intermediate precision ($N = 6$; endogenous level)	1. %RSDs (of the determined amounts) $\leq 20\%$ 2. %Absolute difference (first and second analyst) $\leq 25\%$	%RSD 10.8 %Difference 4.56	%RSD 12.14 %Difference 0.94	%RSD 12.97 %Difference 2.29
Limit of quantitation (LOQ) = 6 µg/canister	Report results as the average of signal-to-noise	164	110	226
Specificity	The mass spectra of the spiked and unspiked samples are the same as the mass spectra of the standard within instrumental precision.	Agree	Agree	Agree
Sample (spiked 200 µg/canister) and standard stability (200 µg/canister)	Report the longest storage condition for which %Recovery is $\geq 90\%$.	Spiked sample 6 days 101% Standard 6 days 102%	Spiked sample 6 days 102% Standard 6 days 100%	Spiked sample 6 days 98.9% Standard 6 days 101%

RSDs, relative standard deviations.

The future holds the promise of new technologies being applied to the analysis of extractables and leachables. A recent report suggests that ion mobility spectrometry (IMS) could be useful in this area.[28] IMS is an analytical technique that separates ions based on their gas-phase mobility at atmospheric pressure, and produces unique patterns for analyte molecules (termed "plasmagrams") based on ion current versus drift time. IMS was shown to have significant sensitivity for a series of model extractables/leachables, including Irgafos 168, oleamide, and erucamide. The latter two compounds are antislip agents used in certain types of plastic. Certain IMS systems used for cleaning verification of pharmaceutical manufacturing equipment are capable of accepting samples from surface wipes,[29] giving IMS the potential for direct analysis of these antislip agents on container closure system component surfaces (e.g., stainless steel and aluminum canisters and MDI valve components). IMS analysis times are also very short (total analysis times are measured in seconds), giving IMS a significant potential for routine quality control and perhaps even in process testing applications.

REFERENCES

1 ICH harmonised tripartaite guideline: Q3A(R) impurities in new drug substances. International Conference on Harmonisation of Technical Requirements for Registration of Pharmaceuticals for Human Use, 2002.
2 Guidance for industry: Q3C impurities: Residual solvents. U.S. Department of Health and Human Services, Food and Drug Administration, Center for Drug Evaluation and Research (CDER), 1997.
3 Guidance for industry: Q3C-tables and lists. U.S. Department of Health and Human Services, Food and Drug Administration, Center for Drug Evaluation and Research (CDER), 2003.
4 Guidance for industry: Q3B(R2) impurities in new drug products. U.S. Department of Health and Human Services, Food and Drug Administration, Center for Drug Evaluation and Research (CDER), 2006.
5 Norwood, D.L., Qiu, F., and Mullis, J.O. Trace level impurity analysis. In: Swarbrick, J., ed. *Encyclopedia of Pharmaceutical Technology*, 3rd ed. Dekker Encyclopedias (a product line from Taylor and Francis Books), New York, 2006; 3797–3813.
6 Qiu, F. and Norwood, D.L. Identification of pharmaceutical impurities. *J Liq Chromatogr Relat Technol* 2007, *30*(5–7), pp. 877–935.
7 Norwood, D.L. and Ball, D. Product Quality Research Institute: Safety thresholds and best practices for extractables and leachables in orally inhaled and nasal drug products. Submitted to the PQRI Drug Product Technical Committee, PQRI Steering Committee, and U.S. Food and Drug Administration by the PQRI Leachables and Extractables Working Group, 2006.
8 Ball, D., Blanchard, J., Jacobson-Kram, D., McClellan, R., McGovern, T., Norwood, D.L., Vogel, M., Wolff, R., and Nagao, L. Development of safety qualification thresholds and their use in orally inhaled and nasal drug product evaluation. *Toxicol Sci* 2007, *97*(2), pp. 226–236.
9 Norwood, D.L., Paskiet, D., Ruberto, M., Feinberg, T., Schroeder, A., Poochikian, G., Wang, Q., Deng, T.J., DeGrazio, F., Munos, M.K., and Nagao, L.M. Best practices for extractables and leachables in orally inhaled and nasal drug products: An overview of the PQRI recommendations. *Pharm Res* 2008, *25*(4), pp. 727–739.
10 Hertz, H. S. and Chesler, S. N., eds. *Trace Organic Analysis: A New Frontier in Analytical Chemistry*. National Bureau of Standards Special Publication 519, U.S. Government Printing Office, Washington, DC, 1979.
11 Beyermann, K. *Organic Trace Analysis*. Ellis Horwood, Chichester, England, 1984.
12 Norwood, D.L., Nagao, L., Lyapustina, S., and Munos, M. Application of modern analytical technologies to the identification of extractables and leachables. *Am Pharm Rev* 2005, *8*(1), pp. 78–87.

13 Valcarcel, M. *Principles of Analytical Chemistry—A Textbook*. Springer-Verlag, Berlin, 2000.

14 Norwood, D.L., Granger, A.T., and Paskiet, D.M. Extractables and Leachables in drugs and packaging. In: Swarbrick, J., ed. *Encyclopedia of Pharmaceutical Technology*, 3rd ed. Dekker Encyclopedias (a product line from Taylor and Francis Books), New York, 2006; 1693–1711.

15 Norwood, D.L. and Mullis, J.O. "Special case" leacheables: A brief review. *Am Pharm Rev* 2009, *12*(3), pp. 78–86.

16 Norwood, D.L., Mullis, J.O., Feinberg, T.N., and Davis, L.K. N-nitrosamines as "special case" leachables in a metered dose inhaler drug product. *PDA J Pharm Sci Technol* 2009, *63*(4), pp. 307–321.

17 Gross, J.H. *Mass Spectrometry—A Textbook*. Springer-Verlag, Berlin, 2004.

18 de Hoffmann, E. and Stroobant, V. *Mass Spectrometry Principles and Applications*. John Wiley & Sons, Chichester, England, 2007.

19 Boyd, R.K., Basic, C., and Bethem, R.A. *Basic Quantitative Analysis by Mass Spectrometry*. John Wiley & Sons, Chichester, England, 2008.

20 Christman, R.F. Guidelines for GC/MS identification. *Environ Sci Technol* 1982, *16*(3), p. 143A.

21 Harrison, A.G. *Chemical Ionization Mass Spectrometry*. CRC Press, Boca Raton, FL, 1983.

22 McLafferty, F.W. and Turecek, F. *Interpretation of Mass Spectra*, 4th ed. University Science Books, Sausalito, CA, 1993.

23 Budzikiewicz, H., Djerassi, C., and Williams, D.H. *Mass Spectrometry of Organic Compounds*. Holden-Day, San Francisco, CA, 1967.

24 Willoughby, R., Sheehan, E., and Mitrovich, S. *A Global View of LC/MS*. Global View Publishing, Pittsburgh, PA, 1998.

25 Norwood, D.L., Mullis, J.O., and Feinberg, T.N. Hyphenated techniques. In: Ahuja, S. and Rasmussen, H., eds. *HPLC Method Development of Pharmaceuticals*. Separation Science and Technology. Elsevier Academic Press, London, 2007; 189–235.

26 Norwood, D.L., Jenke, D., Manolescu, C., Pennino, S., and Grinberg, N. HPLC and LC/MS analysis of pharmaceutical container closure system leachables and extractables. *J Liq Chromatogr Relat Technol* 2009, *32*, pp. 1768–1827.

27 Norwood, D.L., Prime, D., Downey, B.P., Creasey, J., Sethi, S.K., and Haywood, P. Analysis of polycyclic aromatic hydrocarbons in metered dose inhaler drug formulations by isotope dilution gas chromatography/mass spectrometry. *J Pharm Biomed Anal* 1995, *13*(3), pp. 293–304.

28 Mullis, J.O., Granger, A., Qin, C., and Norwood, D.L. The analytical evaluation threshold (AET) concept: Sensitivity and analytical uncertainty. Presented at *Leachables and Extractables Conference, Proceedings of Leachables and Extractables Conference*, Smithers Rapra Technology, Ltd., Dublin, Ireland, March 5–6, 2008.

29 Qin, C., Granger, A., Papov, V., McCaffrey, J., and Norwood, D.L. Quantitative determination of residual active pharmaceutical ingredients and intermediates on equipment surfaces by ion mobility spectrometry. *J Pharm Biomed Anal* 2009, *51*(1), pp. 107–113.

CHAPTER 14

EXTRACTABLES: THE CONTROLLED EXTRACTION STUDY

Thomas N. Feinberg, Daniel L. Norwood,
Alice T. Granger, and Dennis Jenke

14.1 INTRODUCTION AND OVERVIEW

The ultimate goal of pharmaceutical development is to produce safe and effective medicines. Beyond its intrinsic therapeutic value, safety and efficacy of a particular medicine are achieved by controlling drug product purity, stability, and critical quality attributes. Thus, "control" becomes a key aspect of pharmaceutical development, specifically, the quality control (QC) of inputs and the risk control of outputs.

Extractables/leachables studies are a clear example of the fundamental application of this "control" concept. Control of extractables profiles of drug product delivery or container closure packaging systems (i.e., an input) must be achieved so as to avoid any adverse effects that drug product leachables (i.e., an output) may have on product quality, where the leachables represent "any poisonous or deleterious substance which may render the contents injurious to health."[1] Fundamental concepts of manufacturing science suggest that one cannot maintain quality standards by "testing quality into the product." Rather, by applying QC to the inputs of the process and by controlling the process itself, quality outputs can be achieved and quality risks can be minimized. Thus, in order to control the quality of the final drug product relative to leachables, QC tests and quality standards for extractables must be established for the delivery or container closure system and packaging materials. Analytical investigations of the delivery/container closure system and packaging materials that produce the information required for such control are generally referred to as "extractables" studies.

Two questions arise prior to implementing extractables studies:

1. How does one completely and efficiently establish the extent to which container closure system components and materials of construction contain such "deleterious" substances?

Leachables and Extractables Handbook: Safety Evaluation, Qualification, and Best Practices Applied to Inhalation Drug Products, First Edition. Edited by Douglas J. Ball, Daniel L. Norwood, Cheryl L.M. Stults, Lee M. Nagao.
© 2012 John Wiley & Sons, Inc. Published 2012 by John Wiley & Sons, Inc.

2. How are the results of an extractables study used to establish quality standards for container closure system components and materials of construction?

These questions are far from trivial. Obtaining answers to these questions requires a thorough investigation of each individual material of construction and container closure system critical component. This investigation begins with the procurement and evaluation of available information regarding a material's synthetic processes, ingredients, and manufacturing and fabrication processes, and continues with the controlled extraction study (CES) when such available information is found to be incomplete or inadequate for product impact risk assessment.

14.1.1 What is a CES?

Extractables information, specifically the identities and extracted amounts of individual extractables, is essential to the rigorous, effective, and efficient assessment of any suitability for use issue (e.g., safety and efficacy) associated with any contact that may occur between the system and a therapeutic product or patient during the course of the product's manufacture, storage, and/or delivery. Such an impact assessment requires that extractables information be obtained. While it is somewhat of an oversimplification, there are essentially two general processes by which extractables information *can* be obtained. The first process is, in essence, composition based; that is, the list of known compositional ingredients for the materials, components, or system of interest becomes the list of extractables (e.g., Chapter 13, Table 13.1). Thus, for example, if the component manufacturer adds chemical **X** to a base polymer resin as a processing aid and that resin is used to fabricate a critical component that is then incorporated into a drug product's container closure system, then chemical **X** is a *potential* extractable of that system. Presumably, if one could compile a list of all compounds that (1) were intrinsically present in a material's precursor "starting blocks"; (2) were intentionally added to a material at any step of its manufacture and processing into a finished critical component; (3) were unintentionally added to a material in any step of its manufacture and processing into a finished critical component; and (4) were produced via chemical reaction during its manufacture and processing into a finished critical component, then such a list would represent the component's qualitative *extractables profile* in terms of compound identities. A more detailed list, including the levels at which each precursor, chemical additive, and processing aid are present in the finished critical component, would represent the "worst-case" (i.e., the case in which all of the potential extractables are actually extracted in their entirety) quantitative *extractables profile*.

While such a composition-based approach to extractables identification and quantitation is conceptually desirable as it is based on essentially "free" (presumably obtained from the component manufacturer by request) and potentially comprehensive information, it is rarely the case in the current pharmaceutical marketplace that such detailed compositional information is available and freely shared, sufficiently complete, and sufficiently credible to stand alone on its own merits. Furthermore, even if such a comprehensive list was freely available, complete, and credible, it would not necessarily account for reaction chemistries of additives or their reaction products. Thus, it is almost always the case that compositional information *must* be augmented with extraction studies whose purpose is to delineate a component's complete, credible, and relevant extractables profile. In the extreme case where no compositional information is available, the *only means* of obtaining extractables information is to generate the information in the analytical laboratory. This analytical laboratory approach is the second general process by which extractables information can be obtained.

The laboratory characterization of a material, component, or system with respect to extractables requires that appropriate test articles be extracted and the resulting extracts analyzed. This is true because comprehensive and sensitive techniques for the complete characterization of a component or material in its natural solid state do not exist. The process by which extractables information is generated by direct laboratory evaluation of a test article is termed a CES. The objectives of a CES are to systematically and rationally detect and identify all relevant extractables (which can be considered as potential drug product leachables) and establish the worst-case drug product accumulation potential of each detected and identified extractable. It is obvious that the methods used to generate and analytically test the extracts are the essential foundations of a CES, and thus that proper choice and justification of these techniques and methods are the most important factors that influence the validity of a CES. What is less obvious are the critical experimental design parameters that "control" the effectiveness of a CES and what the proper "settings" are for these parameters so that the CES is comprehensive, compelling, cost-efficient, and, most importantly, helps secure regulatory approval for drug products under development.

14.1.2 Regulatory Guidance for Performing Extractions

In certain cases, the regulatory guidance related to the safety qualification of container/closure systems provides clarity in terms of the type of extraction study that should be performed in order to secure regulatory approval to market a particular pharmaceutical product. Thus, for example, it is noted in "Section 4 (Extraction Studies) of the European Medicines Agency (EMEA) Guideline on Plastic Immediate Packaging Materials"[2] that "the aim of extraction studies is to determine those additives that might be extracted by the preparation or the active substance in contact with the material." Furthermore, the EMEA guideline states that "the solvent used for the extraction should have the same propensity to extract substances as the active substance/dosage form as appropriate." The EMEA recommends that the extraction be performed under "stress conditions to increase the rate of extraction." Thus, it is clearly the EMEA perspective that the extraction study should simulate actual product use and that exaggerated extraction conditions should be used only to speed up the extraction process, not to necessarily produce a "super-concentrated" extract. The United States Food and Drug Administration (FDA), in its "Guidance for Industry: Container/Closure Systems for Packaging Human Drugs and Biologics,"[3] offers a similar perspective. Thus, in "Attachment C" of the guidance, it is noted that "the ideal situation is for the extracting solvent to have the same propensity to extract substances as the dosage form, thus obtaining the same quantitative extraction

TABLE 14.1.   Primary Directives for the Extraction Portion of a Controlled Extraction Study

Directive #1	The extraction conditions employed cannot materially change the nature of the extractables profile.
	Corollary 1A. For the qualitative identification of extractables, this means that no new entities are produced as a result of the extraction strategy employed and that no old entities are lost to a level below the identification limit.
	Corollary 1B. For the quantitation of extractables, this means that there can be no change in the number and identity of the entities found as well as no material change in the level of each and every entity.
Directive #2	Any extraction process, other than actual product use, which is used to assess the safety and/or efficacy of a system, must be technically justified in terms of its ability to produce the same extractables profile as the leachables profile that would be obtained by testing the drug product under the worst-case conditions of storage, use, and/or shelf life.

profile." The FDA recommends that a stronger extracting solvent would be used to "obtain a qualitative extraction profile that would be used to establish quality control criteria." The FDA notes that extractions should be performed at elevated temperatures so as to "increase the rate of extraction, so that a short experimental time may simulate a longer exposure time at room temperature, or to maximize the amount of extractables obtained from a sample."

These considerations can be summarized in terms of a two-part extraction directive[4] that reads as follows: An extraction study should be performed with an extracting solvent that minimally has a similar propensity to extract substances as the drug product, and under conditions that accelerate, but do not change, the actual product contact conditions (see Table 14.1).

14.1.3   The "Catch 22" Associated with Extractions

Given the great diversity in both the materials used in pharmaceutical container closure systems and the pharmaceutical applications (products) themselves, it is reasonable to observe that there is no single extraction process (consisting of an extraction solvent and the material/solvent contact conditions) that meets the previously stated requirements for a CES in all potential pharmaceutical situations. This juxtaposition can (and has) lead to considerable fragmentation in extractables survey information because such information is generated using different extraction techniques and different analytical methods. While this fragmentation is not a fatal issue in the context of a single component or material in a single application, it can be a major issue in comparative situations (e.g., when comparing the extractables profiles of two materials or considering the use of a single material in several applications).

Although it may not be possible to establish a universal "standard" CES including protocols/conditions for all pharmaceutical applications, it may be the case that sound scientific principles and best demonstrated laboratory practices can be used to develop (and justify) a set of "standard" CES conditions that focus on specific pharmaceutical applications. These general scientific principles and best demonstrated laboratory practices would not only address the main objectives of the CES as defined previously, but should also address certain "secondary" objectives[5] such as the following:

1. Establish a basis for the development and validation of routine QC methods and acceptance criteria for critical component extractables profiles.

2. Establish a basis for the development and validation of leachables methods suitable for use in drug product leachables studies as well as for potential use as routine QC methods for drug product leachables (should such be required by regulatory authorities).

3. Allow for the "correlation" of extractables and leachables.

It is clear that a comprehensive CES consists of two essential steps: extract generation and extract characterization (analysis). Each of these distinct but inter-related processes is considered in greater detail as follows.

14.2 EXTRACT GENERATION

14.2.1 Overview

There is much discussion of, but little consensus for, establishing universally applied standard extraction protocols and methods. This is true due to the considerable diversity in therapeutic products, their associated container closure systems, the materials and/or components used in those systems, and the composition and processing of those materials. Nevertheless, there are certain concepts that are relevant to the CES, regardless of the component or material, the therapeutic product, and/or the circumstances of contact. For a qualitative CES (i.e., the identification/structural analysis of extractables), the objective is to establish the compound-specific composition of a particular test article. Thus, one must employ a CES extraction process that disrupts the physical integrity of the test article without destroying the structural integrity of its extractables; that is, the extraction must liberate the test article's extractables (in part or in their entirety) but cannot result in their chemical alteration. More specifically, the extraction conditions used must solubilize, but not chemically modify, a sufficient quantity of all potential extractables so that they can be detected and identified. If such a modification were to occur, a flawed or compromised characterization could result. This flaw may be a direct false negative (i.e., the altered extractable may no longer be identifiable or its actual level may not be accurately represented in the extract) or an indirect false positive (i.e., a "new" chemical entity, derived from the extractable, is now present in the extract that was not initially present in the material or component and cannot likely appear as a drug product leachable).

Thus, an "ideal" extraction process for a qualitative CES would

1. extract quantities of the test article's chemical building blocks, chemical additives, processing aids, and so on, sufficient for detection and identification/structural analysis;

2. be simple, straightforward, and safe so as to facilitate laboratory operations;

3. preserve the structural integrity of the extractables;

4. produce an analytically appropriate extract; and

5. be repeatable and reproducible.

Since the intent of the qualitative CES is identification, the extractions used therein need not liberate the total amount of any extractable present in the test article; rather, the extraction must liberate only enough of the extractable so that it can be detected and identified. Thus, the most "mild" extraction conditions that can liberate sufficient extractables for identification are desirable in that such conditions have the least potential for causing the conversion of actual material ingredients to secondary reaction products.

These relatively simple concepts associated with the qualitative aspects of the CES do not directly translate to a CES that addresses the quantitation of individual extractables, specifically, the process by which the test article is extracted to produce the extract for testing. This statement is true because

1. the amount of an extractable that is extracted from a test article will establish the impact that the extractable could potentially exert on a therapeutic product, and

2. the amount of an extractable that is extracted from a test article will depend on the extraction process.

The essential issue involved in establishing the conditions for a quantitative CES is "what is the proper amount of an extractable that should be liberated in the CES?" There are three possible answers to this question:

1. the CES should establish the *total amount of an extractable that is present* in a test article (i.e., "deformulate" the test article);

2. the CES should establish the *absolute amount of an extractable that can be extracted* from a test article; or

3. the CES should establish the *absolute amount of an extractable that will potentially accumulate in a drug product* during its contact with a test article over the course of the drug product's shelf life.

These answers translate to (1) the worst *possible* accumulation ("it all comes out"), (2) the worst *probable* accumulation ("it comes out to the extent that is thermodynamically possible"), and (3) the *actual* accumulation ("it comes out to the extent that is both thermodynamically and kinetically possible"). In a situation where these values are all equal (it all comes out under the actual conditions of contact), then the objective of the CES is clear ("get it all out"; deformulate) and the criteria for judging the acceptability of the extraction process is equally clear (the extraction process is only valid if it has been demonstrated to remove all of an extractable from a test article). However, most product use situations for pharmaceuticals are such that actual product use is considerably less "stressful" than complete extraction, and thus, it is not clear what the proper extraction strategy should be.

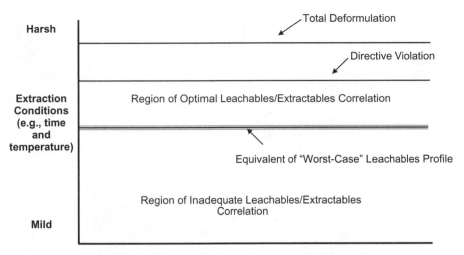

Figure 14.1   Conceptual schematic of a controlled extraction study (CES).

TABLE 14.2.   Comparison of Sample Pretreatment Methods on Amounts of Detected Extractables

Condition	Irganox 1076	Oxidized Irgafos 168	Irgafos 168
Intact	100	158	247
Cryoground	226	227	760

As outlined by the extraction directives in Table 14.1, an effective and accept-able CES is performed under conditions that are more aggressive than the usual conditions of storage, use, and/or shelf life, but not conditions that are so aggressive that the resultant extractables profile is dramatically different than that which would be obtained if the extraction was performed with the drug product under the worst-case conditions of storage, use, and/or shelf life. An illustration of this concept for a typical CES is shown in Figure 14.1. Note that for a given extraction process, as conditions move from "mild" toward "harsh," the overall system moves from a state where there is potential for an inadequate extractables/leachables correlation, past a point which exactly correlates with the worst-case leachables profile into a region of adequate correlation, and finally a point where *directive violation* is reached (i.e., the extractables profile is chemically altered by the extraction process such that extractables/leachables correlation is affected). It is possible, though somewhat unlikely, that total deformulation of the extracted component could be attained without such directive violation. As a relatively simple example, and with reference to Figure 14.1, consider a polypropylene critical component subjected to Soxhlet extraction with methylene chloride (dichloromethane). It is known from the supplier that the component contains 300 µg/g of Irgafos 168 as an antioxidant. Irgafos 168 is designed to react with hydroperoxides to form an oxidized analog as follows (see also Table 14.2 with associated discussion below):

Hydroperoxide

(ROOH)

If the worst-case leachables profile of the drug product contains equal amounts of both intact Irgafos 168 and its oxidized form at a total equivalent level of 100 µg/g of the contacting component, then it would be a directive violation to conduct a CES, which resulted in 250 µg/g of only the oxidized form (likely also including detectable levels of secondary degradation products of the oxidized form, which might not appear as drug product leachables).

To accomplish the goal of generating an appropriate extract within reasonable laboratory time frames, the rates of migration of compounds within the test article matrix must be accelerated. This acceleration can be accomplished by a variety of means including

1. *statistical effects*—by exaggerating the conditions of test article contact (i.e., exaggerated surface area or weight to solution volume, subdivision of the sample);

2. *solvent selection or chemical acceleration*—the use of a more "aggressive" solvent (i.e., a solvent that has a somewhat greater propensity to extract substances than does the drug product);

3. *temperature/time effects*—by an increase in thermal stress (i.e., higher temperature, longer duration); and

4. *physical effects*—by the application of external physical force (e.g., sonication).

14.2.2 Extraction Kinetics

The kinetics of extraction, that is to say the rates at which individual compounds are extracted from test articles (usually measured in units of mass per time), are not completely understood. In the absence of any known mechanisms that form new extractables without some source (even in CESs the law of conservation of mass must be recognized), the limiting amount of any particular extractable in solution is governed by the original amount in the material. The question of how much time must be allowed to obtain useful extracts has no a priori answer. If the only concern is to produce extracts of product importance as described previously, then the total amount of time necessary is limited to the product shelf life (assuming a perfect match in solubility forces between extracting solvent and drug product formulation). However, as discussed earlier the purpose of a CES is to obtain this information in a much shorter period of time. It is left to the experimenter to justify the choices of not only extraction time, but also temperature, solvent, sample size, and physical treatment.

Another guide to the expected results from conducting a CES is the Eley–Rideal model for surface reaction kinetics. For the following mechanism, an additive β migrates to an active surface position, A, to form a surface active compound designated as β_A:

$$\beta + A \rightleftarrows \beta_A,$$

with a forward rate constant given as k_1 and a reverse rate constant as k_{-1}. For such a mechanism,

$$\text{Rate} = \frac{d[\beta_A]}{dt} = -\frac{d[\beta]}{dt} = k_1[\beta][A] - k_{-1}[\beta_A]. \tag{14.1}$$

As the surface active compound interacts with solvent molecules, S, a new solvated species denoted β_S is formed,

$$\beta_A + S \rightarrow \beta_S,$$

with a forward rate constant given as k. For such a mechanism, if it is assumed that the reaction is not reversible, then

$$\text{Rate} = \frac{d[\beta_S]}{dt} = k[\beta_A][S]. \tag{14.2}$$

The square brackets denote the concentration of the species inside. The concentration designated by $[\beta]$ is the concentration of β in the packaging material, while $[\beta_S]$ is the concentration of β in the extraction solvent. The concentrations designated as $[A]$ or $[\beta_A]$ can be thought of as a number density on a surface of unoccupied active sites (designated by $[A]$), or the occupied active sites designated by β_A, respectively. Assuming that the surface area remains fixed, $[A]$ or $[\beta_A]$ can be expressed as the number of unoccupied active sites, N_{UA}, or the number of molecules occupying active sites, $N\beta_A$, respectively. The total number of active sites, N_A, is the sum of N_{UA} and $N\beta_A$. In other words, $N_A = [\beta_A] + [A]$. A surface coverage number, θ, can be defined as the fraction of total active surface sites, N_A, that are occupied by β_A molecules.

If $\theta = [\beta_A]/N_A$, then

$$[\beta_A] = \theta N_A \quad \text{and} \quad [A] = (1-\theta)N_A.$$

Substitution of the β_A concentration into Equation 14.2 gives the following:

$$\frac{d[\beta_S]}{dt} = k\theta N_A[S], \tag{14.3}$$

To ultimately express the rate of extraction of the β molecules in terms of only $[\beta]$ and $[S]$, it is necessary to first determine an analytical solution for θ. Equations 14.1 and 14.2 are combined to give the overall rate of change of β_A:

$$\frac{d[\beta_A]}{dt} = k_1[\beta][A] - k_{-1}[\beta_A] - k[\beta_A][S].$$

Using the steady-state approximation for the occupied active surface sites, such that the rate is zero gives

$$0 = k_1[\beta][A] - k_{-1}[\beta_A] - k[\beta_A][S].$$

Substitution of the concentrations for A and β_A expressed in terms of θ (as described above) allows an analytical solution for θ such that

$$\theta = \frac{k_1[\beta]}{k_{-1} + k_1[\beta] + k[S]}.$$

Substitution of this value into Equation 14.3 gives the desired expression

$$\frac{d[\beta_S]}{dt} = -\frac{d[\beta]}{dt} = \frac{kk_1 N_A[S][\beta]}{(k_{-1} + k[S]) + k_1[\beta]},$$

where $(k_{-1} + k[S])$ represents the rate of decrease of β_A and $k_1[\beta]$ represents the rate of increase of β_A. Two extremes may exist. In the first case, if $k_1[\beta] \gg (k_{-1} + k[S])$, then

$$\frac{d[\beta_S]}{dt} = kN_A[S].$$

Note that this expression is independent of the β species concentration, and the rate of extraction is dependent on the total number of active sites available and the solvent concentration. Since the solvent concentration is typically high compared with the species to be solvated, it is expected that the concentration of the solvent during the extraction will remain high until the situation is reached whereby the species is nearly depleted from the sample. This leads to zero-order kinetics.

For the other case where $k_1[\beta] \ll (k_{-1} + k[S])$, then the rate

$$\frac{d[\beta_S]}{dt} = \frac{kk_1 N_A[S][\beta]}{(k_{-1} + k[S])}.$$

This equation has the properties that at low concentrations of β in the sample, the rate is limited by diffusion and is first order in concentration. Rearrangement of this expression can be used to further elucidate the relative effects of concentration and diffusion. The following equation can be obtained:

$$\frac{d[\beta_S]}{dt} = \frac{k_1 N_A[\beta]}{\left(\dfrac{k_{-1}}{k[S]}+1\right)}.$$

If $k[S] \gg$ or $= k_{-1}$, which describes the case where the rate of extraction is much greater than the rate of leaving an active site and diffusing back into the matrix, then $d[\beta_S]/dt \propto k_1 N_A[\beta]$, where the extraction rate is primarily dependent on the species concentration in the matrix.

Alternatively, if $k[S] \ll k_{-1}$ and $k_{-1} \gg 1$, then $d[\beta_s]/dt \propto k_1 N_A[\beta]/k_{-1}$, where the extraction then becomes constrained by the diffusion rate to the active sites rather than the extraction rate of the solvent. Graphically, what one finds is shown in Figure 14.2.

The implication of this extraction kinetic model is that regardless of the goal of the CES (be it to establish [1] the worst possible accumulation, [2] the worst probable accumulation, or [3] the actual accumulation), the CES should be accomplished to achieve an outcome that agrees with the Figure 14.2 model but stays within the parameters depicted in Figure 14.1. This represents the so-called asymptotic level, which is suggested in various guidance documents.[5]

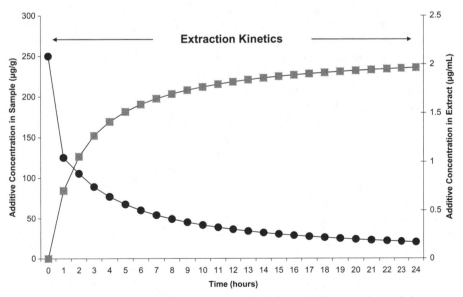

Figure 14.2   Typical extraction kinetics showing initially rapid linear region and slower diffusion limited tail (asymptotic). Note that the additive concentration in the extract (squares) increases while decreasing in the test article sample (circles).

14.2.3   Statistical Effects: Test Article Preparation

Because it is intact components that are contacted by the drug product under use conditions, common sense suggests that a CES should be performed using intact components (e.g., rubber gaskets and seals, dose metering valve components). However, there are compelling reasons why a CES may be performed with "modi-fied" test articles. For example, laboratory extractions can be accelerated by exag-gerating the conditions of contact (e.g., significantly increased surface area, subdivision of the sample). Furthermore, some test articles might be of inappropriate physical size (e.g., too big) for efficient handling in laboratory-scale extractions using any typical extraction apparatus or instrument-based extraction techniques. Lastly, it is important to account for any variability in extractables profiles between individual components. Extractions that involve few, or single, components maxi-mize the potential impact of component-to-component variability on CES results and interpretations.

Any test article sample preparation technique employed must not violate Directive #1 (Table 14.1), which states that "the extraction conditions employed cannot materially change the nature of the extractables profile." The simplest tech-nique for preparing test article samples for extraction that likely complies with this directive is cutting (in the case of elastomers) or breaking (in the case of plastic). Although this method has the virtue of simplicity, it potentially creates an inhomogeneous sample by exposing new surface areas in an irreproducible way. Two techniques that have been employed for plastic are pressing into a thin film (see Chapter 16) and cryogenic grinding. Pressing of plastic test article samples into a thin film is accomplished by the simultaneous application of heat and pres-sure ("hot pressing"). The thin film reduces the distance that individual extractable compounds must migrate within the polymer matrix in order to be extracted into a solvent. Cryogenic grinding uses a "freezer/mill," such as that shown in Figure 14.3. This particular freezer/mill system cools test article samples to cryogenic temperatures (liquid nitrogen) in a closed vial and pulverizes them by magneti-cally shuttling a steel impactor back and forth against two stationary end plugs.[6] The sample vial is immersed in liquid nitrogen throughout the grinding process, keeping the sample at cryogenic temperatures. The low temperature of the process renders the sample brittle and therefore more grindable,[6] as well as preventing the sample from heating excessively due to thermal energy from the grinding process. Note the "before and after" test article samples in Figure 14.3. It is also possible to grind test article samples to a constant particle size by sieving the cryogenically ground samples. Cryogenic grinding significantly increases sample surface area and can also produce a composite test article sample by grinding a number of individual components simultaneously into a homogenized powder. Table 14.2 compares data obtained via a quantitative analysis of extracts from both intact and cryoground polyolefin samples. For the same extraction conditions, the cryo-ground sample showed greater than twofold increase of extracted known additives (Irganox 1076 and Irgafos 168), with a less than 50% increase in the known addi-tive degradation product (oxidized Irgafos 168). The low temperatures used in the cryogrinding process can also prevent the loss of "volatile" organic extractables,

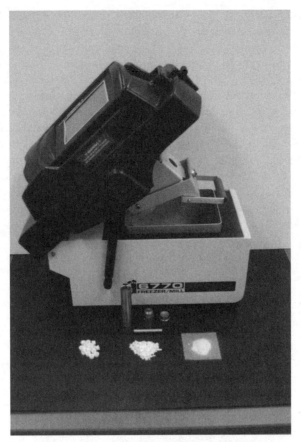

Figure 14.3   A "freezer/mill" used for cryogenic grinding of plastic container closure system components and materials. Note the intact plastic component test article sample (lower left), which is beside a sample that has been crushed to facilitate grinding (lower center). The finished cryoground sample is shown in the lower right.

thus increasing the probability that these potential leachables will be detected and identified during the CES.

Obviously, the more thermal and mechanical energy imparted to the test article sample during a samples preparation process, the more likely it is that Directive #1 will be violated. For plastics, this is particularly true when considering monomers and oligomers (i.e., polymer building blocks), and thermally labile antioxidants as extractables, since thermal and mechanical energy can disrupt and cleave polymer chains potentially releasing additional monomers/oligomers and also possibly additional oligomers, which would not be released from intact test articles. It is therefore recommended that any such sample preparation process be verified by comparison of extractables profiles obtained from both prepared and unprepared test article samples (e.g., ground to intact).

A properly designed CES must consider how to reduce the impact of component-to-component variability through selection of appropriate test article weights (and/or number of components) for extraction, and/or test article sample homogenization through, for example, cryogenic grinding. Two additional factors that influence test article weights for extraction and the final concentration (or dilution) of the resulting extract are the threshold concept (i.e., safety concern threshold [SCT]; analytical evaluation threshold [AET][7]), and the sensitivity requirements of the analytical techniques used for both qualitative and quantitative extractables profiling (the influence of analytical techniques is discussed in the following section).

Regarding the threshold concept, consider a metered dose inhaler (MDI) drug product with 120 labeled actuations per canister, a recommended dose of 8 actuations per day, and a critical component elastomer mass per valve of 250 mg. For an individual organic leachable derived from this elastomer, the estimated AET[7] would be

$$\text{Estimated AET} = \left(\frac{0.15\ \mu g / day}{8\ actuations / day} \times 120\ labeled\ actuations / canister \right),$$

$$\text{Estimated AET} \approx 2.25\ \mu g / canister.$$

Furthermore,

$$\text{Estimated AET} = \left(\frac{2.25\ \mu g / canister \times 1\ canister / valve}{0.25\ g\ of\ elastomer / valve} \right),$$

$$\text{Estimated AET} \approx 9.00\ \mu g / g\ of\ elastomer.$$

This calculation suggests that the selection of test article mass should be guided by the ability of the CES to identify/quantitate individual extractables at levels of at least 9 μg/g (ppm) in the elastomer matrix, assuming 100% extraction efficiency.

In addition to consideration of how a particular test article is prepared for extraction, it is important to ensure that the materials and components that are accepted as test articles for a CES are representative of the actual drug product container closure system. If this is not the case, then the generation of definitive extractables/leachables correlations, either qualitative or quantitative, could be compromised.

14.2.4   Solvent Selection: Chemical Acceleration

The basic physical chemistry involved with selection of appropriate extracting solvents for a CES is conceptually very simple: "like dissolves like." While there are more sophisticated, quantitative, and potentially useful descriptions of this basic concept, such as Hildebrand and Hansen solubility parameters[8] and quantitative measures of hydrophobicity such as the octanol/water partition coefficient (log P),[9] the fundamental concept remains the same for both matrix (e.g., an elastomer) and analytes (e.g., extractable chemical entities). If a suitable solvent is employed that can interact with the matrix and thereby physically facilitate the migration of extractables from the interior toward the surface, then acceleration of extraction will occur. The most common example of this phenomenon is the swelling of elastomers in suitable organic solvents. This swelling has the effect of increasing the "pore size" between cross-linked polymer chains in the elastomer matrix, facilitating migration. Solvent molecules penetrating the elastomer matrix also provide a suitable medium for accelerated migration of analyte species by actually dissolving the analytes.

If the solvent is not able to dissolve a particular analyte or suite of analytes at analytically useful concentrations, then it might not produce a complete extractables profile. One example of such a "poor" solvent for polymer additives, monomers, and oligomers is water, the major component/diluent of many drug product formulations. Although it is important to use water, or other aqueous extracting media such as buffer solutions or mixed aqueous/organic solutions, in CESs involving aqueous drug products (e.g., inhalation solutions, nasal sprays, and parenterals), it may be desirable for a scientifically rigorous CES to use organic solvents with greater potential for accelerated extraction. Furthermore, since the solubility of known elastomer and plastic additives in any particular organic solvent can cover a wide range (e.g., log P ranges from 2 to 12), it is important, as suggested in FDA guidance documents for inhalation drug products,[10,11] to use solvents with a range of polarities. Solvent polarity directly affects the solubility of any particular chemical entity in that solvent, again "like dissolves like." This is graphically illustrated in Figures 14.4–14.6, which show gas chromatography (GC)/mass spectrometry (MS) extractables profiles from Soxhlet extraction of a peroxide-cured elastomer using methylene chloride, isopropanol, and hexane. The relatively dramatic

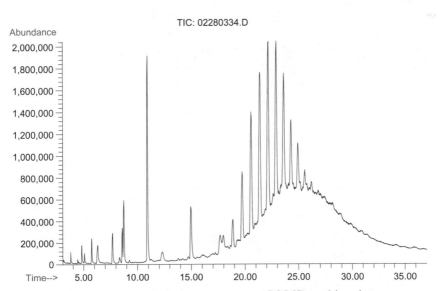

Figure 14.4   Gas chromatography/mass spectrometry (GC/MS) total ion chromatogram (TIC) extractables profile of a peroxide-cured elastomer (16 hours; methylene chloride Soxhlet extraction).

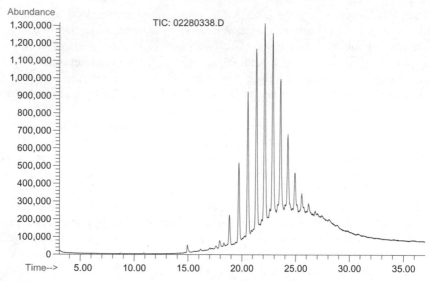

Figure 14.5   Gas chromatography/mass spectrometry (GC/MS) total ion chromatogram (TIC) extractables profile of a peroxide-cured elastomer (16 hours; isopropanol Soxhlet extraction).

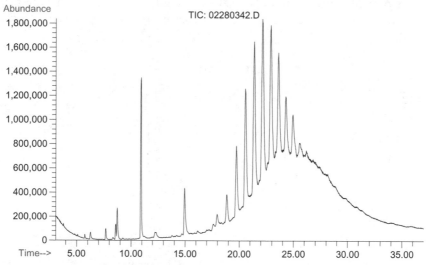

Figure 14.6   Gas chromatography/mass spectrometry (GC/MS) total ion chromatogram (TIC) extractables profile of a peroxide-cured elastomer (16 hours; hexane Soxhlet extraction).

differences in these profiles clearly support the use of varying polarity organic solvents in a comprehensive CES. As stated in the Product Quality Research Institute (PQRI) recommendations,[5] the function of the test article along with knowledge of its composition and drug product's formulation should be used to guide solvent selection. For example, methylene chloride is a good solvent to use for MDI valve components, since it is reasonable to assume that it will have extracting properties similar to those of typically used MDI propellants. It is reasonable (and essential) to use water in a CES of inhalation solution critical components where the drug product formulation is aqueous based. However, water might not be the only extracting solvent used for components from aqueous-based drug products, and would be a curious choice for a test article associated with an organic propellant-based MDI drug product. While knowledge of test article composition is a useful guide, one should never assume that such knowledge can be used to completely define an extractables profile. Solvents with a range of polarities should be selected to cover a wide range of potential extractables. It should be remembered that certain solvents are potentially reactive (e.g., methanol, ethanol) or contain potentially reactive contaminants (e.g., peroxides in ethyl ether and tetrahydrofuran) and their use in a CES should be justified.

14.2.5   Temperature Effects: The Extraction Technique

Simplistically, extraction can be viewed as the removal of analytes from a surrounding matrix by dissolution in a suitable solvent followed by migration out of the matrix to a point where they can be recovered for identification and quantitation. This process can be simplistically modeled as a diffusive "flux" (J) according to Fick's first law of diffusion:

$$J = -D \frac{\partial \phi}{\partial x},$$

where J measures the amount of substance that will flow through a small area during a small time interval, in units of amount/(area^2) time (e.g., mol/(m$^2 \cdot$s)).[12] D is the "diffusion coefficient" (area/time) and is proportional to the squared velocity of the diffusing particles (molecules of analyte), which depends on the temperature and viscosity of the matrix.[12] The term $\emptyset$ is the concentration of analyte in units of amount/volume (e.g., mol/m^3), and x is the position. This flux model states that the rate of material removal through point x is inversely proportional to the concentration gradient along dimension x, which means that chemical entities will move from regions of high concentration to regions of low concentration. Analyte flux also depends on D, which as stated above depends on the absolute temperature and viscosity of the matrix. The surrounding matrix can be either the polymer (very rigid) or the extraction solvent (presumably less rigid).

Increasing extraction temperature has the dual effect of increasing the kinetic energy of the analyte and decreasing the viscosity of the matrix (matrix viscosity represents an opposing force for diffusion). Higher temperatures have the effect of lessening the interactions (as measured by the viscosity) between matrix molecules.

Both of these effects increase analyte flux, and therefore enhance analyte extraction. An additional effect of temperature that is somewhat more difficult to model involves altering the nature of the matrix (e.g., swelling of an elastomer matrix). Consequently, the temperature of an extraction has one of the most significant effects of all extraction parameters on the rate of depletion of analytes from matrices. Solvent selection is also clearly an important experimental parameter, but for many extraction techniques (e.g., reflux, Soxhlet), the boiling point of the solvent is at least as significant as the solubility of analytes in the solvent. Therefore, in practice, it is desirable to use extracting solvents with a range of both polarity *and* boiling point.

There are a number of commonly used solvent extraction techniques that affect the temperature of the extraction. Some, such as the Soxhlet extraction, also favorably control the concentration gradient between the extraction solvent and the material/component sample on a continuous basis. Some extraction techniques are "instrument based" and therefore relatively easy to automate. The reader is cautioned, however, that glassware and other relatively simple laboratory apparatus are likely to be available for many decades and beyond, while instrument-based techniques are continuously upgraded and changed.

14.2.5.1 Maceration Maceration, or the controlled soaking in sealed vessels of test articles in suitable solvents, is the simplest and most ubiquitous example of a solvent extraction technique. It is as simple as making a cup of tea: the solvent (water) is at a defined initial temperature (boiling), the amount of tea is defined by either the tea bag or an external measure such as a teaspoon, the amount of extraction solvent is defined by the volume of the teacup, and the time is controlled by the user depending on the desired strength of the resulting beverage. One additional parameter available for manipulation is stirring during the extraction. Laboratory practice uses an apparatus somewhat more sophisticated than a teacup and spoon. For example, an elastomer may be cut into pieces and soaked in an organic solvent inside a glass bottle placed in a controlled temperature environment (Fig. 14.7). Note that the environment of a controlled temperature/humidity chamber makes it possible to include temperature acceleration in a maceration extraction.

14.2.5.2 Reflux Extraction The acceleration effects of elevated temperature on a controlled extraction suggest the use of the classical laboratory technique of reflux (or hot) extraction in a CES. The boiling point of a solvent is the highest temperature that solvent can attain and still remain a liquid under normal atmospheric conditions. Reflux extraction is accomplished by placing the test article to be extracted into a round-bottom flask charged with solvent, attaching a reflux condenser to recycle the solvent vapor, and then heating the solvent to boiling for a defined period of time (see Fig. 14.8). Extracting solvent volumes for reflux are usually greater than 100 mL with test article amounts typically ranging from 2 to 5 g. Reflux extraction involves considerable temperature acceleration. It is important to point out that since each extracting solvent has a unique boiling point, reflux extractions with different solvents will obviously occur at different temperatures.

Figure 14.7   Maceration extraction of an elastomer. Note the test article elastomer contained in a maceration bottle with solvent (left), along with a number of such bottles contained within a controlled environment chamber (right).

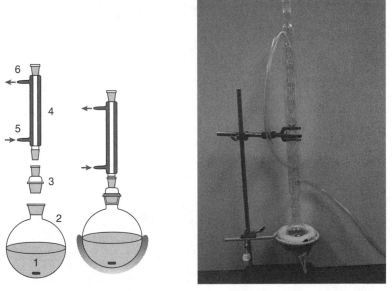

Figure 14.8   Reflux extraction diagram with components (1) stir bar, (2) round-bottom flask, (3) adapter, (4) Liebig condenser, (5) coolant in, and (6) coolant out. Note the photograph of a CES employing reflux extraction (right).

14.2.5.3 Soxhlet Extraction One issue with reflux extraction is that the analyte concentration gradient between the test article matrix and the extracting solvent becomes less favorable as the extraction proceeds. In other words, as more analytes are extracted from the test article matrix and their concentration in the extracting solvent increases, the concentration gradient becomes less favorable for further analyte extraction from the matrix. The Soxhlet extractor (see Fig. 14.9), introduced by Prof. Dr. Franz Von Soxhlet (1848–1926) around the year 1879 (need more be said regarding the longevity of classical laboratory apparatus?) captures distilled (and therefore pure) solvent with a reflux condenser and directs it into a sample holding vessel that holds typically 1–5 g of solid material in a "thimble" made of a nonreactive material such as glass wool or cellulose (note that Soxhlet extractors exist, which can accommodate kilograms of test article and liters of extracting solvent). Periodically, the sample holding area is filled and subsequently siphoned back down into the round-bottom flask (pot still) below. This periodic replenishing of the extracting solvent, which serves to maintain a favorable analyte concentration gradient between matrix and solvent, is called "turnover" and is one of the important laboratory variables to be controlled or reported. Higher energy input to the pot will still increase the rate of boiling and therefore the number of turnovers. From the sample perspective, Soxhlet is akin to repeated maceration at an intermediate

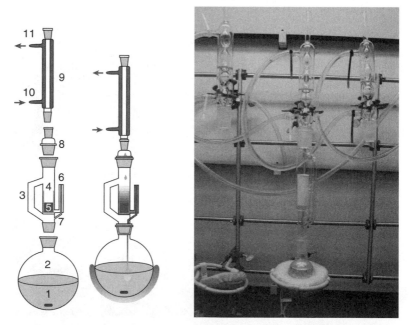

Figure 14.9 Soxhlet extraction diagram with components (1) stir bar, (2) round-bottom flask, (3) vapor path, (4) sample holder/thimble, (5) sample, (6) siphon, (7) solvent return, (8) adapter, (9) Liebig condenser, (10) coolant in, and (11) coolant out. Note the photograph of a CES employing Soxhlet extraction (right).

temperature between that of the condenser and that of the boiling solvent. The favorable analyte concentration gradient achieved with Soxhlet extraction can compensate for the reduced temperature acceleration compared with reflux.

14.2.5.4 Instrument-Based Solvent Extraction

Maceration, reflux, and Soxhlet extraction are all limited to temperatures either at or below the boiling point of a particular extracting solvent. Consideration of the phase diagram for a particular extracting solvent will show combinations of temperature above the boiling point and pressure above ambient at which the solvent remains a liquid. An obvious potential enhancement would be to use extracting systems that are capable of maintaining the solvent at such elevated temperatures and pressures, and therefore achieve increased temperature acceleration. The general term for these techniques is "pressurized fluid extraction," for the obvious reason that the test article and extracting solvent must be placed in pressure-rated vessels to allow for higher extraction temperatures. Differences in heating method and liquid handling have resulted in two common variants. The first of these systems is termed accelerated solvent extraction and is known by the abbreviation ASE®, which is a registered trademark of Dionex Corporation (Sunnyvale, CA). Samples are placed in stainless steel containers (akin to calorimetric "bombs") with pressure-rated seals that allow automated introduction and removal of solvents. The entire instrument is automated, allowing unattended operation after the sample containers have been charged with the test articles to be extracted. After programming, metered volumes of solvent at ambient temperature are automatically transferred into each container, and the containers are then placed into a resistively heated oven. After a preset period of time at a predetermined set temperature, the container is cooled to below the boiling point of the solvent and the extract is allowed to drain into a receiving vessel. The cycle can be repeated for an individual test article sample with different solvents, temperatures, and holding times. The other widely used instrument-based extraction technique utilizes microwave heating in addition to pressurized vessels. To accommodate this more rapid heating mechanism, a major modification to the vessel design was required in order to eliminate any metallic components. Most vessels utilize glass, quartz (for high temperature inorganic digestion) and fluorocarbon polymer wettable surfaces. The other modification for microwave extraction methods is that either the extracting solvent or the test article must be polarizable in order to absorb the microwave energy. The reader is advised that microwave systems (i.e., ovens) suitable for such uses are not to be found in local kitchen appliance stores.

The advantages of instrument-based extraction techniques over classical techniques include significantly lower required volumes of extracting solvents, faster extraction times, and increased sample throughput. These advantages are balanced by the increased cost of the instrumentation, the potential for downtime during repair/maintenance cycles, and the potential for "obsolescence" of the instrumentation. This obsolescence issue is of particular concern when instrument-based extraction techniques are used for routine extractables testing and QC release of container closure system components, as costly and time-consuming change-control processes are usually the result.

14.2.5.5 Direct Thermal Extraction and "Volatiles"   A group of instrument-based extraction techniques is available, which employ purely thermal processes for extraction (i.e., with no extracting solvent), with the thermal process usually coupled directly to an analytical instrument such as a mass spectrometer. Three of the most commonly used techniques are thermal desorption (TD)/MS (or TD/GC/MS), thermogravimetric analysis (TGA)/MS (or TGA/GC/MS), and headspace analysis (HS)/MS (or HS/GC/MS). The TD technique is often utilized for reverse-engineering and QC to determine whether or not certain additives are present in polymers. During the process, test articles are rapidly heated in an inert atmosphere above their breakdown temperatures, and release volatile and semivolatile additives and degradation products. These released substances can be trapped for subsequent introduction into a GC/MS system, or directly into an MS. TGA/MS (GC/MS) combines the capability of the TGA for detecting thermal events (e.g., melting) via measuring test article weight loss with the analytical capabilities of the MS or GC/MS. In this technique, the evolved gas from the TGA oven is sampled either directly into a mass spectrometer, or a GC/MS for separation prior to MS detection (see Fig. 14.10). The extraction/analysis processes for both TD and TGA are automated. As these processes usually do not leave the test article sample structurally unchanged, nor do they remotely resemble how most drug products will interact with their packaging materials, neither of the primary directives (Table 14.1) are satisfied. However, these techniques are effective for rapidly determining primary additives for subsequent experimental confirmation by an appropriate CES.

For situations where the most likely interaction between the drug product and the packaging materials/components will be via a volatilization–deposition mechanism (e.g., blister pack powders for inhalation, or sterile lyophilized powders for injection), the analysis of volatile organic compounds emitted from packaging materials is important. HS/MS and particularly HS/GC/MS are highly useful techniques for this application. The headspace is defined as the gas space in a container (e.g., a chromatography vial) above the test article, and HS is therefore the identification and quantitation of chemical entities present in that gas.[13] In HS, volatile and semivolatile organic compounds in solid and liquid samples are partitioned into the gaseous headspace of the closed sample container, which is then sampled into, usually, a GC/MS system. The partitioning equilibrium can be affected by thermal acceleration at relatively low temperatures (e.g., 50°C) compared with TD and further enhanced by matrix manipulation. HS/GC/MS is effective for low-molecular-weight volatiles in samples that can be efficiently partitioned into the headspace gas volume from the liquid or solid matrix sample, with higher boiling compounds not detectable due to their minimal partitioning into the headspace volume. Headspace analytical techniques also lend themselves to automation and are known to be highly reproducible.

14.2.6 Physical Effects: The Extraction Technique

14.2.6.1 Sonication   Extraction can be accelerated by the application of an external physical force. Arguably, microwave-assisted extraction falls into this category; however, the principal reason for acceleration with this technique would

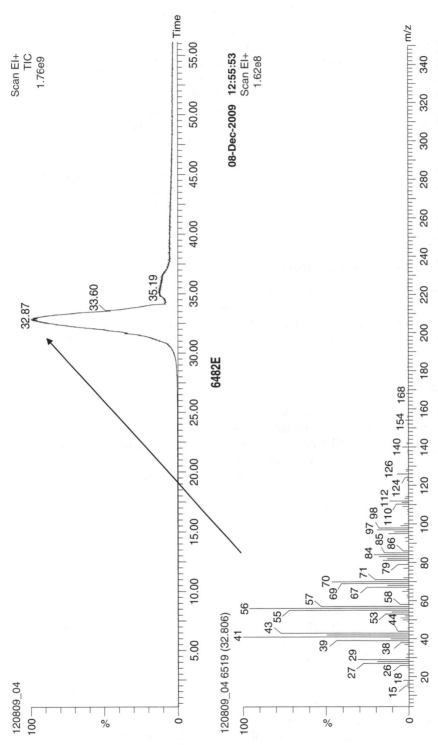

Figure 14.10 A total ion chromatogram (TIC; top) and electron ionization (EI) mass spectrum (bottom) from the themogravimetric analysis/mass spectrometry (TGA/MS) analysis of an elastomer sample. Note that the EI mass spectrum is taken from the apex of the evolved gas from the elastomer sample.

appear to be heating the test article and extracting solvent (i.e., thermal acceleration). The acceleration process facilitated by ultrasonic radiation, however, appears not to be primarily thermal. Ultrasonic extraction is widely applied to the extraction of organic compounds from various matrices[14] and is also applicable to CESs on pharmaceutical container closure system materials and components.[5] In its simplest form, a test article is placed into a glass container, which can be either open or closed, with a quantity of extracting solvent. This container is then placed in a laboratory sonicator with an appropriate level of cooling water, and the sample subjected to sonication for a period of time. Note that a potential issue with this procedure is that most laboratory sonicators have no provision for measuring or controlling the level/intensity of the ultrasonic energy, which can lead to issues related to repeatability/reproducibility and method transfer. The obvious advantages are in speed and simplicity; therefore, sonication like TD is likely most useful for rapidly obtaining a preliminary extractables profile.

14.2.7   Water as an Extracting Solvent

The use of water as a solvent includes a special set of challenges, especially related to the preparation of water extracts for analysis (as discussed in the following section of this chapter). Water is an appropriate extraction solvent when studying aqueous-based drug product formulations (e.g., inhalation solutions and sprays, injectables, opthalmics) or when water is an integral ingredient in the formulation. When the use of water as an extracting solvent is called for in a CES, it is important to consider two points:

1. *Water has a high boiling point (100°C) relative to most other solvents typically used in CESs.* Soxhlet extraction systems, for example, are not designed with water in mind and their use with water would be very difficult. Maceration, sonication, and certain instrument-based extraction techniques (ASE) can be used with water. Reflux with water is also possible; however, the high boiling point requires special attention in order to ensure that solvent loss through the condenser (usually water cooled) is not excessive. Maceration with water can be facilitated by accomplishing extractions in sealed vessels (i.e., leak proof under high internal pressures) within an autoclave or other suitable system. The autoclave uses high pressure steam for heating, and the sealed extraction vessel allows the internal solvent temperature to increase significantly without solvent loss.

2. *Water as an extracting solvent allows for control of pH.* With appropriate buffer systems, the pH (acidity/basicity) of an extraction can be controlled. This is significant for aqueous-based drug products, since it is important to understand the influence of pH on extractables profiles for these container closure systems (corresponding leachables profiles on accelerated drug product stability could also be influenced by pH). In fact, "water" of differing acid/base concentrations (e.g., at pH extremes and containing an organic modifier) can be viewed as different solvents with different "solvating power" in the same way as different organic solvents have different polarities.

14.2.8   Additional Points Regarding Extraction Techniques and Acceleration

1. *The basic principles of trace organic analysis laboratory operations should be considered.* These include the use of only *high-quality solvents*, either purchased as high quality or redistilled and purified in the laboratory. Poor quality solvents will likely include impurities that will be magnified in any extract concentration procedures and can interfere with the analysis of extracts. Likewise, only *thoroughly cleaned glassware and laboratory apparatus* should be used.

2. *Appropriate blanks and controls should be included in any CES study protocol.* The term "blank" in this context refers to analysis of an unused portion of the extracting solvent. The term "control" refers to a portion of the extracting solvent that has undergone the complete extraction process, extract preparation, and analysis process, which includes everything except the actual material/component test article. In Figure 14.11, a control (top) is displayed along with an actual profile from a water extraction (bottom). For both cases, the extracts were exchanged into methylene chloride and concentrated by a factor of 50. Note the relatively large peak in the extraction control around 8 minutes (a direct comparison on the same axis is shown in the bottom profile). Depending on the required analytical threshold, if the control were not available, this constituent would be incorrectly designated as an extractable.

3. *Extracting solvent to test article sample ratios should be relatively high (e.g., 20:1 weight:weight) in order to maintain favorable analyte concentration gradients.* Sample extracts can be concentrated prior to analysis (see discussion below).

4. *CESs should "always" include multiple extracting solvents of different solvating power.* This is to ensure that the widest practical net is cast for potential drug product leachables (note that it is not and can never be the widest possible net).

5. *CESs might incorporate multiple extraction techniques as these can be complementary (see Fig. 14.12).* For example, methylene chloride sonication and methylene chloride reflux are performed at different temperatures, and extraction kinetics are temperature dependent. The use of multiple extraction techniques along with multiple solvents also allows for a more informed decision when choosing an extraction process to optimize for extractables/leachables correlation, development/validation of routine extractables control methods, and so on. (see the later discussion on extraction optimization). Extraction temperature can be a factor affecting both extraction efficiency and the formation of extraction artifacts through thermolysis. Low-temperature extraction techniques such as sonication should be justified regarding their extraction efficiency, while extractables profiles from higher-temperature extraction techniques should be carefully examined for artifacts.

6. *This discussion is not intended to recommend, endorse, or preclude any particular extraction technique or process.* Extraction techniques and processes

(a)

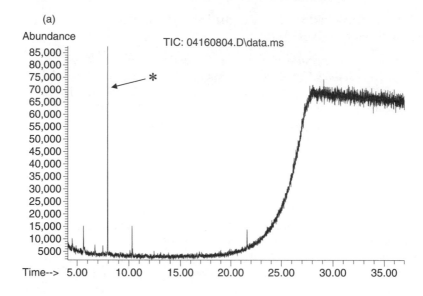

(b)

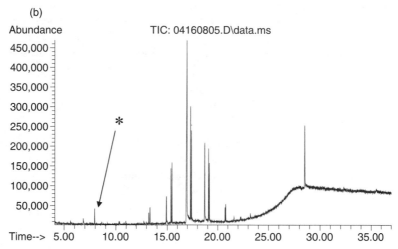

Figure 14.11   Gas chromatography/mass spectrometry (GC/MS) total ion chromatograms (TICs) from (a) a "control" sample and (b) a water extract sample from a container closure system test article.

not discussed in this chapter (e.g., supercritical fluid extraction, high-pressure liquid carbon dioxide extraction) can be used as appropriate in a CES.

14.3   ANALYSIS OF EXTRACT

14.3.1   Overview

Clearly, the CES is not completed by the generation of an extract as accomplishing this objective does not produce the information that defines an extractables profile. Rather, generation of a viable and appropriate extract sets the stage for the second

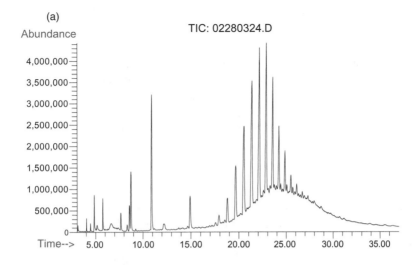

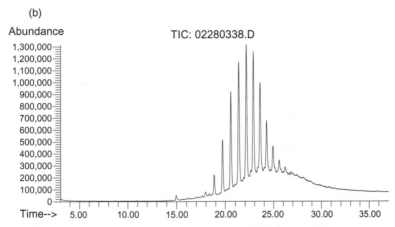

Figure 14.12   Gas chromatography/mass spectrometry (GC/MS) total ion chromatograms (TICs) from (a) a 4-hour isopropanol reflux extract and (b) a 16-hour isopropanol Soxhlet extract of a peroxide-cured elastomer test article.

step of the CES process, the analysis of the extract. It is by analysis of the extract(s) that the extractables profile is revealed and established (the reader is referred back to the discussion in Chapter 13). The goals of the extract analysis process are to generate extractables profiles that

1. are generally consistent with the known ingredients in the material or component test article;

2. can be correlated both qualitatively and quantitatively with end of shelf-life drug product leachables profiles;

3. can be optimized for the potential development and validation of routine extractables QC analytical methods, and the development of appropriate extractables specifications and acceptance criteria;

4. are complete; and

5. are reproducible.

Accomplishing these goals requires extract preparation and analysis procedures that balance considerations of known test article information, analytical technique sensitivity, and the appropriate AET.

As is the case for extract generation (i.e., that there is no universally applicable extraction process), it is the practical reality that no one analytical methodology can universally be applied in all steps of the extractables profiling process. It stands to reason that since the objective of each step is dramatically different, the most effective methods to accomplish those steps are different. That is to say, a method that might be effective in compound discovery might not be nearly as effective in compound identification. Thus, an effective and efficient process for generating an extractables profile would include numerous analytical methods (and modifications thereof), applied at different times in the analytical process.

Extract analysis processes can be classified as follows:

1. *Scouting.* Obtain general chemical information that provides insight into the nature or magnitude of extractables. These include "bulk property" techniques/methods, such as gravimetric (nonvolatile residue [NVR]), total organic carbon (TOC), pH, conductivity, absorbance (ultraviolet [UV], infrared [IR], etc.).

2. *Discovery/detection.* Search for extracted substances (more specifically, the search for instrumental responses that are indicative of these extracted substances). These include techniques/methods with broad scope and range as well as relatively high sensitivity and specificity, such as GC/flame ionization detection (FID), high-performance liquid chromatography (HPLC)/UV detection, and ion chromatography. Note that both GC and HPLC can include other types of detectors, including nitrogen specific, electron capture, evaporative light scattering, and corona assisted discharge. All of these techniques/methods are considered to be noncompound specific (see Chapter 13).

3. *Qualitative/structural.* Provide identities and molecular structures for the chemical entities responsible for the detected responses. These analytical techniques/methods must be compound specific and include GC/MS and LC/MS (see the detailed discussion in Chapter 13). Note that these techniques/methods can also be used for discovery/detection.

4. *Quantitative.* Provide concentration measurements for the identified compounds and for those responses for which identifications could not be achieved. These techniques/methods must be accurate, precise, and sensitive. Note that all of the analytical techniques/methods mentioned above (as well as many others) can be quantitative.

No single analytical technique or method will be sufficient to detect, identify, and quantitate all possible extractables from any particular test article and thus achieve all the goals of the extract analysis process. Therefore, it is recommended that multiple broad-spectrum techniques such as GC/MS and LC/MS be used to ensure complete evaluation of an extractables profile. As noted above, GC/MS and

LC/MS (with the LC/MS system also incorporating diode array detection) are capable of extractables discovery/detection, qualitative/structural analysis, and quantitative analysis. The remaining discussion in this chapter will focus on these two preeminent analytical techniques.

14.3.2 Analytical Technique Sensitivity

A general rule of thumb for instrumental analysis is that between 1 ng and 1 μg of a compound must be introduced into an analytical system (e.g., GC/MS and LC/MS) for reasonable detection and identification. Common introduction volumes for such analyses range from 1 μL (e.g., GC/MS) to 100 μL (e.g., LC/MS), depending on whether a gas chromatograph or liquid chromatograph is being used as an inlet (see Chapter 13). Therefore, the pertinent concentration range for extractables is 1–10 ng/μL. As routine laboratory concentration steps can easily attain a 100-fold concentration increase, the original concentrations in extracts can be as low as 10–100 ng/mL.

14.3.3 Preparation of Extracts for Analysis

In many cases, extracts can be directly analyzed by either GC or HPLC methods. For example, methylene chloride extracts can be injected directly onto GC and GC/MS systems if the concentration of the extract is consistent with the desired AET. However, it is not always appropriate to inject high-boiling or reactive solvents; therefore, it might be necessary to switch solvents prior to extractables profile analysis (see Fig. 14.13). It is usually inappropriate to inject water extracts directly into a GC as water is a very high boiling solvent and is also highly reactive chemically. This chemical reactivity can affect analytes (i.e., extractables) dissolved in the aqueous extract, especially in the high-temperature conditions of the GC injection port, and can also adversely affect the stationary phase of the GC column. Therefore, it is often necessary to remove the organic compounds from the water sample with a more nonpolar (also lower boiling and less chemically reactive) solvent prior to GC analysis (see more detailed discussion below). Other organic solvents such as methanol and ethanol are chemically reactive and can form esters with extracted fatty acids. For example, ethanol can react in a heated GC injection port with stearic acid, a very common additive to rubber and plastic, to form ethyl stearate. Extracted fatty acid esters can also be transesterified; for example, ethanol can react in a heated GC injection port with methyl stearate to form ethyl stearate.

Various laboratory techniques are available and widely used for the concentration of organic solvent extracts prior to GC analysis. The simplest of these techniques is to evaporate the organic solvent, assisted by blowing a gas stream across the surface of the solvent (usually a stream of highly pure and dry nitrogen). Somewhat more sophisticated techniques, such as the Kuderna–Danish (KD, Fig. 14.14) concentrator, are also available. The KD is most applicable to relatively low-boiling solvents, such as diethyl ether and methylene chloride, and works by refluxing the solvent through a three-ball condenser (known as the Snyder column) at ambient temperature. Solvent vapor pressure causes the balls to "bounce" in unison, allowing

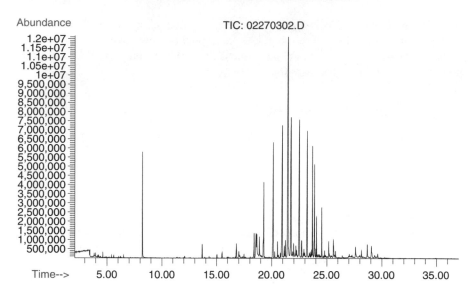

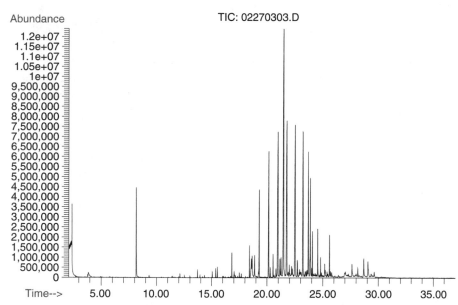

Figure 14.13 Gas chromatography/mass spectrometry (GC/MS) total ion chromatogram (TIC) extractables profile of an elastomer 2-propanol reflux extract. Sample injected "neat" (top) and reconstituted in methylene chloride prior to injection (bottom).

solvent vapor to slowly vent to the atmosphere. Liquid seals forming around each ball help prevent analyte molecules, which vaporize in the KD, from escaping to the atmosphere by redissolving them and returning them to the concentrating liquid extract. Other laboratory concentration techniques, such as rotary evaporation, are also commonly available.

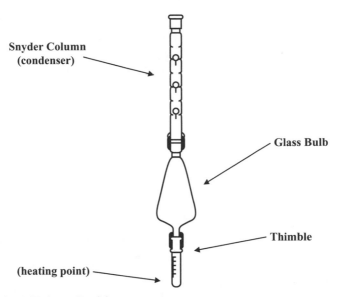

Figure 14.14   A Kuderna–Danish concentrator.

Likewise, when using HPLC-based analytical techniques such as LC/MS, it is usually inappropriate to inject samples in solvents that are not miscible in the mobile phase. For example, methylene chloride extracts should be dried or significantly concentrated and the extractables redissolved in mobile phase prior to analysis. In such cases, however, concentrating the extract to dryness may produce significant analytical issues. Note that for any sample preparation strategy with either GC- or HPLC-based analytical techniques in mind, the implications for recovery and/or loss of extractables should be considered.

14.3.4   Water as an Extracting Solvent: Analytical Considerations

As mentioned previously, aqueous samples present unique challenges for trace organic analysis in general, and for CESs in particular. Water is a problem for GC-based analytical techniques, which usually means that chemical entities extracted into water must be solvent exchanged (e.g., via liquid/liquid extraction) into a solvent more appropriate for GC analysis. Consider the following example.

Problem

During analysis of stability samples for a drug product with a 5-mL aqueous-based formulation, a new leachable was detected at the 3-month time interval using a GC/FID analytical method (see Fig. 14.15 for an example blank profile and Fig. 14.16 for accelerated storage, respectively).

The initial task was to elucidate the molecular structure of this unknown leachable.

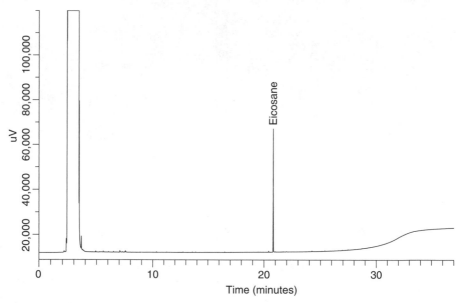

Figure 14.15 Gas chromatography/flame ionization detection (GC/FID) "blank" leachables profile showing internal standard (eicosane).

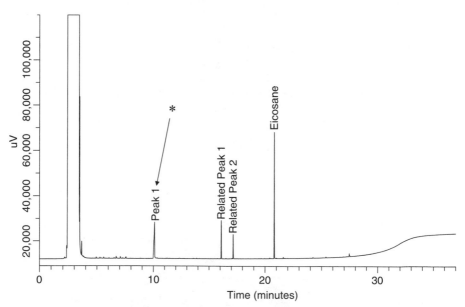

Figure 14.16 Gas chromatography/flame ionization detection (GC/FID) leachables profile of a 3-month, 40°C/75% RH (relative humidity) drug product showing unidentified, new peak.

Solution

1. *Assess the suitability of the existing analytical method for GC/MS.* Figure
    14.17 shows the resulting GC/MS leachables profile for a normal storage
    sample. Note that all leachables peaks observed in the GC/FID profile (Fig.
    14.16) are present in the GC/MS profile.

2. *Utilize computerized library reference mass spectra in an attempt to identify
    the leachable (see Chapter 13).* The unknown electron ionization (EI) mass
    spectrum, along with the best fit reference spectrum (2-ethylhexanoic acid), is
    shown in Figure 14.18. The leachable was unambiguously identified as
    2-ethylhexanoic acid.

3. *Determine the source of the leachable and produce a qualitative leachables/
    extractables correlation.*

The original CES for the drug product's low-density polyethylene (LDPE)
container using both methylene chloride and water as extracting solvents had failed
to identify 2-ethylhexanoic acid as an extractable. Consequently, a modification of
the extracting solvent to facilitate extraction of this organic acid was developed and
applied. The modified extraction procedure used maceration of emptied/rinsed containers, with 0.1 N sodium hydroxide as the extracting solvent, for 24 hours at 70°C.
After extraction, the resulting solution was acidified with 1 N hydrochloric acid to
pH 2 and extracted with an equal volume of methylene chloride. The resulting GC/
FID extractables profile is shown in Figure 14.19 and clearly reveals 2-ethylhexanoic
acid as an extractable, thus allowing a qualitative correlation. Note that the resulting

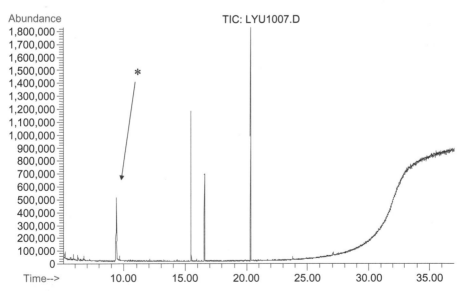

Figure 14.17   Gas chromatography/mass spectrometry (GC/MS) total ion chromatogram
(TIC) leachables profile of a 3-month, 25°C/60% RH drug product showing the
unidentified, new peak.

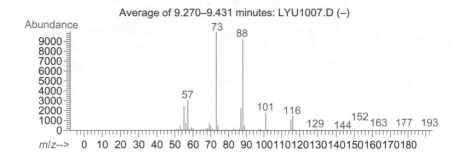

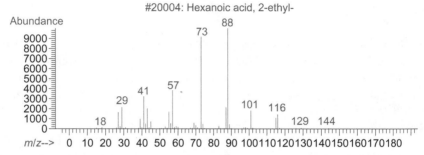

Figure 14.18   Comparison of electron ionization mass spectrum of unknown peak to National Institute of Standards and Testing (NIST) reference library mass spectrum of 2-ethylhexanoic acid.

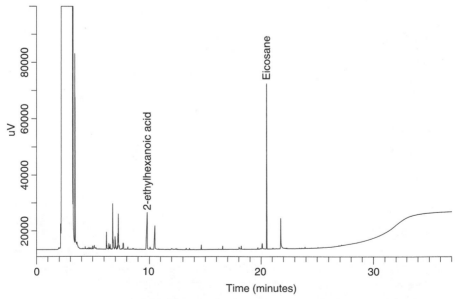

Figure 14.19   Gas chromatography/flame ionization detection (GC/FID) extractables profile of resin sample showing 2-ethylhexanoic acid.

methylene chloride extract could have been easily concentrated for analysis as required to detect the extractable in question.

This "case study" demonstrates the importance of considering pH when accomplishing a CES with aqueous extracting media, as basic pH facilitated the extraction of the aliphatic acid. Using only log P (approximately 2) as a guide to the partitioning of the acid from the plastic into the aqueous product suggested that the polymer *in situ* level was greater than 100 ppm. This particular batch of LDPE had passed the QC specification of not more than 100 ppm of semivolatile organic extractables. What was not considered until later was the role of the pH of the drug product solution. A small (0.1 N) amount of sodium hydroxide added to the extraction solvent (water) was enough to ionize the acid extractable and rapidly solubilize it into the aqueous solvent. The amount found in the resin was never more than 100 ppm, which was consistent with the previous analysis. The issue with the drug product in question was one of uncontrolled pH during production, allowing the extraction of the acid into the drug product at an undesirable level. While the log P was appreciably positive, the log D (at pH levels greater than that of the pKa) was appreciably negative.

While injecting aqueous samples (i.e., extracts) is acceptable with HPLC analytical techniques, the vast majority of which use aqueous mobile phases or mobile phases that are miscible with water, dilute water extracts may require concentration prior to analysis. As noted previously, liquid/liquid extraction with organic solvents can produce extracts that can be relatively easily concentrated. Other concentrating techniques for aqueous samples are available and routinely applied, the most popular of which include solid-phase extraction, vacuum/heat-assisted evaporation, and lyophilization (freeze-drying). The reader interested in additional information on sample preparation for trace organic analysis is referred to the scientific literature, which includes numerous books and manuscripts on this subject.

14.3.5 Considering the AET

Consider a CES of a critical container closure system component where an AET of 20 μg/g (ppm) for any individual organic extractable is determined. Using a sample loading of 2 g into a 50-mL extract volume will result in, at most, 40 μg of a threshold compound in that entire 50 mL extract, which is a final extract concentration of 0.8 μg/mL (0.8 ng/μL). This is at the low end of the range for detection/identification/structural analysis by GC- and HPLC-based analytical techniques such as GC/MS and LC/MS. Since the AET for extractables in a CES is based on the AET for leachables in the associated drug product, which is in turn derived from the SCT (an exposure-based threshold of 0.15 μg/day total daily intake for an individual organic leachable), it is obvious that defining the drug product daily "dose" is the critical element in defining the various AETs. Since drug products and their respective dose definitions differ widely (e.g., consider an MDI with a 200 dose label claim versus a single dose 2-L large volume parenteral), it is clear that extractables AETs will vary widely. Considering both the AETs and the sensitivities of analytical techniques, the scientist designing a CES can control

1. the weight of material/component test article to be extracted;
2. the volume of extracting solvent, keeping in mind that a 20:1 weight ratio of solvent to sample is desirable;
3. the extraction conditions; and
4. the concentration of the final extract.

One must also employ common sense in designing a CES and remember, as in the "case study" described previously, that additional extractions and analyses can always be accomplished to allow for appropriate qualitative and quantitative extractables/leachables correlations.

14.3.6 Optimization of "Definitive" Extraction Processes

After evaluating extractables profiles from various extraction techniques/methods and solvents, it is recommended that one choose a "definitive" extraction technique(s)/method(s) for optimization. An optimized extraction process is defined as one that yields an appropriate number and concentration of extractables, and achieves steady-state levels (i.e., "asymptotic levels") without violating any CES directives discussed previously (see Fig. 14.20 as well as the Case Study sections in this book). Optimization of the extraction technique(s)/method(s) prior to conducting a quantitative CES ensures that the extractables profile represents at least a "worst-case" scenario of potential leachables and their levels. Extractables profiles

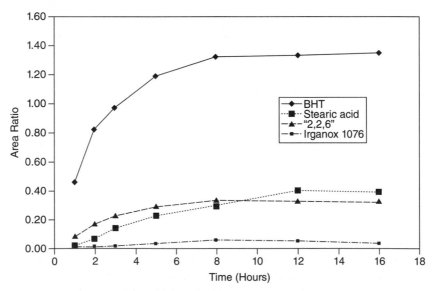

Figure 14.20   An example optimization for a "definitive" extraction process. Note the "asymptotic levels" and compare with Figure 14.2. Note that this optimization is based on GC/MS peak area ratios (analyte/internal standard) for four representative target extractables. BHT, butylated hydroxytoluene.

produced from such optimized technique(s)/method(s) should be thoroughly evaluated both qualitatively and quantitatively. While complete validation is not recommended or necessarily expected for CES extraction processes, it is recommended that appropriate experiments be accomplished to verify that quantitative results are accurate and precise for the definitive extraction process. This is especially true if the quantitative CES results are an integral part of a quantitative extractables/ leachables correlation, and/or the definitive extraction process is intended for use as in routine critical component extractables testing. Appropriate method verification experiments could include evaluations of precision, accuracy, linearity, and selectivity.

14.4   INTERPRETATION OF THE CES

While the outcome of the CES, the extractables profile(s), is actionable information, it is not an action in and of itself. Rather, it is the *interpretation* of the extractables profile that utlimately leads to the decision to either accept or reject materials, components, or systems from the perspective of their suitability for use. The proper interpretation of the extractables profile depends on the overall purpose for having performed the CES. If the purpose of the CES was impact assessment, then the proper interpretation of the extractables profile rests in establishing the impact of the extractables, in whole or in part, on some quality attribute of the product or, less frequently, the tested material. For example, an important, but by no means sole, purpose of a CES is to estimate the impact that contact with a material has on product safety. Thus, a safety assessment would involve establishing the safety impact of the extractables, which is dictated by their identity (which establishes toxicity potential) and their concentration (which establishes the potential patient exposure as leachables). Therefore, it is important that qualitative and quantitative extractables profiles be discussed with and reviewed by toxicologists, so that any potential safety concerns regarding individual extractables (i.e., potential leachables) are identified early in the drug product development process. Early safety review of extractables profiles obtained during CES has significant potential benefit to any pharmaceutical development process. Potential leachables that represent possible safety concerns can be identified and evaluated at a point in the process where corrective changes to the container/closure system would have less effect on the timelines and cost of the development program. Therefore, the results of CES should also be used as a component and material selection tool. In addition to safety, other quality or suitability for use attributes that might be impacted by extractables include compatibility (e.g., stability, formation of particulate matter), container closure system performance, and protection. The comprehensive list of extractables (identified by defined and systematic processes) that could be potential leachables should be compared with available supplier information for the particular material/critical component under consideration. This will help determine if the extraction and analysis methods used are appropriate, and determine the presence of other chemical entities not included in the supplier information (see Fig. 14.21), which is an all too common occurrence.

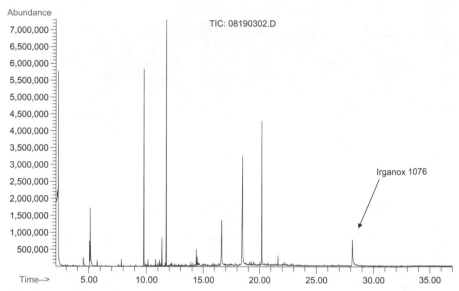

Figure 14.21   Gas chromatography/mass spectrometry (GC/MS) total ion chromatogram (TIC) extractables profile of an elastomer showing the phenolic antioxidant Irganox 1076, which was not reported by the elastomer manufacturer as an elastomer additive. Irganox 1076 was later determined to have been added by the base polymer supplier.

14.5   "SPECIAL CASE" COMPOUNDS

Certain compounds and compound classes of extractables and leachables have been deemed of special safety concern.[5,15] For elastomeric materials, these include N-nitrosamines, polynuclear aromatic hydrocarbons (PAHs or polynuclear aromatics [PNAs]), and 2-mercaptobenzothiazole (2-MBT). Safety thresholds such as the qualification threshold (QT) and SCT do not apply to special case compounds, and these must be evaluated at lower thresholds (i.e., the limits of available analytical technology), along with dedicated methods, appropriate specifications, appropriate qualifications, and risk assessments. N-nitrosamines are associated with sulfur-cured rubber, and CESs are usually accomplished with standard methods, which also include specific analytical techniques and methods.[15,16] PAHs and 2-MBT are also associated with rubber (PAHs with carbon black rubber filler, and 2-MBT with certain sulfur-cured rubber), but no standard methods exist for accomplishing CESs for these compounds. Therefore, the best practice principles for accomplishing CESs described in this chapter apply. Extracts of PAHs are typically analyzed by GC/MS with selected ion monitoring (SIM; see Chapter 13).[5] 2-MBT can be analyzed in rubber extracts with technologies such as LC/MS/MS (see Chapter 13). For phthalate plasticized materials (e.g., polyvinyl chloride [PVC]), "special case" extractables might include "common" phthalates such as di(2-ethyl-hexyl)phthalate (DEHP) and mono(ethyl-hexyl)phthalate (MEHP). Given the recent interest in bisphenol A

(BPA), this compound might be an appropriate "special case" extractable for materials that use BPA as a monomer (e.g., polycarbonate).

14.6 SUMMARY OF RECOMMENDED BEST PRACTICES FOR CESS

Best practice recommendations discussed in this chapter for CESs (also contained in recommendation documents[5] and the scientific literature[17]) on materials and container closure system critical components are as follows:

- CESs should employ vigorous extraction with multiple solvents of varying solvating power.
- CESs may incorporate multiple extraction techniques.
- CESs should include careful sample preparation based on knowledge of analytical techniques to be used.
- CESs should employ multiple analytical techniques.
- CESs should include a defined and systematic process for identification of individual extractables.
- CES "definitive" extraction techniques/methods should be optimized.
- During the CES process, sponsors should revisit supplier information describing component formulation.
- CESs should be guided by an AET that is based on an accepted SCT.
- Special case compounds such as PAHs (or PNAs), N-nitrosamines, and 2-MBT require evaluation by specific analytical techniques and technology-defined thresholds.
- Qualitative and quantitative extractables profiles should be discussed with and reviewed by pharmaceutical development team toxicologists so that any potential safety concerns regarding individual extractables, that is, potential leachables, are identified early in the pharmaceutical development process.

14.7 CONCLUSIONS

The CES is perhaps the most important aspect of any pharmaceutical development program for drug product types at relatively high risk for interaction between the drug product formulation and the container closure system.[3] The CES provides information on the identities and levels of potential drug product leachables, allowing for appropriate assessments of container closure systems and associated materials of construction for safety, compatibility, and other suitability for use criteria.[18] Based on CES results, and the extraction processes and analytical techniques/methods used to produce these results, container closure component selection, drug product leachables studies, and routine extractables testing and release of critical container closure system components are facilitated. Pharmaceutical development scientists are encouraged to perform CESs with the highest regard for good and practical science, assisted by the best practice recommendations discussed in this chapter.

REFERENCES

1 Federal Food Drug & Cosmetic Act. Section 501, Title V. Available at: http://www.fda.gov/RegulatoryInformation/Legislation/FederalFoodDrugandCosmeticActFDCAct/FDCActChapterVDrugsandDevices/ucm108055.htm (accessed October 14, 2009).

2 European Medicines Agency. Guideline on plastic immediate packaging materials. CPMP/QWWP/4359/03, EMEA/CVMP/205/04, 2005.

3 Guidance for industry: Container closure systems for packaging human drugs and biologics. Department of Health and Human Services, Food and Drug Administration, Center for Drug Evaluation and Research (CEDER) and Center for Biologics Evaluation and Research (CBER), 1999.

4 Jenke, D.R. Nomenclature associated with the chemical characterization of, and compatibility evaluations for, medical product delivery systems. *PDA J Pharm Sci Technol* 2003, *57*, pp. 97–108.

5 Norwood, D.L. and Ball, D. Product Quality Research Institute: Safety thresholds and best practices for extractables and leachables in orally inhaled and nasal drug products. Submitted to the PQRI Drug Product Technical Committee, PQRI Steering Committee, and U.S. Food and Drug Administration by the PQRI Leachables and Extractables Working Group, 2006.

6 Operating manual—6770 Freezermill®. SPEX SamplePrep, LLC, Metuchen, NJ, Rev A 7.07, updated July 1, 2008.

7 Ball, D., Blanchard, J., Jacobson-Kram, D., McClellan, R., McGovern, T., Norwood, D.L., Vogel, M., Wolff, R., and Nagao, L. Development of safety qualification thresholds and their use in orally inhaled and nasal drug product evaluation. *Toxicol Sci* 2007, *97*(2), pp. 226–236.

8 Hansen, C.M. Hansen solubility parameters, 2010. Available at: http://www.hansen-solubility.com (accessed January 13, 2010).

9 Castner, J., Anderson, J., and Benites, P. Strategy for development and characterization of HPLC methods to investigate extractables and leachables. *Am Pharm Rev* 2007, *10*(3), pp. 10, 12, 14, 16–18.

10 Guidance for industry: Metered dose inhaler (MDI) and dry powder inhaler (DPI) drug products. Department of Health and Human Services, Food and Drug Administration, Center for Drug Evaluation and Research (CEDER), 1998.

11 Guidance for industry: Nasal spray and inhalation solution, suspension, and spray drug products—Chemistry, manufacturing, and controls documentation. Department of Health and Human Services, Food and Drug Administration, Center for Drug Evaluation and Research (CDER), 2002.

12 Time Domain CVD, Inc. Introduction to diffusion: Physics and math, 2010. Available at: http://www.timedomaincvd.com/CVD_Fundamentals/xprt/intro_diffusion.html (accessed January 13, 2010).

13 LabHat.com. Basic principles of headspace analysis, 2010. Available at: http://www.labhut.com/education/headspace/introduction01.php (accessed January 13, 2010).

14 Sanghi, R. and Kannamkumarath, S. Comparison of extraction methods by Soxhlet, sonicator, and microwave in the screening of pesticide residues from solid matrices. *J Anal Chem* 2004, *59*(11), pp. 1032–1036.

15 Norwood, D.L. and Mullis, J.O. "Special case" leachables: A brief review. *Am Pharm Rev* 2009, *12*(3), pp. 78–86.

16 Norwood, D.L., Mullis, J.O., Feinberg, T.N., and Davis, L.K. N-nitrosamines as "special case" leachables in a metered dose inhaler drug product. *PDA J Pharm Sci Technol* 2009, *63*(4), pp. 307–321.

17   Norwood, D.L., Paskiet, D., Ruberto, M., Feinberg, T., Schroeder, A., Poochikian, G., Wang, Q., Deng, T.J., DeGrazio, F., Munos, M.K., and Nagao, L.M. Best practices for extractables and leachables in orally inhaled and nasal drug products: An overview of the PQRI recommendations. *Pharm Res* 2008, *25*(4), pp. 727–739.

18   Jenke, D. *Compatibility of Pharmaceutical Products and Contact Materials—Safety Considerations Associated with Extractables and Leachables*. John Wiley & Sons, Hoboken, NJ, 2009.

CHAPTER 15

EXTRACTABLES: CASE STUDY OF A SULFUR-CURED ELASTOMER

Daniel L. Norwood, Fenghe Qiu, James R. Coleman,
James O. Mullis, Alice T. Granger, Keith McKellop,
Michelle Raikes, and John A. Robson

15.1 INTRODUCTION

15.1.1 Elastomer Material Overview

Chapter 14 of this book addressed controlled extraction studies (CESs) in detail, including the philosophy of designing and accomplishing CESs in the modern pharmaceutical development environment, and the best demonstrated laboratory practices for producing and analyzing extracts of container closure system critical components. The Product Quality Research Institute (PQRI) developed these best demonstrated laboratory practices by accomplishing CESs guided by defined experimental protocols on custom-made materials of construction, each of which was designed to simulate actual pharmaceutical container closure system components for orally inhaled and nasal drug products (OINDPs).[1,2] This chapter presents a summary of the CES results for one of these custom-made materials of construction, a sulfur-cured elastomer. This summary is divided into four phases, extractables profiling (qualitative CES), selected extraction method optimization, a "simulated" leachables study, and method verification for suitability in routine extractables quality control. The latter phase is appended to the chapter as a case study (Case Study 15.1). Also appended as a case study (Case Study 15.2) is an analytical investigation of the "special case" compound 2-mercaptobenzothiazole (2-MBT) in this elastomeric test article.

A clear advantage to using custom-made materials as test articles for the development of best demonstrated laboratory practices for CESs is that the chemical compositions and compounding processes for these materials are completely known

Leachables and Extractables Handbook: Safety Evaluation, Qualification, and Best Practices Applied to Inhalation Drug Products, First Edition. Edited by Douglas J. Ball, Daniel L. Norwood, Cheryl L.M. Stults, Lee M. Nagao.

TABLE 15.1. Ingredients in Sulfur-Cured Elastomer Test Article[1]

Ingredient	Registry #(s)	Percent (w/w)
Calcined clay	308063-94-7	8.96
Blanc fixe (barium sulfate)	7727-43-7	25.80
Crepe	9006-04-6	38.22
Brown sub MB (ingredients below)	Not available (NA)	16.84
Brown sub loose	NA	33.30
Crepe	9006-04-6	66.70
1722 MB (ingredients below)	NA	2.11
Standard Malaysian rubber (SMR)	NA	60.00
FEF Carbon black (low PNA)	1333-86-4	40.00
Zinc oxide	1314-13-2	4.04
2,2′-Methylene-bis(6-*tert*-butyl-4-ethylphenol)	88-24-4	0.56
Coumarone-indene resin	164325-24-0	1.12
	140413-58-7	
	140413-55-4	
	68956-53-6	
	68955-30-6	
Paraffin	8002-74-2	1.12
	308069-08-1	
Tetramethylthiuram monosulfide	97-74-5	0.11
Zinc 2-mercaptobenzothiazole	149-30-4	0.29
	155-04-4	
Sulfur	7704-34-9	0.84

and can, to a certain extent, be controlled. The chemical composition of the sulfur-cured elastomer[1] was presented in Chapter 13 (Table 13.1), which for the reader's convenience is reproduced here (as Table 15.1 of this chapter). The information contained within this chemical composition, which is useful to the design of a CES and in the subsequent analysis of resulting extracts, was also briefly discussed in Chapter 13. To review and expand:

- The presence of carbon black indicates the likely presence of polyaromatic hydrocarbons (PAHs) in the elastomer (also termed polynuclear aromatics or PNAs). Note that PAHs are the so-called special case compounds requiring specific consideration.[1,3]

- Sulfur-curing agents, such as tetramethylthiuram monosulfide (TMTMS) (see structure below), indicate the likely presence of *N*-nitrosamines, which is a second class of "special case" compound.[1,3]

- 2-MBT (see structure below) is itself a "special case" compound.[1,3]

- Paraffin (an alkane hydrocabon waxy solid mixture of empirical formula C_nH_{2n+2} with generally $20 \leq n \leq 40$) and coumarone-indene resin (a thermoplastic resin obtained by polymerization of indene and coumarone; see structures below) are complex materials, and therefore are likely to produce complex extractables profiles containing many structurally related chemical entities.

- Some of the individual named additives, such as 2,2'-methylene-bis(6-*tert*-butyl-4-ethylphenol) (see structure below), are known to be amenable to analysis by gas chromatography (GC) methods. Hence, GC and GC/mass spectrometry (MS) will be required for separation and analysis of extractables mixtures.

- Calcined clay, blanc fixe (barium sulfate), and sulfur are all inorganic in nature and should not contribute chemical entities to an extractables profile. It should be noted, however, that S_8 (termed "octahedral sulfur") is a stable chemical entity that is amenable to analysis by GC-based techniques and, if present, will likely appear as an extractable.

- "Crepe," "Brown sub MB," "Brown sub loose," "1722 MB," and "SMR" (standard Malaysian rubber) are all designations for various types of unvulcanized rubber.

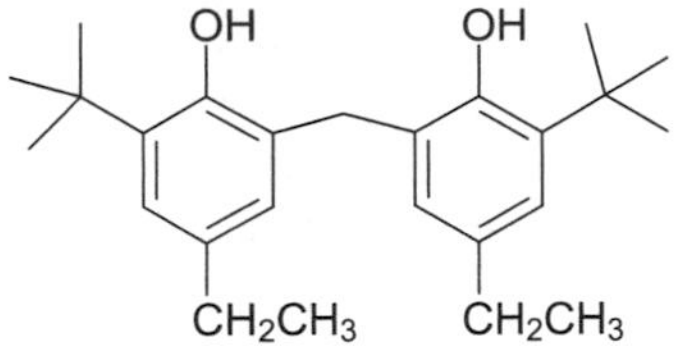

2,2'-Methylene-bis(6-*tert*-butyl-4-ethylphenol)

(Chemical Abstracts Service [CAS] #88-24-4; hindered phenolic antioxidant)

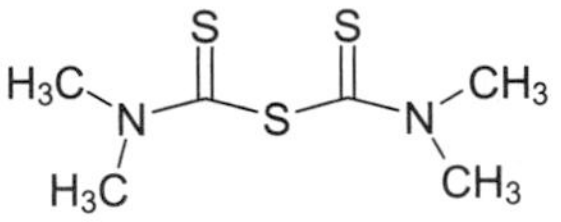

TMTMS

(CAS #97-74-5; sulfur-curing agent)

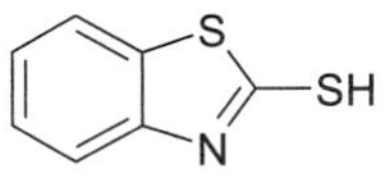

2-MBT

(added as the zinc salt CAS #149-30-4/155-04-4; vulcanization accelerator)

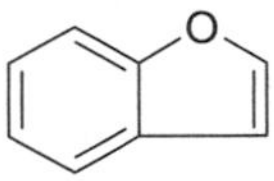

Coumarone

CAS #271-89-6

Indene

CAS #95-13-6

15.1.2   Extraction Considerations

It is known that all three types of "special case" compounds will require attention (i.e., highly specific extraction procedures and analytical methods for their qualitative and quantitative analysis), and that GC-based analytical techniques will be required for more general extractables profiling of this elastomeric test article. In addition, since this material is intended to simulate a metered dose inhaler (MDI) valve component material of construction, and given the fact that MDI drug products contain organic propellants and co-solvents, organic solvents with similar solvating properties should be used in the extraction studies of corresponding container closure system critical components. Elastomers of this type are known to swell in organic solvents, which will serve to release larger quantities of extractables for analysis. With these considerations in mind, methylene chloride (dichloromethane), 2-propanol (isopropanol), and hexane (*n*-hexane) were chosen. These extracting solvents serve to simulate MDI formulation conditions for leaching and represent a range of polarities, which casts a "wide net" for potential leachables (see Chapter 14). Furthermore, these solvents (1) represent a range of boiling points; (2) are relatively nonreactive chemically; (3) are easily and safely handled in a typical analytical laboratory setting; and (4) are readily available in high purity.

Again, since this CES was intended as an MDI critical component study, methylene chloride was chosen to mimic chlorofluorocarbon (CFC) and hydrofluoroalkane (HFA) propellants, and isopropanol was chosen to mimic ethanol (a common co-solvent for MDI drug product formulations). In the case of inhalation solutions and other aqueous-based drug products, water or another aqueous-based medium, for example, aqueous buffer solution, should be used as an extracting solvent. In certain cases, it may be possible to use the actual drug product vehicle as an extracting medium for a CES (note the "simulated" leachables study described later in this chapter).

Three extraction techniques (sonication, reflux, and Soxhlet) were selected for application in these studies. While perhaps not the most efficient techniques for sample throughput and solvent use when compared with automated techniques such as supercritical fluid extraction (SFE) and automated solvent extraction (ASE) (see Chapter 14), these techniques have the advantages of a long history of use in many laboratories (e.g., Soxhlet extraction dates to approximately 1879) and relatively simple requirements for apparatus and equipment. In addition, for routine extractables testing application, these techniques are likely to remain relatively unchanged and thus supportable for many years to come, avoiding the requirement for time-consuming and costly change control processes. The same statement cannot be made for any automated/high-throughput extraction technique that relies on a unique scientific instrument.

It must be stressed that compositional information while necessary is not sufficient to allow for the complete design and conduct of a CES (see Chapter 14), which was taken as a guiding principle by PQRI for the development of CES protocols. For example, even though the information in Table 15.1 would suggest that GC-based analytical techniques are sufficient for the analysis of extracts from this particular material, it was considered necessary to accomplish liquid chromatography (LC) and LC/MS analyses as well. This is because some the elastomer ingredients (e.g., paraffin and coumarone-indene resin) are complex materials and it cannot be assumed that they will only contain chemical entities amenable to GC analysis. Furthermore, the individual named additive 2,2′-methylene-bis(6-*tert*-butyl-4-ethylphenol) might contain higher-molecular-weight impurities and/or degradation products, which might not be amenable to GC analysis.

This chapter and its appended case studies summarize both qualitative and quantitative CESs on the sulfur-cured elastomer test article. The two case studies describe, respectively, (Case Study 15.1) the development of a GC/flame ionization detection (FID) method for the quantitative analysis of extracts, providing the basis for a "routine" extractables test and release method; and (Case Study 15.2) the development of a quantitative method for the analysis of 2-MBT (a "special case" compound) in the elastomer test article.

15.2   EXTRACTABLES PROFILING (QUALITATIVE CES)

As mentioned above, the PQRI Working Group accomplished CESs guided by carefully designed experimental test protocols. These are reproduced as an Appendix to this book, and the interested reader is referred to these study protocols for additional detail and to use as a model for designing other CESs (also see Reference 1). Experimental detail is to be found in these protocols. For the reader's convenience, certain specific experimental details are also summarized in this chapter.

15.2.1   Sample Preparation for Extractables Profiling by GC/MS

15.2.1.1 Sonication Extracts   Test articles of sulfur-cured elastomer were received as sheets approximately 10-cm square, covered with a layer of polymeric release liner. For each extraction, a piece of rubber with release liner attached was cut from the stock, the release liner removed, and the sample weighed on an analytical balance. A 1-g sample of elastomer test article was then cut with scissors into four pieces of approximately equal size and placed into a 40-mL borosilicate I-Chem vial having a Teflon lined cap. The appropriate volume of extraction solvent was then added to each sample vial and accompanying blank vial. The actual volume of solvent added was dependent on the sample weight and was adjusted so as to maintain a solvent-to-sample ratio of 20:1 (v/w) in each sample vial. Twenty milliliters of solvent was added to each blank vial. The vials were tightly capped and placed in a standard laboratory sonicator bath and sonicated for 1 hour. After sonication, the samples were allowed to return to room temperature, and then an aliquot was transferred to a 2-mL glass vial, capped, and analyzed by GC/MS. For isopropanol, the sample was injected directly (neat) onto the GC/MS system, as well as evaporated under stream of dry helium and reconstituted in methylene chloride for a second GC/MS analysis.

15.2.1.2 Reflux Extracts
Test articles of sulfur-cured elastomer were received as sheets approximately 10-cm square, covered with a layer of polymeric release liner. For each extraction, a piece of rubber with release liner attached was cut from the stock, the release liner removed, and the sample weighed on an analytical balance. A 5-g sample of elastomer test article was cut with scissors into 20 pieces of approximately equal size and placed into a 250-mL round-bottom flask containing Teflon boiling chips. A 100-mL volume of extracting solvent was added to each sample flask and accompanying blank flask. Each flask was connected to a water-cooled condenser (tap water) and placed on a heating mantle. The flasks were heated to boiling and refluxed for 4 hours. Fresh solvent was then added as appropriate to account for any solvent lost to evaporation during the extraction. An aliquot of cooled sample extract (or blank) was transferred to a glass vial, capped, and analyzed by GC/MS.

15.2.1.3 Soxhlet Extracts
Test articles of sulfur-cured elastomer were received as sheets approximately 10-cm square, covered with a layer of polymeric release liner. For each extraction, a piece of rubber with release liner attached was cut from the stock, the release liner removed, and the sample weighed on an analytical balance. A 5-g sample was cut with scissors into 20 pieces of approximately equal size and placed into a preconditioned 33×80 mm cotton cellulose thimble. An empty thimble was used in the accompanying extraction blank. A 100-mL volume of extracting solvent was added to each sample flask and blank flask. Each flask was then connected to a water cooled (tap water) Soxhlet apparatus and placed on a heating mantle. The flasks were heated to boiling and extracted for 16 hours. Fresh solvent was then added as appropriate to account for any solvent lost to evaporation during the extraction. An aliquot of cooled sample extract (or blank) was transferred to a glass vial, capped, and analyzed by GC/MS.

15.2.1.4 Nonvolatile Residue Analysis
Nonvolatile residues from the various elastomer test article extracts were subjected to analysis by Fourier transform infrared (FTIR) spectroscopy. These analyses were accomplished by pipetting several drops of each individual extract onto a KBr plate and evaporating the solvent to dryness, leaving a nonvolatile residue for FTIR analysis. FTIR spectra were then acquired with a Nicolet (ThermoFisher Scientific, Madison, WI) Magna 750 FTIR spectrophotometer. Isopropanol extracts prepared in this manner showed significant IR absorbance bands due to isopropanol. To ameliorate this situation, isopropanol extracts on KBr plates were heated on a hot plate for 3 hours at 40°C and reanalyzed.

15.2.1.5 Inorganic Analysis
Each extract was sampled by withdrawing a portion of the fluid with a glass pipette, filtering through Whatman #50 filter paper and drying the material on titanium foil. Examination of samples in the high-vacuum

environment of the scanning electron microscope (SEM) required the removal of all volatile material from the sample. To accomplish this, the sample material was "ashed" at high temperature, up to ~600°C, for an appropriate time period, which varied for each sample. In summary, the liquid extract material and/or wetted solids were placed on laboratory fabricated polished titanium mounts. These were then heated on a hot plate until the volatiles were removed, as evidenced by no further visible change. Once cooled, the titanium samples were adhered to aluminum stubs using double-sided silver tape.

Samples were examined on a Hitachi (Hitachi High Technologies America, Inc., Pleasanton, CA) S-4700 SEM equipped with an Autrata Bse (back-scattered electron) detector and a Noran Vantage EDS spectrometer. Se- (secondary electron) and Bse-SEM examinations were performed on all samples as was qualitative energy dispersive spectrometry (EDS). In addition, the initial samples were analyzed using "spectral imaging" allowing for elemental image maps of the surface.

15.2.2 Preparation of Gas Chromatographic System Suitability Mix

A "system suitability" mixture of reference compounds was prepared in methylene chloride, based on nominal individual compound concentrations stated in the test protocol. The reference compounds and actual concentration values are listed below:

Reference Compound	Concentration (µg/mL)
Pyrene	1.2
2-MBT	6.5
Tetramethylthiuram disulfide	56.0
Butylated hydroxytoluene (BHT)	62.0
Bis(2-ethylhexyl)phthalate	90.5
Bis(dodecyl)phthalate	47.8
Stearic acid	117
2-Ethylhexanol	61.6

15.2.2.1 GC/MS Conditions   The GC/MS conditions used for extract analysis were as follows:

Instrument: Agilent (Agilent Technologies, Santa Clara, CA) 5973 Mass Selective Detector with 6890 plus GC

Column: RTX-1; 30 m × 250 µm; 0.1-µm film

Autosampler: HP7683; 1-µL slow injection (10-µL syringe)

Injector: 280°C

Interface: 280°C

Constant flow/vacuum compensation: Helium

Oven program: 40°C for 1 minute, ramp to 300°C at 10°C/min, hold for 10 minutes.

Source temp.: 230°C

Ionization mode: Electron ionization (EI)

Scanning: 50–650 amu at 4.7 scans/s

Solvent delay: 2.0 minutes

Chemical ionization (CI) GC/MS analysis for molecular weight confirmation of individual extractables was accomplished on a Micromass (Waters Corporation, Milford, MA) Quattro 2 instrument with ammonia as a reactant gas (see Chapter 13). Scan and other conditions were similar to those above for EI. Accurate mass measured EI spectra were acquired on a VG Micromass Autospec double-focusing sector instrument (see Chapter 13) with perfluorokerosene as an internal reference. Scan and other conditions were similar to those above for EI, except that the mass spectrometer was scanned exponentially at approximately 1 second per decade of mass at a resolving power of approximately 5000 (m/Δm; 10% valley definition).

15.2.3   Sample Preparation for Extractables Profiling by LC/MS

For LC/MS analysis, aliquots of individual extracts were evaporated under a stream of dry nitrogen, and the resulting extractables residues dissolved in aliquots of the initial mobile phase (mobile phase A; see the Appendix).

15.2.3.1   LC/MS Conditions   LC/MS analyses were accomplished on a Micromass Quattro Ultima (triple quadrupole) instrument in both positive and negative atmospheric pressure chemical ionization (APCI; see Chapter 13) modes. Conditions for high-performance liquid chromatography (HPLC) are detailed in the study protocol (see the Appendix). Specific instrumental conditions were as follows:

Corona:	2.5 μAmps
Cone:	15 V
Source block temp.:	150°C
APCI probe temp.:	550°C
Cone gas flow:	Off
Desolvation gas flow:	357 L/h
Scan rate:	1 s/scan
Scan cycle:	1.1 seconds
Mass range:	100–1000 amu

15.2.4   Results from Nonvolatile Residue Analyses of Extracts

A representative FTIR spectrum of nonvolatile residue from the methylene chloride Soxhlet extract is shown in Figure 15.1. In general, these FTIR spectra are not very informative regarding the chemical nature of the nonvolatile residue. Significant C–H stretching bands (both aliphatic and aromatic) in the vicinity of 3000 cm^{-1} likely result from paraffin-associated hydrocarbons (see GC/MS results below) and

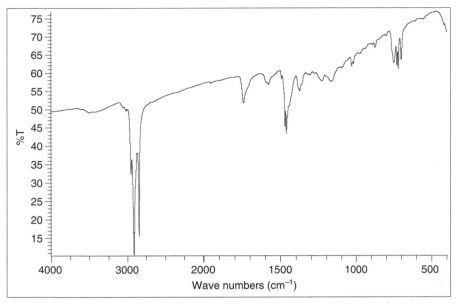

Figure 15.1   Fourier transform infrared (FTIR) spectrum of nonvolatile residue derived from a sample of methylene chloride Soxhlet extract.

coumarone-indene resin-related structures. The absorbance band at approximately $1700 \ cm^{-1}$ suggests the presence of aliphatic fatty acids. There are other bands in the "fingerprint" region of the spectrum suggestive of aromatic structures, of which there are several potential sources. Also apparent is low-intensity OH stretching in the vicinity of $3400 \ cm^{-1}$, which could be consistent with aliphatic fatty acids and/ or phenols. Known potential sources are available to rationalize most of the significant absorbance bands in these spectra.

15.2.5   Results from Inorganic Analyses

Particulate matter was noted in many of the test article extract samples after "ashing." It is probable that these particles formed during the drying and ashing process. In addition, the several steps necessary for sample preparation for SEM/EDS analysis may contribute environmental foreign particulates. The long drying and ashing process could permit environmental particles to settle on the samples. Many particles were observed during the course of this study that did not have an elemental composition consistent with the majority of the particles in a particular sample extract, or with each other. For these infrequent occurrences, no attempt was made to ascertain their origin. However, a general trend was observed, in that all of the samples appeared to have a thin film of material ashed to the titanium stub surface. With the exception of the isopropanol reflux sample, all of the extracted rubber samples contained varying amounts of zinc (Zn) and sulfur (S). The blank samples were predominantly comprised of thin carbon films. Table 15.2 summarizes the result for each sample. The table is broken into three parts:

TABLE 15.2. Summary of SEM/EDS Results for Various Sulfur-Cured Elastomer Extracts

Solvent	Description	Film	Particles	Other
$CHCl_2$	Soxhlet	Cracked, thick film, Zn major, S and O above background	Zn major, S and O above background	
	Blank	Amorphous irregular film, C, O, Si	Few observed	
Isopropyl alcohol (IPA)	Sonication	Heavy deposit with variable surface features and topography, Zn major, S minor	Particles "blobs" plentiful, embedded in film, Zn major, S minor	
	Blank	Sparse deposit, mainly C	C, O, Na, As, Al, Si, Cl, K, and Ca	
	Reflux	Sponge-like film, mainly C	Particles have O, Na, Si, Al in common	
	Blank	Sponge-like film with embedded particles; Si variable in amount but in every particle	Cl associated with particles	
	Soxhlet	Dried as "droplets"; Zn and S present in various amounts but not proportional to each other	Zn and S present in various amounts	Particles adherent to glass container contain Zn and S in variable amounts and apparently inversely proportional
	Blank	Thin C Film	Particles/flakes contain NaCl	
Hexane	Soxhlet	Thick, smooth film; C, O, Zn, S present	Many small particles: Na, Mg, Al, Si, C, O, Fe	
	Blank	Heavy deposit, C present	Some C particles observed	
	Reflux	Zn minor, C, O major, S occasionally above background	Few particles: 1. C, O, Zn, Mg, Al, Si, P, S, Ca 2. Na, Al, Si, K 3. Na, Zn, Mg, Al, Si	
	Blank	Sponge-like film, C, O present	Few particles	

1. the composition of the film, which identifies elements that were ubiquitously observed;

2. frequently observed particle species; and

3. other—solid material that was adhered to the sample submission vial.

Figure 15.2 shows a representative SEM micrograph illustrating the surface morphology of the methylene chloride Soxhlet extract.

The predominance of zinc and sulfur is completely consistent with the known composition of the elastomer test article. The zinc obviously results from the addition of the zinc salt of 2-MBT, and the sulfur results from both 2-MBT and TMTMS, along with elemental sulfur. It is important to note that no significant unexpected (and potentially toxic) inorganic species were observed.

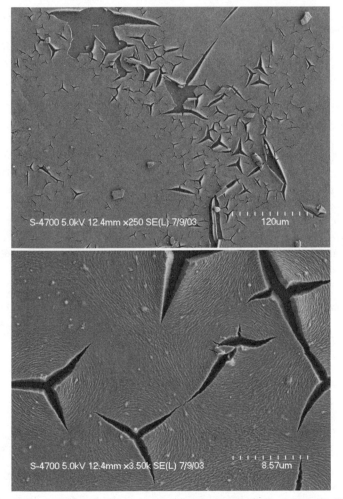

Figure 15.2   SEM micrograph illustrating the surface morphology of an "ashed" sample of methylene chloride Soxhlet extract.

15.2.6 Results from Extractables Profiling by GC/MS

Figures 15.3–15.5 show extractables profiles (in the form of GC/MS total ion chromatograms [TICs]) from the 2-propanol, hexane, and methylene chloride reflux, and Soxhlet and sonication extracts of the sulfur-cured elastomer test article. A detailed analysis was accomplished on the methylene chloride Soxhlet extract in order to identify as many individual extractables as reasonably possible. Table 15.3 summarizes the results of this detailed analysis, with reference to the expanded and partially numbered TIC in Figure 15.6. Identifications were accomplished following the general procedures and identification categories presented and described in Chapter 13 (also reference Table 15.4 in this chapter). Key points and conclusions from the qualitative extractables profiling study by GC/MS are as follows:

1. *All of the extractables profiles are similar.* However, note that the profiles differ somewhat in number and intensity of peaks depending on the particular solvent and extraction technique employed, an observation which supports the use of multiple solvents and extraction techniques in comprehensive CESs. The notable differences that can be observed between various extractables profiles are discussed in detail below. In choosing an extraction technique for optimization and possible use for routine control of extractables profiles for critical container closure system components, it is clear from examination of these profiles that many of the solvent/extraction technique combinations are viable candidates. An important observation is that sonication appears to be a viable extraction technique for this purpose. Sonication extractions produce representative extractables profiles in significantly less time and with lower volumes of extracting solvent, than either reflux or Soxhlet. However, sonication extraction efficiency likely depends on the energy produced by the ultrasonic bath which could be difficult to calibrate, control, and reproduce. Furthermore, as stated above, sonication necessarily involves a device (e.g., the ultrasonic bath) that will have a limited service lifetime and therefore potentially require change control processes to update.

   When using GC-based analytical techniques, it is not always appropriate to inject high-boiling or reactive solvents; therefore, it might be necessary to switch solvents prior to extractables profile analysis by either GC or GC/MS. Figure 15.7 shows extractables profiles (TICs) of sulfur-cured elastomer extracts from reflux in 2-propanol. In Figure 15.7A, the 2-propanol was evaporated from the sample, and the sample was then reconstituted in methylene chloride. Figure 15.7B shows results of "neat," that is, 2-propanol, sample injection. In this case, there appears to be no significant differences in the GC/MS extractables profiles based on extract preparation for injection (however, careful data analysis would be required to confirm this observation).

   At this point, it is appropriate to briefly discuss the preparation of elastomer and plastic test articles for extraction in a comprehensive CES. It is the opinion of many pharmaceutical development scientists, including the authors of this chapter, that CESs are best accomplished on intact components since it is the intact component that the drug product formulation can potentially

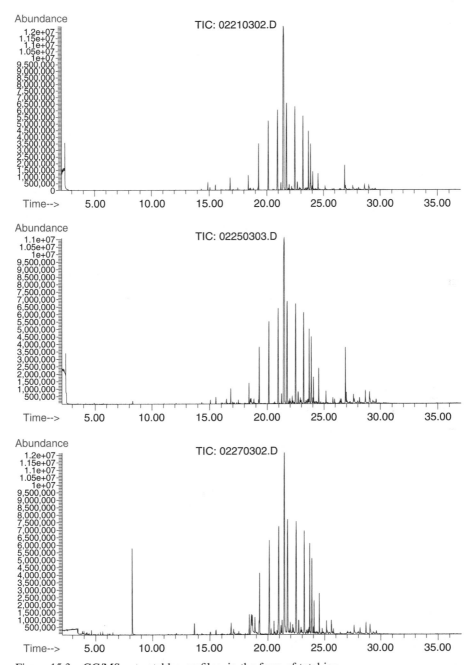

Figure 15.3   GC/MS extractables profiles, in the form of total ion chromatograms (TICs), from the methylene chloride (top), hexane (middle), and 2-propanol (bottom) reflux extracts.

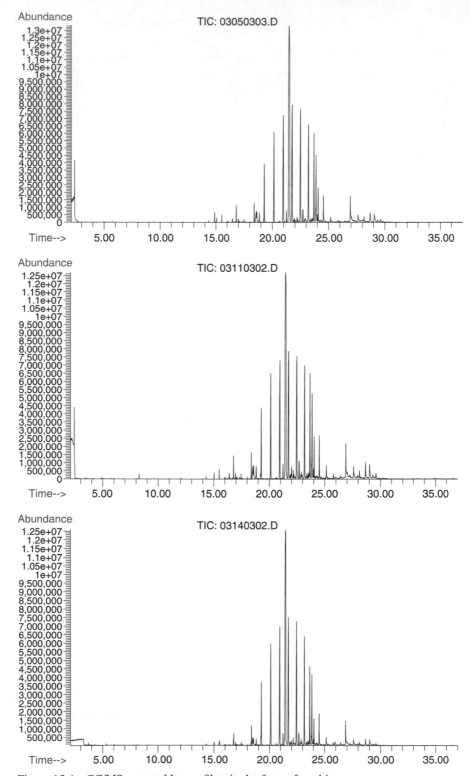

Figure 15.4   GC/MS extractables profiles, in the form of total ion chromatograms (TICs), from the methylene chloride (top), hexane (middle), and 2-propanol (bottom) Soxhlet extracts.

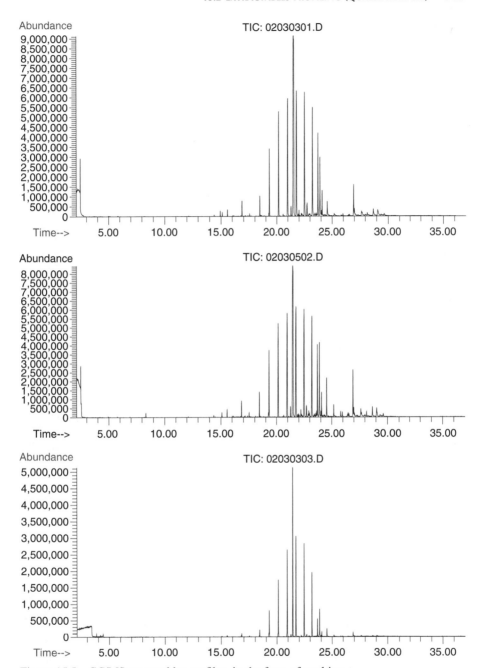

Figure 15.5   GC/MS extractables profiles, in the form of total ion
chromatograms (TICs), from the methylene chloride (top), hexane (middle) and
2-propanol (bottom) sonication extracts.

TABLE 15.3.   Extractables Identified from the GC/MS Analysis of a Methylene Chloride Soxhlet Extract of the Sulfur-Cured Elastomer Test Article

Peak #	Identification	Retention time (minutes)	Identification categories	Additional information	Identification level
1	α-Methylstyrene	4.90	A, C, D, E		Confirmed
2	Indene	5.70	A, C, D, E		Confirmed
3	Naphthalene	7.65	A, B, D		Confirmed
4	Tetramethylthiourea	8.05	A, C, D, E		Confirmed
5	Benzothiazole	8.15	A, C, D, E		Confirmed
6	Ethyl-4-*tert*-butyl phenyl ether	11.05	A, D		Confident
7	2,5-Di-*tert*-butylphenol	12.10	A, D		Confident
8	2-(Methylthio)benzothiazole	12.86	A, D		Confident
9	Coumarone-indene resin related	14.35	A, C, D		Confirmed
10	2-(Chloromethylthio)benzothiazole	14.86	A	Injection artifact	Confident
11	Coumarone-indene resin related	15.05	A, C, D, E		Confirmed
12	Coumarone-indene resin related	15.52	A, C, D		Confirmed
13	Coumarone-indene resin related	15.97	A, C		Tentative
14	Coumarone-indene resin related	16.07	A, C	9 with one additional double-bond or a related structure; MW 234; $C_{18}H_{18}$	Tentative
15	Coumarone-indene resin related	16.24	A, C	Isomer of 14	Tentative
16	2-Mercaptobenzothiazole	16.40	A, C, D		Confirmed
17	Coumarone-indene resin related	16.80	A, C		Tentative
18	Hexadecanoic acid	16.98	A, C, D, E		Confirmed
19	3,5-Bis(1,1-dimethylethyl-4-hydroxy)benzoic acid	17.04	A, D		Confident
20	Isomer of peak 19	17.11	A, D		Confident
21	Coumarone-indene resin related	17.31	A		Tentative

No.	Compound	RT	Sources	Notes	Confidence
22	n-Eicosane	17.47	A, D		Confident
23	Bis(4-methylphenyl)disulfide	17.53	A, D		Confident
24	Unknown (possible coumarone-indene resin related)	18.03	A		Tentative
25	Heneicosane	18.39	A, B, D, E		Confirmed
26	Linoleic acid	18.52	A, D		Confident
27	(E)-octadecenoic acid	18.60	A, D		Confident
28	Stearic acid	18.84	A, C, D, E		Confirmed
29	1-Octadecene	19.22	A, D		Confident
30	n-Docosane	19.28	A, B, D, E		Confirmed
31	Tricosane	20.12	A, B, D, E		Confirmed
32	Unknown (MW 366)	20.53	–		–
33	Tetracosane	20.94	A, B, D, E		Confirmed
34	Coumarone-indene resin related	21.24	A, C		Tentative
35	2,2'-Methylene-bis(6-tert-butyl-4-ethylphenol)	21.47	A, B, D, E		Confirmed
36	Pentacosane	21.73	A, B, D, E		Confirmed
37	Coumarone-indene resin related	21.88	A, C		Tentative
38	Unknown (possible coumarone-indene resin related)	21.96	A		Tentative
39	n-Alkane	22.17	A		Tentative
40	Unknown	22.24	–		–
41	Hexacosane	22.48	A, B, D, E		Confirmed
42	Coumarone-indene resin related	22.68	A, C		Tentative
43	Coumarone-indene resin related	22.71	A, C	Similar to 34; MW 354; $C_{27}H_{30}$	Tentative
44	Coumarone-indene resin related	22.86	A, C	Similar to 42; MW 352; $C_{27}H_{28}$	Tentative
45	Heptacosane	23.20	A, B, D		Confirmed

TABLE 15.3. *(Continued)*

Peak #	Identification	Retention time (minutes)	Identification categories	Additional information	Identification level
46	Coumarone-indene resin related	23.40	A, C	MW 366; $C_{27}H_{26}O$	Tentative
47	Coumarone-indene resin related	23.45	A, C	Mixture of related structures	Tentative
48	Coumarone-indene resin related	23.53	A, C		Tentative
49	Coumarone-indene resin related	23.68	A, C		Tentative
50	Coumarone-indene resin related	23.83	A, C		Tentative
51	Octacosane	23.88	A, B, D, E		Confirmed
52	Coumarone-indene resin related	23.99	A, C		Tentative
53	Coumarone-indene resin related	24.06	A, C		Tentative
54	Coumarone-indene resin related	24.15	A, C	Mixture of related structures	Tentative
55	Nonacosane	24.54	A, B, D, E		Confirmed
56	Triacontane	25.17	A, D, E		Confident
57	*n*-Alkane	25.80	A		Tentative
58	B-sitosterol	26.93	A, D		Tentative
59	Coumarone-indene resin related	27.05	A	Mixture of related structures	Tentative
60	Coumarone-indene resin related	27.63	A, C		Tentative
61	Coumarone-indene resin related	28.01	A	Mixture of related structures	Tentative
62	Coumarone-indene resin related	28.16	A	Mixture of related structures	Tentative
63	Coumarone-indene resin related	28.69	A, C		Tentative
64	Coumarone-indene resin related	29.07	A, C	Isomer of 63	Tentative
65	Coumarone-indene resin related	29.35	A, C	Mixture of related structures; MW 466 and 468; $C_{36}H_{34}$	Tentative
66	Coumarone-indene resin related	29.63	A, C	MW 466; $C_{36}H_{34}$	Tentative

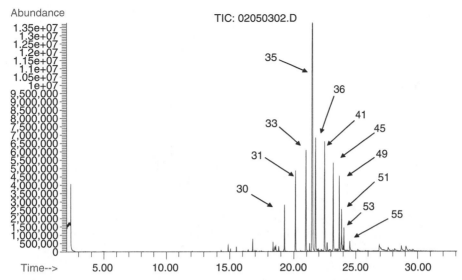

Figure 15.6   Expanded and numbered (reference Table 15.3) GC/MS extractables profiles, in the form of a total ion chromatogram (TIC), from the methylene chloride Soxhlet extract.

TABLE 15.4.   Identification Categories for the Identification of Extractables and Leachables by GC/MS and LC/MS

Identification category	Mass spectral data element
A	Mass spectral fragmentation behavior
B	Confirmation of molecular weight
C	Confirmation of molecular formula
D	Mass spectrum matches automated library/database or literature reference
E	Mass spectrum and chromatographic retention index match authentic reference compound

- A *confirmed* identification means that identification categories A, B (or C), and D (or E) have been fulfilled.
- A *confident* identification means that sufficient data to preclude all but the most closely related structures have been obtained.
- A *tentative* identification means that data have been obtained that are consistent with a class of molecule only.

leach over its shelf life. However, this does not necessarily preclude the use of additional sample preparation procedures (e.g., grinding or pressing in the case of plastic components, cutting or grinding in the case of elastomeric components), provided such procedures are scientifically justified and demonstrated not to produce artifacts.

2. *The major peak in all extractables profiles was confirmed to be the phenolic antioxidant 2,2′-methylene-bis(6-*tert*-butyl-4-ethylphenol).* This compound is one of the primary elastomer formulation ingredients (see Table 15.1).

(A)

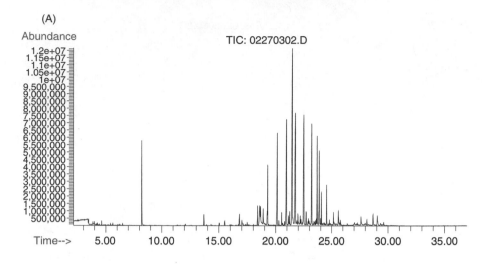

(B)

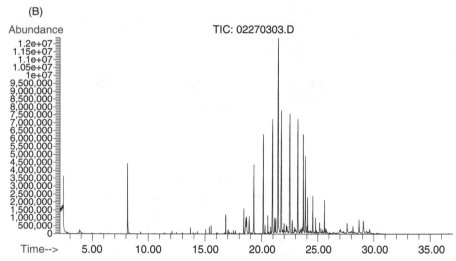

Figure 15.7 GC/MS extractables profiles, in the form of total ion chromatograms (TICs), the 2-propanol reflux extract. (A) Solvent switch to methylene chloride prior to injection. (B) "Neat" injection in 2-propanol.

3. *The* n-*alkanes identified as extractables, and appearing as a homologous series in all extractables profiles, are likely components of paraffin.* This substance is also one of the primary elastomer formulation ingredients (see Table 15.1). Figure 15.8 shows a mass chromatogram (m/z 71) computer generated from the methylene chloride Soxhlet extractables profile, which clearly shows this *n*-alkane (or hydrocarbon) "envelope." Note that m/z 71 is a unique fragment ion in the EI mass spectra of *n*-alkanes.

4. *A series of structurally related extractables identified in all extractables profiles appears to be related to coumarone-indene resin.* This substance is also

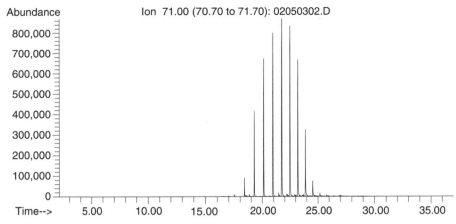

Figure 15.8   Mass chromatogram (*m/z* 71) computer generated from the extractables profile of the methylene chloride Soxhlet extract showing paraffin-derived *n*-alkanes.

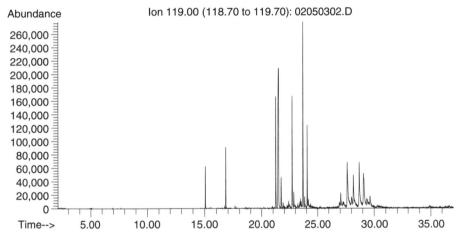

Figure 15.9   Mass chromatogram (*m/z* 119) computer generated from the extractables profile of the methylene chloride Soxhlet extract showing coumarone-indene resin-derived structures.

one of the primary elastomer formulation ingredients (see Table 15.1). Figure 15.9 shows a mass chromatogram (*m/z* 119) computer generated from the methylene chloride Soxhlet extractables profile, which shows this series of structurally related extractables. Note that *m/z* 119 is a unique fragment ion in the EI mass spectra of these compounds. The confirmed and proposed structures for many of these compounds are shown in Figure 15.10 (along with additional selected structures). The origin of these compounds was confirmed by the analysis of an authentic sample of coumarone-indene resin (see Fig. 15.11).

5. *The known formulation ingredient 2-MBT appears as a relatively minor peak in the extractables profiles.* Figure 15.12 shows a mass chromatogram (*m/z*

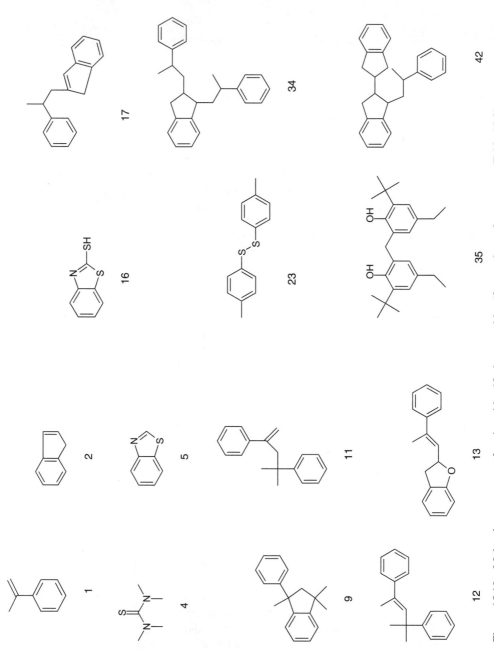

Figure 15.10 Molecular structures of various identified extractables (for number references, see Table 15.3).

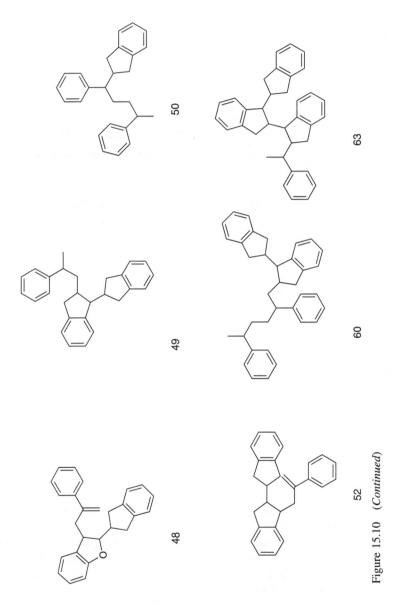

50

63

49

60

48

52

Figure 15.10 (*Continued*)

353

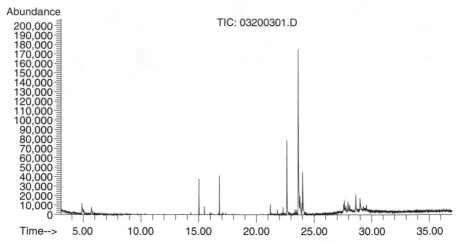

Figure 15.11 GC/MS profile, in the form of a total ion chromatogram, of an authentic sample of coumarone-indene resin.

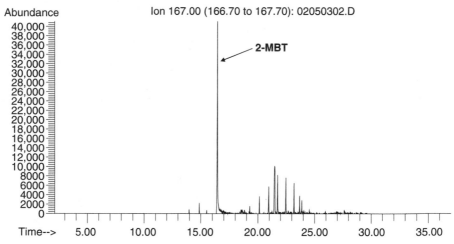

Figure 15.12 Mass chromatogram (*m/z* 167) computer generated from the extractables profile of the methylene chloride Soxhlet extract showing 2-mercaptobenzothiazole.

167) computer generated from the methylene chloride Soxhlet extractables profile, which shows 2-MBT. Note that *m/z* 167 is the molecular ion of 2-MBT in its EI mass spectrum.

However, in Figure 15.3, note the peak at approximately 8 minutes' retention time, which is not as significant in any of the other GC/MS extractables profiles. This extractable was identified as benzothiazole, and its presence in the 2-propanol reflux extract at this relatively high level is likely the result of thermolysis of 2-MBT. The boiling points of the extracting solvents are, respectively, methylene chloride 40.1°C, 2-propanol 82.3°C, and *n*-hexane 69.0°C. It is attractive to hypothesize that the higher temperature at which the

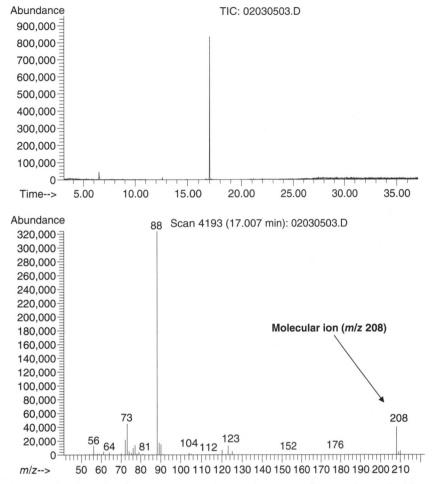

Figure 15.13   GC/MS profile, in the form of a total ion chromatogram, of an authentic sample of tetramethylthiuram monosulfide.

protic solvent 2-propanol refluxes is responsible for this high level of benzothiazole:

At first consideration, this might be interpreted as an extraction artifact; however, it is important to point out that comparison of all GC/MS extractables profiles, along with a basic understanding of organic chemistry and chemical reactivity, would alert the analytical chemist to the potential presence of the special case extractable 2-MBT.

An interesting "extractable" was observed in various methylene chloride extracts of the sulfur-cured elastomer:

2-(Chloromethylthio)benzothiazole

This chemical entity was determined not to be an extractable from the sulfur-cured rubber, but is in fact a reaction product between methylene chloride and 2-MBT, which is likely promoted by the relatively high temperatures in the GC injector, and is clearly an artifact:

However, the identification of this GC injection artifact also suggests that 2-MBT is present in the various extracts of the sulfur-cured elastomer, perhaps at higher levels than suggested by the relative amounts of actual 2-MBT detected. The pharmaceutical development scientist should always be vigilant for extraction and analytical artifacts, which could affect the interpretation of extractables profiles.

6. *TMTMS, a known elastomer formulation ingredient (see Table 15.1), was not detected in any of the GC/MS extractables profiles.* TMTMS is a vulcanization accelerator and known N-nitrosamine precursor, making it a potential leachable of some interest. The reaction of TMTMS with a "nitrosating agent" (NOX) to form N-nitrosodimethylamine is as follows:

An authentic reference standard of TMTMS was analyzed by GC/MS under the same analytical conditions used to characterize elastomer extracts. Figure 15.13 shows a TIC from the GC/MS analysis of authentic TMTMS, indicating that this additive would likely be detected in GC/MS profiles of sulfur-cured elastomer extracts. Based on these results, it is reasonable to assume that TMTMS was significantly consumed during the elastomer polymerization/ cross-linking process (however, it should be noted that the laboratory experienced difficulties with the GC/MS analysis of tetramethylthiuram disulfide in the system suitability mixture). However, this result does not relieve the burden of N-nitrosamine testing for this elastomer, as N-nitrosamines are reaction products of TMTMS, which could form during the elastomer curing process.

15.3   RESULTS FROM EXTRACTABLES PROFILING BY LC/MS

The LC/MS extractables profiles for the sulfur-cured elastomer test article serve to complement the GC/MS extractables profiles. A representative example that best illustrates the complementary nature of the LC/MS extractables profiling results is that for the 2-propanol Soxhlet extract. Figure 15.14 shows a series of chromatograms from two LC/MS analyses of this particular extract. The top chromatogram is a representative UV trace (280 nm; taken from diode array data) with the middle and bottom chromatograms being respectively, the TIC from the positive ion APCI analysis (middle) and the TIC from the negative ion APCI analysis (bottom). The TICs are dominated by HPLC mobile-phase background ions, so the in-line UV trace is used to assist in locating individual extractable peaks.

The major extractable (retention time of 17.02 minutes in the UV trace) was confirmed to be 2,2′-methylene-bis(6-*tert*-butyl-4-ethylphenol), which is the major known additive already identified by GC/MS. The positive and negative ion APCI mass spectra for this extractable are shown in Figure 15.15. Another major extractable (retention time 3.71 minutes in the UV trace) was determined to be 2-MBT (see mass spectra in Fig. 15.16). Other lower-level extractables observed in the LC/MS analyses were not readily identifiable. This result confirms what was hypothesized from GC/MS, that 2-MBT is present in detectable quantities in the sulfur-cured elastomer. Furthermore, the LC/MS results confirm the idea that multiple complementary analytical techniques should be used for analysis of extracts in comprehensive CESs.

15.4   OPTIMIZATION OF A SELECTED GC/MS EXTRACTABLE PROFILING METHOD

After evaluating extractables profiles from various extraction techniques/methods and solvents, a pharmaceutical development team should choose a "definitive" extraction technique(s)/method(s) to optimize. An optimized extraction method is defined as one that yields a high number and concentration of extractables, for example, steady-state or "asymptotic levels," without violating any of the established CES principles (see Chapter 14). This is not meant to imply that 100% of all known additives must be recovered, and it is therefore not a "deformulation analysis." Optimization of the extraction technique(s)/method(s) prior to conducting quantitative CESs ensures that the extractables profile represents at least a "worst-case" scenario of potential leachables and their levels. Extractables profiles produced from such optimized technique(s)/method(s) should be thoroughly evaluated both qualitatively and quantitatively. Adequate experimental studies, for example, accuracy, precision, linearity, and selectivity, should be accomplished in order to verify the accuracy of the quantitative results, should these results become an integral part of an extractables/leachables correlation. An optimized extraction technique/method can also serve as the basis for development and validation of routine extractables control methods. These fully validated routine extractables control methods can then be used to produce qualitative and quantitative databases of component extractables information, which can facilitate correlation of extractables and leachables. Based on an objective evaluation of all extractables profiles, Soxhlet extraction in methylene chloride was selected for optimization experiments for the sulfur-cured elastomer test article. A timed Soxhlet extraction with a fixed mass of rubber (7 g cut into 20–30 approximately uniform pieces), and with 200 mL of methylene chloride

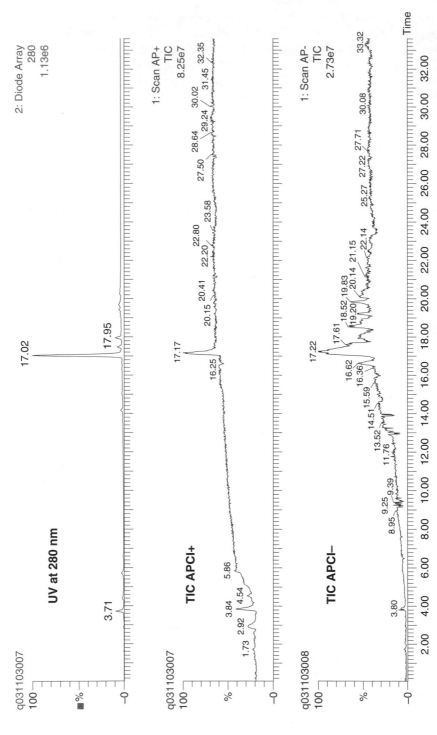

Figure 15.14 Chromatograms (extractables profiles) from two LC/MS analyses of the isopropanol Soxhlet extract. The top chromatogram is a representative UV trace (280 nm; taken from diode array data) with the middle and bottom chromatograms being, respectively, the total ion chromatogram (TIC) from the positive ion APCI analysis (middle) and the TIC from the negative ion APCI analysis (bottom).

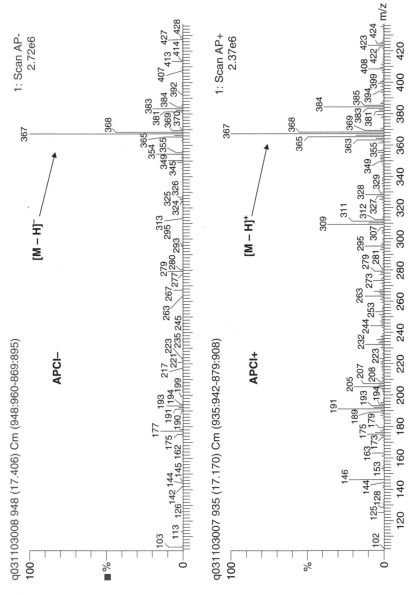

Figure 15.15 Positive (top) and negative (bottom) ion APCI mass spectra of 2,2'-methylene-bis(6-*tert*-butyl-4-ethylphenol). Note the "hydride ion abstraction product" $[M - H]^+$ at m/z 367 (molecular weight 368), in the positive ion APCI spectrum (see Chapter 13).

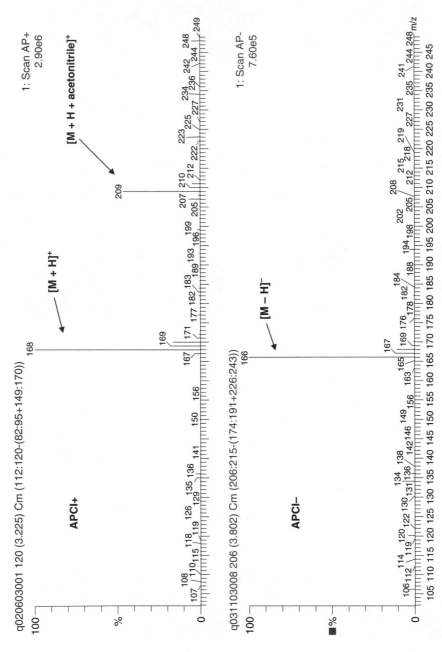

Figure 15.16 Positive (top) and negative (bottom) ion APCI mass spectra of 2-mercaptobenzothiazole.

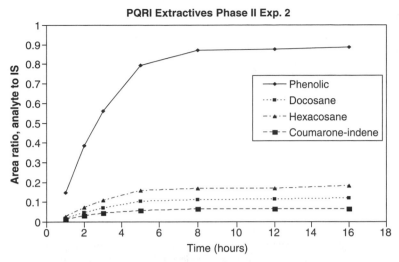

Figure 15.17 Optimization of methylene chloride Soxhlet extraction method ("asymptotic" levels). The individual extractables followed for optimization were 2,2'-methylene-bis(6-*tert*-butyl-4-ethylphenol), docosane, hexacosane, and one of the major coumarone-indene resin components.

spiked with an internal standard (2-fluorobiphenyl), was performed. The drop rate of methylene chloride in the Soxhlet extractor was approximately 20 per minute. Samples of methylene chloride extract (1.0 mL taken through the sidearm when boiling stopped) were collected at time intervals of 1, 2, 3, 5, 8, 12, and 16 hours. These samples were then diluted 10:1 with fresh methylene chloride and analyzed by GC/MS. Selected ion peak area ratios (A_{ion}/A_{is}) were monitored for four of the most significant and representative extractables over the time course. These peak area ratios were then plotted versus extraction time (see Fig. 15.17). Based on the data, it was determined that a 16-hour extraction is suitable (see Fig. 15.18), with asymptotic levels for all four monitored extractables clearly reached after approximately 8 hours of extraction time.

Note that this is but one example of how extraction technique/method optimization studies might be accomplished during the overall CES process. Obviously, there are other study designs that could accomplish the same purpose, and the reader should not infer that the study designs described in this chapter are the only ones acceptable. However, the end result of extraction technique/method optimization studies should always be the achievement of asymptotic levels of extractables with high overall extractables yields in order to facilitate qualitative and quantitative correlation of extractables and leachables, and to allow for development and validation of appropriate analytical methods for routine control of extractables.

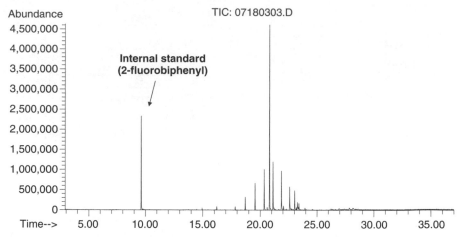

Figure 15.18 GC/MS extractables profile, in the form of a total ion chromatogram (TIC), from the 16-hour time point in the methylene chloride Soxhlet extraction optimization study, with added internal standard (2-fluorobiphenyl).

15.5 SIMULATED LEACHABLES STUDY (QUALITATIVE EXTRACTABLES/LEACHABLES CORRELATION)

To investigate the concept of qualitative extractables/leachables correlation, a "simulated" leachables study was accomplished with the sulfur-cured elastomer. Quantities of elastomer test article were cut into relatively small pieces and placed into glass bottles, which were then filled near the top with CFC-11 (Sigma-Aldrich Chemical Company, Milwaukee, WI), which is a liquid under standard environmental conditions. The sample bottles were then sealed with a crimped metal cap that incorporated a thermoplastic elastomer seal of known composition. Appropriate control bottles were prepared, which did not include any test article, in order to correct for any "leachables" associated with the thermoplastic elastomer seal. Sample and control bottles were stored in a laboratory stability chamber (VWR Scientific Products, West Chester, PA/USA, Model 9000) under accelerated conditions (40°C/75% relative humidity), and samples were pulled for GC/MS analysis (extractables profiling) over a period of 3 months. Figure 15.19 shows pictures of a sample bottle (note that the solvent swollen elastomer pieces float on the CFC-11) and the stability chamber. An internal standard (2-fluorobiphenyl) was added to each sample prior to GC/MS analysis, and samples in CFC-11 were injected directly into the GC/MS system.

A GC/MS "leachables profile" (1 week time point) for the sulfur-cured elastomer test article is shown in Figure 15.20. Even without a detailed analysis of this profile, the qualitative similarity with any of the GC/MS extractables profiles is readily apparent. This result supports the choice of extracting solvents for simulation of a typical MDI drug product.

Figure 15.19   Simulated leachables study sample bottle and stability chamber.

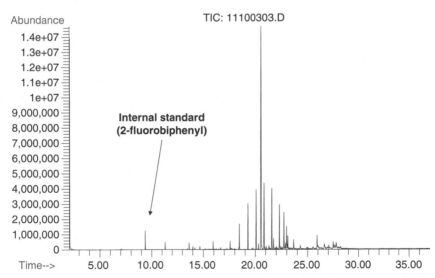

Figure 15.20   GC/MS leachables profile, in the form of a total ion chromatogram (TIC), from the 1 week time point, with added internal standard (2-fluorobiphenyl).

15.6   CONCLUSIONS AND RECOMMENDATIONS

The intention of this study was not to produce a CES suitable for inclusion in a regulatory submission, but to accomplish CES-related work that would enable the understanding and development of "best practice" recommendations for the conduct of CESs in support of regulatory submissions. To that end, this study was successful, in that it facilitated the previously discussed and reported[1,2] recommendations. Furthermore, all results of the CES are totally consistent with the known formulation of the sulfur-cured elastomer test article.

We consider this CES to be, in spite of the extensive and detailed work that was accomplished, little more than a "first pass." Use of this CES in a regulatory submission for an MDI drug product would potentially require the following:

- additional data analysis of both GC/MS and LC/MS extractables profiles in order to (1) upgrade *tentative* and *confident* GC/MS identifications to *confirmed*; (2) find and identify additional less abundant extractables in the LC/MS data, which are either hidden in the TICs or lack UV chromophores;

- a complete leachables/extractables correlation, both qualitative and quantitative. This assumes that an appropriate drug product leachables study, including accelerated storage conditions, has been accomplished. Note that if this were a real OINDP container closure system critical component, then the drug product would have dosing parameters from which an analytical evaluation threshold (AET) for leachables could be calculated (based also on the safety concern threshold [SCT]; 0.15 µg/day total daily intake for an individual organic leachable). This leachables AET could be used to guide further evaluation of CES data;

- a thorough search for any detectable levels of the TMTMS, which is the only known elastomer formulation ingredient which could not be detected in any extractables profile;

- complete validation of the GC/FID quantitative extractables method (Case Study 15.1) according to established validation parameters, for use as a routine extractables control method;

- development and complete validation of a second routine extractables control method capable of quantifying and controlling extractables not amenable to GC (i.e., TMTMS);

- development and complete validation of analytical methods for "special case" extractables, including PAHs, *N*-nitrosamines, and 2-MBT (the latter would require complete validation of the analytical method discussed in Case Study 15.2).

REFERENCES

1 Norwood, D.L. and Ball, D. Product Quality Research Institute: Safety thresholds and best practices for extractables and leachables in orally inhaled and nasal drug products. Submitted to the PQRI Drug

Product Technical Committee, PQRI Steering Committee, and U.S. Food and Drug Administration by the PQRI Leachables and Extractables Working Group, 2006.

2 Norwood, D.L., Paskiet, D., Ruberto, M., Feinberg, T., Schroeder, A., Poochikian, G., Wang, Q., Deng, T.J., DeGrazio, F., Munos, M.K., and Nagao, L.M. Best practices for extractables and leachables in orally inhaled and nasal drug products: An overview of the PQRI recommendations. *Pharm Res* 2008, *25*(4), pp. 727–739.

3 Norwood, D.L. and Mullis, J.O. "Special case" leachables: A brief review. *Am Pharm Rev* 2009, *12*(3), pp. 78–86.

APPENDIX: CASE STUDY 15.1

DEVELOPMENT AND EVALUATION OF A ROUTINE EXTRACTABLES TEST METHOD FOR A SULFUR-CURED ELASTOMER

David Olenski, John Hand, Sr., and Melinda K. Munos

Introduction

The objective of this case study was to develop and evaluate an analytical method for the quantitative extractables profiling of the sulfur-cured elastomer test article. It was determined based on the results of the qualitative controlled extraction study (CES) that the analytical method should be based on gas chromatography (GC) and incorporate the optimized extraction process (i.e., Soxhlet extraction for 16 hours with methylene chloride). The evaluation of the method was designed to determine its suitability for use as a routine extractables control method for any OINDP critical components manufactured from the sulfur-cured elastomer.

During the conduct of the qualitative CES, it was determined that GC/mass spectrometry (MS) was capable of separating and individually detecting extractables associated with all of the known ingredients in the sulfur-cured elastomer test article. However, since advanced analytical techniques such as GC/MS are not always desirable in a quality control environment, GC/flame ionization detection (FID) was chosen as a potentially viable alternative. GC/FID is known to correlate well with GC/MS (see Fig. CS 15.1.1) and is generally considered to be a sensitive, selective, and robust analytical technique. GC/FID is routinely used by quality control laboratories in pharmaceutical manufacturing for such applications as quantitation and control of residual solvents in active pharmaceutical ingredients. Evaluation of the analytical method, including the extraction procedure and GC/FID analysis of resulting extracts, was accomplished by investigating the parameters: system suitability (instrument precision, chromatographic resolution, chromatographic tailing factor), linearity and range, precision (repeatability), specificity, accuracy, limit of quantitation (LOQ), and standard and sample stability.[1,2] Test article extracts were employed in the evaluation, along with certain selected authentic reference compounds identified as significant extractables in the qualitative CES. These included 2,2'-methylene-bis(6-*tert*-butyl-4-ethylphenol), *n*-docosane, *n*-tricosane, *n*-tetracosane, *n*-pentacosane, *n*-hexacosane, and *n*-octacosane.

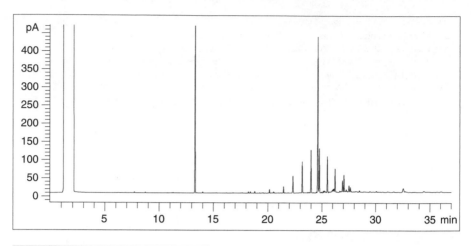

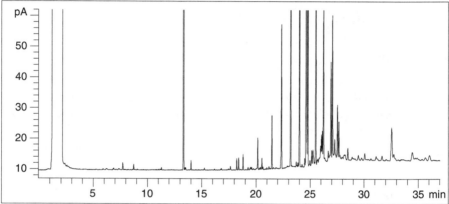

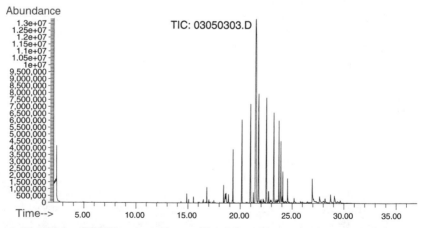

Figure CS 15.1.1 GC/FID extractables profile of the sulfur-cured elastomer test article (top); vertically expanded (middle); GC/MS total ion chromatogram (TIC; bottom). Note that the GC/FID profiles contain an internal standard (2-fluorobiphenyl) at approximately 13.2 minutes (retention time).

Experimental Summary

General Procedure

- All extractions were accomplished using methylene chloride spiked with an internal standard (2-fluorobiphenyl) at a nominal concentration of 100 μg/mL.

- All standard and sample materials were dissolved and diluted as needed in methylene chloride spiked with the internal standard.

- For all Soxhlet extractions, the cellulose thimbles were pre-extracted with methylene chloride for ~2 hours.

- The elastomer sample was cut into small squares of approximately 5 mm. Approximately 7 g of sample was then placed in a thimble and extracted with 200 mL methylene chloride for 16 hours.

- After 16 hours of extraction, the methylene chloride was collected and diluted 1:10 with the extraction solvent described above.

- The diluted material was analyzed by GC/FID.

GC/FID Instrumental Parameters

- Instrument: Agilent 6890 with ChemStation software
- GC column: Restek 0.32 mm × 30 m DB-1; 1.0-μm film thickness
- Injector temperature: 280°C
- Injection mode: Splitless
- Injection volume: 1 μL
- Purge valve: On at 1.00 minute; off initially
- Temperature program: 40°C for 1 minute, then linear ramp to 300°C at 10°C/min; hold for 10 minutes
- Carrier gas: Helium; flow rate, 2.7 mL/min constant flow
- Detector: FID; air 400 mL/min, hydrogen 30 mL/min

Additional experimental details relative to the various validation criteria are summarized and discussed below.

Results and Discussion

System Suitability   System suitability parameters were evaluated using the selected authentic reference compounds representing major extractables identified in the qualitative CESs, and the internal standard. Six replicate injections of a test solution containing approximately 10 μg/mL 2,2′-methylene-bis(6-*tert*-butyl-4-eth-ylphenol) and *n*-pentacosane, spiked with 2-fluorobiphenyl (internal standard) in methylene chloride were accomplished. Peak areas and area ratio measurements of reference compounds and the internal standard were determined, and means and percent relative standard deviations (% relative standard deviations [RSDs]) of area ratios and relative response factors (RRFs) were calculated. Table CS 15.1.1 shows the calculated area ratios, RRFs, means, and %RSDs. Utilizing the second

TABLE CS 15.1.1. System Suitability: Instrument Precision

Trial	Area 2,2′-MBTBE	Area ratio	RRF	Area n-pentacosane	Area ratio	RRF
1	98.03	0.84	1.01	48.14	0.41	0.43
2	96.06	0.85	1.03	46.50	0.41	0.43
3	94.73	0.83	1.00	45.03	0.40	0.42
4	91.89	0.84	1.01	45.87	0.42	0.44
5	98.12	0.86	1.03	48.78	0.43	0.45
6	97.06	0.85	1.03	47.76	0.42	0.44
Mean	96.0	0.85	1.02	47.0	0.41	0.43
%RSD	2.48	1.25	1.25	3.08	2.54	2.54

2,2′-MBTBE, 2, 2′-methylene-bis(6-*tert*-butyl-4-ethylphenol).

TABLE CS 15.1.2. System Suitability: Tailing and Resolution of 2,2′-Methylene-Bis(6-*tert*-Butyl-Ethylphenol); 2,2′-MBTBE; and n-Pentacosane

Trial	Resolution 2,2′-MBTBE	Tailing[a] 2,2′-MBTBE	Resolution n-pentacosane	Tailing[a] n-pentacosane
1B	8.411	–	2.132	–
2B	8.336	–	2.136	–
3B	8.441	–	2.135	–
4B	8.375	–	2.097	–
5B	8.324	–	2.133	–
6B	8.446	–	2.113	–
Mean	8.39	–	2.12	–
%RSD	0.62	–	0.75	–

[a] Tailing was not observed in the chromatographic analysis of these compounds. The peaks appeared to be fronting and the software would not calculate a tailing factor. The hand calculated USP tailing factor is 0.9.

2,2′-MBTBE, 2, 2′-methylene-bis(6-*tert*-butyl-4-ethylphenol).

injection of each mixed standard in the instrument precision section, resolution and tailing for 2,2′-methylene-bis(6-*tert*-butyl-ethylphenol) and n-pentacosane were determined using the GC/FID system software (Agilent ChemStation). Percent relative standard deviations and means were also calculated. These results are shown in Table CS 15.1.2.

Linearity and Range Linearity and range were evaluated by analyzing 2,2′-methylene-bis(6-*tert*-butyl-4-ethylphenol) and n-pentacosane in duplicate at six different concentrations ranging from 1.0 to 20.0 μg/mL in methylene chloride spiked with 2-fluorobiphenyl internal standard. Relative response factors were calculated and regression analyses were performed. The analyte concentrations, peak area ratios, and relative response factors are shown in Table CS 15.1.3. The regres-

TABLE CS 15.1.3. Linearity Study: Analyte Concentrations, Peak Area Ratios, and Relative Response Factors (RRFs)

Trial	Internal standard area ratio n-pentacosane	RRF n-pentacosane	Internal standard area ratio 2,2′-MBTBE	RRF 2,2′-MBTBE
1.0 μg/mL A	0.027	0.003	0.060	0.007
1.0 μg/mL B	0.028	0.003	0.061	0.007
2.0 μg/mL A	0.054	0.011	0.129	0.028
2.0 μg/mL B	0.053	0.011	0.129	0.028
5.0 μg/mL A	0.154	0.081	0.349	0.188
5.0 μg/mL B	0.155	0.081	0.350	0.188
10.0 μg/mL A	0.331	0.347	0.742	0.801
10.0 μg/mL B	0.322	0.338	0.739	0.797
15.0 μg/mL A	0.549	0.864	1.187	1.919
15.0 μg/mL B	0.540	0.850	1.179	1.907
20.0 μg/mL A	0.727	1.525	1.598	3.446
20.0 μg/mL B	0.734	1.539	1.592	3.435

2,2′-MBTBE, 2,2′-methylene-bis(6-*tert*-butyl-4-ethylphenol).

sion analyses, slopes, y-intercepts, and coefficients of determination (R^2) and provided in Figures CS 15.1.2 and CS 15.1.3.

Additionally, single-point relative response factors were determined for the remaining authentic reference compounds (n-docosane, n-tricosane, n-tetracosane, n-hexacosane, and n-octacosane). The results are shown in Table CS 15.1.4.

Precision Repeatability was assessed by performing six separate extractions of the sulfur-cured elastomer test article using the optimized Soxhlet extraction procedure. Seven major extractables identified in the qualitative controlled extraction studies were quantified using the optimized GC/FID method. The means and %RSDs of these analyses were calculated for each analyte. These results are provided in Table CS 15.1.5.

Specificity Specificity was evaluated as part of the qualitative CES. A mixed standard containing 2,2′-methylene-bis(6-*tert*-butyl-4-ethylphenol), n-docosane, n-tricosane, n-tetracosane, n-pentacosane, n-hexacosane, n-octacosane, and 2-fluorobiphenyl as the internal standard was analyzed by GC/MS. The chromatogram from this analysis was compared with chromatograms of GC/MS analysis of the test article to confirm peak identification and verify that there were no co-eluting peaks. GC/FID chromatograms of a method blank (with internal standard), a mixed standard, and sample extract are shown in Figure CS 15.1.4. GC/MS total ion chromatograms from an extraction blank and sample extract are shown in Figure CS 15.1.5.

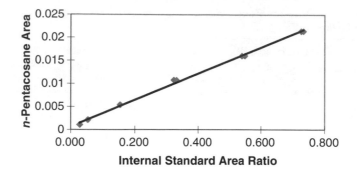

Regression Statistics	
R^2	0.999
Standard Error	0.0004
Observations	12
y-Intercept	0.0007
Slope	0.0286

Figure CS 15.1.2   Linearity study and regression statistics for *n*-pentacosane.

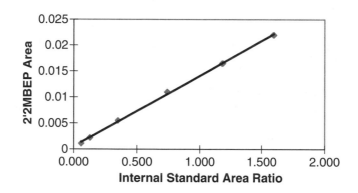

Regression Statistics	
R^2	0.999
Standard Error	0.0003
Observations	12
y-Intercept	0.0006
Slope	0.0135

Figure CS 15.1.3   Linearity study and regression statistics for 2,2'-methylene-bis(6-*tert*-butyl-4-ethylphenol).

TABLE CS 15.1.4.   Linearity Study: Single-Point Relative Response Factors (RRFs) for Selected Major Extractables

Analyte ID	Analyte (µg/mL)	Analyte area	Internal standard area	Internal standard (µg/mL)	Area ratio	RRF
n-Docosane	10.0	102.9126	142.8876	10.2	0.72	0.71
n-Tricosane	10.6	81.86713	126.4398	10.2	0.65	0.63
n-Tetracosane	10.3	73.93506	125.9505	10.2	0.59	0.58
n-Hexacosane	10.4	52.41956	127.3931	10.2	0.41	0.40
n-Octacosane	10.4	38.76235	132.3707	10.2	0.29	0.29

TABLE CS 15.1.5.   Precision: Quantitation of Selected Major Extractables

Trial	n-Doco (µg/g)	2,2'-MBTBE (µg/g)	n-Trico (µg/g)	n-Tetra (µg/g)	n-Penta (µg/g)	n-Hexa (µg/g)	n-Octa (µg/g)
1	35	501	60	72	56	45	11
2	35	503	57	72	53	43	11
3	37	543	58	71	52	40	9
4	32	492	52	64	47	36	8
5	29	464	45	53	37	38	12
6	30	464	46	55	39	34	13
AVG	33	494	53	64	47	39	11
%RSD	4.49	2.98	6.15	6.72	8.21	5.38	7.46

n-Doco, n-docosane; 2,2'-MBTBE, 2,2'-methylene-bis(6-*tert*-butyl-4-ethylphenol); n-Trico, n-tricosane; n-Tetra, n-tetracosane; n-Penta, n-pentacosane; n-Hexa, n-hexacosane; n-Octa, n-octacosane.

Acceptance criteria for specificity were that peak identifications should be confirmed and there should be no significant co-eluting peaks for each target extractable.

Accuracy   Accuracy was assessed by preparing standard additions, in triplicate, of approximately one, two, and three times the target extractable concentration in 50 mL of a Soxhlet extract test solution. N-pentacosane and 2,2'-methylene-bis(6-*tert*-butyl-4-ethylphenol) were selected as the target extractables. The individual spiking levels were chosen to represent the appropriate range of extractable concentrations expected based on the method development experiments. Spiked samples were analyzed by the optimized GC/FID method and individual mean recoveries were determined for each spiking level. The spiking levels and percent recovery are provided in Tables CS 15.1.6 and CS 15.1.7.

LOQ   LOQ was assessed by analyzing a standard solution of n-pentacosane and 2,2'-methylene-bis(6-*tert*-butyl-4-ethylphenol) designed to produce a response of approximately 10 times the LOQ (i.e., a response that provides a signal-to-noise

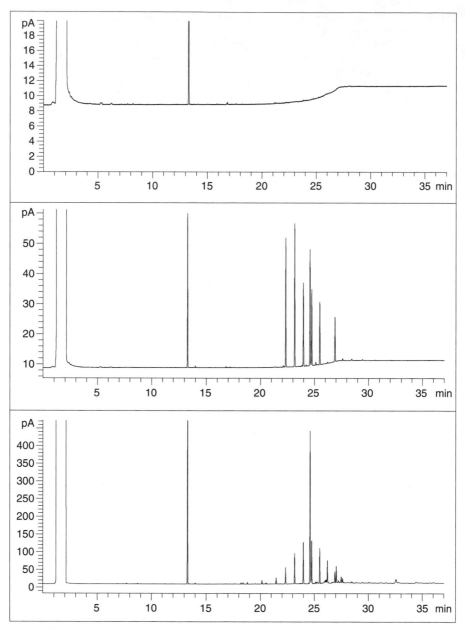

Figure CS 15.1.4   GC/FID specificity chromatograms from an extraction blank (top), mixed standard (middle), and elastomer test article (bottom).

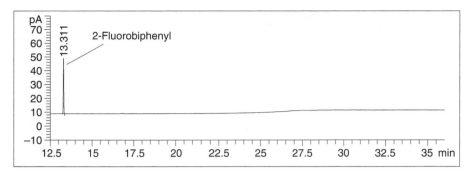

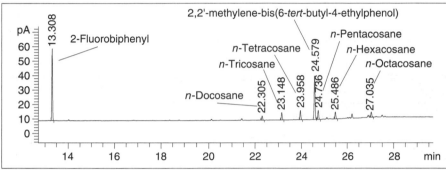

Figure CS 15.1.5   GC/MS specificity total ion chromatograms (TICs) from an extraction blank (top) and elastomer test article (bottom).

TABLE CS 15.1.6.   Accuracy: Spiking and Recovery Data for *n*-Pentacosane

Trial ID	Spiked (µg)	Recovered (µg)	% recovery
Low spike A	352	380	107.9
Low spike B	352	398	113.0
Low spike C	352	411	116.9
Mid spike A	704	687	97.6
Mid spike B	704	680	96.6
Mid spike C	704	694	98.6
High spike A	1056	980	92.8
High spike B	1056	980	92.8
High spike C	1056	1017	96.3

[Root mean square (RMS)] ratio [S/N] of approximately 100:1). The standard was analyzed six times by the optimized GC/FID method. Based on the average S/N ratios for these target extractables, LOQs were estimated by extrapolation (S/N 10:1). Based on these extrapolated LOQs, a solution of target extractables was prepared and analyzed six times for LOQ confirmation. Table CS 15.1.8 provides the S/N ratios for analysis of 0.5 µg/mL of both extractables in test solution.

**TABLE CS 15.1.7.   Accuracy: Spiking and Recovery Data for
2,2'-Methylene-Bis(6-*tert*-Butyl-4-Ethylphenol)**

Trial ID	Spiked (μg)	Recovered (μg)	% recovery
Low spike A	554	511	92.2
Low spike B	554	528	95.3
Low spike C	554	538	97.1
Mid spike A	1108	1080	97.5
Mid spike B	1108	1059	95.5
Mid spike C	1108	1110	100.2
High spike A	1662	1579	95.0
High spike B	1662	1591	95.7
High spike C	1662	1605	96.6

**TABLE CS 15.1.8.   Limit of Quantitation: Signal-to-Noise (S/N) Ratios for 0.5 μg/mL Target
Extractables in Test Solution**

Trial ID	S/N *n*-pentacosane	S/N 2,2'-methylene-bis(6-*tert*-butyl-4-ethylphenol)
1	8.8	13.6
2	8.0	13.7
3	8.2	13.3
4	8.6	13.8
5	9.9	16.2
6	9.3	11.3
AVG S/N	8.8	13.7

Standard and Sample Stability   Standard and sample stability were evaluated over a period of 5 days by analyzing on each day an appropriate mixed standard of selected major extractables (*n*-pentacosane and 2,2'-methylene-bis(6-*tert*-butyl-4-ethylphenol)) as in the system suitability section, and an extract of the sulfur-cured elastomer test article, as in the "Precision" section. 2-Fluorobiphenyl was used as the internal standard. Area ratios of major extractable to internal standard were calculated. The area ratios are provided in Tables CS 15.1.9 and CS 15.1.10.

Conclusion

A GC/FID method was developed for extractables profiling and the quantitative analysis of target extractables in methylene chloride Soxhlet extracts of a sulfur-cured elastomer test article. The method was evaluated with respect to system suitability (instrument precision, chromatographic resolution, chromatographic tailing factor), linearity and range, precision (repeatability), specificity, accuracy, LOQ, and standard and sample stability. Quantitative levels of target extractables from duplicate analyses of six test article extracts are shown in Table CS 15.1.11.

TABLE CS 15.1.9. Standard Stability: Ratios of Major Extractable (in a Mixed Standard) to Internal Standard over a 5-Day Period

Trial ID	Internal standard area ratio n-pentacosane	Internal standard area ratio 2,2'-MBTBE
1A	0.43	0.85
1B	0.41	0.85
1C	0.45	0.87
1D	0.42	0.86
1E	0.43	0.86
1F	0.42	0.85
Average	0.43	0.86
2A	0.41	0.84
2B	0.42	0.84
2C	0.42	0.85
2D	0.42	0.84
2E	0.42	0.84
2F	0.42	0.85
Average	0.42	0.84
3A	0.46	0.88
3B	0.44	0.82
3C	0.45	0.82
3D	0.43	0.82
3E	0.47	0.84
3F	0.48	0.85
Average	0.46	0.84
4A	0.46	0.85
4B	0.46	0.84
4C	0.43	0.81
4D	0.47	0.85
4E	0.48	0.85
4F	0.44	0.81
Average	0.46	0.83
5A	0.51	0.87
5B	0.48	0.85
5C	0.48	0.83
5D	0.52	0.87
5E	0.51	0.85
5F	0.44	0.80
Average	0.49	0.85

2,2'-MBTBE, 2,2'-methylene-bis(6-*tert*-butyl-4-ethylphenol).

TABLE CS 15.1.10.   Sample Stability

Trial ID	Internal standard area ratio n-pentacosane	Internal standard area ratio 2,2'-MBTBE
Sample stability A	0.18	0.81
Sample stability B	0.17	0.80
Sample stability C	0.17	0.79
Sample stability D	0.17	0.78
Sample stability E	0.18	0.81
%RSD	2.9	1.6

2,2'-MBTBE, 2,2'-methylene-bis(6-*tert*-butyl-4-ethylphenol).

TABLE CS 15.1.11.   Quantitative Levels of Major Extractables

Trial ID	n-Doco (µg/g)	2,2'-MBTBE (µg/g)	n-Trico (µg/g)	n-Tetra (µg/g)	n-Penta (µg/g)	n-Hexa (µg/g)	n-Octa (µg/g)
1A	35	501	60	72	56	45	11
1B	35	494	56	70	52	42	11
2A	35	503	57	72	53	43	11
2B	34	496	55	68	50	40	10
3A	37	543	58	71	52	40	9
3B	36	542	57	70	51	40	9
4A	32	492	52	64	47	36	8
4B	33	491	52	63	46	37	8
5A	29	464	45	53	37	28	12
5B	30	470	46	55	39	30	13
6A	30	464	46	55	39	31	13
6B	30	466	46	56	40	31	13
AVG	33	494	53	64	47	37	11
%RSD	4.05	2.73	5.25	5.84	7.03	7.75	7.78

n-Doco, n-docosane; 2,2'-MBTBE, 2,2'-methylene-bis(6-*tert*-butyl-4-ethylphenol); n-Trico, n-tricosane; n-Tetra, n-tetracosane; n-Penta, n-pentacosane; n-Hexa, n-hexacosane; n-Octa; n-octacosane.

The developed GC/FID method was deemed suitable for use in CESs and for validation as a routine extractables testing method.

REFERENCES

1 ICH harmonised tripartite guideline: Q2A(R1), text on validation of analytical procedures. International Conference on Harmonisation of Technical Requirements for Registration of Pharmaceuticals for Human Use, 2006.
2 ICH harmonised tripartite guideline: Q2B(R1), validation of analytical procedures: Methodology. International Conference on Harmonisation of Technical Requirements for Registration of Pharmaceuticals for Human Use, 2006.

APPENDIX: CASE STUDY 15.2

ANALYSIS OF 2-MERCAPTOBENZOTHIAZOLE (2-MBT) FROM SULFUR-CURED RUBBER BY LIQUID CHROMATOGRAPHY–TANDEM MASS SPECTROMETRY (LC/MS/MS) METHOD

Tianjing Deng, Xiaochun Yu, Derek Wood, Shuang Li,
Song Klapoetke, and Xiaoya Ding

Introduction

2-MBT and other benzothiazoles, such as benzothiazole disulfide (MBTS), are common vulcanization accelerators that have been used in the manufacture of elastomeric components incorporated into certain pharmaceutical container closure systems, including those for certain orally inhaled and nasal drug products (OINDPs). 2-MBT is of particular concern since it is considered a "special case" leachable with respect to OINDP. The designation of "special case" leachable means that defined safety thresholds such as the qualification threshold (QT) and safety concern threshold (SCT) (and therefore the analytical evaluation threshold [AET], see earlier chapters) do not apply.[1–3] Therefore, for critical components of an OINDP container closure system, 2-MBT must be investigated, and potentially controlled, with specific and highly sensitive analytical methods. The molecular structures of 2-MBT and MBTS are as follows:

2-MBT

MBTS

The special safety concerns and potential leachability of 2-MBT (and other benzothiazoles such as MBTS) have resulted in the development and application of various specific and sensitive analytical methods. Literature related to the analysis of 2-MBT and other "special case" compounds and compound classes has been reviewed by Norwood and Mullis.[4] Application areas reported by these authors include rubber glove extracts, water and wastewater, urine, and an iodinated intravenous contrast medium. The methods developed for these applications include gas chromatography (GC) and GC/mass spectrometry (MS), high-performance liquid chromatography/ultraviolet (HPLC/UV) detection, HPLC with amperometric detection, capillary electrophoresis (CE) with amperometric detection, and capillary zone electrophoresis (CZE) with UV detection. Also reported[4] are applications incorporating liquid chromatography (LC)/MS and LC/MS/MS.

As discussed in the comprehensive review,[4] 2-MBT can potentially be analyzed by GC-based methods. However, as demonstrated in the qualitative controlled extraction study of the sulfur-cured elastomer test article (Chapter 15), 2-MBT is thermally labile under certain extraction conditions and potentially reactive in the heated GC inlet. A GC/MS-based method for 2-MBT determination in urine reviewed by Norwood and Mullis[4] includes chemical derivatization with pentafluorobenzyl-bromide, presumably to improve GC performance.[5] This case study describes the development and evaluation of a method using LC/MS/MS for the quantitative analysis of 2-MBT and MBTS in the sulfur-cured elastomer test article. The method uses positive ion electrospray ionization (ESI) combined with selected reaction monitoring (SRM) on a triple quadrupole mass spectrometer. The reader is referred to Chapter 13 of this book for detailed discussions of ESI, SRM, and other technical details related to LC/MS/MS. The capabilities of the LC/MS/MS method, and its accompanying extraction procedure, were assessed relative to the parameters of selectivity/specificity, repeatability, accuracy, limit of detection (LOD)/LOQ, and linearity. These results of the case study demonstrate the feasibility of developing a highly specific and sensitive analytical method, incorporating an analytical technique such as LC/MS/MS, for application in quantitative controlled extraction studies and routine extractables testing (as well as any required leachables studies) of this sulfur-cured elastomer test article for 2-MBT.

Method Summary

Analytical Method Conditions
HPLC Parameters

Mobile phase A: acetonitrile : water : formic acid (20:80:0.05)

Mobile phase B: acetonitrile : water : formic acid (90:10:0.05)

Flow rate: 0.2 mL/min

Column: Waters Symmetry C18, 3.5 μm, 2 × 50 mm

Column temperature: 40°C

Autosampler temperature: Ambient

Injector volume: 1 μL

MS

Instrument: PE Sciex API 2000/API365 Triple Quadrupole Mass Spectrometer

Ionization mode: Electrospray (positive ion)

The collision-induced dissociation (CID) transitions used for 2-MBT and MTBS both involve fragmentation of the respective protonated molecular ions ($[M + H]^+$) to their primary product ions. With reference to the CID spectra in Figures CS 15.2.1 and CS 15.2.2, these transitions are m/z 168 → 135 (2-MBT) and m/z 333 → 167 (MBTS).

Extraction Method Hansson et al.[6] studied the extraction of 2-MBT/MBTS using different solvents. These workers determined methyl-*tert*-butyl ether (MTBE) to be

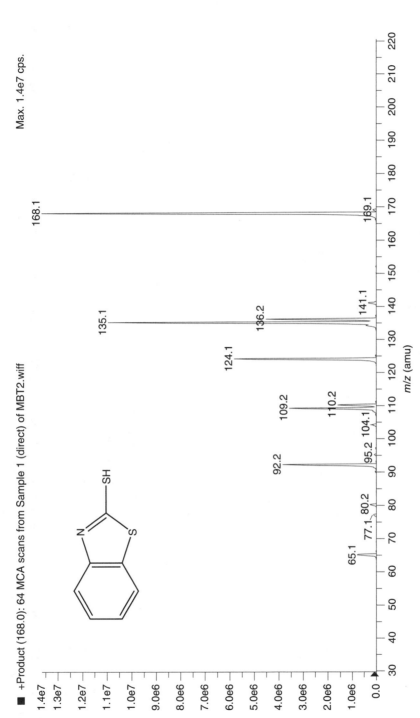

■ +Product (168.0): 64 MCA scans from Sample 1 (direct) of MBT2.wiff

Max. 1.4e7 cps.

Figure CS 15.2.1 Collision-induced dissociation (CID) product ion (MS/MS) mass spectrum of 2-mercaptobenzothiazole (2-MBT) in positive ion ESI mode. Note the $[M + H]^+$ at m/z 168.

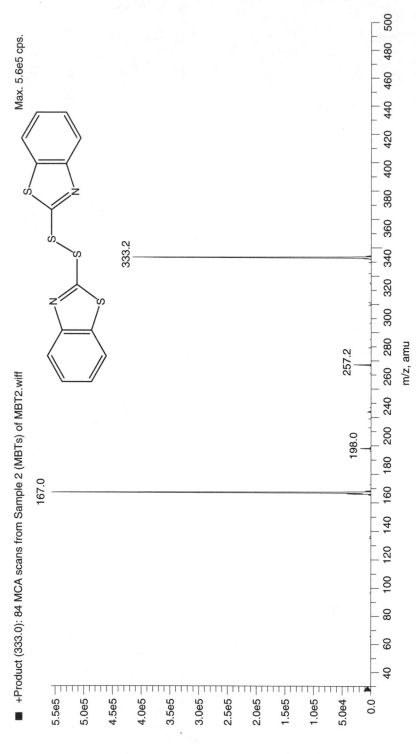

+Product (333.0): 84 MCA scans from Sample 2 (MBTs) of MBT2.wiff

Max. 5.6e5 cps.

Figure CS 15.2.2 Collision-induced dissociation (CID) product ion (MS/MS) mass spectrum of benzothiazole disulfide (MBTS) in positive ion ESI mode. Note the $[M + H]^+$ at m/z 333.

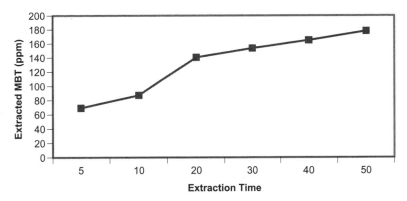

Figure CS 15.2.3   Extraction time optimization study for 2-MBT.

a good extracting solvent for MBT/MBTS due to its powerful extraction potential, low toxicity, inertness to MBT/MBTS, and high volatility. In this study, samples of elastomer test article were cut into 3 × 3 mm pieces. One gram of this cut elastomer was then extracted with 10 mL of MTBE for 30 minutes by sonication. After extraction, each resulting extract was diluted with methanol : water (50:50) and filtered using glass fiber syringe filters prior to LC/MS/MS analysis. The 30-minute extraction time was chosen based on an extraction time study shown in Figure CS 15.2.3.

Results and Discussion

Selectivity/Specificity   Selectivity/specificity was assessed by the analysis of extraction blanks and standard solutions of 2-MBT and MBTS, the results of which are shown in Figures CS 15.2.4–CS 15.2.6, respectively. Note that the extraction blank (Fig. CS 15.2.4) shows no detectable 2-MBT (upper trace) or MBTS (lower trace), while the 2-MBT standard shows a definitive peak for 2-MBT (Fig. CS 15.2.5). The MBTS standard (Fig. CS 15.2.6) shows two peaks for MBTS as well as some detectable levels of 2-MBT, suggesting that MBTS is chemically unstable in solution (degrading to 2-MBT).

Repeatability   Four replicate samples of the sulfur-cured elastomer test article were extracted and analyzed by the LC/MS/MS method. The mean concentration of 2-MBT was 56.4 ppm and the %RSD ($n = 4$) was 7.1% (see Table CS 15.2.1). An overlaid chromatogram from a representative elastomer test article extract is shown in Figure CS 15.2.7. Note that no MBTS was detected in this extract.

Accuracy: Filter Study   A filter study was conducted to verify that the syringe filter used in the sample preparation did not reduce the recovery of 2-MBT and MBTS. Three 500 ng/mL standards were analyzed before and after the filtration, and the area responses of 2-MBT and MBTS were compared. The percent differences between the filtered and nonfiltered samples were less than 2.5%, suggesting that filtration does not affect the method accuracy (see Table CS 15.2.2).

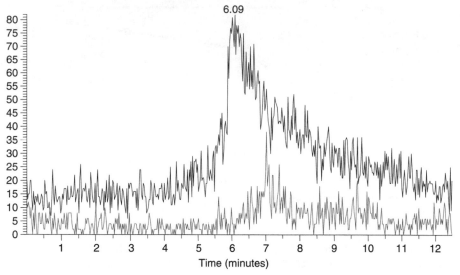

Figure CS 15.2.4   Multiple reaction monitoring (MRM) chromatograms of an extraction blank (2-MBT, upper trace; MBTS, lower trace).

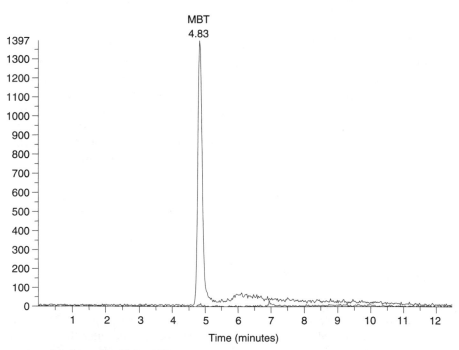

Figure CS 15.2.5   Multiple reaction monitoring (MRM) chromatograms of a 2-MBT standard (250 ng/mL; 2-MBT, upper trace; MBTS, lower trace). Note that no MBTS was detected in the 2-MBT standard.

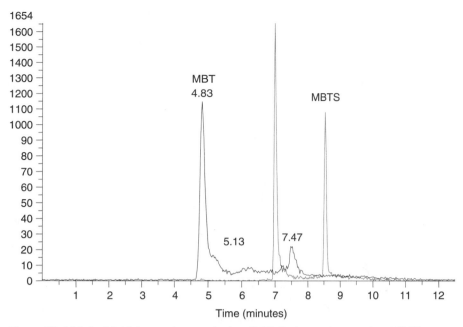

Figure CS 15.2.6   Multiple reaction monitoring (MRM) chromatogram of an MBTS standard (500 ng/mL; 2-MBT, upper trace; MBTS, lower trace). Note that 2-MBT was detected in the MBTS standard.

TABLE CS 15.2.1.   Calculated MBT Concentration (Parts per Million) in Four Replicate Extracts (Repeatability)

Calculated concentration (ppm)				Mean	%RSD ($n = 4$)
Extract 1	Extract 2	Extract 3	Extract 4		
61.2	52.7	53.4	58.0	56.4	7.1

Accuracy: 2-MBT Recovery   Approximately 360 ng/mL of 2-MBT was spiked into an extract of elastomer test article. Three replicates of spiked sample were prepared and analyzed by LC/MS/MS. The mean recovery of spiked 2-MBT was 87.3% (see Table CS 15.2.3).

Accuracy: Spiking Study of MBTS   MBTS (410 ng/mL) was spiked into an extract of elastomer test article. Three replicates of spiked sample were prepared and analyzed by LC/MS/MS. No MBTS was recovered, but ~204 ng/mL of 2-MBT was observed (see Table CS 15.2.4). It was hypothesized that MBTS was not chemically stable under the experimental conditions and underwent homolytic cleavage of the S–S bond to form one molecule of 2-MBT and a second unidentified hydrolysis product.

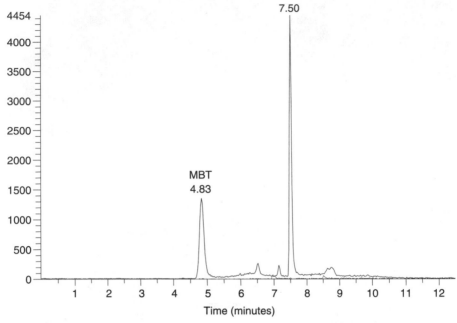

Figure CS 15.2.7   Multiple reaction monitoring (MRM) chromatograms of a representative rubber extract (2-MBT, upper trace; MBTS, lower trace). Note that no MBTS was observed in the extract.

TABLE CS 15.2.2.   Results of the Accuracy—Filter Study (Before and After Filtration)

	Mean peak area responses	
	2-MBT	MBTS
Before filtration	4082	2130
After filtration	3990	2151
% difference	2.3	1.0

TABLE CS 15.2.3.   Recovery Results for 2-MBT

	Extract (ng/mL)	Replicate 1 (ng/mL)	Replicate 2 (ng/mL)	Replicate 3 (ng/mL)
Calculated	219.5	567.8	552.6	553.2
%RSD	3.5 ($n = 7$)	2.3	2.4	3.0
%Recovery	N/A	89.9	86.0	86.1

N/A, not applicable.

TABLE CS 15.2.4.   Spiking Study of MBTS (2-MBT Concentration [ng/mL])

	Extract[a] (ng/mL)	Replicate 1 (ng/mL)	Replicate 2 (ng/mL)	Replicate 3 (ng/mL)
Calculated (ng/mL)	219.5	428.8	422.4	421.6
2-MBT recovered (ng/mL)	N/A	209.2	202.9	202.1
Average 2-MBT recovered (ng/mL)	204.7			
MBTS spiked (ng/mL)	410.7 (equivalent[b] to 206.7 ng/mL 2-MBT)			
% recovery	99.0			

[a] Endogenous level of 2-MBT in the extract.

[b] Assuming hydrolysis of MBTS to one molecule of 2-MBT along with one molecule of a second unidentified reaction product.

N/A, not available.

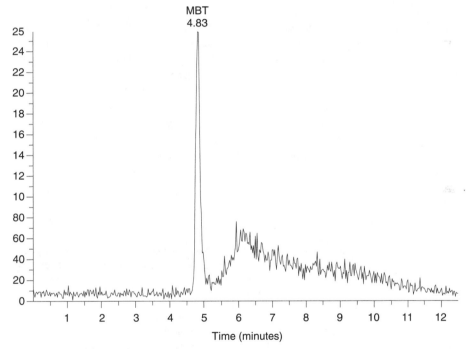

Figure CS 15.2.8   MRM chromatograms of 2-MBT standard (50 ng/mL) used for LOD estimation.

LOD   The LOD of 2-MBT was estimated from a 50 ng/mL standard (see Fig. CS 15.2.8) using an S/N ratio of 3. The LOD was estimated to be 6 ng/mL in solution or 12 pg on-column.

Linearity   Linearity was determined over the range of 50–1000 ng/mL for 2-MBT (see Fig. CS 15.2.9). Note the excellent correlation ($R^2 = 1$).

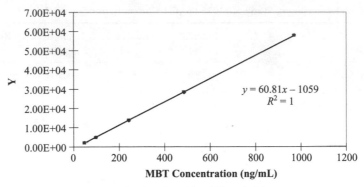

Figure CS 15.2.9   Linearity plot for 2-MBT (50–1000 ng/mL).

Conclusions

An LC/MS/MS method was developed for the quantitative analysis of 2-MBT in the MTBE extracts of a sulfur-cured elastomer test article. The method was demonstrated to be selective and specific for 2-MBT, linear over the range of 50–1000 ng/mL in elastomer extracts, and repeatable. The percent recovery of 2-MBT was determined to be 87.7% at the 400 ng/mL spiking level. The LOD for 2-MBT was estimated to be 6 ng/mL in solution or 12 pg on-column. MBTS was found to not be chemically stable under the optimized elastomer extraction and analysis conditions suggesting that that the method is not applicable to this analyte.

The developed LC/MS/MS method is deemed suitable for use in controlled extraction studies and for validation as a routine extractables testing method.

REFERENCES

1   Norwood, D.L. and Ball, D. Product Quality Research Institute: Safety thresholds and best practices for extractables and leachables in orally inhaled and nasal drug products. Submitted to the PQRI Drug Product Technical Committee, PQRI Steering Committee, and U.S. Food and Drug Administration by the PQRI Leachables and Extractables Working Group, 2006.

2   Ball, D., Blanchard, J., Jacobson-Kram, D., McClellan, R., McGovern, T., Norwood, D.L., Vogel, M., Wolff, R., and Nagao, L. Development of safety qualification thresholds and their use in orally inhaled and nasal drug product evaluation. *Toxicol Sci* 2007, *97*(2), pp. 226–236.

3   Norwood, D.L., Paskiet, D., Ruberto, M., Feinberg, T., Schroeder, A., Poochikian, G., Wang, Q., Deng, T.J., DeGrazio, F., Munos, M.K., and Nagao, L.M. Best practices for extractables and leachables in orally inhaled and nasal drug products: An overview of the PQRI recommendations. *Pharm Res* 2008, *25*(4), pp. 727–739.

4   Norwood, D.L. and Mullis, J.O. "Special case" leachables: A brief review. *Am Pharm Rev* 2009, *12*(3), pp. 78–86.

5   Manninen, A., Auriola, S., Vartainen, M., Liesivuori, J., Turunen, T., and Pasanen, M. Determination of urinary 2-mercaptobenzothiazole (2-MBT), the main metabolite of 2-(thiocyanomethylthio)benzothiazole (TCMTB) in humans and rats. *Arch Toxicol* 1996, *70*(9), p. 579.

6   Hansson, C., Bergendorfe, O., Ezzelarab, M., and Sterner, O. Extraction of mercaptobenzothiazole compounds from rubber products. *Contact Dermatitis* 1997, *36*(4), pp. 195–200.

CHAPTER 16

CASE STUDY OF A POLYPROPYLENE: EXTRACTABLES CHARACTERIZATION, QUANTITATION, AND CONTROL

Diane Paskiet, Laura Stubbs, and Alan D. Hendricker

16.1 INTRODUCTION

A polypropylene (PP) material was employed as a test article for an extractables characterization, quantitation, and control study. The project entailed three phases: profile (qualitative controlled extraction studies), method optimization, and qualification for control of extractables (routine extractables testing) in the test material.

For the first phase, the PP was extracted using three different solvents and three different extraction techniques. An analytical survey was performed in order to examine the extracts to indicate whether or not the extraction techniques and analysis methods were adequate to profile the material. The extraction techniques used were reflux, Soxhlet, and sonication. The analysis methods involved the use of gas chromatography/mass spectrometry (GC/MS), high-performance liquid chromatography with diode array (ultraviolet) detection (HPLC/UV), and liquid chromatography/mass spectrometry (LC/MS). Fourier transform infrared spectroscopy (FTIR) and optical microscopy/electron microprobe (OM/EM) were also employed on the nonvolatile residues (NVRs) remaining after extraction.

The second phase of this project was to develop an optimized quantitative method for routine control of extractables. Three target analytes, Millad 3988® (Milliken, Spartanburg, SC), Irganox® 1010 (Ciba Specialty Chemicals, Tarrytown, NY), and Ultranox® 626 (Chemtura, Philadelphia, PA), were selected to be analyzed based on the results obtained in the profile study. Extraction of the PP was performed

Leachables and Extractables Handbook: Safety Evaluation, Qualification, and Best Practices Applied to Inhalation Drug Products, First Edition. Edited by Douglas J. Ball, Daniel L. Norwood, Cheryl L.M. Stults, Lee M. Nagao.

using reflux, and separation and detection by HPLC/UV. The method extraction time, solvent, and LC conditions were optimized for the targets selected.

The final phase of the project involved qualifying the optimized method for extraction and measurement of the target analytes. The method was qualified based on the following parameters: specificity, system suitability, linearity, range, precision, accuracy, limit of quantitation (LOQ), and stability.

This chapter presents the results for the extractables characterization of the PP via various means, optimization of extraction conditions, selection of an analytical methodology for control of extractables, and assessment to determine suitability of the method for meeting validation criteria.

This work was performed as part of the Product Quality Research Institute (PQRI) Leachables and Extractables Working Group's development of best practices for orally inhaled and nasal drug products (OINDPs).[1] A threshold concept for extraction studies was ultimately determined by the Working Group, and this process is outlined following presentation of this case study.

16.2   CONTROLLED EXTRACTION STUDY

Plaques of a PP base resin material were obtained specifically to demonstrate the conduct of a controlled extraction study on this material. Additives in polymers are necessary to prevent degradation as well as to broaden the application range by enhancing functionality, performance, and appearance. Additives are also used to facilitate compounding, processing, and assembly. In this case study, the resin plaques were considered as a surrogate for a final container used in nasal sprays, to illustrate extractables characterization and control methodologies.

PP is a linear hydrocarbon with a methyl group on every alternate carbon of the hydrocarbon chain. The structure of a PP is illustrated in Figure 16.1.

The PP resin was supplied by Phillips Sumika Polypropylene Company (Pasadena, TX). The resin was molded into 2.5-in.-diameter $\times$ 0.040-in.-thick disks by Owens-Illinois Closure, Inc. (Maumee, OH). Identity and amount of additives in the material formulation were known to the laboratories and targeted during the testing. Names and structures of the additives are shown below in Figure 16.2. Ultranox 626 and Irganox 1010 serve as antioxidants, Pationic® 901 (Patco Additives, Kansas City, MO) is an antistatic and mold release agent, Millad 3988 is a clarifier, and calcium stearate is a stabilizer.

16.2.1   Extraction Considerations

Controlled extraction studies provide a detailed understanding of the materials tested in order to determine potential leachables. The materials are subjected to "worst-case" scenarios to determine potential leachables levels by using various

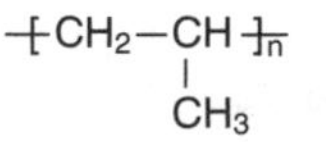

$$+CH_2-CH+_n$$
$$|$$
$$CH_3$$

Figure 16.1   Structure of a polypropylene (fragment of polypropylene chain).

Ultranox 626
Bis(2,4-di-*t*-butylphenyl)pentaerythritol diphosphite
CAS #26741-53-7
0.05% in formulation

Pationic 901
Glycerol Monostearate
CAS #31566-31-1
0.3% in formulation

Millad 3988
1,3:2,4-Bis(3,4-dimethylbenzylidene)sorbitol
CAS #135861-56-2
0.2% in formulation

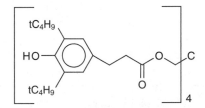

Irganox 1010
Pentaerythritol tetrakis(3,5-di-*tert*-butyl-4-hydroxyhydrocinnamate)
CAS #6683-19-8
0.08% in formulation

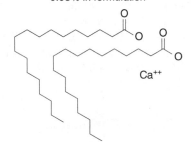

Calcium Stearate
Stearic Acid Calcium Salt
CAS #1592-23-0
0.03% in formulation

Figure 16.2   Structures of various known polypropylene additives.

solvents and harsh extraction conditions meant to far exceed standard scenarios that the materials will experience as part of the drug product. In this case, the PP disks were aggressively extracted by reflux, sonication, and Soxhlet techniques to explore the impact of these different extraction techniques. For each technique, individual samples were extracted with 2-propanol (IPA), methylene chloride, and hexane.

16.2.2 Sample Preparation for Extraction Profiles

Approximately 2 g of the PP plaques were hot-pressed into a thin film to obtain a total surface area of 100 cm^2 for each extraction. The thin film was then cut into strips to enable insertion into the extraction vessels. The sample-to-solvent ratio was based on the known amount of additives in the plaques to ensure the extractables concentrations were suitable for the analytical methods of choice. For reflux extraction, 50 mL of solvent was pipetted into a 125-mL Erlenmeyer flask containing the sample strips. The flasks and contents were weighed and connected to a water-cooled condenser and placed on a hot plate. The flasks were heated to boiling and refluxed for 2 hours. After cooling, extracts were decanted. For sonication extraction, 40 mL of each solvent was pipetted into a 125-mL Erlenmeyer flask containing the sample strips. The flasks and contents were weighed and connected to a water-cooled condenser and placed into a Bronson 2510 (Bronson Ultrasonics, Division of Emerson, Danbury, CT) sonicator bath for 30 minutes. After cooling, extracts were decanted. For Soxhlet extraction, the sample strips were placed into a pre-extracted 25 × 80 mm Whatman cellulose thimble. Pre-extraction of the thimble was performed for 2 hours using the appropriate solvent and then discarding the liquid. The sample in the thimble was placed in the apparatus, and 200 mL of solvent was added to the flask. The flasks were connected to a water-cooled Soxhlet apparatus and placed on a hot plate. The flasks were heated to boiling for 8 hours, turned off overnight, and then heated to boiling the next day for 8 hours for a total of 16 hours. The extract liquid was decanted.

16.2.3 Analysis of Extractables

Liquid extracts were analyzed for organic extractables using chromatographic techniques HPLC/UV, LC/MS, and GC/MS. NVRs were obtained by evaporating known volumes of the extracts to dryness. Portions of the NVR samples were evaluated using the FTIR for general characterization and evaluated for inorganic compounds using OM/EM techniques. The known additives were identified and measured along with evaluation of any unspecified extractable.

16.2.4 System Suitability for Chromatographic Techniques

During the material characterization phase of work, system suitability standards were prepared, which consisted of typical extractables found in a variety of types of materials. The purpose of these standards was to demonstrate gas and liquid chromatographic characteristics as well as sensitivity for the profiling methods

TABLE 16.1. LC System Suitability Standards

Compound	Target concentration (μg/mL)	Amount injected on-column (HPLC method) (ng)
Pyrene	1.0	10
2-Mercaptobenzothiazole (2-MBT)	50	500
Tetramethylthiuram disulfide (TMTDS)	50	500
Butylated hydroxytoluene (BHT)	50	500
Bis(2-ethylhexyl)phthalate (DEHP)	50	500
Diphenyl amine	50	500
Diisononyl phthalate	50	500
Irganox 1010	50	500

TABLE 16.2. GC System Suitability Standards

Compound	Target concentration (μg/mL)	Amount injected on-column (GC method) (ng)
Pyrene	1.0	1
2-Ethylhexanol	50	50
2-Mercaptobenzothiazole (2-MBT)	50	50
Tetramethylthiuram disulfide (TMTDS)	50	50
Butylated hydroxytoluene (BHT)	50	50
Bis(2-ethylhexyl)phthalate (DEHP)	50	50
Stearic acid	100	100
Diisononyl or diisodecyl phthalate	50	50

employed. The compounds selected as standards are listed in Tables 16.1 and 16.2, along with the concentrations in solution and amount injected. Peaks should be readily detectable to demonstrate appropriate instrument sensitivity. Additional acceptance criteria with regard to reproducibility may also add value, but were not investigated in detail for the semiquantitative materials characterization portion of work.

No formal specifications were placed on the system suitability standards, since the profiling methods are general methods not validated for any specific test article. However, they represent a good measure of chromatographic performance and were valuable in terms of understanding the methods' ability to separate and detect typical extractables compounds, including ones that are considered "special case compounds" by the PQRI recommendations (e.g., polynuclear aromatics [PNAs] *N*-nitrosamines, 2-mercaptobenzothiazole) and require lower thresholds of toxicological concern.

16.2.5 Results from Extractables Profiling Analysis by HPLC/UV

16.2.5.1 Instrumental Conditions HPLC/UV Instrument: Agilent 1100 (Agilent Technologies, Santa Clara, CA) with diode array detector (DAD)

Eluent A: Acetonitrile

Eluent B: Water

Gradient (linear)

Time (minutes)	% A	% B
0	50	50
11.00	100	0
19.00	100	0

Post time: 5 minutes at 50% A : 50% B

Flow rate: 1.0 mL/min

Injection volume: 10 µL

Column temperature: 60°C

Column: Vydac (Grace Davison Discovery Sciences, Deerfield, IL), C18, 150 mm × 4.6 mm P/N #201TP5415

Detector: DAD

Signals: 200 nm, bandwidth 4 nm; 220 nm, bandwidth 4 nm

Reference signal: 550 nm, bandwidth 100 nm

16.2.5.2 Reference Standard Preparation and Extract Analysis (for Retention Time and UV Spectral Confirmation) A solution of additive reference standards was prepared to identify and measure the known additives in the PP formulation. Calcium stearate was excluded due to limited solubility and detectability. Irganox 1010, Ultranox 626, and Pationic 901 were prepared in IPA. Millad 3988 was not completely soluble in IPA and was prepared separately in 50:50 IPA : tetrahydrofuran (THF). The reference standard compounds and concentrations are listed in Table 16.3. Responses of the sample extract peaks were compared with those observed in the reference standards to provide semiquantitative results. The Pationic 901 responded poorly due to lack of UV activity and was not targeted using this technique for a quantitative result. Reference standard chromatograms are shown in Figures 16.3 and 16.4. The Ultranox 626 reference material standard produced four peaks under the HPLC conditions used in this study. Only one of the four peaks (peak B) was quantitated in the extract chromatograms. This peak is identified as 2,4-di-*tert*-butylphenol, an Ultranox 626 fragment.

The chromatograms were evaluated at 200 and 220 nm. The 200-nm signal was initially evaluated for detection and quantitation of all analyte peaks. However, it was noted that better selectivity of Millad 3988 occurred using 220 nm; thus, this wavelength was chosen for Millad 3988 quantitation in subsequent activities (using the DAD). Representative chromatograms for sample extracts are shown in Figures

TABLE 16.3.   Reference Standards of Known Additives and Concentrations Used for Confirmation and Quantitation

Compound	Reference standard concentration (µg/mL)
Millad 3988	20
Irganox 1010	17
Ultranox 626	21
Pationic 901	503

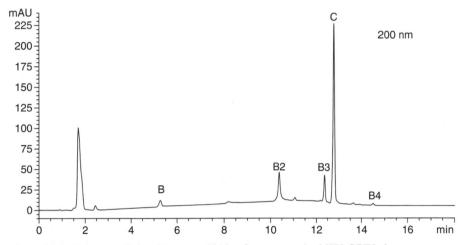

Figure 16.3   Ultranox 626 and Irganox 1010 reference standard HPLC/UV chromatogram. Note: B2, B3, and B4—Ultranox 626 related peaks 2, 3, and 4; these peaks are degradants of the parent compound and only observed in reference material standard.

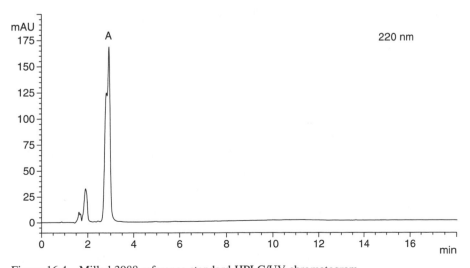

Figure 16.4   Millad 3988 reference standard HPLC/UV chromatogram.

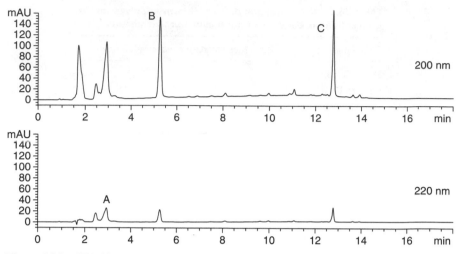

Figure 16.5   HPLC/UV chromatogram 2-propanol reflux extraction.

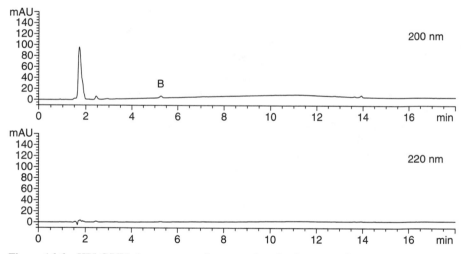

Figure 16.6   HPLC/UV chromatogram 2-propanol sonication extraction.

16.5–16.9. The chromatograms of each extraction technique using IPA are shown in Figures 16.5–16.7. Representative chromatograms of methylene chloride and hexane extracts are shown in Figures 16.8 and 16.9, respectively. Peaks are identified as follows:

A—Millad 3988,

B—Ultranox 626 related (2,4-di-*tert*-butylphenol),

C—Irganox 1010.

Peaks observed with peak height above 5 mAU were evaluated. Peaks were confidently identified by retention time and UV spectral match to peaks in

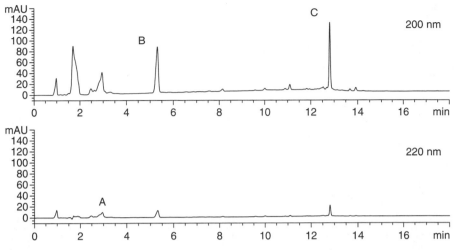

Figure 16.7   HPLC/UV chromatogram 2-propanol Soxhlet extraction.

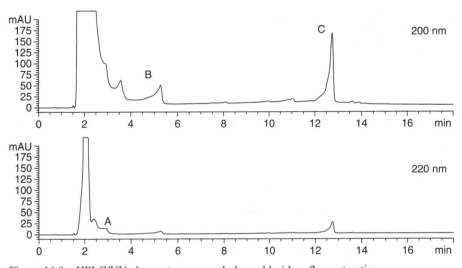

Figure 16.8   HPLC/UV chromatogram methylene chloride reflux extraction.

the reference standards. The peaks are listed in Tables 16.4–16.6 with respect to extraction solvent (not extraction technique). The unidentified peaks were of minor height (<20 mAU) with absorbance maxima less than 225 nm. These peaks may be related to PP oligomers (low-molecular-weight polymer). LC/MS and GC/MS were employed to confirm and further characterize the extracts. The semiquantitative results are shown in Table 16.7. All results were based on the response of the reference compound.

16.2.5.3   *Comparison of Extraction Results*   The reflux extraction using IPA solvent produced the most consistent extraction for all targeted analytes. Reflux in

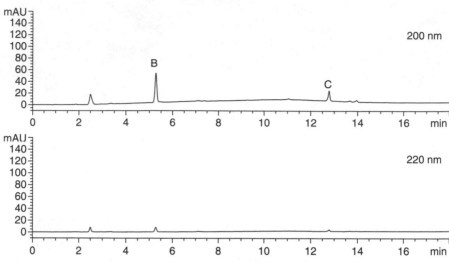

Figure 16.9   HPLC/UV chromatogram hexane sonication extraction.

TABLE 16.4.   HPLC/UV Peak Identification Summary: Isopropanol Extractions

Retention time (minutes)	Identification
1.7	2-Propanol
2.7	Related to Millad
2.9 and 3.2	Millad 3988
5.6	Di-*tert*-butylphenol
8.4	Pationic 901
10.3	Unknown
11.3	Related to Irganox 1010
12.9	Irganox 1010

TABLE 16.5.   HPLC/UV Peak Identification Summary: Methylene Chloride Extractions

Retention time (minutes)	Identification
1.8–3.5	Methylene chloride
2.9 and 3.9	Millad 3988
5.6	Di-*tert*-butylphenol
12.9	Irganox 1010

methylene chloride extracted a greater amount of analytes overall; however, the solvent absorbs at <250 nm causing interferences and poor peak shape when a detection wavelength of 200 nm is employed, as in this study. Hexane was not effective in extracting Millad 3988 by any of the techniques. Sonication generally achieved poor extraction of all analytes. Soxhlet extraction produced favorable results in some

TABLE 16.6. HPLC/UV Peak Identification Summary: Hexane Extractions

Retention time (minutes)	Identification
2.7	Also in hexane blank
3.6	Unknown
5.6	Di-*tert*-butylphenol
8.4	Pationic 901
10.3	Unidentified
11.3	Related to Irganox 1010
12.9	Irganox 1010

TABLE 16.7. Semiquantitative Levels of Targets Detected Based on Extraction Solvent and Condition

	Solvents	μg/g		
		Millad 3988	Ultranox 626	Irganox 1010
Reflux	2-Propanol	350	380	250
	Methylene chloride	450	230	550
	Hexane	ND	390	590
Sonication	2-Propanol	ND	12	ND
	Methylene chloride	84	250	110
	Hexane	ND	140	21
Soxhlet	2-Propanol	140	200	140
	Methylene chloride	660	57	760
	Hexane	20	76	730

ND, not detected.

cases, but not for all three targets in methylene chloride or hexane. Favorable extraction of all three target analytes was achieved in IPA using Soxhlet, but reflux extracted greater amounts. These results indicate that IPA reflux is a favorable extraction technique to consider for material control of the PP if HPLC is the analytical method of choice. For this portion of work, attempts were not made to correlate amounts extracted to the amounts present in the formulation.

16.2.6 Results from Extractables Profiling Analysis by HPLC/MS

The methylene chloride reflux extract sample was also analyzed using HPLC/MS in order to attempt to confirm identity of peaks detected in the extract and test for species without chromophores. For this portion of work, a screening method was employed designed to elute all species using a strong organic gradient. Attempts were not made to obtain similar chromatography to the HPLC/UV portion of analytical work.

16.2.6.1   *Instrumental Conditions for HPLC/MS*   HPLC/MS conditions used are presented below:

HPLC instrument: Agilent 1100

Eluent A: 75:25 acetonitrile:water

Eluent B: 50:50 acetonitrile:THF

Gradient (linear)

Time (minutes)	% A	% B
0	100	0
10	60	40
20	0	100
30	0	100
32	100	0
45	100	0

Post time: None

Flow rate: 1.0 mL/min

Injection volume: 50 μL

Column temperature: Ambient

Column: Alltech Alltima C18 (Alltech Associates, Deerfield, IL), 5-μm particle size, 250 mm length × 4.6 mm inner diameter

UV detector: 280 nm

Mass spectrometer: Micromass Platform II (Micromass, Beverly, MA)

Ionization mode: Positive and negative ion atmospheric pressure chemical ionization (APCI)

Scan mode: Scanning; m/z 50–1350

Scan rate: 1.3 s/scan

A summary of the peaks detected (combining all solvents and extraction techniques as well as both ionization methods) is presented in Table 16.8. Identification of species present was accomplished by the use of authentic reference standards, correlation with GC/MS data, and manual interpretation of obtained spectra. In addition to targets and related compounds, LC/MS confirmed the presence of fatty acids, which could not be observed in the HPLC/UV analysis, since no chromophore is present on these species. These species may have been derived from the calcium stearate additive. Not all peaks were observed in all detection schemes (UV, positive ion MS, and negative ion MS). Individual peaks detected are labeled in each figure with numbers corresponding to assignments in Table 16.8. HPLC/UV chromatograms and total ion chromatograms (TICs) are shown in Figures 16.10 and 16.11 for the methylene chloride reflux extract analysis, which produced the most peaks.

TABLE 16.8.   Peaks Detected and Proposed Identifications Using LC/MS Analysis

Peak number	Approximate retention time (minutes)	Identification
1	4.9	Bis(dimethylbenzylidene)sorbitol isomer (Millad 3988 related)
2	5.3	Bis(dimethylbenzylidene)sorbitol isomer (Millad 3988 related)
3	8.6	Unknown
4	10.6	Di-*tert*-butylphenol (Ultranox 626 fragment)
5	13.6	Unknown
6	15.6	Tetradecanoic acid
7	16.0	Hexadecanoic acid
8	18.4	Glycerol monopalmitate/glycerol monostearate (Pationic 901)
9	19.0	Irganox 1010 fragment
10	19.4	Irganox 1010 related
11	20.3	Octadecanoic acid
12	21.0	Irganox 1010
13	22.3	Tris(2,4-di-*t*-butylphenyl)phosphate (Ultranox 626 fragment)

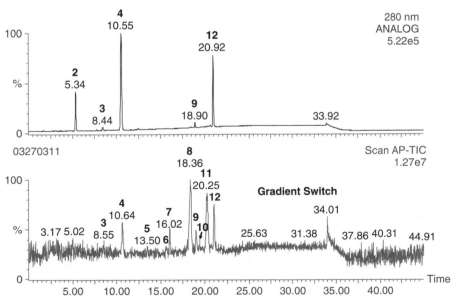

Figure 16.10   HPLC/UV chromatogram and negative ion APCI total ion chromatogram. Methylene chloride reflux extract analysis.

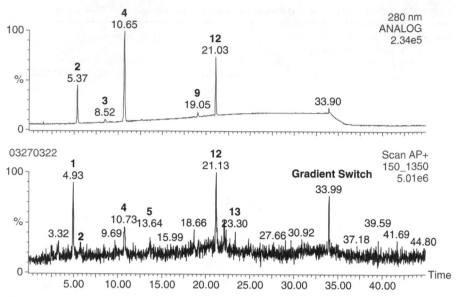

Figure 16.11   HPLC/UV chromatogram and positive ion APCI total ion chromatogram. Methylene chloride reflux extract analysis.

16.2.7   Results from Extractables Semivolatile Profiling Analysis by GC/MS

Extracts were prepared as described in Section 16.2.2. The same extracts used for HPLC/UV and HPLC/MS were used for the GC/MS analysis.

Profiling GC/MS conditions are defined below:

Flow rate: 2.69 mL/min helium, constant flow

Injection volume: 1.0 μL

Column temperature: 40°C, hold 1 minute, heat 10°C/min to 300°C, hold 10 minutes

Column: Agilent HP-1, 30 m × 0.32 mm ID, 0.25-μm film

Solvent delay: 3 minutes

Injector temperature: 280°C

Detector: Agilent Mass Spectrometer (MSD), electron ionization (EI)

Detector temperature: 280°C

Run time: 37 minutes

A material reference solution of known additives was prepared in IPA to be used as a semiquantitative reference. The reference standard components and concentrations are listed below. Irganox 1010 and calcium stearate were excluded from this reference standard because they are not suitable for analysis using this technique. Responses of the sample extract peaks were compared with those observed in the reference standard to provide semiquantitative results.

Compound	Reference concentration (µg/mL)
Pationic 901	203
Ultranox 626	207
Millad 3988	207

Representative chromatograms for the methylene chloride extraction analyses are shown in Figures 16.12–16.14. Chromatograms of the IPA and hexane extracts appeared very similar to the methylene chloride chromatograms and are not shown. The TICs were evaluated for the presence of the target analytes in the reference

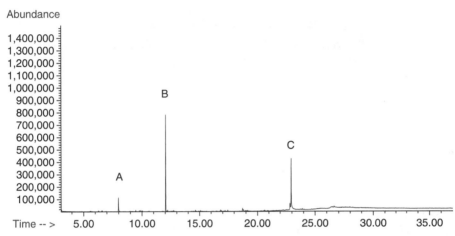

Figure 16.12 GC/MS chromatogram for the methylene chloride reflux extract analysis.

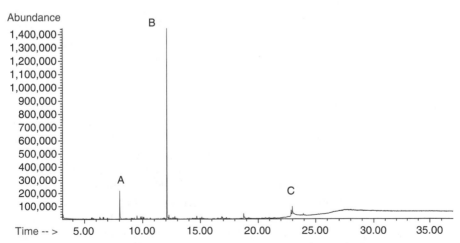

Figure 16.13 GC/MS chromatogram for the methylene chloride sonication extract analysis.

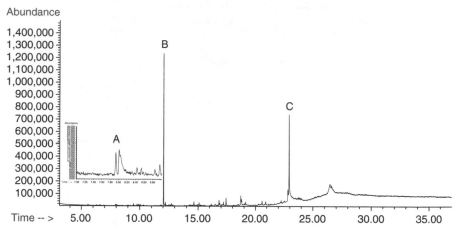

Figure 16.14 GC/MS chromatogram for the methylene chloride Soxhlet extract analysis.

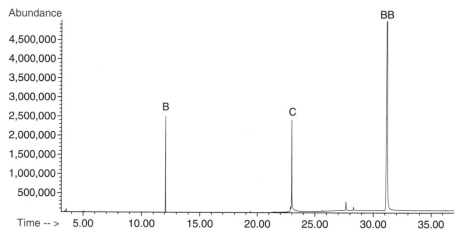

Figure 16.15 GC/MS chromatogram for the target reference material standard analysis. BB = Ultranox 626, intact molecule.

standard as well as other unknown peaks. Minor peaks observed were indicative of PP homologs. Peaks were identified by retention time and mass spectral match to peaks in the reference standard. Two of the additives were not detected in the sample extracts as intact molecules. The peak at approximate retention time of 12 minutes was identified as 2,4-di-*tert*-butylphenol, a breakdown product of Ultranox 626. Additionally, the intact molecule was observed in the reference material at an approximate retention time of 31.2 minutes under the GC/MS conditions used in this study. The relevant mass spectra are shown in Figures 16.15 and 16.16. Only the breakdown product, 2,4-di-*tert*-butylphenol, was observed in the extract chromatograms. The peak identified at approximate retention time of 8 minutes was

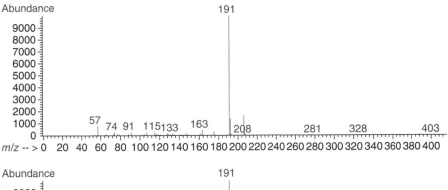

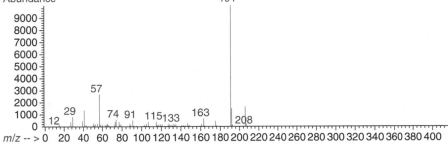

Figure 16.16   EI mass spectrum of peak B from Figure 16.15, attributed to 2,4-di-*tert*-butylphenol (Ultranox 626 related). Top spectrum—Ultranox 626 reference standard 12.0 minute peak. Bottom spectrum—library reference spectrum of 2,4-di-*tert*-butylphenol.

identified as 3,4-di-methylbenzaldehyde, a breakdown product of Millad 3988. Additionally, the intact molecule was observed in the reference material at approximate retention time of 27 minutes. The relevant mass spectra are shown in Figures 16.17 and 16.18. Only the breakdown product of Millad 3988 was present in the extract chromatograms. The following peaks are labeled in the figures:

A—3,4-dimethylbenzaldehyde (degradation product of Millad 3988),

B—2,4-di-*tert*-butylphenol (degradation product of Ultranox 626),

C—Pationic 901.

The peak identifications along with the semiquantitative results are listed in Table 16.9 with respect to extraction solvent and extraction technique. For each extract, the additive or degradant was quantified based on comparison of the response of the respective peak in the extract chromatogram to the response of that same peak in the reference chromatogram.

Methylene chloride appears to be the most favorable solvent for GC analysis, yielding the greatest amount of target analytes of all the tested extraction techniques. The other solvents showed inconsistencies since some analytes were not detected in the extracts. Again, sonication produced poor extraction overall. Reflux and Soxhlet could be viable choices for a method development study. HPLC was selected as the

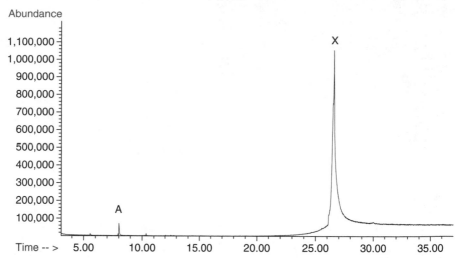

Figure 16.17   GC/MS chromatogram for the target reference material Millad 3988 analysis. X = Peak attributed to Millad 3988: 1,3:2,4-bis(3,4-dimethylbenzylidene)sorbitol.

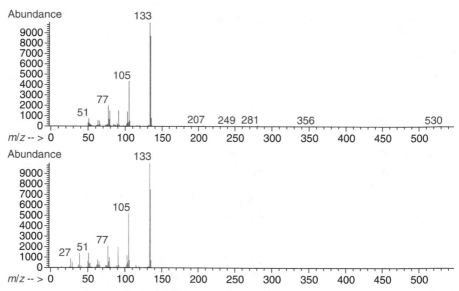

Figure 16.18   EI mass spectrum of peak A, from Figure 16.17, and reference library match, attributed to 3,4-dimethylbenzaldehyde (Millad 3988 fragment). Top spectrum—Millad 3988 reference standard 8.0-minute peak. Bottom spectrum—library reference spectrum of 3,4-dimethylbenzaldehyde.

TABLE 16.9. GC/MS Peak Identification and Semiquantitative Results

Extraction type	Solvents	3,4-Dimethyl benzaldehyde (fragment of Millad 3988)		2,4-Di-*tert*-butylphenol (fragment of Ultranox 626)		Pationic 901	
		RT (minutes)	µg/g	RT (minutes)	µg/g	RT (minutes)	µg/g
Reflux	2-Propanol	8.02	170	12.04	3500	22.95	1100
	Methylene chloride	7.99	220	12.04	3800	22.94	990
	Hexane	7.98	190	12.04	3900	22.94	50
Sonication	2-Propanol	ND	ND	12.06	30	ND	ND
	Methylene chloride	8.00	240	12.06	220	22.97	150
	Hexane	8.00	42	12.06	500	ND	ND
Soxhlet	2-Propanol	8.04	31	12.05	1500	22.94	160
	Methylene chloride	8.02	640	12.05	1700	22.95	1600
	Hexane	ND	ND	12.05	380	ND	ND

RT, retention time; ND, not detected, peak not observed.

analytical technique as the material control method in this case study. However, in a real-world situation, GC/MS may also be considered, especially for Pationic 901, which is not particularly suited for HPLC/UV detection.

16.2.8 NVR, FTIR, and Microscopy

The NVR results showed significant variation in the amount of residue obtained from each extraction technique and solvent. An extrapolated residue of less than 0.1 µg/g of PP was found for the IPA sonication, while 50,000 µg/g was found for the hexane Soxhlet extraction. The NVR was ultimately used to gain additional profile information by FTIR and microscopy. A small portion of each NVR was cast onto a KRS-5 crystal with chloroform and an infrared spectrum obtained using attenuated total reflectance. One hundred scans were collected in the 4000–400 cm^{-1} region. The major component of the NVR was interpreted based on the functional groups observed. The FTIR data indicated the residue from all extracts was a mixture of Pationic 901 additive and PP. Based on the differences in relative size of the bands at 1720 cm^{-1} (attributed to Pationic 901) and 720 cm^{-1} (attributed to PP), the IPA and methylene chloride extracts appear to contain more Pationic 901, and the hexane extracts appear to contain more PP. A spectrum representative of the IPA extract sample is shown in Figure 16.19. The methylene chloride extract sample FTIR spectrum, although not shown, was similar. A spectrum representative of the hexane extracts is shown in Figure 16.20.

A small portion of each NVR was examined using a stereo-binocular microscope for general appearance. A small portion of each NVR was mounted for qualitative analysis by electron microprobe for elemental composition. An energy dispersive

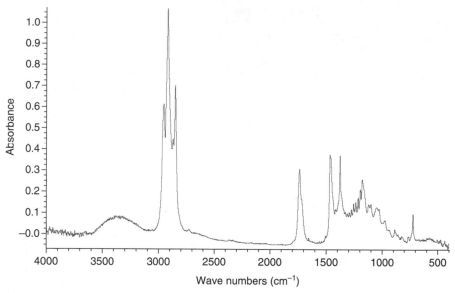

Figure 16.19 FTIR spectrum of NVR from 2-propanol Soxhlet extraction.

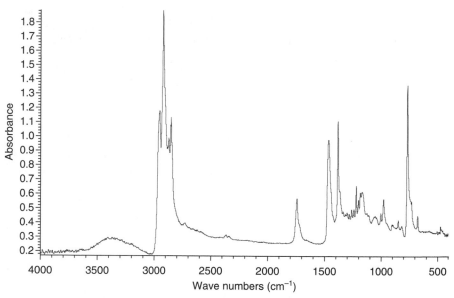

Figure 16.20 FTIR spectrum of NVR from hexane reflux extraction.

X-ray (EDX) detector was used to survey elements from atomic number 11 though 92, and a wavelength dispersive X-ray (WDX) detection system was used to resolve elemental overlaps and to detect carbon. The detection of calcium was specifically investigated with this technique since calcium stearate could not be detected in the HPLC or GC/MS techniques. In addition to the NVR samples, a charred IPA extract

NVR and a charred piece of the PP material were analyzed by electron microprobe.

The NVR samplings of all extracts were evaluated for general appearance. All had the appearance of a translucent gel. The hexane appeared to add a stiff texture to the gels (the stiffer texture is likely a result of the hexane extracting more of the polymer than the other solvents). Carbon was the major element detected by electron microprobe in all extracts. The elemental analysis detected the presence of calcium in the IPA sonication extract residue and in the charred PP material.

16.3 DEVELOPMENT OF ANALYTICAL METHOD FOR ROUTINE CONTROL OF EXTRACTABLES

Based on the above results using a variety of detection methods, the HPLC/UV method was chosen as the method for detection and quantitation of the extractables from the PP material. Irganox 1010 was not suitable for analysis via GC/MS, precluding use of that technique to fully control extractables in the material. Calcium, from calcium stearate, was not observed directly via the HPLC/UV or GC/MS analyses since it is ionic in nature. Stearic acid was observed via GC/MS, and also by negative ion LC/MS. The stearic acid is possibly originating from the calcium stearate, typically used as a mold release agent, and is a low-level additive. Alternatively, it could also be originating from the Pationic 901 (glycerol monostearate) additive. Since the stearic acid was not a direct additive, it was potentially created from the harsh extraction process and was deemed less toxicologically significant, it was not chosen as a target. Millad 3988, Ultranox 626, and Irganox 1010 are known additives and were selected as targets to be monitored by HPLC/UV for quantitative control. A second method to monitor the Pationic 901 would be needed for comprehensive control of a component profile. The following sections outline the process for the optimization of the extraction and HPLC method as well as challenges to the method to ensure successful validation.

16.3.1   Optimization of HPLC Conditions

The chromatography conditions used in the profile analysis were optimized for better separation of Ultranox 626 and Irganox 1010 peaks. The conditions were also made suitable for analysis of a 50:50 IPA:THF solvent, which would be needed for improved solubility of Millad 3988.

To address the separation issues, a 250 mm × 3.2 mm Columbus C18 column (Phenomenex, Torrance, CA) was selected to improve resolution and sensitivity. The organic portion of the initial mobile phase was reduced to 30%, which produced a single Millad 3988 peak instead of a split peak as observed in the profile. The gradient to 100% acetonitrile was extended to 12 minutes, and the total run time was extended from 19 to 25 minutes due to use of the longer column.

The column and mobile phase composition optimization produced better chromatography overall of a standard mix of the target analytes. The new conditions also adequately detected and resolved all target peaks at 200 nm, which eliminated the

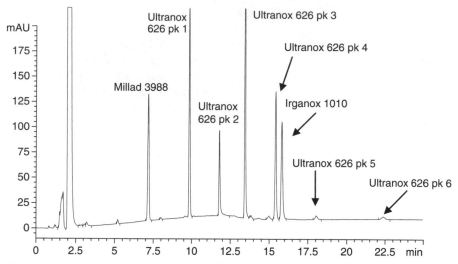

Figure 16.21 Representative working standard HPLC/UV chromatogram. Six peaks were observed, which were attributed to Ultranox 626 degradants.

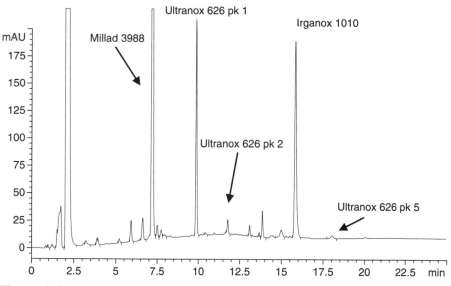

Figure 16.22 Representative polypropylene sample extract HPLC/UV chromatogram. Three peaks related to Ultranox degradants were observed.

need for collection of data at 220 nm. A sample and standard chromatogram under optimized conditions are shown in Figures 16.21 and 16.22.

16.3.2 Optimization of Extraction Method

Reflux extraction was selected as the technique to optimize for the PP routine extraction method, based on results of the materials characterization study. Reflux extraction demonstrated higher levels of extractables for the analytes of interest and provided a more complete extraction in the times assessed. It was also the most rapid in terms of extractables recovery from the materials.

The Millad 3988 target showed limited solubility when prepared as a standard. In this case, an alternative extraction solvent was chosen for optimization, which would fully solubilize the targets. A 50:50 IPA:THF solvent was employed as the extraction solvent upon which optimization experiments were performed. This allowed a thorough extraction as well as complete solubilization of the Millad 3988 target. It also provided a means for direct analysis of extract liquid without solvent exchange, which would have been required for hexane or methylene chloride extract liquids, since those extraction solvents are not compatible with reversed-phase mobile phases.

16.3.3   Optimization of Extraction Time

One of the most critical factors involved in extraction method optimization is the determination of an appropriate extraction time. Sufficient time is needed to reach equilibrium and obtain reliable results. Insufficient time will provide an incomplete extraction and may contribute to variability in observed extractables levels. Excessive time may cause sample or target degradation and decrease method precision and recovery, as well as add unnecessary time during release testing of materials. For this material, two extraction time studies were carried out to determine the optimal length for extraction by reflux. The first extraction study consisted of six separate extractions of 50 cm^2 total surface area PP weighing approximately 1 g in 25 mL of 50:50 IPA:THF at six different time intervals. Extracts from reflux times of 1, 2, 4, 6, 8, and 10 hours were analyzed under the optimized HPLC conditions. The amounts found for each target were calculated at each time interval and are shown in Figure 16.23.

Based on these results, asymptotic levels of the extractables occurred rather quickly. To further refine the optimal extraction time and verify these results, a

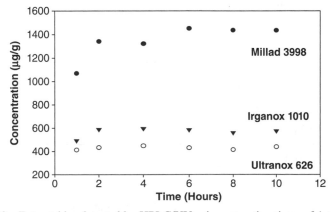

Figure 16.23   Extractables detected by HPLC/UV using extraction times of 1 to 10 hours.

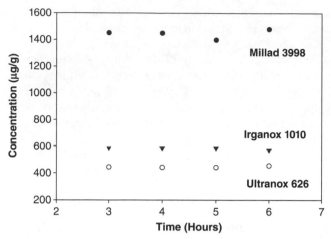

Figure 16.24   Extractables detected by HPLC/UV using extraction times of 3 to 6 hours.

second extraction study was performed consisting of four separate extractions over time intervals of 3, 4, 5, and 6 hours. The results are shown in Figure 16.24.

These results confirmed the time needed to achieve extraction of a maximum concentration of targets under the given conditions, and were consistent with the first extraction time study. A 3-hour extraction time appears sufficient and was chosen as a conservative time for which complete extraction was achieved.

16.3.4   Method Optimization Summary

The developed and optimized method employed a 3-hour reflux extraction of 50 cm^2 (approximately 1 g) of PP material in 25 mL of 50:50 IPA:THF using a modified HPLC/UV assay with detection at 200 nm. This method was subsequently challenged as part of the validation process against traditional acceptance criteria.

16.3.4.1   Final Optimized HPLC/UV Conditions
Eluent A: acetonitrile

Eluent B: water

Gradient (linear)

Time (minutes)	% A	% B
0	30	70
12.00	100	0
25.00	100	0

Post gradient time: 5 minutes

Flow rate: 1.0 mL/min

Injection volume: 10 µL

Column temperature: 60°C

Column: Phenomenex Columbus, 5 μ, C18, 250 mm × 3.2 mm

Detector: DAD

Signal: 200 nm, bandwidth 4 nm

Reference signal: 550 nm, bandwidth 100 nm

16.4   METHOD VALIDATION FOR EXTRACTABLES CONTROL IN PP MATERIAL

A method was developed for control of target extractables in PP plaques and the method was tested with respect to typical validation parameters to ensure suitability. The method was tested for specificity, system suitability, linearity range, precision, accuracy, LOQ, and stability.

Specificity was determined using diode array detection to provide spectral information for each analyte. Individual standards were prepared for the target extractables and analyzed via the optimized method to ensure adequate separation. A sample was also prepared via the optimized extraction method and analyzed contemporaneously to ensure that peaks identified had identical diode array spectra to standards, and that no other peaks were observed, which co-eluted with the targets.

System suitability was established for the target extractables. Standards were injected six times using the optimized method and evaluated for repeatability, tailing, capacity factor, resolution, and sensitivity. Method linearity and range was established by preparation of standards of each target at various levels spanning the ranges observed in the extraction studies and relative to the known amounts in the PP.

Method precision was established by analysis of six replicate extractions of PP material, followed by analysis with the analytical method. The repeatability was evaluated based on % relative standard deviation (RSD) for each target for the replicate analyses. Intermediate precision was established by a second analyst using a different instrument and column and analyzing six replicate extract preparations of PP. The results were compared between two analysts for percent difference for each target. The second analyst's results were also measured versus the proposed repeatability criteria for the method.

Method accuracy and recovery were established by spiking the targets into the solvent and refluxing for the time specified in the method. The samples were then extracted and analyzed via the HPLC method. Three different spiking levels were chosen and each triplicate samples were prepared for each spiking level. Recovery was assessed by comparison of detected levels of targets versus spiking level. LOQ was assessed by preparation of low-level standards and assessing signal to noise. The levels that produced an approximate signal-to-noise ratio of 10 were determined for each target. The LOQ was verified by replicate injection of the determined LOQs, and recovery at LOQ was assessed by calculating recovery after spiking targets at LOQ levels into samples.

The stability of both the standards and the samples was assessed by comparison of target levels in prepared standards and samples over a 7-day period. The absolute difference versus the day zero samples was used to establish the method stability characteristics.

TABLE 16.10. Summary of Results from Routine Extraction Method Validation Experiments

Validation parameter	Acceptance criteria	Results	
Linearity/range	**Linear ranges tested**		
	Millad 3988: 73.5–0.39 μg/mL	Millad 3988	$r^2 = 0.99960, 0.99961$
	Ultranox 626: 20.6–1.00 μg/mL	Ultranox 626	$r^2 = 1.00000, 0.99966$
	Irganox 1010: 30.6–1.10 μg/mL	Irganox 1010	$r^2 = 0.99878, 0.99885$
System suitability	**Instrument precision:**	**Instrument precision**	%RSD
	%RSD of each target extractable is not more than 5%.	Millad 3988	0.41
		Ultranox 626	0.34
		Irganox 1010	0.35
	Resolution:	Resolution	2.49 (mean)
	The most critical pair of extractables shall have a half-width resolution of not less than 1.5.	Tailing factor (*T*)	*T*
	Tailing factor:	Millad 3988	0.94
	The target extractables shall have a tailing factor of not more than 2.	Ultranox 626	1.13
		Irganox 1010	1.11
	Capacity factor:	Capacity factor (k′)	k′
	The target extractables shall have a capacity factor not less than 2.	Millad 3988	6.20
		Ultranox 626	8.87
		Irganox 1010	14.85

TABLE 16.10. *(Continued)*

Validation parameter	Acceptance criteria	Results			
Sensitivity	The signal-to-noise ratio for the target extractables in the working LOQ standard not less than 10.	Sensitivity LOQ Millad 3988 0.39 μg/mL Ultranox 626 1.00 μg/mL Irganox 1010 1.00 μg/mL	S/N ratio 19.9 20.1, 26.9 29.9		
Precision (repeatability)	%RSD of each target extractable is not more than 15%.	Extractable Millad 3988 Ultranox 626 Irganox 1010	%RSD 1.69 2.18 1.34		
Intermediate precision	%RSD of each target extractable is not more than 15%. The % absolute difference between means is not more than 15%.	Extractable Millad 3988 Ultranox 626 Irganox 1010	%RSD chemist 1 1.69 2.18 1.34	%RSD chemist 2 3.11 3.67 1.90	% absolute difference 1.80 13.7 0.30
Specificity	Confirms peak identifications and confirms no co-elution peaks for each target extractable.	No peaks were present in the 2-propanol/THF blank or standards that cause interferences with the quantitation of the target extractables.			

TABLE 16.11. Summary of Routine Method Validation Results—Spiking and Stability Experiments

Validation parameter	Acceptance criteria	Results		

Accuracy

Acceptance criteria: The mean recovery for each target extractable at each spiking level should be within 80%–120% of the calculated concentration.

Concentration level	Extractable	% mean recovery
Low	Millad 3988	118
	Ultranox 626	95.2
	Irganox 1010	50.3[a]
Mid	Millad 3988	104
	Ultranox 626	95.9
	Irganox 1010	93.0
High	Millad 3988	98.4
	Ultranox 626	95.6
	Irganox 1010	101

Limit of quantitation (LOQ)

Acceptance criteria: Report results based on extrapolated LOQs. The quantitation limit (LOQ) is the lowest level of the range that has acceptable accuracy, linearity, and precision. The signal-to-noise ratio for each injection at the LOQ must be greater than or equal to 10.

Extractable	LOQ	% RSD
Millad 3988	0.39 µg/mL	0.39
Ultranox 626	1.00 µg/mL	1.00
Irganox 1010	1.00 µg/mL	1.00

Standard-sample stability

Acceptance criteria: For information only; no defined acceptance criteria. Reported as % absolute difference.

Extractable	Day								
	Sample			Working standard			LOQ solution		
	1	3	7	1	3	7	1	3	7
Millad 3988	2.2	0.33	2.1	3.5	2.3	0.11	3.1	0.24	1.7
Ultranox 626	1.0	1.1	3.2	5.6	7.8	0.97	7.4	5.0	4.3
Irganox 1010	1.1	0.045	1.6	0.11	11	4.9	4.5	12	12

[a] Failed proposed criterion.

414

Tables 16.10 and 16.11 summarize the validation parameters, acceptance criteria and results from the validation experiments. The method met all proposed acceptance criteria except for one. The low-level spike of Irganox 1010 during the accuracy determination was the only result to fail proposed criteria. This indicated that the measurement range would need to be adjusted to meet the proposed acceptance criteria.

16.5 SUMMARY AND CONCLUSIONS

This chapter provides an example of the variables to be considered in the design of a comprehensive extraction study for a container closure system material used for inhalation products. The study incorporates the PQRI best practice recommendations for materials characterization as well as development of a routine material control method. The method was challenged to illustrate that measurement range, accuracy, and precision were within acceptable limits. The results of the method challenges indicated that the optimized method would be suitable, within a given range, to meet typical specifications associated with a comprehensive and rugged validation.

Extractions for the materials characterization were performed using multiple techniques, multiple solvents, and exaggerated conditions. The resulting extract liquids were profiled for species that migrated from the material into solution using a variety of techniques including GC/MS and HPLC/UV/MS; FTIR and OM/EM were used on the NVRs from the extracts. Detected extractables included known additives such as Pationic 901, Millad 3988, Ultranox 626, and Irganox 1010. Other species detected could be related to low-molecular-weight PP oligomeric species (branched aliphatic hydrocarbons, saturated and unsaturated), fatty acids, degradants, or reaction products of known additives. Extractables were quantitated using authentic reference standards where possible.

An additional goal of this work was to demonstrate the process for assuring a validated method could be produced for material control in the PP test article. The targets selected were Ultranox 626, Irganox 1010, and Millad 3988—known additives and detected extractables in the material. The extraction method was optimized to ensure that asymptotic levels of targets were obtained. The method assessment consisted of a variety of standard tests of method performance including specificity, system suitability, linearity range, precision (repeatability), intermediate precision, accuracy, LOQ, and stability. This process can be used to ensure accurate measurements and to generate a database of results for eventual development of specification setting on materials for release.

REFERENCE

1   Norwood, D.L. and Ball, D. Product Quality Research Institute: Safety thresholds and best practices for extractables and leachables in orally inhaled and nasal drug products. Submitted to the PQRI Drug Product Technical Committee, PQRI Steering Committee, and U.S. Food and Drug Administration by the PQRI Leachables and Extractables Working Group, 2006.

CHAPTER 17

ANALYTICAL LEACHABLES STUDIES

Andrew D. Feilden and Andy Rignall

17.1 INTRODUCTION

The identification and quantification of leachables and the establishment of a suitable strategy for their control are key aspects of pharmaceutical container closure system/delivery system development, as the patient may be exposed to any species that migrate from these container closure/delivery systems into the drug product. This chapter covers the critical areas that should be considered when embarking on a leachables testing program, which include the selection of analytical methods, including the limits of detection required, the planning and execution of leachables studies, and specific challenges associated with leachables testing. The correlation between extractables and leachables is an important element in any control strategy, and ways of achieving this correlation will be discussed.

Leachables are the chemical species that migrate into a pharmaceutical product under normal conditions of storage and use. Potential leachables can be identified by performing controlled extraction studies on the individual material of construction and critical container closure/delivery system components associated with the product. Theoretically, leachables should be a subset of the extractables observed during the controlled extraction study (see Chapter 14); however, there are some exceptions, and examples are provided. The separation and quantification strategies developed during the controlled extraction studies typically form the basis for the subsequent leachables methods. The diversity of potential materials of construction and their extraction profiles used in container closure/delivery system component manufacture results in a large number of potential leachables, creating challenges for any separation/detection strategy. In addition, leachables analysis has the added complication of potential interferences from the active ingredient(s) and its impurities and/or the formulation excipients.

To outline the main attributes of a successful leachables program, this chapter contains historical data from actual drug products to highlight these attributes along with a summary of current regulatory expectations in this area. The chapter will end

Leachables and Extractables Handbook: Safety Evaluation, Qualification, and Best Practices Applied to Inhalation Drug Products, First Edition. Edited by Douglas J. Ball, Daniel L. Norwood, Cheryl L.M. Stults, Lee M. Nagao.

with a discussion of some novel approaches to increase the efficiency of leachables testing.

17.2 ANALYTICAL METHODS

Analytical methods used for leachables are typically based on the methods used during the controlled extraction studies, usually with further optimization to ensure that the specificity challenges associated with interfering responses originating from other materials used in the container closure system components, the active ingredient(s), and any excipients, are overcome. The methods should be sufficiently sensitive to detect specific leachables at the analytical evaluation threshold (AET) limits. The AET is drug product specific and plays a critical part in any extractables and leachables analytical technique selection and subsequent method development, as it acts as a guide for the minimum sensitivity required. The AET is defined as the threshold at or above which identification of a particular leachable and/or extractable should occur followed by potential toxicological assessment. See Chapter 5 for further details on the AET and how it can be calculated. Leachables methods should be validated to appropriate standards and should be robust under typical use conditions. The selection of suitable methods for leachables separation, detection, and quantification is driven by matching the known specificity and detection capability of specific analytical techniques with the calculated AET.

As controlled extraction study analytical methods can form the basis of leachables methods, a brief review of controlled extraction studies is useful. Controlled extraction studies employ a wide range of analytical techniques to extract, concentrate, separate, quantify, and identify the extractables species. The selection of techniques depends on a number of factors. These factors would include the following points to consider:

- solute characteristics (what is the target compound);
- what is the sample matrix (very important for leachables);
- concentration level of solute (%—ppb);
- molecular weight (impact on volatility);
- polarity (determines volatility, solubility, adsorption, etc.)
- Log P (Log D);
- functional groups;
- thermal stability and reactivity;
- detectability (e.g., the presence or absence of a UV chromophore).

The number of points relevant to the potential leachables methods and their individual complexity is dictated by the number of extractables species observed and their chemical and physical diversity.

A wide variety of analytical techniques are routinely used in the trace analysis of potential extractables (see Fig. 17.1); however, not all techniques are appropriate for extractables and leachables studies. For example, size-exclusion chromatography

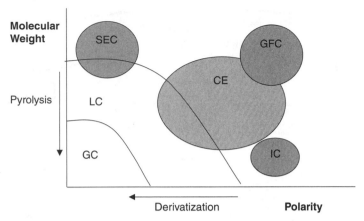

Figure 17.1   Possible analytical techniques based on molecular weight and polarity of the potential leachables species.

(SEC) and gel filtration chromatography (GFC) are generally not used since extractables and leachables studies focus on the migration of low-molecular-weight species and not the polymer itself. Despite being theoretically applicable, capillary electrophoresis (CE) is not widely used as it suffers from low loadability, resulting in significant interference from the relatively high amounts of active ingredient(s) and excipient present. Pyrolysis can reduce the molecular weight allowing species of higher molecular weight to be analyzed by gas chromatography (GC), and by derivatization, species can be analyzed by liquid chromatography (LC), including high-performance liquid chromatography (HPLC), ultra performance liquid chromatography (UPLC) (and ultra high performance liquid chromatography [UHPLC]), or GC. An example is the derivatization of fatty acids to the more volatile and less polar methyl esters, which are easily analyzed by GC-based analytical methods. LC is also employed for the analysis of species not amenable to any GC-based analytical methods. GC- and LC-based analytical methods are considered to be complementary for extractables/leachables analysis (see Chapters 13–16) due to their alternate modes of separation and detection. Hyphenated analytical techniques are widely used for both leachables and extractables analyses (see Chapter 13). These techniques include GC and LC (often run in gradient mode) with mass spectrometry (MS) (i.e., GC/MS and LC/MS) but could also include various other detector types where particular selectivity and specificity are required. GC is compatible with a number of detectors, including the flame ionization detector (FID), nitrogen phosphorous detector (NPD), thermal energy analyzer (TEA), nitrogen chemiluminescence detector (NCD), and electron capture detector (ECD). In addition to mass spectrometers, LC-based techniques can include UV detectors (and photodiode array detectors), charged aerosol detectors (CADs), and evaporative light-scattering detectors (ELSD). Bulk analytical techniques, such as total organic carbon (TOC) or ion mobility spectrometry (IMS) can be used to confirm that all the potential organic extractables species have been separated for further identification. Often, the wide diversity of extractables species associated with the engineering plastics and elastomers used in the construction of container closure and delivery systems means that

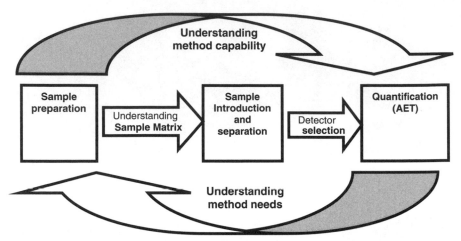

Figure 17.2   Diagram showing how an understanding of the AET and sample requirements interact to aid the development of leachables analytical methods.

no single technique and/or detector can be used to perform a complete suite of organic extractables and leachables analyses. For inorganic species, analytical techniques can include inductively coupled plasma (ICP) and ion chromatography (IC). For routine extractables testing, simplified or more routinely used detectors may be employed, such as LC/UV and GC/FID (see Chapter 18).

The leachables method development process should be informed by the various operational requirements for the method along with knowledge of potential limitations associated with sample preparation or reagent/sample matrix compatibility (see Fig. 17.2). Dialogue and information sharing with component suppliers, and possibly the subsuppliers of the component raw materials, will facilitate the compilation of a list of starting materials, intermediates, additives, processing aids, or other residual species related to the manufacturing process that may potentially be extracted and leached. Component suppliers may also be a potential source of reference standards for certain species if they are not readily commercially available. The physical and chemical properties of these species can be researched to help choose the most appropriate sample preparation, analysis, and detection technique.

Semivolatile/nonvolatile species are typically more amenable to LC/MS or LC/UV. This technique may be more prone to chromatographic interference from the active ingredient(s) or excipients and/or ion suppression/enhancement effects if co-elution occurs.

17.2.1   Quantification Targets and Application of the AET

As has already been discussed, the AET is drug product specific and is calculated using the following information:

- the labeled number of doses in a container/delivery system;
- the number of actuations required per dose;

- the dosing regime (i.e., number of doses per day);
- the weight of each component material in the container/delivery system;
- the safety concern threshold (SCT).

Note that the SCT is defined as the threshold below which a leachable would have a dose so low as to present negligible safety concerns from carcinogenic and noncarcinogenic toxic effects (see the detailed discussions in earlier chapters). The threshold recommendations are meant to provide general guidance for orally inhaled and nasal drug products (OINDPs). The approaches used to derive the SCT are based on calculated lifetime exposure (chronic). In general, a leachable with a total daily intake (TDI) at or below the SCT would

- have a dose so low as to present negligible safety concerns from noncarcinogenic toxic effects;
- be considered qualified, so no toxicological assessment would be required;
- have a low lifetime cancer risk of 1:1,000,000 (10^{-6}).

For certain classes of potential leachables compounds with special safety concerns, for example, N-nitrosamines, polynuclear aromatics (PNAs or polyaromatic hydrocarbons [PAHs]), and 2-mercaptobenzothiazole (2-MBT), appropriate specifications and appropriate qualification and risk assessments are required along with dedicated methods to underwrite the challenging lower detection and quantification thresholds required. Potential interference with active ingredient(s) and/or excipients may present specific difficulties for these methods. The detection required limits for these species can be defined as ALARP (as low as reasonably practicable) limits.

Examples of AET calculations for some different dosage forms are provided below.

17.2.2   Examples of AET Calculations

17.2.2.1   Example 1: A Metered Dose Inhaler (MDI)   This example is for a typical OINDP following the Product Quality Research Institute (PQRI) guidance. The following assumptions are made:

- The dosing regimen for the MDI product is 12 actuations/day.
- The product contains 120 doses.

This allows the AET for an individual organic leachable to be calculated as follows:

$$\text{Estimated AET} = \left(\frac{0.15\,\mu\text{g/day}}{12\,\text{actuations/day}} \times 120\,\text{labeled actuations/MDI} \right),$$
$$\text{Estimated AET} \approx 1.5\,\mu\text{g/MDI}.$$

Assuming that the MDI incorporates a 0.5-g rubber gasket, then the AET calculation for an individual organic extractable is

$$\text{Estimated AET} = \left(\frac{1.5\ \mu g/\text{MDI}}{0.5\ g\ \text{rubber/MDI}} \right),$$

$$\text{Estimated AET} \approx 3.0\ \mu g/g.$$

Note that the estimated AET for an individual organic extractable is useful as a guide for analyzing information from controlled extraction studies and for developing routine extractables quality control analytical methods.

17.2.2.2   *Example 2: Disposable Single-Use Plastic Syringe*   Example 2 is provided for another drug product type that constitutes a relatively high risk for container–dosage form interaction. While the PQRI guidance is currently recommended for OINDPs, the concept of the SCT (and qualification threshold [QT]) has been used to demonstrate how this guidance might be applied to other dosage forms. Note that for this particular delivery system, the main critical component is the plastic syringe barrel. We assume that the dose = 1.05 mL, maximum of three times per day.

The AET calculation for an individual organic leachable is

$$\text{Estimated AET} = 0.15\ (\mu g/\text{day}) \times \left(\frac{1\ \text{day}}{(3) \times 1.05\ \text{mL}} \right) = 0.0476\ \mu g/\text{mL or } 47.6\ \text{ng/mL}.$$

For this particular example drug product, 47.6 ng/mL = 50.0 ng/syringe. Assuming a critical component mass of 6 g, the estimated AET for an individual organic extractable is

$$\text{Estimated AET} = \left(\frac{50.0\ \text{ng/syringe}}{6\ g/\text{syringe}} \right) = 8.33\ \text{ng/g}.$$

Note that these estimated AETs are well into the range of trace organic analysis (see Chapter 13).

In both of the above examples, the *estimated AET* does not take into account any uncertainties in either the analytical technique or the method, and this is where the so-called *final AET* is used. This analytical uncertainty is especially applicable to extractables and leachables where there is a wide range of chemical classes, molecular weights, and analytical techniques used. There are two main approaches used to evaluate analytical uncertainty, one of which is through response factors (RFs). An RF is defined as

$$RF = A_a / C_a,$$

where

A_a = response of an individual analyte, for example, chromatographic peak area, and

C_a = concentration (or mass) of the individual analyte.

A more precise uncertainty evaluation can be obtained through the use of relative response factors (RRFs), which are defined as follows:

$$RRF = C_{is}A_a/A_{is}C_a,$$

where

C_{is} = concentration (or mass) of an internal standard,

A_{is} = response of the internal standard,

A_a = response of an individual analyte, and

C_a = concentration of the individual analyte.

The RRF normalizes individual RF to the RF of an internal standard. The use of internal standards is a well-established procedure for improving the accuracy and precision of trace organic analytical methods.

The other approach to account for analytical uncertainty is to use a "worst-case" RF of 50%. Analytical uncertainty in the estimated AET can be defined as %relative standard deviation (%RSD) in an appropriately constituted and acquired RF database or a factor of 50% of the estimated AET, whichever is greater. Typically, the 50% factor is most commonly used. The estimated AET is then reduced by the uncertainty factor to yield the final AET for the particular extractables/leachables profile. At an early stage of controlled extraction studies and/or leachables studies, an assumption of a 50% conversion of the estimated AET to the final AET should be sufficient before an RF database can be developed. It should be noted that the application of the AET concept to LC/MS methods suffers from limitations due to the relatively high background of "chemical noise," and is more applicable to methods involving GC and LC/UV, which have relatively low chemical noise background levels. The reader interested in additional detail regarding the AET concept is referred to Chapter 5.

17.2.3 Can Extractables Methods be Applied to Leachables Analysis?

Assessing the suitability of the optimized conditions used in the controlled extraction studies for use as leachables methods involves assessment of analyte preparation and confirmation of specificity. To check whether any interferences from the other species present occur (e.g., active ingredients and/or excipients), the optimum solvent, selected during the controlled extraction studies, should be tried to see if it can dissolve the leachables in the presence of the formulation constituents. If the formulation constituents are not soluble, then additional work will be needed to test for losses during the physical separation of the insoluble species. This may simply be a filtration step. Ideally, aged samples should be used to perform this suitability assessment so that the maximum levels of leachables are present. If aged samples are not available, another approach would be to spike the formulation with known extractables.

The AET limits will define the method requirements, such as the volume used to dissolve the example MDI contents to obtain an appropriate concentration of analytes. Consider, for example, if the final AET is 2 µg/MDI and the limit of quantification of the analytical technique is 0.2 µg/mL; then dissolving the contents of the MDI in a volume greater than 10 mL would make quantifying a species at the AET more challenging but could help resolve solubility issues of the excipient(s) and/or active ingredient(s). As has been previously discussed, an understanding of trace organic analysis is required for the development of any leachables methods. If specificity for the required leachables is confirmed and a suitable sample preparation process is developed, the method must be formally validated (as described in detail later in this chapter).

17.2.4 Assessment of an Extractables Method for Use as a Leachables Method: Case Study for an MDI

Dichloromethane is an excellent solvent for GC. It has a low boiling point (solvent focusing can be used to improve the chromatography) and has a low expansion volume (large volumes can be injected). During a controlled extraction study undertaken for an MDI product, dichloromethane was found to be the optimum solvent for generating GC extractables profiles from critical container closure system components. It was then assessed as a potential solvent and diluent for the MDI residue containing drug product leachables, following evaporation of the volatile propellant. The solvent was found to dissolve all of the required leachables and formulation components without requiring filtration. The volatile species present could be separated, detected, and quantified using GC/MS. Inlet and column temperatures were set as low as possible to avoid thermal degradation of either the active ingredient(s) and/or the excipients. Care had to be taken during the propellant removal step as this was observed to have a critical effect on the level of a hindered aliphatic leachable with the boiling point of 180°C (aliphatic a in Fig. 17.3), relative to another more polar leachable with a boiling point of 170°C (leachable b in Fig. 17.3). The time gap between venting of the propellant to the addition of the dichloromethane affected the levels of these leachables as determined with the method. As can be seen from Figure 17.3, a 75% loss in the level of the aliphatic leachable is observed. The data generated allowed an optimum vent time to be stipulated in the method.

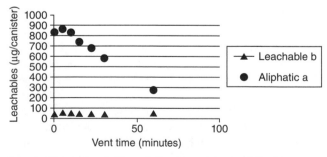

Figure 17.3   Level of target leachables relative to canister vent time.

17.3 ANALYTICAL METHODS FOR SPECIAL CASE LEACHABLES

Where species such as volatile *N*-nitrosamines, PAHs (or PNAs), and 2-MBT are identified as potential extractables/leachables, specific analytical methods able to achieve detection at ALARP levels (typically low to sub parts per billion levels) are required. For *N*-nitrosamines, this typically involves the use of GC/TEA and NCD, or more recently, the use of GC/MS and LC/MS has shown suitable limits of detection. PAHs or PNAs originate from the historical use of carbon black as an additive in elastomers, particularly sulfur-cured elastomers. Even though carbon black is typically no longer added to elastomers used in medical applications, confirmation of the absence of PNAs is still currently expected unless a suitable (and rigorous) control strategy is put in place. GC/MS is typically used to detect and quantify target PNAs at or below parts per million levels in elastomeric components, but HPLC with a range of detectors can also be used to quantify target PNAs. For leachables analysis, where the formulation is less suited to HPLC analysis, GC-based analytical methods are favored as they can more easily remove any matrix effects and potential interferences.

17.3.1 Case Study: The Determination of PNAs in an MDI Product

Table 17.1 shows levels of target PNA leachables determined in an MDI drug product over extended storage conditions. The analysis was performed by GC/MS

TABLE 17.1. Levels of Target PNA Leachables in an MDI Drug Product

Target PNAs	Total micrograms detected per MDI	
	3 months	12 months
Acenapthene	<LOD	<LOD
Acenaphthylene	<LOD	<LOD
Anthracene	<LOD	<LOD
Benzo[a]anthracene	<LOD	<LOD
Benzo[a]pyrene	<LOD	<LOD
Benzo[k] and [b] fluoranthene	<LOD	<LOD
Benzo[g,h,i]perylene	<LOD	<LOD
Chrysene	<LOD	<LOD
Dibenzo[a,h]anthracene	<LOD	<LOD
Fluoranthene	<LOD	<LOD
Fluorene	<LOD	<LOD
Indeno[1,2,3-cd]pyrene	<LOD	<LOD
Naphthalene	<LOD	<LOD
Phenanthrene	<LOD	<LOD
Pyrene	<LOD	<LOD

with a limit of detection (LOD) of 0.1 μg/MDI for each target analyte. These results are typical for MDI drug products with dose metering valve elastomeric critical components manufactured without added carbon black.

17.4 SETTING UP LEACHABLES STUDIES

Leachables studies can be prolonged and resource intensive since the studies last over the shelf life of the drug product as determined during appropriate real-time and accelerated stability studies. In these studies, drug product is stored under a variety of controlled environmental conditions and analyzed for leachables (both qualitatively and quantitatively) at multiple time points over the anticipated shelf life. Ideally, materials that represent the full extent of their manufacturing process design space (International Conference on Harmonisation [ICH] Q8(R2))[1] for each critical container closure/delivery system component should be used in the leachables study to enable the most representative leachables data set (see Fig. 17.4 and Table 17.2). This will facilitate the development of a representative specification (leachables specifications will be discussed later in this chapter). Leachables methods should be validated according to ICH Q2(R1),[2] and this will also be discussed later in this chapter. The storage conditions described in ICH Q1A(R2)[3] are typically used (see Table 17.2). Sample orientation should be carefully considered and controlled

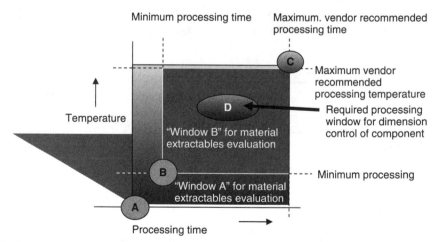

Figure 17.4   A representation of a material process design space.

TABLE 17.2.   Potential Storage Conditions and Testing Time Points for a Leachables Study

Condition (temperature/relative humidity)	Time points (months)
25/60	3, 6, 12, 18, 24, 36 to end of shelf life
30/65	3, 6, 12, 18, 24, 36 to end of shelf life
40/75	3, 6

since the storage orientation can have a significant effect on the leachables levels observed. For MDI products, valve down storage produces the highest levels of leachables as the formulation is in greater contact with the dose metering valve critical components. There should be minimal delay between batch completion and stability set-down to ensure that any data generated are fully representative of marketed product. A typical study may involve testing at all the recommended conditions along with multiple orientations, such as upright and inverted, along with multiple product strengths.

The stability of batches that are fully representative of the drug product used and assessed during clinical studies, and intended for commercial use, should be assessed. These drug product batches should contain a diversity of critical component batches that are typical of component supply variability and, if possible, explores the manufacturing process design space for each major critical component. The utilization of components produced over the full range of their manufacturing process design space forms part of the overall quality-by-design (QbD) approach for inhalation product development. However, to achieve this requires an excellent partnership with the component manufacturer and potentially even further back through the supply chain. If the drug product has multiple presentations, for example, different formulation volumes to achieve different numbers of label claim doses, the stability of each presentation should be assessed.

17.4.1 Reporting Leachables Data

To illustrate how leachables data might be reported, example data from 60 and 120 actuation MDI drug products are presented in the Tables 17.3 and 17.4. These data

TABLE 17.3. Leachables (µg/Inhaler) in 60 Dose Inhalers

Storage time	Initial	18 months	24 months
	Mean (µg/inhaler)	Mean (µg/inhaler)	Mean (µg/inhaler)
Diphenylamine	40	30	37
Isopropyl diphenylamine	<LOQ	<LOQ	<LOQ
9,10-Dihydro-9,9-dimethyl acridine	33	23	21
Dibutyl phthalate	<LOQ	<LOQ	<LOQ
2,2′-Methylene-bis(6-*tert*-butyl-4-methylphenol)	70	62	69
Ziram	<LOQ	<LOQ	<LOQ
Triethyleneglycol dicaprylate	<LOQ	51	84
Triethyleneglycol caprate-caprylate	33	118	167
Triethyleneglycol dicaprate	<LOQ	38	61
Total unassigned peaks	38	45	72
Most abundant unassigned peak	20	12	16

TABLE 17.4. Leachables (μg/Inhaler) in 120 Dose Inhalers

Storage time	Initial	18 months	24 months
	Mean (μg/inhaler)	Mean (μg/inhaler)	Mean (μg/inhaler)
Diphenylamine	47	41	39
Isopropyl diphenylamine	<LOQ	<LOQ	<LOQ
9,10-Dihydro-9,9-dimethyl acridine	41	34	32
Dibutyl phthalate	<LOQ	<LOQ	<LOQ
2,2'-Methylene-bis(6-*tert*-butyl-4-methylphenol)	67	62	68
Ziram	<LOQ	<LOQ	<LOQ
Triethyleneglycol dicaprylate	33	70	50
Triethyleneglycol caprate-caprylate	75	135	95
Triethyleneglycol dicaprate	20	46	25
Total unassigned peaks	34	61	54
Most abundant unassigned peak	18	16	13

show the general trend of leachables increasing over time but, as apparent in Table 17.3, some target leachables reach their equilibrium almost immediately with little change over storage time.

17.5 ASSESSMENT OF LEACHABLES DATA

All leachables species detected at levels that exceed the SCT need to be reported for some sort of safety assessment. The controlled extraction study data can be used to calculate a theoretical or worst-case maximum concentration that will be present in the formulation, which can provide early insight into potential safety concerns and potentially aid material selection for critical components. It is only when the leachables studies have been completed and the maximum levels of leachables observed during the stability study (up to and including the proposed shelf life of the product) are determined that a comprehensive toxicological assessment can be finalized.

17.5.1 Case Study Comparing Leachables with Extractables

Table 17.5 shows some extractables data from an MDI elastomeric critical component, determined by Soxhlet extraction for 24 hours with dichloromethane, along with the maximum expected leachables levels based on the weight of the material used (0.5 g).

TABLE 17.5.   Correlating Leachables and Extractables

Extractable	Amount (μg/g)	Maximum expected leachables (μg/MDI)	Leachables (μg/MDI) 12 months 25°C/60% RH
Abietic acid	83	42	ND
1-Chloro-4-(1-chloroethynyl)cyclohexene	104	57	125
1,6-Dichloro-1,5-cyclooctadiene	221	122	226
Diphenylamine	6	3	3
Hexadecanoic acid	17	9	50
2,2-Methylene-bis(4-methyl-6-*t*-butylphenol)	1582	791	271
Oleic acid	46	23	18

ND, not detected.

These data show the importance of testing for leachables, as a number of species are above the maximum expected level, suggesting either asymptotic levels were not achieved during the controlled extraction study and/or the solvent used in the study was not as efficient as the MDI propellant at extracting the potential leachable.

17.6   SPECIFICATION SETTING

Establishing leachables specifications, which include quality attributes, validated test methods, and acceptance criteria, is required by many regulatory authorities for OINDP. The leachables acceptance criteria most representative of future large-scale commercial manufacture of drug product can be proposed when a diverse range of input materials and critical components are used during development. To enable statistically robust specifications and acceptance criteria, larger sample sizes are desirable, but as the sample size becomes ever larger, the cost of the leachables testing increases, both in terms of time and money, suggesting the need for a cost–benefit analysis. Leachables specifications also take time to establish, as leachables tend to increase over the shelf life of the drug product and tend to reach equilibrium either at or beyond the shelf life of the product. The large number of leachables species combined with wider unit-to-unit variation can make this process more challenging compared with other drug product impurities. Using the data presented previously, example acceptance criteria can be derived using a "mean plus 4 standard deviation" approach. This approach is typically used early in the development life cycle. When additional data are available, the number of standard deviations used can be reduced, for example, a "mean plus 3 standard deviation" approach may be more appropriate (see Table 17.6).

TABLE 17.6.   Example Specification Using the Data Generated from Three Drug Product Batches

	Drug product leachables (µg/canister)			Mean	SD	Derived specification
	Lot A	Lot B	Lot C			
Peroxide-related leachables	2.4	0.9	0.7	1	1	4
Myristic acid	107	95	87.2	96	10	126
Ethyl myristate	44.4	16.3	15.3	25	17	75
Palmitic acid	188	123	124	145	37	257
Ethyl palmitate	111	67.9	54.5	78	30	166
Stearic acid	289	184	165	213	67	413
Ethyl stearate	113	60.9	70.2	81	28	165
PBT dimer	192	185	178	185	7	206
PBT trimer	20.6	20.1	19.2	20	1	22
PBT tetramer	0.4	<0.4	0.6	1	0	1
PBT pentamer	<LOD	<LOD	<LOD	<LOD	<LOD	<LOD

PBT, polybutylene terephthalate.

Note that the approach presented here for the establishment of leachables acceptance criteria is but one of several possibilities. For example, upper limits for leachables could be established considering safety assessment of individual leachables species, rather than with just historical stability data. It is also possible to employ more statistically rigorous process capability analysis, perhaps combined with individual leachables safety evaluation.

Note further that any leachables specifications should include acceptance criteria for

- identities of known leachables detected and identified during primary stability studies;
- quantitative levels of these known target leachables; and
- quantitative limits for "unspecified" leachables based on the calculated AET for the particular drug product.

17.7   LEACHABLES CONTROL STRATEGY

Leachables are most easily controlled by

1. selecting the appropriate component materials, taking formulation constituents into account; and
2. developing extractables control specifications with appropriate acceptance criteria for these component materials, and/or the critical components fabricated from these materials.

To achieve this control, routine extractables testing to verify the continuing suitability of the input materials is required, including extractables specifications with appropriate acceptance criteria. Demonstrating that this testing effectively controls drug product leachables requires an understanding of the levels of leachables predicted from the available extractables data via a suitable extractables/leachables correlation. This can involve simply comparing observed identities and quantitative levels of leachables and extractables. Extractables profiles from quantitative studies should be compared with leachables profiles to determine extractables and leachables correlations. To establish a qualitative correlation between profiles, analysts must show that compounds detected in the leachables studies were also present in the controlled extraction studies. To establish a quantitative correlation between profiles, analysts must show that levels of leachables obtained from leachables studies are generally less than the levels of extractables obtained from quantitative controlled extraction studies. An example data set showing the leachables/extractables correlation for three MDI drug product lots incorporating a development valve with prewashed elastomeric components is presented in Table 17.7.

The data in Table 17.7 show the importance of leachables testing since some leachables are not a subset of the extractables but are related due to reaction with a formulation component resulting in esterification of fatty acid leachables to the corresponding ethyl esters (indirect qualitative correlation). The data also show the importance of testing multiple drug product lots/batches as some batch-to-batch differences are apparent. This point is reinforced by the data in Table 17.8, which presents extractables/leachables data from a second MDI drug product.

TABLE 17.7. Predicted Maximum Levels of Extractables and the Corresponding Leachables Following Storage for 18 Months

Target leachables	Maximum predicted total extractables (μg/valve)	Drug product leachables (18 months inverted; 30°C/70% RH) (μg/canister)		
		Lot A	Lot B	Lot C
Peroxide-related leachables	3.2	2.4	0.9	0.7
Myristic acid	332	107	95.0	87.2
Ethyl myristate	<LOD	44.4	16.3	15.3
Palmitic acid	710	188	123	124
Ethyl palmitate	<LOD	111	67.9	54.5
Stearic acid	1230	289	184	165
Ethyl stearate	<LOD	113	60.9	70.2
PBT dimer	5620	192	185	178
PBT trimer	1810	20.6	20.1	19.2
PBT tetramer	270	0.4	<0.4	0.6
PBT pentamer	227	<LOD	<LOD	<LOD

TABLE 17.8. Predicted Maximum Level of Extractables and the Corresponding Leachables Following Storage for 3 and 12 months

Extractable	Amount (μg/g)	Maximum expected leachables (μg/ MDI)	Leachables 3 months, 40°C/75% RH (μg/MDI)	Leachables 12 months, 25°C/60% RH (μg/MDI)
1-Chloro-4-(1-chloroethynyl) cyclohexene	104	57	78	68
1,6-Dichloro-1,5-cyclooctadiene	221	122	154	122
Diphenylamine	6	3	2	2
Hexadecanoic acid	17	9	<LOD	<LOD
2,2-Methylene-bis(4-methyl-6-*tert*-butylphenol)	1582	791	272	367
Oleic acid	46	23	<LOD	<LOD

17.8 VALIDATION OF LEACHABLES METHODS

Validating leachables methods can pose additional challenges relative to "standard method validation" due to the number of potential analytes (i.e., leachables), their diversity in terms of physical and chemical attributes (i.e., amines and phenols), their trace levels, and their wide endogenous levels (e.g., from the data in Table 17.8, the leachables range from 2 to 367 μg/MDI canister). As with standard method validation, the following parameters typically need to be investigated (see ICH Q2(R1))[2]:

- linearity
- range
- accuracy
- system precision (instrument precision)
- method precision
- repeatability
- intermediate precision
- specificity
- detection limit/quantitation limit
- standard and sample stability
- robustness
- system suitability.

As leachables tend to increase over time to asymptotic maxima, the validation of leachables methods can only be completely accomplished once the end of shelf life has been reached for pilot studies, unless the methods are validated over very wide linear ranges based on worst-case scenario extractables data. The use of such wide dynamic ranges for drug product leachables concentrations will tend to increase analytical uncertainty at the lowest concentrations. Experience with various drug products on stability suggests that leachables tend to reach an equilibrium concentration between 12 and 18 months, so it would only be at that point that the range for the analytical method could be accurately determined. It is also certain that the concentration ranges will be significantly different for all potential leachables. Examples of a calibration standard and leachables profile following storage for 18 months at 30°C are shown in Figures 17.5 and 17.6.

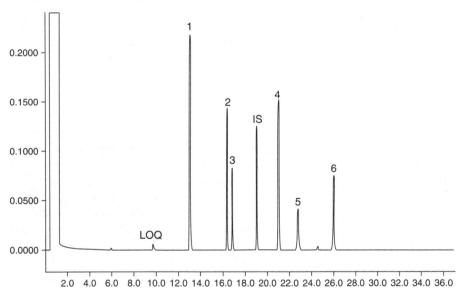

Filename: T:\QC\SANDRINE\NONPAH\MA5286 RUN Channel: A = A

Label	Compound
LOQ	2-fluorobiphenyl
1	Diphenylamine
2	9,10-dihydro-9,9-dimethylacridine
3	Dibutylphthalate
IS	4-terphenyl-d$_{14}$
4	2,2′-methylene-bis(6-*tert*-butyl-4-methylphenol)
5	Ziram
6	Triethyleneglycol dicaprate

Figure 17.5   GC/FID chromatograms of a calibration standard showing six target analytes, together with internal standard (IS; 80 µg/mL) and 2-fluorobiphenyl (LOQ standard).

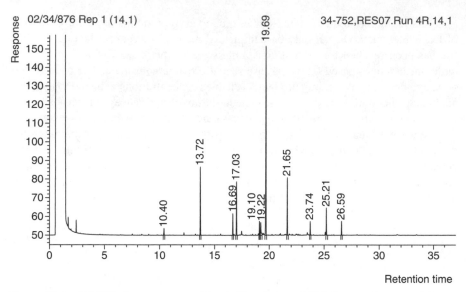

Figure 17.6   GC/FID chromatogram of leachables from an MDI drug product stored for 18 months at 30°C.

17.8.1   Linearity and Range

Linearity can be defined as "the ability of the method to elicit test results that are directly proportional to analyte concentration within a given range,"[4] with *range* being defined as "the (inclusive) interval between the upper and lower levels of analyte that have been demonstrated to be determined with precision, accuracy, and linearity using the method."[4] A linear detector response over the required concentration range should be confirmed for all target leachables for which the method is being validated. The required linear range for each analyte will typically span from the LOD to 150% of the anticipated leachables acceptance criteria depending on the individual analyte and its anticipated level. Note that since many leachables increase in concentration significantly over the course of an accelerated stability study, relatively wide linear dynamic ranges could be required. Note that in some cases, it might be acceptable to verify linearity and range for one representative leachable from a class of closely related leachables. For example, ethyl stearate could be used in a validation exercise to represent other closely related aliphatic ethyl esters (see Table 17.7). Linearity should be assessed relative to specific acceptance criteria, which usually involves a statistical linear correlation.

17.8.2   Accuracy

Accuracy "is the measure of exactness of an analytical method, or the closeness of agreement between the measured value that is accepted either as the conventional, true value, or an accepted reference value."[4] This is achieved by the use of spiked samples at levels typically expected, either based on predicted maximum levels from the controlled extraction studies or from known levels at the end of drug product shelf life. The spiking matrix is typically the drug product formulation (which could also contain a high level of endogenous leachables), or a spiking matrix created in the laboratory from the known drug product formulation constituents. The laboratory-created spiking matrix is often less appropriate as many drug product formulations can be quite complex; however, this approach has the advantage of including no endogenous levels of target leachables, so lower spiking levels can be accomplished with minimum statistical error. Another possible approach is to use a specific batch of drug product for leachables validation, which is manufactured as close to the time of validation protocol execution as possible so as to minimize endogenous leachables content. This approach also facilitates spiking to lower target levels with minimal analytical bias. Three spiking levels are recommended: 50%, 100%, and 150% of a nominal leachables concentration, which of course will differ for each target leachable or class of target leachables. Acceptance criteria for accuracy typically involve percentage recovery of standard spike at all three spiking levels.

17.8.3 System Precision (Instrument Precision)

Precision is "the measure of the degree of repeatability of an analytical method under normal operation."[4] The precision of the analytical system (i.e., the analytical instrument) should be independently assessed for the analytical method. For example, a GC/MS (or GC/FID) system can be assessed by making multiple repeat injections of a single sample solution at the nominal level (e.g., near the middle of the linear concentration range).

At a minimum, the following parameters should be determined for each target leachable (analyte) and assessed relative to specific acceptance criteria:

1. mean (i.e., target leachable peak area, or area ratio relative to an internal standard);
2. standard deviation (i.e., of target peak areas or area ratios);
3. %RSD;
4. 95% confidence interval about each mean.

17.8.4 Method Precision

17.8.4.1 Repeatability *Repeatability* "refers to the results of the method operating over a short time interval under the same conditions (inter-assay precision)."[4] Method repeatability can be challenging unless care is taken regarding the composition of materials and critical components used in the fabrication of container closure /delivery systems for the batches of drug product employed in the assessment. If multiple critical component batches created from multiple batches of raw materials are used, then there could be significant variability in this validation parameter due to the variability in extractables levels in the various critical component batches. Repeatability should be assessed by performing at least six complete determinations according to the analytical method on a single drug product sample set. Repeatability assessment can be combined with accuracy assessment.

TABLE 17.9. Leachables Data from Different MDI Batches

Species	Total micrograms detected per MDI		
	MDI (1)	MDI (2)	MDI (3)
Hexadecanoic acid (palmitic acid)	<LOQ	<LOQ	<LOQ
Octadecenoic acid (oleic acid)	<LOD	<LOD	<LOD
Octadecanoic acid (stearic acid)	<LOD	<LOD	<LOD
Diphenylamine	2	2	2
2,2-Methylene-bis(4-methyl-6-*tert*-butylphenol)	304	413	384
1,6-Dichloro-1,5-cyclooctadiene	114	126	128
1-Chloro-4-(1-chloroethynyl)-cyclohexene	59	73	72

At a minimum, the following parameters should be determined for each target leachable (analyte) and assessed relative to specific acceptance criteria:

1. mean (i.e., target leachable quantitated amounts);
2. standard deviation (i.e., of target leachable quantitated amounts);
3. %RSD;
4. 95% confidence interval about each mean.

The potential variability within a drug product batch is demonstrated in Table 17.9, which shows the levels of leachables from an MDI batch stored at 25°C/60% relative humidity (RH) for 12 months.

17.8.5 Intermediate Precision

Intermediate precision "refers to the results from within-lab variations due to random events such as differences in experimental periods, analysts, equipment, and so forth."[4] Intermediate precision should be determined using the same drug product sample (i.e., the same batch) between days, analysts, instrumentation, columns (including different lots of packing material), reagents, and so on, as deemed relevant to give a total of not less than six analysis occasions. A "design of experiment" (DOE) process is often used to create an experimental plan to assess intermediate precision. An example of such an experimental plan is shown in Table 17.10. Each study should be operated as a separate run, and at least duplicate sample preparations should be employed for each run.

17.8.6 Specificity

Specificity is "the ability to measure accurately and specifically the analyte of interest in the presence of other components that may be expected to be present in the sample matrix."[4] Specificity can be assessed by the analysis of control samples, which could be (for example) the laboratory-prepared accuracy spiking matrix without any spiked

TABLE 17.10. **Experimental Design for Method Development**

Test run	Operator	Equipment	Column	Results
1	1	A	1	1, 2
2	1	B	2	3, 4
3	2	A	1	5, 6
4	2	B	2	7, 8
5	1	A	2	9, 10
6	2	B	1	11, 12
7	1	B	1	13, 14
8	2	A	2	15, 16

analytes. Acceptance criteria would include a lack of any interference with individual analyte specific identification and quantification. Furthermore, each target leachable in a spiked or actual sample should be resolved (either chromatographically or spectrally) from other target leachables. Acceptance criteria for specificity are necessarily dependent on the properties of the analytical technique used.

17.8.7 Detection Limit/Quantitation Limit

The LOD "is defined as the lowest concentration of an analyte in a sample that can be detected, though not necessarily quantitated."[4] The *limit of quantitation* (LOQ) "is defined as the lowest concentration of an analyte in a sample that can be determined with acceptable precision and accuracy under the stated operational conditions of the method."[4] LOD is usually expressed as a concentration at a signal-to-noise ratio of 3 to 1. Also recognized as appropriate for LOD assessment are visual noninstrumental methods and a means of calculation.[4] Likewise, LOQ is usually expressed as a concentration at a signal-to-noise ratio of 10 to 1, and visual noninstrumental methods and a means of calculation are also recognized.[4] For leachables LOD/LOQ assessments, it is common to analyze a standard or spiked sample at a concentration sufficient to produce a signal-to-noise ratio of approximately 100 for each target analyte, and then extrapolate to the LOD/LOQ levels. LOD/LOQ confirmation standards can then be prepared and analyzed.

17.8.8 Standard and Sample Stability

These parameters define the length of time that standard solutions and prepared drug product samples can be stored prior to analysis. Assessment usually involves percent recovery acceptance criteria relative to an initial analysis with appropriate acceptance criteria. Standard stability can be accomplished by analysis of at least three standard solutions at an initial time point and then, for example, at 24, 48, and 72 hours. At each succeeding time point, the original standards should be analyzed as unknowns, with freshly prepared standard solutions. If room temperature storage is desired and stability is considered to be an issue, refrigerated storage may need to be evaluated, as well as protection from light.

Stability of prepared drug product samples can be assessed by analysis of at least three sample solutions at an initial time point, and again at, for example, 24, 48 and 72 hours (remember to include any sample preparation time). Freshly prepared standards should be used at each sample analysis time point. Again, if stability is considered to be an issue, refrigerated storage, protection from light, and so on may need to be evaluated. Examine the results for indications of instability, including the appearance of any new peaks in the sample chromatogram over time.

17.8.9 Robustness

Robustness is "the capacity of a method to remain unaffected by small deliberate variations in method parameters."[4] Considering a leachables analysis by GC/MS, a robustness assessment might include small deliberate variations in injector port temperature, GC oven temperature program rate, mass spectrometer scan time, and so on. Acceptance criteria are typically expressed as percentage recovery relative to a sample analyzed by the method-specified instrumental parameters. As with intermediate precision, DOE statistical approaches can be used to develop and appropriate experimental matrix.

17.8.10 System Suitability

System suitability tests are integral parts of any analytical method,[4] and leachables methods are no exception. Typical system suitability tests for leachables methods could include verification of LOD/LOQ, chromatographic resolution, linearity, range, and so on, relative to appropriate acceptance criteria determined during the method validation exercise.

17.9 LEACHABLES ARISING FROM MATERIAL CONTACT DURING MANUFACTURING

Another potential source of leachables that requires consideration is elastomeric and polymeric materials associated with reaction vessels and other materials that the formulation contacts during normal processing (e.g., filters). These can be especially important for biological systems with single-use bags (reaction vessels) and other processing aids with prolonged contact times, which can be much longer than small-molecule drug products. The key factors that affect the risk associated with leachables from reaction vessels and/or other processing steps are as follows:

Product composition. In general, if a product or a formulation includes higher levels of organics, high or low pH, or solubilizing agents such as surfactants (detergents), the regulatory and safety concern for potential leachables will increase.

Surface area. As can be predicted, the higher the surface area, the higher the risk. This can be especially important for filters where the internal surface area can be very high.

Contact time and temperature. The longer the contact time, the higher the likelihood that potential leachables can be removed from a material until equilibrium is reached. Also, the higher the temperatures, the faster the migration of leachables from materials into a process stream or formulation.

Pretreatment steps. These can either potentially increase or decrease the levels of leachables. Rinsing may pre-extract species, reducing the levels of potential leachables. Treatments such as sterilization by either steam autoclave and/or gamma irradiation may result in higher levels of leachables.

Leachables arising from material contacts during processing are controlled in the same manner as those arising from the container closure system, that is, selection of the most appropriate materials, potentially using information generated by controlled extraction experiments, and ensuring appropriate controls are in place.

17.10   THE FUTURE: NOVEL APPROACHES FOR LEACHABLES TESTING IN MDIS

A number of novel experimental approaches are being developed with the overall goal of increasing the efficiency of leachables testing.

An approach for liquid systems is to determine the maximum solubility of each potential leachable in the drug product formulation and then assess this limit toxicologically as a worst-case scenario for drug product leachables levels. For example, for an MDI product, the solubility of a given extractable can be determined in the propellant system.

17.10.1   Nuclear Magnetic Resonance (NMR) Solubility

Various approaches exist to determine such solubility, for example, the potential leachables species can be added to a suitable NMR (NMR spectroscopy) tube (Fig. 17.7) and propellant added to produce a saturated solution. By either using an eretic signal generated by the spectrometer or an internal standard of known solubility, the solubility of the potential leachable can be determined. If this level is determined to be safe toxicologically, then the maximum theoretical concentration for the leachable would be its solubility limit, as the MDI is a closed system.

However, the relatively common leachables species analyzed so far have solubilities far higher than their determined extractables levels (Table 17.11). If individual maximum possible concentration and therefore maximum potential patient exposure are considered safe, then there would be no need to routinely test for these particular species as a drug product leachable.

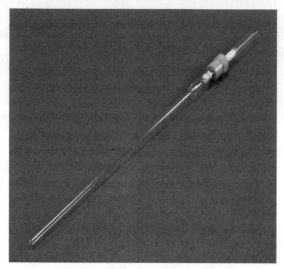

Figure 17.7   High-pressure NMR tube.

TABLE 17.11.   Solubility of Common Potential Leachables in HFA 227 at 293°K

Reference standard	Solubility (ppm)
Antioxidant 2246	600
Nonylphenol	2500
Oleic acid	930
Irgafos 168	35

HFA (hydrofluoralkane) 227, 1,1,1,2,3,3,3-heptafluoropropane.

17.10.2   Sub-Ambient Soxhlet Extraction

It is well known that leachables take time to reach asymptotic equilibrium in a drug product, particularly under ambient temperature storage conditions. One approach to quickly and efficiently assess if known extractables are likely to be observed as leachables is to carry out critical component controlled extraction studies using the MDI propellant as the extraction solvent rather than the standard extraction solvents (such as dichloromethane). This involves accomplishing Soxhlet extraction at −30°C. A typical apparatus is shown in Figure 17.8.

17.10.3   Ultrafast Analysis of MDIs

Direct analysis of the plume emitted from an MDI for its leachables content obviates the need for volumetric work-up of analyte solutions, and therefore offers a rapid alternative to conventional leachables methods. Rather than "extract" (i.e., dissolve) the leachables back into solution to form the analyte mixture, there are a number of

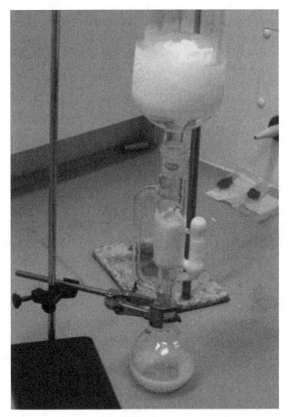

Figure 17.8 Apparatus used in sub-ambient Soxhlet extraction.

approaches that can be used to assess the emitted dose for leachables. This approach offers the added advantage that leachables are assessed using the drug product delivery system in the same manner as the patient would use it.

To assess the semivolatile leachables content of an MDI, one approach is to actuate the MDI directly into a spray chamber of a mass spectrometer (see Fig. 17.9). Any leachables that can be ionized by the mass spectrometer, making the approach especially suited to acids and antioxidants, can be identified and quantified. This approach also allows for the actual complete drug delivery system to be tested in a nondestructive manner, which then allows other critical tests to be accomplished on the same unit. This could facilitate a correlation between leachables levels and other delivery system performance attributes.

For example, eight commercially available MDIs were actuated twice into the spray chamber and a specific leachables species (a hindered phenolic antioxidant) was detected and quantified. As shown in Figure 17.10, four MDI drug products were found to contain readily measurable levels of the antioxidant, with one MDI drug product containing no antioxidant.

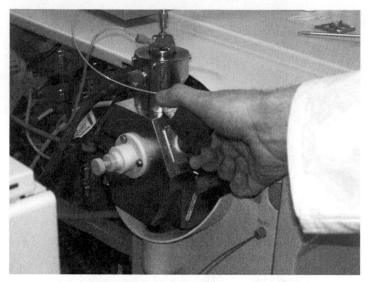

Figure 17.9   The process of actuating an MDI into a spray chamber of a mass spectrometer.

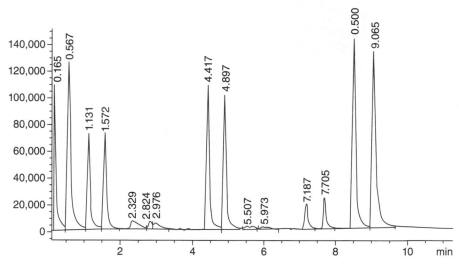

Figure 17.10   Total ion chromatogram (TIC) showing relative levels of a common hindered phenolic antioxidant found in eight MDI drug products.

17.10.4   Direct Addition GC

Volatile leachables in an MDI drug product can be assessed by actuating the MDI into a cryocooled inlet interfaced to a gas chromatograph (GC) and then introduce the sample onto the GC column as a large-volume injection with the instrument in "solvent venting mode" (Fig. 17.11).

Figure 17.11   Direct GC injection via a cryocooled inlet.

The sensitivity of the leachables analysis can be significantly increased by this approach. For example, conventional volatiles analysis typically involves the dilution of the contents of an MDI into, for example, 10 mL of solvent followed by injecting 1 μL of this solution into the GC. Using this approach with a 100-dose MDI means that only 1% of a dose is being analyzed. Using the direct injection approach, an entire dose is analyzed. However, this can cause potential problems such as reduced chromatographic resolution and column overloading if the levels of leachables are high. Some examples from commercially available MDIs are included in Figure 17.12.

As MDIs commonly use a diversity of critical component materials, including various elastomers, in prolonged contact with organic propellants, the likelihood of leachables occurring is among the highest for any container/closure system. However, dry powder inhaler (DPI) drug products (as well as all other OINDP types) can contain leachables and therefore need to be assessed accordingly. Example leachables data from a DPI drug product can be seen in Figure 17.13. In this example, the powder formulation is contained in a sealed blister. The blister material is an aluminum–aluminum (Alu–Alu) type of construction and contains a heat seal coating and polyvinyl chloride (PVC) laminate that is in direct contact with the dosage form. Based on the dosing scheme and mass of powder in the blister, the AET was calculated at 1 ppm based on an SCT of 0.15 μg/day.

17.11   CONCLUSION

Leachables are species that migrate from the drug product container closure/delivery system and can be found in medicinal product under normal storage conditions, and as such have the potential to be detrimental to patient safety and optimal drug product performance. To understand and control specific leachables require a

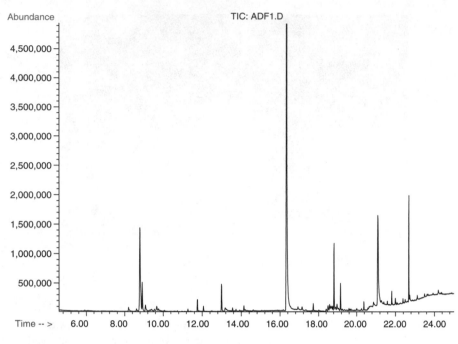

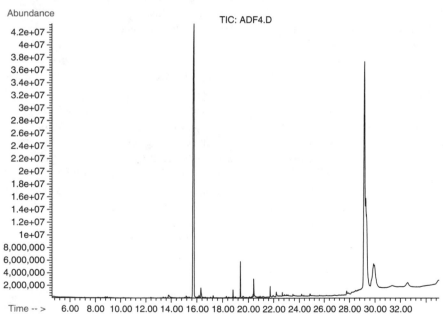

Figure 17.12   Chromatograms obtained from commercially available MDIs using the cryocooled inlet.

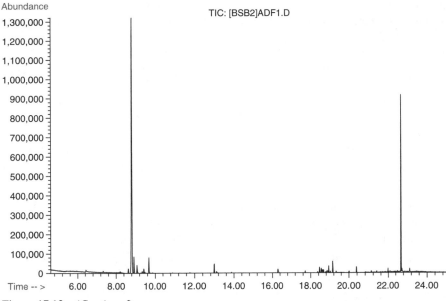

Figure 17.12   (*Continued*)

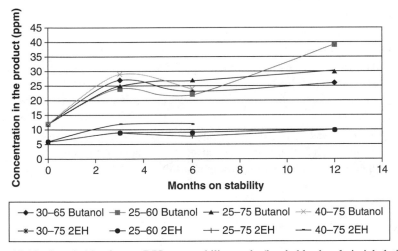

Figure 17.13   Leachables from a DPI on a stability study (leachables levels in inhalation powder: 1-butanol and 2-ethyl-1-hexanol).

considerable amount of effort and a high degree of skill in trace-level analytical chemistry. This requires a detailed understanding, knowledge and upstream control of the extractables associated with each individual component of the container closure system. Upstream control is aided by communication and knowledge sharing throughout the complete container/closure system supply chain. This is particularly important when assessing the impact of potential changes to component materials and/or their manufacturing process. Adherence to these principles facilitates a more complete understanding of container closure/delivery systems and how these impact product quality and performance. Applying this enhanced understanding will ensure that product quality, safety, and efficacy are fully optimized.

Leachables testing can be extensive, but the application of advanced analytical approaches can help to increase efficiency. As further analytical technology advances are applied to this type of testing, more efficient analytical methods that deliver the additional information required to develop optimized control strategies may become available.

The presentation and discussion in this chapter are generally consistent with the recommendations developed and issued by the PQRI.[5] These are the following:

- Analytical methods for the qualitative and quantitative evaluation of leachables should be based on analytical technique(s)/method(s) used in the controlled extraction studies.

- Leachables studies should be guided by an AET that is based on an accepted SCT.

- A comprehensive correlation between extractables and leachables profiles should be established.

- Specifications and acceptance criteria should be established for leachables profiles in OINDP as required.

- Analytical methods for leachables studies should be fully validated according to accepted parameters and criteria.

- PAHs (or PNAs), N-nitrosamines, and 2-MBT are considered to be "special case" compounds, requiring evaluation by specific analytical techniques and technology defined thresholds for leachables studies and routine extractables testing.

- Qualitative and quantitative leachables profiles should be discussed with and reviewed by the pharmaceutical development team toxicologists so that any potential safety concerns regarding individual leachable are identified as early as possible in the pharmaceutical development process.

ACKNOWLEDGMENTS

The authors would like to thank the appropriate pharmaceutical companies for supplying the data presented in this chapter.

REFERENCES

1 Guidance for industry: ICH Q8(R2): Pharmaceutical development. International Conference on Harmonization of Technical Requirements for Registration of Pharmaceuticals for Human Use, August 2009. Current *Step 4* version. Available at: http://www.ich.org/fileadmin/Public_Web_Site/ICH_Products/Guidelines/Quality/Q8_R1/Step4/Q8_R2_Guideline.pdf (accessed September 1, 2011).

2 Guidance for industry: ICH Q2(R1): Validation of analytical procedures: Text and methodology. International Conference on Harmonization of Technical Requirements for Registration of Pharmaceuticals for Human Use, Parent Guideline, October 27, 1994. Current *Step 4* version. Available at: http://www.ich.org/fileadmin/Public_Web_Site/ICH_Products/Guidelines/Quality/Q2_R1/Step4/Q2_R1__Guideline.pdf (accessed September 3, 2011).

3 Guidance for industry: ICH Q1A(R2): Stability testing of new drug substances and products. International Conference on Harmonization of Technical Requirements for Registration of Pharmaceuticals for Human Use, February 6, 2003. Current *Step 4* version. Available at: http://www.ich.org/fileadmin/Public_Web_Site/ICH_Products/Guidelines/Quality/Q1A_R2/Step4/Q1A_R2__Guideline.pdf (accessed September 3, 2011).

4 Swartz, M.E. and Krull, I.S. *Analytical Method Development and Validation.* Marcel Dekker, New York, 1997.

5 Norwood, D.L. and Ball, D. Product Quality Research Institute: Safety thresholds and best practices for extractables and leachables in orally inhaled and nasal drug products. Submitted to the PQRI Drug Product Technical Committee, PQRI Steering Committee, and U.S. Food and Drug Administration by the PQRI Leachables and Extractables Working Group, 2006.

CHAPTER 18

DEVELOPMENT, OPTIMIZATION, AND VALIDATION OF METHODS FOR ROUTINE TESTING

Cheryl L.M. Stults and Jason M. Creasey

18.1 INTRODUCTION

Routine leachables or extractables testing is associated with control of materials or processes at a microscopic level. The impetus for examination of the materials at such low levels has come from cases like EPREX, where adverse events were observed due to unanticipated consequences of a formulation change.[1] Materials or processes may create a concern because of the chemicals that can be unintentionally introduced into the product. These unexpected chemicals may be organic or inorganic and may result in reaction products that form particulates.[2] The first point of entry of leached chemicals may come from the processing equipment used to manufacture the dosage form. Additionally, chemicals may leach from either the container closure system (CCS) or the delivery device (DD) into the dosage form. Patients may then be exposed to the leached chemicals ("leachables"), which at particular concentrations may pose safety concerns. Therefore, the management of leachables has been a topic of increasing concern.

Control of leachables can be addressed either by emphasizing process controls throughout the CCS and DD manufacturing supply chain or by application of appropriate tests at strategic points in the supply chain. A risk-based evaluation specific to the product can be employed to determine the appropriate balance between process controls and testing. Routine leachables or extractables methods may be used for different testing purposes throughout the product life cycle.

As product development advances, the chemical testing of materials progresses from general safety assessment to chemical characterization and, finally, to control. This progression is outlined in Figure 18.1. Early in development, compendial methods may be used for general assessment, while later in development,

Leachables and Extractables Handbook: Safety Evaluation, Qualification, and Best Practices Applied to Inhalation Drug Products, First Edition. Edited by Douglas J. Ball, Daniel L. Norwood, Cheryl L.M. Stults, Lee M. Nagao.

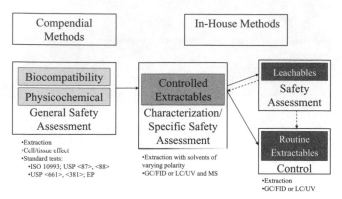

Figure 18.1 Chemical testing progression for material evaluation.

trace-level customized chemical analysis is undertaken to regularly evaluate leachables in the drug product and/or extracted chemicals (extractables) from the CCS/DD materials. The information gained during the controlled extraction study is utilized to develop routine test methods for leachables and extractables. The information gained from leachables testing during product stability studies may indicate a need to perform further characterization of one or more materials and can be utilized to guide the development of routine control methods.

The development of routine leachables or extractables methods employs approaches that are similar to those taken in trace-level or impurity analysis.[3] To achieve the goal of a qualitative and quantitative profile, consideration must be given to the complexity of sample matrices and sensitivity of the analytical techniques. Currently, there are no compendial methods available. This is primarily due to the fact that each pharmaceutical dosage form and CCS/DD has a unique composition when examined at parts per million levels. Just as is the case with pharmaceutical impurity methods, depending on its end use and phase of development, there may be varied expectations for the method. These may include, but are not limited to, the following: instrument qualification, method validation, restricted access to testing equipment, material traceability (e.g., individual chemicals, component, resin), certified reference materials, control charts, and formal reporting criteria. Additional practical considerations include cycle time, cost, and automation potential. Thus, it is important to carefully consider both the material characteristics and the end use of the method. Although many articles[4–7] and books[8–14] have been written on the general topic of analytical method development and validation, it is the intent of this chapter to provide the reader with a comprehensive approach to the development, optimization, and validation of leachables and extractables routine methods.

18.2 PURPOSES OF ROUTINE METHODS

A routine method is performed on a regular basis to characterize several lots of the same material/product, to release material/product, or to perform periodic testing of one or more lots during stability studies. Routine methods may be used at various

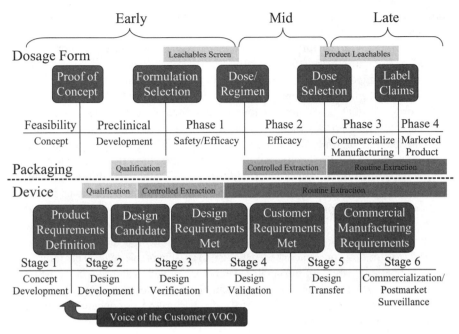

Figure 18.2   Testing associated with drug product development progression.

stages of development for a variety of intents that may dictate the method type: ID, content, impurities or another specific type of chemical measurement. The major focus of this chapter is extractables and leachables routine methods. As discussed below and illustrated in Figure 18.2, these methods are used for a variety of purposes throughout development and manufacturing.

18.2.1   Material Selection and Safety Evaluation

Extractables or leachables testing is performed for characterization of CCS/DD or the drug product, respectively. Early in the product development process, extractables profiles are generated by a variety of techniques with a range of solvent polarities. This characterization process, controlled extraction, is described in detail in Chapter 14. Although the same general starting parameters may be used for sample preparation and analysis, these often need to be adjusted for individual sample types. The result is a one-time implementation of an extraction method. A method that is used for a one-time characterization, for example, controlled extraction, is not considered a routine method.

Early in development, the focus is on compound identification and semiquantitative estimates. When a characterization method is used to compare samples, it is important that any differences observed are a result of the sample and not the method. Although these methods do not undergo the same level of development or validation applied to routine methods, extra care must be taken to ensure the reliability and accuracy of the results. The chemical profile of a CCS or container closure components of a drug delivery system is used at the early stage of product development to make material selection decisions. A preliminary safety assessment based on this information is utilized in conjunction with other safety information obtained from suppliers and selected biocompatibility testing to qualify CCS/DD materials.

The chemical profile generated in a controlled extraction study forms the foundation for further development activities in which leachables may be evaluated in the drug product. This progression is illustrated in Figure 18.2. Leachables methods that are used to characterize multiple lots or several samples stored for different lengths of time and/or under varied conditions are considered routine methods. Early in development, a routine leachables method (leachables screening) may be performed on a drug product with CCS/DD made from different materials to guide drug product CCS/DD design. It is preferable to test leachables in a drug product, or an early representation of the drug product, or a simulated drug product as described by Jenke,[15] early in the drug development process so that if unacceptable leachables are detected, there is opportunity to redesign the container closure materials to eliminate or reduce the leachables to an acceptable level. This minimizes the risk to patients who will eventually use the developed drug product. If the CCS/DD can be designed with demonstrated leachables levels less than the defined safety concern threshold (SCT; see Chapter 4 for a discussion of this concept), then further study of leachables can justifiably be limited and possibly no routine control for leachables will be warranted.

However, it is not always possible to limit all leachables to less than the SCT. The chemical profiles obtained with early-phase extractables or leachables methods may contain peaks representing compounds of specific safety concern (CSSC) or compounds at levels above the SCT. Taken together, the results from the leachables and extractables routine methods can be used to establish a leachables/extractables correlation. From the SCT and dosage information, an analytical evaluation threshold (AET) can be derived and used with the correlation information to develop later-phase routine leachables or extractables methods for safety assessment or control. The application of thresholds to analytical methods is discussed in detail in Chapter 5.

During the later stages of product development, it is expected that routine methods will be used to monitor inhalation and parenteral product leachables during registration stability studies. These methods are focused on known compound quantification and detection of new species. This study of leachables is usually included in the market application made in support of a new drug product. Information on the effect of storage time, temperature, and relative humidity on leachables is normally included. An estimate of maximum leachables levels likely to be dosed to a patient is made together with a safety assessment of leachables at this level. Information on the source of leachables and a strategy for their control in the marketed product can also be included.

18.2.2 Control of Product Leachables

Typically, leachables require weeks or months to appear in the drug product at a significant level, since they diffuse slowly from the materials of the CCS. Thus, it is normally inappropriate to control leachables levels with release tests for the drug product. It is more often the case that controls are placed on the compounds that can be extracted from the CCS materials as part of routine analytical tests. The information gained from the controlled extraction study can be used to develop an appropriate routine extractables test. A correlation between the leachables and extractables levels can usually be demonstrated and thus an extractables specification provides good control over leachables in a drug product. If a leachables/extractables correlation cannot be made, then it may be necessary to implement a leachables routine test. These methods are focused on detection and quantification of analytes above a specified threshold. These specifications are only appropriate if the leachables significantly affect the safety or quality of the drug product.

18.2.3   Verification of Material Consistency

Although leachables and extractables methods have been regularly used to evaluate the pharmaceutical dosage form and its CCS/DD, there is an increased emphasis on upstream controls. This has led to the application of routine methods to evaluate the impact of the manufacturing process and assessment of single-use equipment.[7,16,17] The processing components are different from those in CCS/DD, but the materials of construction are similar. A routine extractables test can be used to release raw materials that are used to manufacture either processing or CCS/DD components. These methods rely on composition information or controlled extraction study results, and are focused on quantification of specific analytes, qualitative evaluation of the profile, and detection of new compounds.

18.3   A RISK-BASED APPROACH TO THE ROUTINE CONTROL DECISION

The principles outlined in International Conference on Harmonisation (ICH) Q8[18] and Q9[19] are used to develop a risk-based approach to the evaluation of leachables and extractables. This approach principally seeks to build an extensive knowledge of the factors affecting the potential for leachables in a drug product. Through this exercise, it is possible to define a control strategy that identifies critical parameters and attributes that are relevant to minimize the effect of potential product leachables. An appropriate suite of routine methods is developed based on the defined strategy.

The need for development of routine methods may be guided in part by the nature of the drug product. The drug product type will heavily influence the probability that leachables will be present at a level that presents a safety concern. The United States Food and Drug Administration (FDA) document "Guidance for Industry: Container Closure Systems for Packaging Human Drugs and Biologics"[20] summarized its view of the relative "likelihood of packaging component–dosage form interactions"; this may be considered a surrogate for the likelihood of leachables from these different product types. Important mechanisms to consider include solid/liquid transfer where the container closure materials are in direct contact with a solvent vehicle, which will promote leaching. This is the case with metered dose

Degree of Concern with Route	Likelihood of Packaging Component–Dosage Form Interaction		
	High	Medium	Low
Highest	Inhalation: Aerosols, Solutions Injection: Liquids	Inhalation: Powders Injection: Powders	
High	Ophthalmics Transdermal Nasal		
Low	Topical: Liquids, Aerosols Oral: Liquids	Topical: Powders Oral: Powders	Oral: Tablets Capsules

Figure 18.3   Likelihood of packaging component–dosage form interactions for different classes of drug products. Adapted from FDA "Guidance for industry: Container closure systems for packaging human drugs and biologics," 1999.

inhalers (MDIs) that contain organic liquids or parenteral products where aqueous solutions may be in contact with plastic components, glass containers, or an elastomeric stopper. Alternative mechanisms to consider include (1) solid/solid transfer where the CCS is in prolonged direct contact with a powder dosage form and a compound migrates from one to the other, and (2) solid/gas transfer where, via vaporization, volatile compounds are released from the CCS/DD that subsequently adsorb to solids or remain in trapped air, which are then dosed to patients. These might be relevant mechanisms where classical leaching via a solvent vehicle is impossible, but leaching remains a possibility, for example, dry powder inhalation systems. As illustrated in Figure 18.3, the route of administration must be considered when determining whether or not a leachable is of safety concern; parenteral and inhalation products share the highest risk category. In addition to the route of administration, it is also relevant to consider duration over which exposure is likely. Chronic versus acute dosing may lead to different levels of safety concern. All of these factors are to be taken into account when considering development of routine methods to determine leachables levels.

The need for routine control via extractables testing ideally can only be made after enough information is available from leachables stability studies on the developed product. However, it is often not feasible to wait for all of the leachables information to be gathered prior to making such a decision. The controlled extraction study information can be used in conjunction with material composition, biocompatibility test results, and knowledge of the component manufacturing process to perform an appropriate risk analysis to determine whether or not routine extractables testing is an appropriate control for leachables.

This element of risk analysis may be addressed as part of a component or assembly failure modes and effects analysis (FMEA) by the DD design authority in

TABLE 18.1. Risk Analysis Rating Examples for Mitigation by Routine Control

Risk rating factor	Material safety (detectability)	Component criticality (severity)	Component manufacturing process controls (occurrence)
1	USP class VI, food compliance—all regs, no known CSSCs, used in predicate device	Functionally important	Well-controlled/ cleanroom environment
5	USP class I/III, food compliance—some regs, CSSCs below AET	Drug path/mucosal contact	Some controls/monitored clean environment
10	No USP class, food compliance—none, CSSCs above AET	Functionally important and drug path/ mucosal contact	Few controls/clean environment

collaboration with the pharmaceutical sponsor. Alternatively, the pharmaceutical sponsor may perform a separate risk analysis that is focused on the chemistry associated with a particular component or assembly. This may be accomplished by following a risk assessment scheme similar to that used for FMEA or failure modes and effects criticality analysis (FMECA).[21,22] The material composition, component type, and manufacturing environment can be rated in terms of safety, criticality, and control, respectively. The safety of the material composition can be thought of as detectability—The more characterization that has been performed and standards to which the material is compliant, the more is known. The criticality of the component can be likened to severity. This, in practical terms, can be related back to the component's functional importance and nature of contact with either the dosage form or the patient.

Finally, control of the manufacturing environment can be thought of as indirectly relating to occurrence. For example, the occurrence of unspecified extractables would be more likely in an environment that is not tightly controlled. Brief descriptions of these factors and the respective rating numbers are given in Table 18.1. Risk rating numbers for each of these factors, referred to as risk rating factors (RRFs), can then be combined by multiplication to obtain a risk priority number (RPN). Thus, with an RRF range of 1 (low risk) to 10 (high risk), the lowest RPN obtainable is 1 ($1 \times 1 \times 1$), and the highest is 1000 ($10 \times 10 \times 10$). Based on the RPN, a determination can be made regarding the need for routine testing; the higher the number, the greater the need for mitigation. Prior to the risk analysis, the pharmaceutical sponsor would determine the RPN above which routine testing is required. A second threshold RPN may be established above which the recommended routine control would be an extractables method. This type of determination is unique to the component in question and should be handled on a case-by-case basis. A few examples are provided below that demonstrate how this risk analysis process may be used to determine the need for one or more routine control methods:

Example 1

- mouthpiece for DPI
- material safety (RRF = 5)—USP class I, Code of Federal Regulations Title 21 (21 CFR) compliant, no CSSCs above AET
- component type (RRF = 10)—functionally important and drug path/mucosal contact
- component manufacturing (RRF = 5)—process controlled/monitored clean environment
- RPN = 5 × 10 × 5 = 250
- routine controls
  - material: Fourier transform infrared spectroscopy (FTIR), visual
  - component: visual, dimensional, routine extraction for lot release
- rationale: high RPN, component is both patient and drug contacting, extractables testing will ensure consistency of material

Example 2

- molded part that forms the mechanical link between assemblies during inhaler use
- material safety (RRF = 5)—no USP class, 21 CFR compliant, no known CSSCs
- component type (RRF = 1)—functionally important, requires specific polytetrafluoroethylene (PTFE) level
- component manufacturing (RRF = 5)—process controlled/monitored clean environment
- RPN = 5 × 1 × 5 = 25
- routine controls
  - material: PTFE content, FTIR, melt viscosity, visual
  - component: visual, dimensional
- rationale: low RPN, component is not patient or drug contacting, PTFE level is better measured at the compounding stage than via an extractables method after fabrication in a clean environment

Example 3

- molded elastomeric part that holds pressure during inhaler use
- material safety (RRF = 5)—21 CFR compliant, no CSSCs above AET
- component type (RRF = 10)—functionally important/drug path
- component manufacturing (RRF = 10)—few controls/clean environment
- RPN = 5 × 10 × 10 = 500
- routine controls: extractables, FTIR, visual, dimensional, cure level (*R*-value)
- rationale: very high RPN, extractables implemented to ensure chemical integrity of material for proper functionality and mitigate low level of control in

manufacturing environment to ensure material consistency for drug path component

As can be seen from the above examples, it is not always the case that an extractables method is put in place as a routine control for a functionally important OINDP component. A risk-based, rational approach to the routine control of materials will, in some cases, depending on the level of risk and nature of the mitigation required, warrant the use of alternate methods or may invoke the use of extractables testing. Similar constructs could be developed for other types of routine methods— these examples are provided for illustrative purposes only.

18.4 ROUTINE METHOD REQUIREMENTS

If it is determined that a routine testing method is required, the next step should be to determine the requirements (design intent) for the method. Crafting requirements for a routine method is an activity that involves several stakeholders. Unlike one-time characterization methods, there are several considerations for routine methods beyond the scientific realm that include practical concerns of time and resources, for example, personnel and capital. The team responsible for performing the method may or may not be directly involved in the early stages of development or even validation. The final method must meet its intended use and ultimately the expectations of the regulatory authorities. For this reason, it is important to capture what is known in process excellence programs (e.g., six sigma) as the "voice of the customer (VOC)."[23] This can be accomplished with a systematic process to evaluate the desired inputs and outputs prior to commencing development of the method. In six sigma terms, this process involves suppliers and customers (the stakeholders) and is known as SIPOC, which stands for suppliers–inputs–process–outputs–customers.[24] An input–process–output (IPO) diagram can be created that depicts the desired inputs and outputs. Inputs may include method targets for selectivity, sensitivity, precision, and any considerations from customers (project teams and regulators) for method results. Also included in inputs may be slightly less obvious items such as training and capital purchase requirements. Outputs may include a documented procedure to perform the analysis; a method which is fit for purpose (see the "Validation" section); and a set of system suitability criteria to apply in routine use. By documenting the design intent using an IPO process and sharing this with all stakeholders, the aims of the method development are clear from the beginning. An example of an IPO diagram for an MDI leachables method is shown in Figure 18.4.

When method development begins, the IPO can be used as a guide for development. It can define the minimum requirements for the method and highlight the areas that need focus during the development process. Other documents will also be relevant, such as guidelines on acceptable practices with regard to safety. If the data will be part of a regulatory submission, controls to ensure that work is conducted in a controlled environment, such as working to good manufacturing practice (GMP) standards, may be necessary.

Method development activities are best performed only after considering the identified critical aspects of the method and then translating those into specifics. The

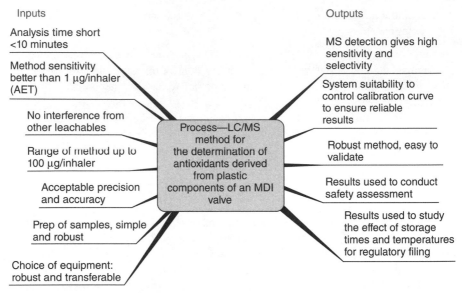

Figure 18.4   Method development IPO.

specifics may include method scope; potential target molecules or analytes of interest; requirements for system suitability, stability, or other method properties; practical requirements; technical constraints; and proposed analytical figures of merit. To facilitate this process, a method development worksheet (see Fig. 18.5 for an example) can be completed with the involvement of all stakeholders. Each of these method development considerations is discussed below.

18.4.1   Method Scope

To develop an appropriate method, the purpose of the test must be clearly understood. For routine control of materials, several tests may be performed. Upon receipt of incoming materials, it is commonly expected that the identity of a material will be confirmed (21 CFR211 and other regulations). Depending on the applicable regulation and the material of construction, this may be accomplished by receipt of a certificate of analysis or performance of an identity test. The definition of an identity test per ICH Q2(R1)[25] is as follows:

> *Identification tests are intended to ensure the identity of an analyte in a sample. This is normally achieved by comparison of a property of the sample (e.g., spectrum, chromatographic behavior, chemical reactivity) to that of a reference standard.*

Identity tests are typically spectroscopic in nature. For organic materials, plastics, and elastomers, this most often involves the infrared (IR) region of the electromagnetic spectrum. For example, FTIR can be used quite readily by placing a sample directly onto an attenuated total reflectance (ATR) crystal and performing a collection of scans to generate a spectrum. Typically, this type of test can be used to confirm the presence of analytes that are greater than or equal to 5% of the total composition. Alternatively, for the detection of metals, X-ray fluorescence can be used to determine the elemental composition. These types of methods can be used to evaluate chemical composition differences at the macroscopic level.

To evaluate chemical compositions at the microscopic level, an impurity test is used. Most routine extractables and leachables tests fall into this category. Per ICH Q2, the definition of an impurity test is as follows:

Testing for impurities can be either a quantitative test or a limit test for the impurity in a sample. Either test is intended to accurately reflect the purity characteristics of the sample. Different validation characteristics are required for a quantitative test than for a limit test.

Method Scope	Preclinical: ☐ Clinical phase: I ☐ II☐ III☐ Commercial: ☐ Purpose: Characterization ☐ Release ☐ Stability ☐ Type: ID ☐ Content ☐ Impurity-Limit ☐ Impurity-Quant ☐ Specific ☐		
Method Description	INSERT statement regarding general goal of the method and any specifics that are not captured in the other sections, e.g., control of extractables in a particular grade of plastic		
Potential Quantitative Targets	INSERT list of chemicals and CAS # (if known) that may be quantified		
Potential Qualitative Targets	INSERT list of chemicals and CAS # (if known) that may be relevant for specificity		
General Requirements	System suitability:	☐ INSERT requirements, e.g., method must have minimum resolution of 1.5 between two antioxidants analyzed by method	
	Method specific properties:	☐ INSERT requirements, e.g., method is expected to utilize HPLC/UV	
	Standard stability:	☐ INSERT requirements, e.g., should show validated standard solution stability of greater than 3 months at room temperature and in the light	
	Sample stability:	☐ INSERT requirements, e.g., should show validated sample solution stability of no less than 1 week at room temperature and in the light	
	Other:	☐ INSERT requirements, e.g., previous experience of this material type showed sample plastic must be ground to a minimum particle size prior to extraction, need to confirm for this grade	
Technical Constraints (of routine test lab)	Instrumentation:	No ☐ Yes ☐ INSERT constraints, e.g., lab developing procedure only has one type of LC/MS that is different from routine lab, need to show equivalence	
	Materials:	No ☐ Yes ☐ INSERT constraints, e.g., lab developing procedure only has access to one type of mill to pregrind sample that is different from routine test lab, need to show equivalence	
	Other:	No ☐ Yes ☐ INSERT constraints	
Practical Considerations	Run time:	☐ INSERT requirements, e.g., run time for analytical finish will need to be minimized to enable sequence of prepared samples to be analyzed prior to expiry of samples due to degradation of antioxidant in solution	

Figure 18.5 Method development requirement worksheet.

	Std prep time:	☐ INSERT requirements, e.g., solubility of antioxidants is poor in mobile phase, need to introduce a 10-minute sonication step to ensure dissolution
	Sample prep time:	☐ INSERT requirements, e.g., must be able to process at least 10 batches per day to keep up with production release requirement
	Other:	☐ INSERT requirements
Proposed Analytical Figures of Merit	Accuracy:	☐ INSERT requirements, e.g., should show validated accuracy of recovery of better than 90% in the first extraction cycle
	Repeatability:	☐ INSERT requirements, e.g., RSD of analyte amount extracted should be less than 10% for replicate identical samples during sample extraction sequence
	Intermediate precision:	☐ INSERT requirements, e.g., RSD of analyte amount extracted should be less than 20% for replicate identical samples during at least two different extraction sequences regardless of operator, instrument used, analysis employed, or day of analysis undertaken
	Specificity:	☐ INSERT requirements, e.g., method must be able to determine the amount of antioxidants in the presence of other formulation constituents and also able to quantify accurately antioxidants in the presence of antioxidant synthetic impurities that may be present by showing the required chromatographic separations and peak purity
	LOD:	☐ INSERT requirements, e.g., LOD should be lower than ½ AET level
	LOQ:	☐ INSERT requirements, e.g., LOQ should be lower than AET level
	Linearity:	☐ INSERT requirements, e.g., coefficient of determination > 0.990
	Range:	☐ INSERT requirements, e.g., linear response should cover expected range of samples and from LOQ to 150% nominal level for sample
	Reproducibility:	☐ INSERT requirements, e.g., any additional laboratories performing method should meet precision requirements outlined above.
	Other:	☐ INSERT requirements, e.g., robustness of the procedure will be explored

Figure 18.5 (*Continued*)

The amounts of analytes typically measured in a routine extractables or leachables method are at the parts per million or parts per billion level. The component extract derived from the material being tested may be analyzed using gas chromatography (GC), liquid chromatography (LC), inductively coupled plasma (ICP), atomic emission spectroscopy (AES), or atomic absorption spectroscopy (AAS), depending on the type of analyte to be determined in the material. The choice of analytical technique will be discussed below.

Although most extractables and leachables routine methods can be categorized as impurity methods, there are some methods that are developed to measure a single type of analyte (e.g., nitrosamines) in a matrix. These types of methods are very specific and may require unique techniques to reach the levels of sensitivity and selectivity required.

It is important to be specific in setting the scope of the method. The typical questions that should be answered in considering the scope are the following:

1. What is the sample to be analyzed?
2. What is to be detected?
3. At what level?

These method properties are specific to the type of analysis desired. For example, it would be important to clarify if a leachables method is to be designed to measure only volatile compounds. For an extractables method, it may be desirable to have a method that can be used for molded components as well as raw material, or for other components made of the same raw material, or more broadly, for molded components and raw material of the same chemical type, for example, polypropylene (PP), polybutylene terephthalate (PBT), and polyvinyl chloride (PVC). There may also be properties of the method that are specific to the type of sample preparation desired. For example, a particular extraction technique—for example, Soxhlet, reflux, or accelerated solvent extraction (ASE)—may be specified for business or regulatory reasons. These may include instrumentation purchase or maintenance costs or regulatory requirements for environmental discharge. In some cases, there may be technical constraints that dictate this selection. Such constraints may include available equipment at the testing site or other factors like analyte or drug formulation availability, solubility, temperature, or light sensitivity.

18.4.2 General Requirements

After the method scope is defined and the analytical method is identified, the general requirements for system suitability may be formulated. For results obtained from any analytical instrumentation to be considered valid, it is essential that proper system performance be verified at the time of use. System suitability is the means or method of verifying that the integrated instrument and data system works according to the performance expectations and criteria set forth in the method. This test or group of tests assures that the "system" met an acceptable performance standard at the time of the testing. It must be performed no later than the end, preferably before and possibly during the analyses. For example, biochemical and immunological assays may incorporate test controls within the samples being analyzed. The results of the analysis are accepted or rejected on the basis of the measured value of the internal control. In special cases, the incorporation of such internal controls may be necessary for leachables or low-level extractables methods. This type of test is one form of system suitability. For chromatographic assays, system suitability criteria are commonly comprised of precision, theoretical plates, tailing, and resolution.[8,26] For other assays, suitable criteria must be established.

In the pharmaceutical industry, it is a common practice for quantitative impurity methods to use a "check standard" as a part of system suitability requirements.[26] This involves preparation of a second standard using the same procedure as that used to prepare the method working standard. The concentration of the check standard is expected to be within an acceptable range of the working standard. Unfortunately, this approach, at best, measures an analyst's ability to repeat the preparation but really gives no indication of whether or not the instrumentation is operating appropriately. An alternative to this approach is to compare the measured concentration to the theoretical concentration, which is calculated from the amount per volume and includes application of purity factors and percent moisture correction. At a minimum, this ensures the accuracy of the standard preparation and may also confirm system performance.

A more comprehensive measure of the overall measurement system performance can be evaluated by using a control chart. A general discussion of the application of control charts to analytical methods can be found in Nunnally and McConnell[27] or Konieczka and Namiesnik.[28] Control charts may be used to establish that a system meets the performance standard. Additionally, it has been shown that drift in analytical results may be predicted from the drift observed in a system suitability control chart.[29] It should be noted that this implementation requires several data sets before it can be utilized as a system suitability check; this will likely be in the later stages of method development or possibly after validation.

18.4.2.1 *Control Chart Example*

As an example of control chart usage, consider two different quantitative routine extractables methods (method 1 and method 2), where there are two analytes (analyte 1 and analyte 2) in each of the methods. System suitability includes a baseline check, precision measurement (six injections), resolution and tailing factor calculations for selected chromatographic peaks, and a calculation of theoretical plates. While this system suitability determination provides adequate information to assess the quality of a particular set of chromatograms generated in one sequence for one analyst, it does not allow the opportunity to monitor the system from sequence to sequence or analyst to analyst. In this particular example, a response factor for the precision injections was calculated from the peak area normalized to the standard analyte amount. The results from several sequences were used to determine σ. A control chart for each analyte in each method was generated according to the principles of Shewhart[30] and is shown in Figure 18.6. The mean values are plotted with lines representing the grand mean (solid line), $\pm 1\sigma$ (dotted line), $\pm 2\sigma$ (dashed line), and $\pm 3\sigma$ (dotted dashed line). The control limits are $\pm 3\sigma$ and the alert (warning) limits are $\pm 2\sigma$. To interpret the control charts, the following rules from the Western Electric handbook[31] were adopted:

1. one or more points outside of the control limits,
2. two of three consecutive points outside the two-sigma warning limits but still inside the control limits,
3. four of five consecutive points beyond the one-sigma limits, and
4. a run of eight consecutive points on one side of the center line.

If one or more of these rules is broken an appropriate action must be taken. If the control limit is exceeded for one sequence, immediate action must be taken and the system suitability result is "fail." Other trends require investigation. For method 1, none of the rules was broken for either analyte 1 or analyte 2. However, it is noted that on two different occasions, analyte 1/sequence run #3 and analyte 2/ sequence run #10, the system suitability precision means were outside the alert limits, while the corresponding precision means for the alternate analytes, analyte 2/sequence run #3 and analyte 1/sequence run #10, were well within the 1σ limit.

Since the two standard analytes are prepared in one solution, it is likely that the response factor variation is a preparation error and not an analytical system error. This root cause has been confirmed for similar occurrences with other control charts. The converse situation is illustrated for method 2. Two different analyte response factors were observed to vary in the same direction for each sequence. As shown in Figure 18.6, there were more than eight consecutive points on one side of the center line for each analyte and rule #4 was broken. These two facts taken together made the instrumentation the first point of investigation. The decline in response factor

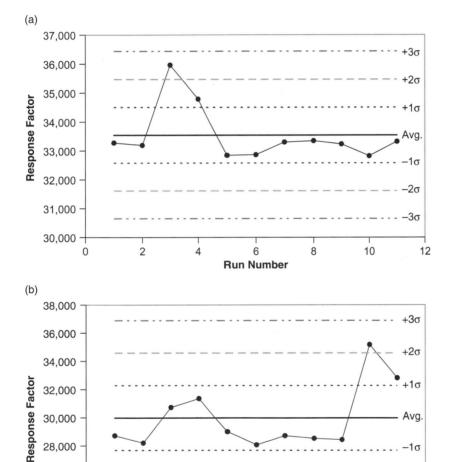

Figure 18.6   Control chart examples: (a) method 1–analyte 1; (b) method 1–analyte 2.

(c)

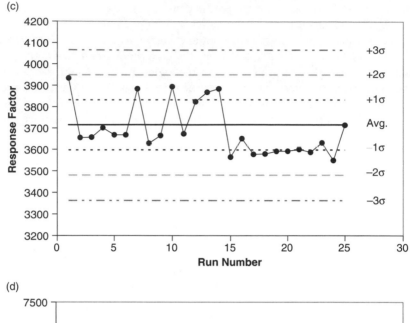

(d)

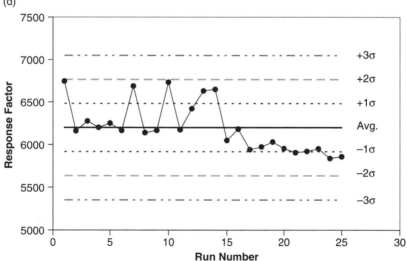

Figure 18.6 Control chart examples (*Continued*): (c) method 2–analyte 1; (d) method 2–analyte 2.

corresponded to a loss in sensitivity. Cleaning the cell of the ultraviolet (UV) detector restored the response factor. Although the instrument was qualified and within its calibration window, the control chart provided an early signal that otherwise would have been missed and may have led to suboptimal results.

Another aspect of the general requirements for a method is the sample and standard stability. For samples that will undergo testing over a period of days, it is critical that the sample, sample preparation, standards, and any associated solutions be stable over the period required for testing. Often, stability testing is only performed to cover the initial testing period. However, if additional one time testing is performed to extend the stability period, many hours spent on sample and standard preparation can be saved several times over for a routinely used method in the event that a retest is required. For chromatographic methods a 7- to 10-day stability period is beneficial.

In addition to the general requirements for a method, there are practical issues that must be addressed.[32] Methods used to run many samples are best suited to have short run times. This is often a balance between the amount of information needed and the time available to perform the study. Similarly, the number of standards and amount of time required to prepare each must be addressed. Sample preparation time may or may not be an item of concern depending on whether the routine method is intended for release of material or simply for a one-time leachables study. In any case, the time considerations need to be addressed with the purpose and frequency of the method in mind. It is important to discuss the method specifics and practical considerations with the team who will be performing the testing prior to development of the method.

18.4.3 Identification of Targets

After the controlled extraction study is performed as described in Chapter 14 and information is obtained from the supplier, a list of potential target molecules can be generated. Any CSSCs are expected to be tracked as quantified analytes in leachables or routine material control methods. It is assumed that method development activities will ensure that these compounds are reliably recovered from the sample and measurable at toxicologically relevant levels.

For routine control of materials, in addition to any CSSCs, it may be useful to identify target molecules that are suitable indicators of material consistency. To be considered as a suitable analyte for an extractables method, these target molecules should possess the following characteristics: expected and relevant to the material, extraction repeatability, present at readily measurable concentrations in the parts per million range, solution stability for the analysis turnaround time. A list of potential target types is compiled in Table 18.2 based on additives that are typically present in materials commonly used in OINDPs. These targets, in large part, originate from material composition requirements as well as processing steps that may introduce chemical changes associated with a particular component. Detailed lists of chemicals that may be expected in different materials can be found in Chapters 11, 13, 15, and 16. Commonly extracted chemicals have been identified by Jenke,[33,34] Yu et al.,[7] Fliszar et al.,[35] and Wang.[36] To ensure consistency of materials, the target may be quantified or tracked qualitatively.

In addition to quantified targets, there may be other analytes that should be tracked qualitatively during method development. These are chemical species that are present at detectable levels and ideally would have been identified from the controlled extraction study or supplier information; they may be known constituents of the material of composition, process residuals, or degradants thereof. These compounds may interfere with the quantification of targets due to having the same absorption maxima, mass, or chromatographic behavior. Alternatively, they may be

TABLE 18.2. Potential Targets Typically Present in OINDP Materials

Polymer	Potential extractables
PP	Antioxidants, lubricants, oligomers
HDPE	Clarifying agents, oligomers
LDPE	Oligomers
PET	Oligomers, volatile impurities
PC	Bisphenol A, antioxidants
PBT	Oligomers, THF
PVC	Monomer (vinyl chloride), phthalates
PS	Monomer, volatile impurities
ABS	Oligomers
PMMA (acrylic)	Oligomers
FKM (fluoroelastomers)	Solvents, oligomers
Silicone	Oligomers, catalyst residues
EPDM	Antioxidants, cure agents
Butyl	Oligomers
Glass	Metals
Stainless steel (grades 304, 316)	Drawing oils, surfactants

HDPE, high-density polyethylene; LDPE, low-density polyethylene; PET, polyethylene terephthalate; PC, polycarbonate; PS, polystyrene; ABS, acrylonitrile butadiene styrene; PMMA, polymethylmethacrylate; FKM, ASTM D1418 designation for fluoroelastomers that contain vinylidene fluoride; EPDM, ethylene-propylene-diene monomer; THF, tetrahydrofuran.

present at levels that would make them significant contributors to the qualitative profile of the material. This is particularly true for chromatographic methods that are used to ensure consistency of materials via an extractables profile. For example, there may be three measurable stabilizers (e.g., antioxidants, UV stabilizers) present in a particular material. For purposes of routine control, a decision could be taken to quantify the one that is most relevant to the desired characteristic of the material (long-term stability, yellowness, etc.) and treat the other two as qualitative targets. In some cases, it may turn out that a qualitative target evaluated during method development may need to be quantified in the future due to changes in the regulatory environment (e.g., phthalates content now required to be known).[37] To minimize the opportunity for unexpected impact on target quantification and to be prepared for alternate quantification needs, it is important to identify all analytes of interest at the outset of method development.

18.4.4 Selection of Routine Methodologies

The method will include methodologies for sample preparation, analytical separation and/or detection, and data interpretation. While the latter two will be dictated primarily by the nature of the analytes, the former will be primarily dictated by the nature of the materials to be analyzed. The exception to this is volatile analytes where the nature of the analyte plays an equally important role in sample preparation. For sample preparation in routine extractables methods, consideration must be given to the materials of construction for the CCS or DD; additionally, for leachables, consideration must be given to the properties of the dosage form.

18.4.4.1 Sample Preparation: Extraction Container closures typically consist of blown glass ampoules or vials, formed or stamped metal components, molded plastic or elastomeric components, and sealable foil laminates. To be effective and efficient, it is important that sample preparation methodologies be complementary to the physical form and properties of the material to be analyzed. An awareness of the material properties can also minimize sample loss as a result of handling due to evaporation, nonquantitative transfers, or interaction with other materials.

18.4.4.2 Component Considerations The first general consideration pertaining to the physical form of a component is the size and shape of the component. For large or irregularly shaped components, a reduction in size is often necessary so that minimal solvent or container volumes can be used. This can be accomplished by cutting, chopping, or cryogrinding, and depends on the final size desired. When the particle size is reduced to microns for materials such as plastics, the extraction time can be minimized because of the greatly increased surface area per gram of material.

Alternatively, for components that are very small, it is sometimes difficult to obtain enough of the components for an extraction. In such cases, it may be necessary to use microlevel extraction techniques to obtain appropriate concentrations of extractables. These may include the use of miniature extraction vessels, concentration of the extract, or surface sampling techniques.

Second, for some materials, the physical construction of the material may require selective extraction of only the drug contacting side to be relevant. This is particularly significant for foil laminates, which may have an asymmetrical construction where there are a number of layers of different chemical composition on the drug contacting side versus the nondrug contacting side. For specific applications such as MDIs, the interior drug contacting surface of the container may be coated and present a similar issue. Additionally, elastomeric closures may have a PTFE coating. In such cases, it would be inappropriate to chop or grind the component prior to extraction. These cases may necessitate the development of an apparatus configured for extracting only the relevant surface of the material.

18.4.4.3 Material Considerations First, the physical state of the material is important to consider in the development of the extraction conditions. Crystalline materials have lower dissolution rates than amorphous materials, which should be considered when determining liquid contact times. Since the goal of the extraction experiment is not deformulation, extraction times that avoid polymer dissolution are preferred. Elastomeric materials are porous and tend to swell because of solvent absorption. The R-value (amount of cross-linking) of a material should be considered when selecting the amount and type of solvent for the extraction. A solvent that results in minimal swelling is preferred so that repeatability is maximized.

Several different forms of liquids may be applied to device or packaging components. These include inks, adhesives, lubricants, lacquers, and coatings. Most often, these are tested in the finished form of the component since the quantities are generally very small relative to the size of the component and there may be a change in physical phase from liquid to solid after application, for example, as in the case of inks. Additionally, there may be some chemical change after application related to processing of the finished form of the component such as heat sealing.

Other physical properties of the material may dictate sample preparation conditions. The glass transition temperature is an important consideration for grinding and extraction. To enable efficient size reduction, particularly for elastomers, it is preferable to grind materials at temperatures below the glass transition; this is often accomplished by cryogrinding with liquid nitrogen. To enhance the mobility of the analytes, it is preferable to extract materials above the glass transition temperature. There are several practical reasons that it is not prudent for the extraction conditions to exceed the melt temperature of a material. These include excessive polymer precipitation in liquid extracts, contamination of the extraction apparatus, and carryover in the analytical method. The boiling points of potential extractables from a material must also be considered when selecting the extraction temperature to ensure that the extract preparation does not inadvertently cause the molecules of interest to evaporate unmeasured. Volatile extractables are typically captured using some type of gas-phase sampling technique such as a cartridge, loop, and purge/trap.

Second, in addition to the physical form of the material, the chemical composition of the material must also be considered in developing proper sample preparation conditions. A general measure of chemical compatibility is solubility. For known compositions, this parameter is expected to inform proper extraction solvent choice. Solvents or temperatures that are known to degrade the primary chemical constituents or known additives in a material should be avoided. The chemical compatibility of several polymeric materials with a variety of solvents can be evaluated using online tools like the one found at http://flw.com/datatools/compatibility/ (accessed September 9, 2011). Several examples are detailed in Table 18.3. Where the

TABLE 18.3. Examples of Solvent/Plastic Compatibility from Online Searches

Polymer (ref)	Water	Ethanol	Propanol	Ethyl acetate	Hexane	Cyclohexane
ABS (1,3)	Good	Good	Good	Poor	Poor	Excellent
HDPE (3)	Excellent	Excellent	Excellent	–	Good	Fair
LDPE (1, 3)	Excellent	Good	Excellent	Excellent	Poor	Good
PBT (2)	Good	Good	–	Fair	–	–
PC (1, 3)	Good	Good	Poor	Poor	Poor	Good
PP (1, 3)	Excellent	Excellent	Excellent	Excellent	Good	Poor
PS (3)	Good	Good	Excellent	–	Poor	Poor
PVC (1)	Good	Fair	–	Poor	Good	Poor

(1) http://flw.com/datatools/compatibility/ (accessed September 9, 2011); (2) http://www.ides.com/resources/plastic-chemical-resistance.asp (accessed September 10, 2011); (3) http://www.psyche.ee/pdf/Plastiku%20sobivus%20keemiliste%20ainetega.pdf (accessed July 31, 2010).

compatibility is listed as "poor," the solvent is expected to interact severely with the material and would not be suitable for extraction. For example, ethyl acetate (EtOAc) causes chain scission and is therefore inappropriate for use with polycarbonate. A similar table for elastomers is shown in Table 18.4. In general, solvents that are listed as "good" or "fair" may be suitable for extraction. However, the reactivity of the solvent must also be considered. For example, ethanol (EtOH) or methanol is not considered a suitable solvent because of the tendency to form reaction products with many molecules. Low-temperature antioxidants such as tris(2,4-di-*tert*-butylphenyl) phosphite (TBPP) are easily oxidized to a varied extent depending on the solvent that is used (see Fig. 18.7). These types of chemical reactions can be sources of irreproducibility and should be avoided.

Some extraction techniques may not be suitable for specific chemical composi-tions. Thermoplastic elastomers and low-melting-point plastics readily reach their softening point when microwave extraction is used. Metal components are generally not compatible with microwave extraction.

18.4.4.4 Sample Preparation: Leachables
Leachables methods must be able to detect analytes at parts per million levels in the presence of a complex drug product matrix. The greatest challenge is eliminating or minimizing the effect of this matrix. Often, the enrichment of the sample solutions is undertaken to present more analyte to the detector. For example, in the analysis of MDIs, it is frequently possible to enhance detection limits by combining the contents of individual inhalers from the same batch. MDIs contain a volatile propellant in which the leachables become dissolved. By allowing the volatile propellant to evaporate, the remaining nonvola-tile contents contain the leachables that can then be redissolved in a smaller volume of solvent for analysis, resulting in an enhancement in sensitivity. For other sample types, alternative strategies may be necessary. For example, in the analysis of aqueous-based systems, enrichment of the sample might be possible by analysis via headspace sampling of volatile leachables that can be captured if the sample is heated. Semivolatiles may be captured by cold filtration.[38] Another potential approach is a liquid–liquid extraction (LLE) employing an organic solvent, which is not mis-cible in water, but into which hydrophobic leachables will readily migrate. If the organic phase volume is smaller than the aqueous phase volume, then enrichment of the analyte concentration is possible. Other enrichment techniques include solid-phase extraction (SPE),[39] solid-phase microextraction (SPME),[40] and more novel approaches such as dispersive liquid–liquid microextraction (DLLME).[41] In all of these techniques, the aim is the same—to minimize matrix effects and enhance the concentration of leachables so they are most readily detected with good precision and accuracy.

In some cases, it may not be possible to remove an analyte from a complex matrix, and other approaches must be taken to eliminate matrix effects. Rather than focus on sample preparation, it may be necessary to focus on analyte separation during analysis. For example, chromatographic conditions may be chosen so that the effect of viscous surfactants is minimized.[42]

For all these techniques to be successful, the solubility of the leachables in the selected solvents must be understood. For successful LLE, the solubility of the

TABLE 18.4. Solvent Compatibility Chart for Elastomers

Polymer	Nitrile elastomer	Fluoro-elastomer	Platinum-cured silicone elastomer	Peroxide-cured silicone elastomer	Ethylene-propylene-diene monomer	Polytetrafluoroethylene
ASTM D-1418 designation	NBR	FKM	VMQ	VMQ	EPDM	PTFE
Common trade names	Buna-N	Viton, Fluorel	Silastic, Elastosil	Silastic, Elastosil	Dutral, Nordel	Teflon
Acids, dilute	Good	Excellent	Good	Good	Good	Excellent
Alkalis	Good	Good	Excellent	Excellent	Excellent	Excellent
Alcohol—glycols	Excellent	Good	Excellent	Excellent	Excellent	Excellent
Animal oils and fats	Excellent	Excellent	Good	Good	Fair	Excellent
Soaps, bleaches, detergents	Fair/good	Good	Fair	Fair	Excellent	Excellent
Steam	Fair	Fair	Good	Good	Excellent	Good
Vegetable oils	Excellent	Excellent	Excellent	Excellent	Good	Excellent
Water	Good	Excellent	Fair	Fair	Excellent	Excellent

Adapted from http://www.stiflow.com/FS_RefGuide.htm.

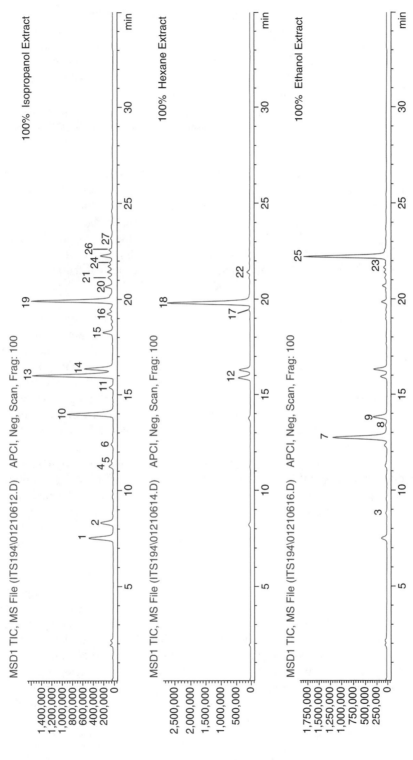

Figure 18.7 Chromatograms of a polypropylene sample extracted with three different solvents.

leachables in the organic solvent selected must be greater than in the sample matrix. If the leachable is not freely soluble in the solutions chosen for sample preparation or analysis, it is highly likely that the method will give poor results. Solubility can be assessed in a number of ways including, but not limited to, comparison of literature values, evaluation of partition coefficients[43] or physical measurements of solubility, for example, determining concentrations of solutions with and without a filtration step to monitor for potential solubility issues. Determination of conditions appropriate to dissolve the leachables of interest is an important part of method development.

18.4.4.5 Analytical Methodology Selection The selection of an analytical technique for a routine method depends on a variety of factors that include, but are not limited to, the following: analyte type—physical form and chemical nature, sensitivity required, and complexity of the sample matrix. These factors can be incorporated into a decision tree such as the one shown in Figure 18.8. The first level of concern is what needs to be measured. The second level of concern is the nature of the matrix that will be used to present the analyte to the measurement system. The third level of concern is the level at which it must be measured.

First, in consideration of the nature of the analyte, it is important to know the volatility of the compound of interest. Non- and semivolatile compounds are readily measured in the liquid phase, whereas volatiles are more readily measured in the gas phase. For a given sample, more than one analytical technique may be needed to quantitatively measure all analytes of interest. If a compound is to be measured in the liquid phase, it is important to know the solubility of the compound in various solvents. Additionally, it is important to consider the ionic nature of the compound along with its pKa. Based on these considerations a suitable solvent system for the measurement can be identified.

Second, in consideration of the matrix in which the analyte will be presented, it is important to know whether physical or chemical separation is required. A physical separation may be performed utilizing a boiling point (themogravimetric analysis [TGA]) for a volatile, or density (centrifugation) or molecular weight (gel chromatography) for a liquid. A chemical separation is typically accomplished by some type of chromatography utilizing a stationary phase that retains the analyte until it is eluted either by boiling point or chemistry of a mobile phase as in GC and high-performance liquid chromatography (HPLC), respectively. If a routine method is performed utilizing dissolution of a solid formulation or aggressive extraction of a solid sample, it may be necessary to use centrifugation or filtration of a sample prior to presenting the sample to the chromatographic system. The chemical complexity of the sample will not only dictate the level of separation required, but will also influence the detector specificity required as shown in Figure 18.8. For example, if two volatile peaks co-elute and the chromatography cannot be adjusted to separate them, a mass spectrometric or FTIR detector would be required instead of the less specific flame ionization detector. Liquid chromatographic detection systems most commonly rely on the absorbance of a chromophore (UV/Vis or fluorescence).

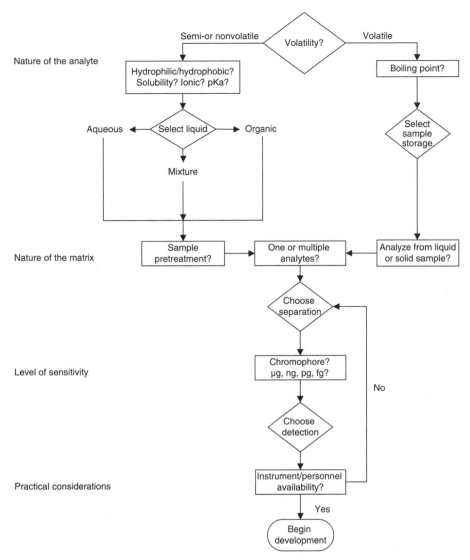

Figure 18.8   Analytical technique decision tree.

Third, in consideration of the sensitivity required, the detection options will depend on whether the analyte will be in a solid, liquid, or gas. The sensitivities of a variety of detection systems are given in Figure 18.9. Typically, for leachables and extractables routine methods, the measurements are at the parts per million or parts per billion level. To make an appropriate selection for a detection system, the chemical nature of the analyte must be considered. For example, if a nonvolatile analyte with no chromophore is to be measured in a liquid form at the picogram level, the choices are possibly electrochemical, or more likely, mass spectrometric detection. Alternatively, the analyte could be derivatized with a chromophore or detected by

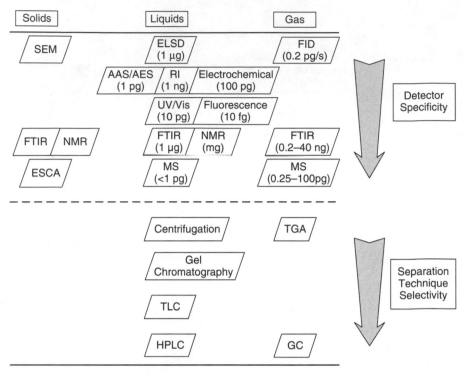

Figure 18.9 Specificity, selectivity, and sensitivity of analytical systems. Detection limits in parentheses are adapted from Skoog et al.[80] and Ewing.[81] SEM, scanning electron microscopy; RI, refractive index; ESCA, electron spectroscopy for chemical analysis.

immunoassay, which may introduce greater variability into the measurement. In some cases, the choice of the detection system may also be based on the dynamic range that is required.[36] For example, if the analyte concentration varies over two orders of magnitude, evaporative light-scattering detector (ELSD) or IR may have limited usefulness, whereas flame ionization detection (FID), mass spectrometry (MS), or UV/Vis may be more appropriate.

The last consideration in the selection of an analytical technique is the practical aspect. It is important to consider the lab at which the method will be performed and whether or not such instrumentation is readily available or can be qualified for routine use. The level of expertise required to perform the technique may be a stronger consideration, since the more complex the instrumentation, the more time may be required to perform the analysis and the more opportunities for issues to arise. Thus, it is important to balance both practical and scientific considerations.

18.4.4.6 Standard Selection After the selection of the analytical technique, it is important to select the standard compounds that will be used in method development. Standards will typically include the identified analytical targets and possibly others. A standard may be used for the following purposes:

- establish a calibration curve or response factor for quantification of analytes;
- a retention time marker for calculating relative retention time;
- to determine system suitability—tailing, resolution, repeatability, sensitivity (limit of quantitation [LOQ]);
- to determine accuracy—may be used for spiking; and
- confirm identification.

A standard compound is expected to have a known purity, which should be near 100%. Standards ideally should be sourced from a reliable, high-quality supplier able to certify purity of the standards. Impurities may interfere with the analytical measurement due to a difference in molar absorptivity or co-elution with a peak of interest. This aspect is important regardless of whether the standard is used for qualitative or quantitative purposes. Standards that are used for quantitative purposes are expected not only to have chemical integrity but also to have known physical attributes that would allow calculation of concentrations from first principles using properties such as molar absorptivity. When added to the sample at the beginning of the preparation, an internal standard can be used to calculate the recovery of the analyte through the process. In this case, the internal standard must be sufficiently chemically similar to the analyte of interest to ensure that the calculation of recovery is meaningful. For quantification of a specific analyte by using a response factor or calibration curve, it is preferable to use certified reference materials.

Specific leachables or extractables may or may not have readily available standards to allow accurate quantitative measurements. If not, consideration will have to be given to alternative approaches. A surrogate standard could be used if its suitability to provide accurate results could be demonstrated. This might be accomplished by evaluating the chromatographic behavior and detector response to the surrogate. This demonstration would be dependent on chemical similarity and, of necessity, have to be done on a case-by-case basis. Alternatively, one may have to consider manufacture of a suitable standard. This might be achieved through extraction and purification of the materials that are the source of the analytes, or it might be achieved by chemical synthesis. If a purified or synthesized compound is to be used as a standard reference material, it is important to characterize that material. The following are the requirements for standard reference materials[44]:

- Confirm structure by spectroscopic methods—UV/Vis, FTIR, and nuclear magnetic resonance (NMR).
- Confirm molecular mass by MS.
- Evaluate impurities—inorganic (ICP), organic (LC/UV, LC/MS), and volatile (Karl Fischer [KF], GC/MS).

The preparation and characterization of a standard reference material can be difficult and time-consuming, and is not considered trivial. Depending on the particular use of the standard, a partial characterization may suffice.

18.4.4.7 Selection of Response to be Measured and Reporting Format   After selection of the sample preparation and analytical/detection

methodologies, it is necessary to define the response that is to be measured. While this may seem straightforward, it is not necessarily the case. For example, standards are usually prepared in units of mass per volume of solution. However, samples are typically provided in units of mass or numbers of components. Thus, for a quantitative method, there will be some type of manipulation of the signal from the analytical instrument that involves correlation with a standard and subsequent conversion to a meaningful value for the sample.

Typically, for leachables and extractables, the goal is to report their mass per component or per gram of component. After the analytical signal (microvolts, counts per second, etc.) is transformed into an analytical measurement (transmittance, peak height, peak area, etc.), a standard calibration curve or response factor may be used to convert the measurement into a concentration. Through the use of standard purity correction factors, sample dilution/concentration factors, and accurate measurements of volumes and masses, this concentration can be converted to the desired mass per gram of component. An example of a standard calculation is as follows:

$$\text{Response factor} = \frac{\text{Average peak area for analyte over run}}{\text{Concentration of analyte in working standard } (\mu g/mL) \times \text{purity}},$$

$$\text{Conc analyte } (\mu g/g, \text{ppm}) = \text{Peak area} \times \frac{\text{Sample volume (mL)}}{(\text{RF}) \times \text{sample mass (g)}}.$$

It should be pointed out that to use a response factor in place of a calibration curve, it must be demonstrated that the response is linear over the range of use and that the intercept is not significantly different from the value obtained with no analyte present. If a calibration curve is used, it is not necessary that the curve is linear, but that an equation can be derived to reliably predict the values of the standard response over the measured range. Most often, these equations are first- or second-order polynomials. The reported values can be used subsequently to assess safety or consistency of the material of construction.

18.5 DEVELOPMENT AND OPTIMIZATION CONSIDERATIONS FOR ROUTINE METHODS

After selection of the methodology for sample preparation and analysis, there are several steps to be taken. Generally, these can be grouped as follows: development, optimization, assessment, and validation. The overall progression involves discovery of the experimental parameter boundaries, determination of the range of conditions that will produce acceptable results, and finally, confirmation that the selected range of conditions will regularly produce reliable results.

At the outset of method development, it is common to rely on previous experience, published information, standardized methods (United States Pharmacopeia [USP], American Society of Testing and Materials [ASTM], European Pharmacopoeia [Ph. Eur.]), technical notes from instrument suppliers and internal routine method requirements to define the boundaries of the experimental space that will be explored. A list of useful information to obtain prior to experimentation can be found in Crowther et al.[45] The establishment of a "knowledge space" can require more time than is available for experimentation and oftentimes is not fully explored. Prioritization of specific aspects of the methodology can be assisted by using an Ishikawa diagram/FMEA approach as described by Borman et al.[46]

The traditional approach to method development is univariate and involves examining one factor at a time (OFAT) to determine its effect on the measurement. Typically, one experimental parameter, such as time or temperature, is varied while all others are held constant. Sequentially, each parameter of the method is varied until all parameters that are considered important have been examined. A second set of experiments may be performed by combining the levels for each parameter that produced an optimal response. By a sequential iterative approach, a set of conditions is identified that produces an optimal experimental output, for example, maximum signal, minimum background, minimal degradation, and minimal variability.

For example, consider the development of an HPLC method with UV detection. Typical parameters to be investigated include column—length, particle size, stationary phase composition, and temperature; mobile phase—solvent composition, ion pairing agents, pH, flow rate, and gradient/isocratic; sample—injection volume, matrix; and detector—wavelength, cell path length, and volume. The normal progression as outlined by Snyder et al.[13] includes (1) selecting a detector, (2) choosing "best guess" conditions for preliminary runs, and (3) optimizing the separation conditions through trial and error. Krull et al.[47] discussed a similar simplified approach that involves (1) column screening with one mobile-phase/pH combination; (2) mobile-phase selection with chosen column; and (3) selection of other parameters with chosen column and mobile phase. This process can be very tedious and result in an incomplete data package due to co-elution and peak exchange. The OFAT approach, while systematic, does not provide the opportunity to characterize parameters that interact and produce a different effect than any of the parameters when characterized individually.

An alternate approach to method development is multivariate and involves examining multiple factors at one time with a design of experiment (DOE).[48,49] Typically, a screening design is used to determine which of the selected parameters has the most significant impact on the response; these are called main effects. An optimization design is used to determine the main effect levels that will produce the optimal response. These results can be used to map a response surface. From this, a confirmation design can be used to test the model's accuracy and the robustness of the method. For example, in addition to optimizing the chromatography, it is important to optimize the detection system. This could involve the use of chemical understanding of the analytes and performing a series of experiments to select parameters for a mass spectrometer that will ensure that the response is maximized and consistent. Many instruments come with software to aid in the DOEs to accomplish this.

The difference between the DOE approach and the OFAT approach is illustrated in Figure 18.10, where the diamond shape represents one combination of experimental parameters. For example, consider two experimental parameters and a total of 11 experiments to be performed. The OFAT approach involves sequential

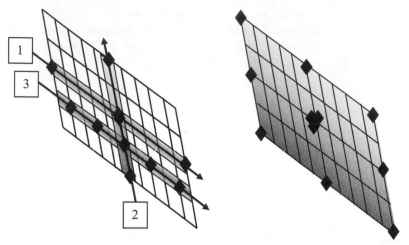

Figure 18.10   Graphical representation of OFAT (left) and DOE (right) approaches.

sets of experiments (identified as 1, 2, and 3 in Fig. 18.10) that explore a progressively smaller space to identify an optimum combination. This iterative approach relies on simple linear y versus x types of analysis. The DOE approach involves one set of 11 experiments to create a response surface across the entire space that is then used to identify an optimal region. This approach relies on matrix analysis of polynomial equations that contain xy terms. If there is any curvature to the response surface, it is possible that the sequential approach may not reveal a truly optimal combination, while the DOE approach is more likely to do so because of its treatment of the interaction between two or more factors in the mapping process.

The practical use of DOE may be illustrated by a few examples. The first is one in which the critical experimental parameters are determined—the screening DOE. A headspace GC method was to be developed for measuring extractables volatiles. A Plackett–Burman design was utilized to determine which parameters had the greatest effect on the signal intensity. In this case, nine factors were evaluated at two different levels. If a traditional orthogonal experiment had been used to investigate these factors, it would have required 512 (2^9) experiments instead of the 12 that were performed. While it is true that there is no complete representation of the effect of each factor and many of the factor combinations are confounded, this approach provided enough information to construct a second set of experiments using the most significant factors. The factors investigated and results obtained are listed in Table 18.5. Statistical analysis of the results showed that for EtOH, all factors were significant, and that for EtOAc, seven out of nine factors were significant. Rank ordering of the factors by significance showed that the top three factors were incubation temperature, mass, and oven temperature. From this, an optimization DOE could be designed where the three factors were varied at three levels. From this, the response surface for both compounds could be mapped.

A second example is discussed to illustrate the use of an optimization DOE. A method for extracting plastic samples was to be developed. From a six-factor/

TABLE 18.5. Screening DOE for Headspace GC Volatile Extractables Method

Pattern[a]	Experiment number	Split ratio	Inlet temperature (°C)	Incubation time (minutes)	Incubation temperature (°C)	Solvent volume (μL)	Mass (mg)	Detector temperature (°C)	Initial oven temperature (°C)	Column film thickness (μm)	EtOH (ppm)	EtOAc (ppm)
−+++−−−−+	1	0.1	180	10	140	20	500	220	30	1.4	5.23	199.77
+−+++−−−+	2	20	140	30	140	20	100	220	30	3.0	9.84	220.78
+−+−+−++−	3	20	140	10	140	10	500	300	45	1.4	7.64	203.62
−++++−−+−	4	0.1	180	30	140	10	100	220	45	1.4	ND	228.43
+++++−−+−	5	20	180	30	100	10	100	300	30	1.4	3.54	53.21
+++++++++	6	20	180	30	140	20	500	300	45	3.0	10.32	199.65
−+−−+−−+−	7	0.1	140	10	140	10	100	300	30	3.0	0.67	22.79
++++−−+−+	8	20	180	10	100	10	500	220	30	3.0	1.45	14.98
+−+−−+−−+	9	20	140	10	100	20	100	220	45	1.4	ND	23.89
−++−−++++	10	0.1	180	10	100	20	100	300	45	3.0	2.36	35.64
−+−++−+++	11	0.1	140	30	100	10	500	220	45	3.0	2.81	2.47
−+−+++++−	12	0.1	140	30	100	20	500	300	30	1.4	2.24	58.34

[a] For each factor, two levels are chosen (e.g., incubation time = 10 or 30), and one (e.g., 30) is characterized as "high" (+) and the other (e.g., 10) as "low" (−). The combination of the nine factors is expressed as a string of these symbols representing the level for each factor. This shorthand notation (pattern) allows the user to quickly inspect the design.
ND, not detected.

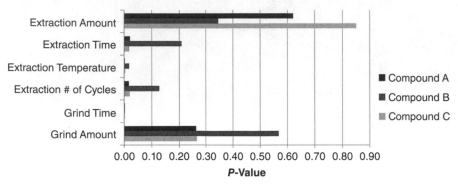

Figure 18.11 Statistical analysis of results from screening DOE.

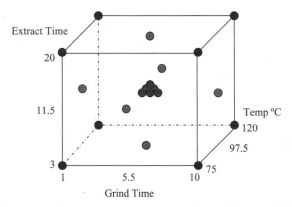

Figure 18.12 Face-centered central composite design for optimization DOE.

two-level screening DOE, it was determined that the statistically significant ($P < 0.05$) experimental parameters were grind time, temperature, and extraction time as shown in Figure 18.11. An optimization DOE was constructed utilizing the face-centered central composite design shown in Figure 18.12. In this design, there are a total of 20 experiments: eight at the corner points, six at the face-centered points, and six replicates at the center point. Each point in this design is a unique combination of the three parameters. The response surfaces for the three compounds measured by the HPLC method are shown in Figure 18.13. Inspection of these surfaces showed that, to measure all three compounds in a reliable manner, there is a limited operating space where the maxima overlap. The area of overlap is governed primarily by compound B, which has a diagonal ridge on either side of which the response decreases. The optimum conditions were identified and upon performing triplicate experiments—at each of the following conditions: extraction time, extraction time plus 4 minutes, and extraction time minus 2 minutes—the models for each compound were confirmed.

The DOE approach to method development as described and illustrated here is broader than that typically undertaken in the pharmaceutical industry. Several

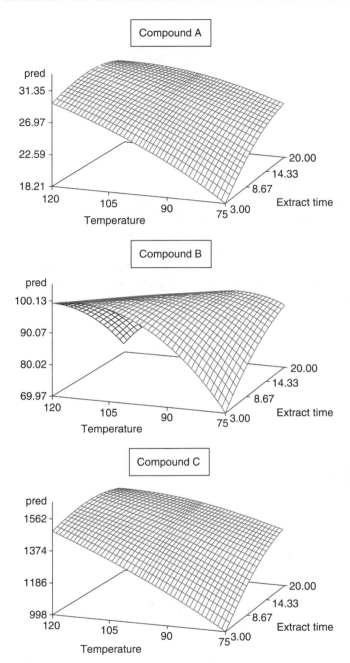

Figure 18.13   Response surfaces for extract preparation.

authors have described the use of DOE to optimize subsets of analytical parameters,[50] to assess robustness at the end of development[10,51–53] or to validate accuracy and precision.[54] The application of DOE to the entire range of method development activities has been applied primarily to chromatographic separations.[47,55,56] The broader approach described here begins with a screening design, which may be a two-level Plackett–Burman type of design that is only intended to identify the main effects. The upper and lower limits of each parameter explored provide the boundary conditions for the knowledge space that is created. The optimization design is usually a three-level fractional factorial* for two to four main effects where higher-order interaction terms are explored. Due to the fractional nature of the design not all higher-order interaction terms can be statistically evaluated individually, but are treated as combinations of an interaction term and main effect. This type of design was used in this example. In this type of design, it is important to consider the order assigned to the factors so that a significant effect of one factor is not confounded with a significant effect of the interaction of two other factors; results from screening DOE can help guide this assignment. More than one response may be monitored; typically, for leachables and extractables methods, these may be analyte levels, chromatographic separation, or quality of prepared sample. A design space is defined by the results from the optimization DOE. In this case, the fitting of a model to the data produces an equation, or set of equations for multiple responses, that can be solved simultaneously through an iterative approach to obtain optimal parameter settings. Alternatively, for multiple responses a systematic grid search over the factor space similar to that described by Murphy[57] can be used to identify the optimum parameters. The third and final use of DOE for method development is a confirmation DOE. This DOE is typically a three-level design for one or more factors that are main effects with multiple replicates. The results from this set of experiments confirm the response model, demonstrate robustness, and provide estimates of variability. By using fractional factorial designs, the number of experiments required is greatly reduced, and the development time can be reduced from months to weeks.

At the end of the method development process, the results are documented and a draft method is prepared. Depending on the phase of product development that the analytical method supports, documentation of the results may be (1) only recorded in a notebook, (2) summarized in a memo, or (3) written in a detailed report. For late-phase or registered products, a detailed report is appropriate. The draft procedure should include the following:

- summary of the scope—include description of materials to which method applies;

- list of chemicals with Chemical Abstracts Service (CAS) numbers: reagents with grade specified, standards;

---

* A full factorial is the complete number of experiments that would need to be done if every combination of parameters was tested; a fractional factorial is a statistically derived representative portion of the factorial.

- list of supplies: glassware with class designation; filters, thimbles, guard columns, and so on, with purchasing description;
- list of instrumentation and equipment with model or type;
- procedures for sample, standard, and solution preparation—include safety precautions;
- instrumentation settings and relevant operation procedures;
- proposed system suitability with acceptance criteria and calculation method specified;
- sequence of analysis for blanks, standards, samples;
- calculations to be performed including mathematical formulas to be used.

The draft method is used for experiments performed during assessment and revised to the final form after validation is complete. The final document will include the analytical figures of merit obtained during validation and final acceptance criteria for system suitability. For chromatographic or spectroscopic methods, the final method document should include representative scans for blanks, standards, and samples.

18.6   ASSESSMENT AND VALIDATION OF ROUTINE METHODS

After a routine testing method is developed, it is important to confirm that the method can be reliably used to generate accurate results and that it meets the expectations for its intended use. The topic of method validation has been addressed by several authors, and books have been written on the subject.[8–10,12,58] General step-by-step guides to analytical method validation have been published.[59,60] More specifically, Jenke published a detailed guide to the steps involved in analytical method validation for leachables determinations.[61]

Historically, the term "validation" may have been used interchangeably with "verification" and "qualification." As of 2003, it was established that processes, for example, analytical methods, are validated, and equipment, for example, analytical instruments, is qualified.[62] "Verification" is a term used by the USP to describe the process of confirming that a compendial method is suitable under specific conditions of use.[63] Since there are no compendial methods for the type of leachables and extractables determinations described here, only the term "validation" is appropriate and will be used in this discussion.

The goal of this discussion of method assessment and validation is to review the basic concepts and point out the salient features of method validation that are relevant to leachables and extractables routine methods. The variety of different analytical methodologies that may be used to routinely measure extractables or leachables can be found in the USP and ICH Q2(R1) guidelines on method validation and are summarized in Table 18.6. The attributes of the method that must be assessed are dependent on the purpose of the method. For example, a method that is used for identification only needs to be assessed for specificity, whereas a

TABLE 18.6. Validation Attributes per ICH Q2(R1) and USP

	I		II		III	IV
	Content (pure materials)	Content (products or intermediates)	Impurity quantitative	Impurity limit test	Specific tests	Identity
Accuracy	R	R	R	N[a]	R[a]	N
Precision	R	R	R	N	R	N
Specificity	R	R	R	R	R[a]	R
LOD	N	N	P	R	N[a]	N
LOQ	N	P	R	N	N[a]	N
Linearity/ range	R	R	R	N[a]	P[b]	N
Robustness[c]	R	R	R	P	P	N
Sample stability	R	R	R	R	P	P
Standard stability	R	R	R	R	P	P

[a] USP indicates "P."

[b] ICH Q2 indicates "R."

[c] May be performed during method development.

R, required; N, not required; P, perform only if appropriate.

quantitative impurity method will require assessment of nearly all attributes listed. Most often, leachables and extractables fall into this latter category of quantitative impurity methods.

To ensure successful completion of a formal validation activity the most critical attributes of the method may be tested in an assessment activity. This assessment allows the user to get a preview and confirm the capabilities of the method prior to validation. For example, a quantitative method that measures one target compound and is used to evaluate the chromatographic profile for new chemical entities may be assessed for precision, linearity, and specificity prior to the formal validation. Detailed examples of attribute selection and assessment strategy are provided by Jenke et al.[64,65]

18.6.1 Phase-Appropriate Validation

The concept of risk management integrated into the realm of validation combined with the fact that ICH guidelines for validation apply only to registered products has spawned the use of the term "phase-appropriate" validation. Several authors have provided suggestions for simplification of the validation of analytical methods that support products in early phases of pharmaceutical development.[10,66,67] The validation of analytical methods is principally driven by the end use. That is, the degree and type of validation will depend on the expectations for the method. For example, an extractables method may be required at the start of a selection process for a container closure component. At this stage, the requirements for the method will be different than at a later stage. Later, an extractables method might form part of a control strategy for a container closure material or component, which has been developed and finalized for use and submitted as part of a drug product application to a regulatory authority. These are just two examples that may be used to illustrate the two ends of the development continuum.

Consider these two examples further to elucidate the differences in use and thus in validation requirements for these methods. At the material selection stage, the extractables method will typically be a general screening method, aimed at collecting the maximum amount of information about the material(s) under consideration. Some of the materials may change as development progresses; thus, the method validation requirements may well focus on elements such as selectivity and sensitivity (limit of detection [LOD]) rather than on a method that is precise and accurate with good linearity. Therefore, if the only requirement is a demonstration of extractables below a preset amount, it could be appropriate to validate the extractables method as a limit test at this stage. For example, the limit could be derived from a general knowledge of the product and the Product Quality Research Institute (PQRI) SCT of 0.15 μg/day. Therefore, if a material can be demonstrated to be absent of extractables at this same limit, then there is a very good case for selecting the material for further study. At this early phase of development, a minimal validation effort could involve notebook documentation of the validation performed by only one analyst using one set of equipment.

At the opposite end of the continuum, there are routine methods that must be fully validatable, transferable, robust, reliable, accurate, and precise (V-TR^{2}AP, an acronym coined by Crowther, Salomons, and Callaghan).[45] It is this type of quantitative method that will form part of the control strategy for the life of the product. Typically, such a method will be used for routine release of materials and may be performed at multiple testing sites. As alluded to above, an extractables method may fall into this category. It could happen that with minimal effort, the extractables limit test described above could be modified and validated to be a quantitative test, especially if the method requirements are set with this as the long-term goal. The validation package for this type of method at the point of regulatory submission would need to include assessments as specified by the ICH guidelines: accuracy, precision (repeatability, intermediate precision, and reproducibility), specificity, LOD, LOQ, linearity, range, robustness, standard, and sample stability. Reproducibility and robustness are particularly important if the method is transferred to another laboratory to perform routine analysis against agreed specifications for the material, which may well be the case for long-term testing to support a product. This type of validation should be performed per protocol with acceptance criteria and potentially include a set of designed experiments for robustness and precision to efficiently use multiple instruments, personnel, and test sites. The testing environment should be compliant with GMPs including trained analysts, data security, qualified instruments, and sample traceability. The results and validation status of the method should be captured in a detailed report that includes the following:

- component name, lot number, and manufacturer's part number with revision level;
- test dates;
- CAS numbers and expiry dates for reagents;
- confirmation that instrumentation is within calibration period;
- details of sample storage, preparation, and storage after preparation;
- description of visual appearance of samples pre- and postpreparation;
- deviations from the protocol;
- representative chromatograms or spectra, and CAS numbers for identified compounds;
- all other relevant experimental observations, details, and results; and
- summary of acceptance criteria and pass/fail status.

As with any continuum, there may be several points where routine methods are required between the early development phase and marketed product. The type of validation activity is dictated not only by the type of method as categorized in the ICH guidelines but also by the phase of development of the product that is supported by the routine method. Table 18.7 contains a list of items to consider with respect to the validation of a routine extractables or leachables quantitative method and some options for the magnitude of effort required. Typically, less time and

TABLE 18.7. Clinical Development Phase Appropriate Activities to Support Validation

	Low	Mid	High
	Phase I	Phase II	Phase III/IV
	One-time study	Extended one-time study	Ongoing testing
	Material evaluation	Lot testing	Material release, registration stability
Testing environment	Standard lab practices	GMP	GMP
Documentation	Lab notebook	Lab notebook, report	Protocol, lab notebook, report
Instrumentation	Calibrated	Calibrated, PM	Calibrated, PM, change control
Qualification performed	IQ/OQ	IQ/OQ	IQ/OQ/PQ
Precision	Repeatability	Repeatability, intermediate precision	Repeatability, intermediate precision, reproducibility

PM, preventive maintenance; IQ/OQ/PQ, installation qualification, operational qualification, performance qualification.

resources are committed to early-phase methods than to later-phase methods. Since early work is generally more qualitative, the validation acceptance criteria for early-phase methods may be broader than those for a late-phase method. The combination of the intended use and phase of product development supported must be considered together to determine the level of control applied. Where there is a mismatch between the phase of product development and the method purpose and frequency of testing the more stringent approach should be taken.

18.6.2 Analytical Figures of Merit

Prior to commencing validation, it is important to finalize the acceptance criteria for successful completion. The requirements for the method and information gained during assessment can be used to establish suitable acceptance criteria; examples can be found in McPolin,[10] Bloch,[67] and Ermer.[68] Each of the attributes that are part of a final validation package are discussed, and typical acceptance criteria are presented below. Many of the considerations presented are relevant to quantitative impurity method validation with respect to leachables and extractables.

18.6.2.1 Accuracy

The accuracy of an analytical procedure expresses the closeness of agreement between the value which is accepted either as a conventional true value or an accepted reference value and the value found. ICH Q2(R1)

Accuracy validation for leachables and extractables methods is a particular challenge. Unlike drug impurity validation, it is not normally possible to obtain a ready supply of all the analytes that may be present in a sample. Additionally, it is not possible to produce an authentic blank matrix spiked with known quantities of analytes, which is normally the case in impurity validation. Therefore, accuracy needs to be determined in a different way. Spiking and recovery can still be employed to determine accuracy in leachables and extractables methods, but it is typically necessary to use samples that already contain some of the analytes. Thus, it is necessary to subtract this residual amount from any spike added to obtain the correct value for the recovery of the spiked component. Spiking of dry samples can be particularly challenging since the volumes used to spike are small to minimize wetting of the sample and subsequently render the sample nonhomogeneous. An alternative approach to an accuracy experiment is to vary the quantity of sample analyzed. For example, if normally 1 g of sample is analyzed for leachables in the accuracy experiment, additional samples of 0.5 and 2 g can be used. This is the equivalent of spiking with 50% and 200% of the nominal leachables amount. The accuracy is then determined by confirming that the leachables levels in the 0.5- and 2-g samples are 50% and 200%, respectively, of the expected level in the 1-g sample. A refinement of this experiment is to analyze a range of sample masses, to plot the actual result versus the expected result, and to use linear regression analysis to evaluate accuracy.

The use of percent recovery as a measure of accuracy can be problematic when processing steps in the method compromise the recovery of the analyte. Where a leachables method is developed that includes an enrichment step, such as LLE, it may not be possible to develop a method where 100% of the analyte is extracted from the sample matrix during sample preparation. It is also possible that recoveries may be low due to degradation or evaporation of the analyte. If low recoveries are observed during validation or development, it might be appropriate to introduce a known amount of surrogate standard that can be used to correct for low recoveries due to sample processing. A complex sample matrix and multiple processing steps increase the likelihood that there will be low recovery. Trace-level methods are expected to have recoveries with a broader range than content methods; for analyte levels at 0.1–10 ppm, the range is estimated to be 80%–110%.[11]

Typical acceptance criteria: recovery = 70%–130% with relative standard deviation (RSD) = 20% for headspace-GC, 15% for GC or 10% for HPLC.

18.6.2.2 Precision

The precision of an analytical procedure expresses the closeness of agreement between a series of measurements obtained from multiple sampling of the same homogeneous sample under the prescribed conditions. Precision may be considered at three levels: repeatability, intermediate precision and reproducibility.

Precision should be investigated using homogeneous, authentic samples. However, if it is not possible to obtain a homogeneous sample it may be investigated using artificially prepared samples or a sample solution. The precision of an analytical procedure is usually expressed as the variance, standard deviation or coefficient of variation of a series of measurements. ICH Q2(R1)

18.6.2.3 Repeatability

"Repeatability expresses the precision under the same operating conditions over a short interval of time. Repeatability is also termed intra-assay precision" (ICH Q2(R1)).

18.6.2.4 Intermediate Precision

"Intermediate precision expresses within laboratories variations: different days, different analysts, different equipment, etc." (ICH Q2(R1)). Note: This can be thought of as long-term variability[69] of the measurement process and has been referred to as intralaboratory precision or ruggedness as described in the USP.

18.6.2.5 Reproducibility

"Reproducibility expresses the precision between laboratories (collaborative studies are usually applied to standardization of methodology)" (ICH Q2(R1)). Note: This can be thought of as interlaboratory precision.

The precision associated with a result from the method is a cumulative total that results from the intrinsic variability of the method and extrinsic variability from both short-term and long-term contributions as outlined in Table 18.8. While instrumentation repeatability is not part of the formal validation requirement, it is acknowledged as a contributor to method repeatability and typically assessed as part of system suitability. Analytically, taking a propagation of error approach, the overall variability of the method could be represented as $\sigma^2_{method} = \sigma^2_{instrument} + \sigma^2_{short\text{-}term} + \sigma^2_{long\text{-}term}$. Each of these terms could be further broken down

TABLE 18.8. Extrinsic Sources and Levels of Variability That Contribute to Method Precision

	Short term		Long term	
	System repeatability	Method repeatability	Intermediate precision	Interlaboratory precision
Testing laboratory	Single	Single	Single	Several
Test environment	Single	Single	Several	Several
Analyst	Single	Single	Several	Several
Day	Single	Single	Several	Several
Instrument	Single	Single	Several	Several
Reagents/standards lots	Single	Single	Several	Several
Number of replicate preparations	One	Multiple	Multiple	Multiple
Number of replicate tests	Multiple	Multiple	Multiple	Multiple
Typical RSD	1%–5%	5%–15%	10%–20%	15%–20%

into cumulative sums of contributing variabilities so that, if each of the individual variability was known, the overall variability could be computed. Practically, these are not readily determined, but conceptually, it is important to keep the cumulative sum approach in mind when setting acceptance criteria for each type of precision. For example, this would mean that the RSD for intermediate precision would not be equal to, but greater than the RSD for repeatability. Typical RSD values obtained for leachables and extractables methods are provided in Table 18.8.

The short- and long-term factors associated with method precision are distinguished from those associated with robustness in that they are external to the analytical system. These variabilities can be viewed as "noise factors" that contribute to the random distribution of results and are generally not possible to control directly or fix. Since leachables and extractables methods tend to be somewhat complex, it is prudent to consider reducing the human noise factor contribution by appropriate training and skill set of the analysts.

It is appropriate to validate the method to the level at which it will be performed so that all of the appropriate contributions to variability are taken into consideration. For example, if a method will be performed by several analysts at only one test site, it would not be relevant to assess the variability associated from multiple labs, but it would be appropriate to validate intermediate precision. An efficient way to validate intermediate precision could be to use a DOE. One type of design for this is based on the Kojima design, used for registration in Japan and is represented in Table 18.9. Other types of designs could be used for intermediate precision and reproducibility as well.

Properly designed reproducibility studies will ensure that method transfer is successful in the short term and that the method is likely to perform in the longer term.

TABLE 18.9.   Kojima-Based DOE for Intermediate Precision

Experiment	1	2	3	4	5	6
Date	A	B	C	D	E	F
Analyst	1	2	1	2	1	2
Instrument	1	2	2	1	1	2
Column	1	2	2	1	2	1

Often, in the pharmaceutical industry, the acceptance criteria for precision and intermediate precision are set differently depending on the type of sample matrix and the analytical measurement technique used. Reproducibility has been measured for over 10,000 interlaboratory analytical studies and a model was fit to the data.[70] The Horwitz "trumpet" curve may be used to guide the selection of appropriate acceptance criteria. This curve describes the relationship between the concentration of an analyte in its matrix and the relative interlaboratory reproducibility and is expressed as

$$RSD = Q \cdot e \wedge (1 - 0.5 \log f_a),$$

where Q is 2 for trace substances in food and f_a is the fraction of analyte in the sample.[11,71,72] Interestingly, this relationship has been found to hold true regardless of sample matrix or analytical technique. For leachables and extractables quantitative methods that measure analytes in the range of 1 ppm, the expected interlaboratory RSD is estimated at 16%.

Typical acceptance criteria: six replicates, at least one different analyst, instrument, day, column; see RSD values in Table 18.8.

18.6.2.6   Specificity

Specificity is the ability to assess unequivocally the analyte in the presence of components which may be expected to be present. Typically these components might include impurities, degradants, matrix, etc. Lack of specificity of an individual analytical procedure may be compensated by other supporting analytical procedure(s). ICH Q2(R1)

Method specificity for extractables methods is normally demonstrated by showing absence of interference from other compounds with the target analyte within the solvent blank or sample matrix, and adequate resolution of all analytes. This is not always easy to demonstrate since even the purest solvents may contain trace levels of analytes that may be targets in an extractables method, or compounds may be extracted into the blank from the thimble that holds the sample. The latter interference can be eliminated by prewashing the extraction thimble. Standards can be used to confirm that there is no co-elution of identified analytes. However, it is possible that for some analytes, standards may not exist or the analyte may not be identified. Alternative approaches, such as peak purity determinations, are therefore sometimes necessary. For example, for a GC/FID method, it may be necessary to run the same method on a GC/MS to evaluate the homogeneity of the peak with respect to consistency of the mass profile throughout the peak. For an LC/UV method, it may be possible to use a diode array detector (DAD) to assess peak purity, or it may be necessary to use LC/MS in a manner similar to that for GC to confirm peak homogeneity.

Leachables methods present slightly different challenges in the confirmation of specificity. Leachables methods typically measure low-level analytes in the presence of much higher quantities of drug or excipient. Therefore, leachables methods usually employ a selective detector such as a mass spectrometer to minimize interferences. For example, many leachables methods employ mass selective detection in which the mass spectrometer monitors only ions of the target leachable to ensure specificity. If a nonselective detector is used, such as an FID fitted to a GC or a UV detector fitted to an HPLC, it is particularly important to demonstrate that the response due to leachables is free from interference from complex sample matrix. This may be achieved by running the method in a one-time study utilizing an MS detector as described above.

Obtaining a suitable blank for a leachables method can be problematic. It might be achieved by analysis of a sample that has limited opportunity to leach because it is stored for a short time at low temperature, or by analysis of samples that are similar but known not to contain the leachable of interest. Depending on the outcome of the specificity experiments, it may be important to add some additional type of specificity check to the system suitability requirements for a routine leachables method. For leachables or extractables methods, it is common to include a resolution check utilizing two closely eluting standards.

Typical acceptance criteria: no interfering peaks in blank or extraction blank (DAD peak purity threshold > 995).

18.6.2.7 *Detection Limit/Quantitation Limit*

The detection limit of an individual analytical procedure is the lowest amount of analyte in a sample which can be detected but not necessarily quantitated as an exact value.

The quantitation limit of an individual analytical procedure is the lowest amount of analyte in a sample which can be quantitatively determined with suitable precision and accuracy. ICH Q2(R1)

For leachables and extractables methods, it is essential that the LOD is at or below the AET and more preferably that even the LOQ is below the AET. This follows the same rationale as that used for drug impurity tests.[73] ICH standard definitions are as follows:

Detection limit

  3:1 signal to noise (S/N)

  3.3σ/slope

Quantitation limit

  10:1 S/N

  10σ/slope

These determinations are typically made from standard solutions prepared to have concentrations close to the expected LOD and LOQ obtained from extrapolation of the calibration curve results. The concentrations are adjusted until the respective S/N ratios are obtained. Since it is important that the LOQ be reliably measured, it is possible that the concentration of the LOQ standard may need to be raised slightly to achieve adequate precision and recovery. Precision is assessed by preparation of six replicate standards at the LOQ concentration. Where no suitable standards are available, standards that are expected to elicit a detector response similar to the target analytes are sometimes used as a substitute.

For a leachables method, the precision and recovery of the target response at the LOQ concentration may be quite different in the presence of the sample matrix than in the presence of only standard diluent. This may be assessed by preparing a target spiked placebo that contains an appropriate amount of freshly prepared drug formulation. The same assessment as described above would then be performed using a calibration curve constructed with sample matrix as the diluent rather than a simple solvent diluent. In any case, if measurements are made at a level close to the LOQ concentration, a system suitability requirement for the method may include use of an LOQ solution, sometimes referred to as a "sensitivity standard," to confirm an adequate S/N and appropriate signal response.

Typical acceptance criteria: LOD has S/N = 3; LOQ has S/N = 10 with RSD ≤ 10% (*n* = 6, LC/UV).

18.6.2.8   Linearity/Range

The linearity of an analytical procedure is its ability (within a given range) to obtain test results which are directly proportional to the concentration (amount) of analyte in the sample.

The range of an analytical procedure is the interval between the upper and lower concentration (amounts) of analyte in the sample (including these concentrations) for which it has been shown that the analytical procedure has a suitable level of precision, accuracy and linearity. ICH Q2(R1)

Leachables and extractables methods commonly employ detection systems where a linear relationship between response and concentration of analyte can be expected. Thus, a linearity measurement is normally made, which covers the working range of the method. In some cases, the working range for these methods can be over two orders of magnitude. For example, leachables concentrations in the drug product can vary quite considerably over the study period in which the method is employed. A similarly broad range may be encountered for specific analytes in elastomeric components where the sources of that analyte are many and not controlled prior to the compounding step. It is standard practice in some methods, such as those utilizing GC, to include an internal standard to normalize the response where precision of the volume of sample introduced is likely to be variable and may affect linearity. To accommodate the extended range of analysis, it may be necessary to use a weighted linear, or possibly a polynomial, fit to the data obtained from a calibration curve that contains at least five equally spaced points over the range. In some cases, the response may be nonlinear due to the use of a detector whose intrinsic response function is not linear, for example, fluorescence or evaporative light scattering, and the relevant fit to the data should be applied. The goodness of fit is often judged by the value of the correlation coefficient, r, coefficient of determination, r^2, relative residual standard deviation, or the residual sum of squares. For a detailed discussion of the merits of these and other fit indicators, the reader is referred to Ermer.[74]

For a standard impurity method, the range is the reporting limit to 120% of nominal. However, due to the broader ranges typically encountered in leachables and extractables routine methods, the range may be from a measured blank, "zero," up to 1000% of nominal. Typically, the lowest concentration measured will be LOQ or zero, and the highest will be 300% of nominal. In any case, it is important that the expected range be covered with at least five standards with concentrations equally spaced over the range. For nonlinear responses, eight or more standards may be needed to adequately fit a model to the data. Alternatively, a sample can be diluted so that its concentration is bracketed by the linear range. Where the precision of the measurement varies, for example, GC, it may be necessary to perform replicate measurements at each concentration.

Typical acceptance criteria: $r = 0.98-1.00$, $r^2 \geq 0.995$, relative residual standard deviation $\leq 20\%$, random scatter of the residuals with no systematic trend.

18.6.2.9 Robustness "The robustness of an analytical procedure is the measure of its capacity to remain unaffected by small, but deliberate, variations in method parameters and provides an indication of its reliability during normal usage" (ICH Q2(R1)).

Robustness testing should not be restricted to chromatographic parameters but should include all parameters of the analytical procedure, for example, sample preparation. Robustness can be defined as the ability of the method to withstand small changes in method conditions (parameters) and still deliver precise and accurate results. These changes may be predictive of variations that could be introduced either due to equipment, material, or reagent changes in the long term over which the method may be performed. Method development utilizing DOE will enable a determination of which parameters should be investigated for robustness testing. The decision on whether to fix a parameter or explore its design space should also take account of a possible need for future flexibility.

Typical acceptance criteria: no significant difference between results obtained with slight variations in method parameters.

18.6.3 Sample and Standard Stability

Sample and standard solution stability should be confirmed if the method is expected to analyze samples or standard solutions that may be stored prior to analysis. The stability of these stored solutions can be determined by assaying them against freshly prepared solution to confirm that their expected concentrations remain after storage. Some analytes are more likely to degrade in solution than others. Additives, such as antioxidants or UV stabilizers, are liable to degrade in solution if exposed to light and/or heat. Other analytes, especially at low concentration, adsorb onto the surface of glassware and thus are lost. It is expected that the stability period of the sample and standard solutions, at the very least, will bracket the analysis period.

18.6.4 Triggers for Revalidation

After validation is complete and the routine method is used regularly for testing, it is important to regularly review the procedure. At some point in the life cycle of the product, it may become necessary to revise or change a routine method that supports it. Methods that are used for material release are typically under a change control system and may require revalidation if a change is made. Listed below are some of the events that may trigger a revalidation[75,76]:

- change in material formulation or synthetic process for the material,
- change in component fabrication process,
- new instrumentation that operates by a different principle,
- loss of supply of reagents,
- change in product dosing scheme resulting in a change in the AET, and
- new component made from same material.

If a change needs to be made to the method, then the development process would restart with the stakeholders clarifying the requirements followed by development, optimization, assessment, and validation. If the validation package can be thought of as a set of modules,[77] it may be possible to consider revalidation of the module that is impacted by the change. One example of a change might be a case in which a different formulation is packaged in the same container closure as that used in an already marketed product. In this case, since the same leachables could be expected if the formulation matrix is similar, it may be possible to modify a previously validated leachables method so that only the sample preparation step requires revalidation. Another example of a change might be a case in which a plastic resin used to mold a drug path component undergoes a formulation change. The routine extractables method used for that drug path component would be assessed for suitability with the new material. Depending on the nature of the change, it may be necessary to modify and revalidate sample preparation and/or some or all of the analytical parameters. In contrast, if the plastic resin were to remain the same but was used to mold a different drug path component with similar processing conditions, it is likely that the extractables profile would remain unchanged and only the sample preparation would require revalidation. If target levels in the new component are found to exceed the validated range, then a revalidation of the linear range would be required. Thus, the validation activities may be a subset of what was initially required depending on the nature of the change.

18.7 PQRI RECOMMENDED BEST PRACTICES FOR ROUTINE TESTING

The concepts and approaches discussed in this chapter generally follow the recommendations for routine testing addressed in the PQRI recommendations and the scientific literature[78,79]:

1. Analytical methods for routine extractables testing should be based on the analytical technique(s)/method(s) used in the controlled extraction studies.

2. Routine extractables testing should be performed on critical components using appropriate specifications and acceptance criteria.

3. Analytical methods for leachables studies and routine extractables testing should be fully validated according to accepted parameters and criteria.

18.8   CONCLUSION

The approach to development and optimization of methods for routine measurement of extractables or leachables is based on the approaches used for trace impurity analysis. There are several opportunities to incorporate principles of quality by design and risk management into the development and optimization process. Development and optimization can proceed by a structured (DOE) or random (OFAT) process or a combination of the two. Validation of these methods can be performed in a phase-appropriate manner and may also utilize designed experiments. The development of analytical methods to support pharmaceutical combination products is inherently linked to the life cycle of the product.

ACKNOWLEDGMENTS

The authors gratefully acknowledge Matt Ingram, William Duffield, Mitch Rosner, Mary Ann Smith, and Rong Ming Liu for their contributions to this work.

REFERENCES

1   Pang, J., Blanc, T., Brown, J., Labrenz, S., Villalobos, A., Depaolis, A., Gunturi, S., Grossman, S., Lisi, P., and Heavner, G.A. Recognition and identification of UV-absorbing leachables in EPREX pre-filled syringes: An unexpected occurrence at a formulation-component interface. *PDA J Pharm Sci Technol* 2007, *61*(6), pp. 423–432.

2   Castner, J., Williams, N., and Bresnik, M. Leachables found in parenteral drug products. *Am Pharm Rev* 2004, *7*(2), pp. 70–75.

3   Norwood, D.L., Nagao, L., Lyapustina, S., and Munos, M. Application of modern analytical technologies to the identification of extractables and leachables. *Am Pharm Rev* 2005, *8*(1), pp. 78–87.

4   Patel, R.B. and Patel, M.R. Analytical methods development and validation play important roles in the discovery, development, and manufacture of pharmaceuticals, 2008. Available at: http://www.pharmainfo.net/reviews/introduction-analytical-method-development-pharmaceutical-formulations (accessed September 5, 2011).

5   Krause, S.O. Validating analytical methods for biopharmaceuticals, part I: Development and optimization. *BioPharm Int* 2004, *17*(10), pp. 52–61.

6   Krause, S.O. Validating analytical methods for biopharmaceuticals, part II: Formal validation. *BioPharm Int* 2004, *17*(11), pp. 46–52.

7   Yu, X., Wood, D., and Ding, X. Extractables and leachables: Study approach for disposable materials used in bioprocessing. *BioPharm Int* 2008, *21*(2), pp. 42–51.

8   Swartz, M.E. and Krull, I.S. *Analytical Method Development and Validation*. Marcel Dekker, New York, 1997; 74.

9 Chan, C.C., Lee, Y.C., Lam, H., and Zhang, X.-M., eds. *Analytical Method Validation and Instrument Performance Verification*. John Wiley & Sons, Inc, New York, 2004.

10 McPolin, O. *Validation of Analytical Methods for Pharmaceutical Analysis*. Mourne Training Services, Northern Ireland, United Kingdom, 2009; 60–64, 75–89.

11 Huber, L. *Validation and Qualification in Analytical Laboratories*, 2nd ed. Informa Healthcare USA, New York, 2007; 144, 146.

12 Ermer, J. and Miller, J.H.M., eds. *Method Validation in Pharmaceutical Analysis: A Guide to Best Practice*. Wiley-VCH Verlag GmbH & Co., Darmstadt, 2005.

13 Snyder, L.R., Kirkland, J.J., and Glajch, J.L. *Practical HPLC Method Development*, 2nd ed. John Wiley & Sons, Inc, New York, 1997; 2.

14 Krause, S.O. *Validation of Analytical Methods for Biopharmaceuticals: A Guide to Risk-Based Validation and Implementation Strategies*. DHI Publishing, River Grove, IL, 2007.

15 Jenke, D. *The Prototype Stage in Compatibility of Pharmaceutical Products and Contact Materials: Safety Considerations Associated with Extractables and Leachables*. John Wiley & Sons, New York, 2009; 159–190.

16 Extractables and Leachables Subcommittee of the Bio-Process Systems Alliance. Recommendations for extractables and leachables testing: Part 1—Introduction, regulatory issues, and risk assessment. *BioProcess Int* 2007, *5*(11), pp. 36–44.

17 Extractables and Leachables Subcommittee of the Bio-Process Systems Alliance. Recommendations for extractables and leachables testing: Part 2—Executing a program. *BioProcess Int* 2008, *6*(1), pp. 44–52.

18 ICH harmonised tripartite guideline: Q8(R2) pharmaceutical development. International Conference on Harmonisation of Technical Requirements for Registration of Pharmaceuticals for Human Use, 2009.

19 ICH harmonised tripartite guideline: Q9 quality risk management. International Conference on Harmonisation of Technical Requirements for Registration of Pharmaceuticals for Human Use, 2005.

20 Guidance for industry: Container closure systems for packaging human drugs and biologics. Department of Health and Human Services, Food and Drug Administration, Center for Drug Evaluation and Research (CDER), 1999.

21 Tague, N. *The Quality Toolbox*, 2nd ed. ASQ Quality Press, Milwaukee, WI, 2005; 236–242.

22 Mil-Std-1629A: Procedures for performing a failure mode, effects and criticality analysis. US Department of Defense, 1980.

23 Tague, N. *The Quality Toolbox*, 2nd ed. ASQ Quality Press, Milwaukee, WI, 2005; 16–19.

24 Tague, N. *The Quality Toolbox*, 2nd ed. ASQ Quality Press, Milwaukee, WI, 2005; 475–476.

25 ICH harmonized tripartite guideline: Q2(R1) validation of analytical procedures: Text and methodology. International Conference on Harmonization of Technical Requirements for Registration of Pharmaceuticals for Human Use, 2005.

26 Crowther, J.B., Jimidar, M.I., Niemeijer, N., and Salomons, P. Qualification of laboratory instrumentation, validation, and transfer of analytical methods. In: Miller, J.M. and Crowther, J.B., eds. *Analytical Chemistry in a GMP Environment: A Practical Guide*. John Wiley & Sons, New York, 2000; 423–458.

27 Nunnally, B.K. and McConnell, J.S. *Six Sigma in the Pharmaceutical Industry*. Taylor & Francis Group, LLC, Boca Raton, FL, 2007; 167–184.

28 Konieczka, P. and Namiesnik, J. *Quality Assurance and Quality Control in the Analytical Chemical Laboratory*. Taylor & Francis Group, LLC, Boca Raton, FL, 2009; 29–35.

29 Krause, S.O. A guide for testing biopharmaceuticals part 2: Acceptance criteria and analytical method maintenance. *BioPharm Int* 2006, *19*(10), pp. 42–54.

30 Montgomery, D.C. *Introduction to Statistical Quality Control*, 5th ed. John Wiley & Sons, New York, 2005; 153.

31 Western Electric Co. *Statistical Quality Control Handbook*. Mack Printing Company, Eaton, PA, 1982; 208.

32 Crowther, J.B., Salomons, P., and Callaghan, C. Analytical method development for assay and impurity determination in drug substances and drug products. In: Miller, J.M. and Crowther, J.B., eds. *Analytical Chemistry in a GMP Environment: A Practical Guide*. John Wiley & Sons, New York, 2000; 342–344.

33 Jenke, D. Appendix: Materials used in pharmaceutical constructs and their associated extractables. In: Jenke, D. *Compatibility of Pharmaceutical Products and Contact Materials: Safety Considerations Associated with Extractables and Leachables*. John Wiley & Sons, New York, 2009; 347–370.

34 Jenke, D. Extractable/leachable substances from plastic materials used as pharmaceutical product containers/devices. *PDA J Pharm Sci Technol* 2002, *56*(6), pp. 332–371.

35 Fliszar, K.A., Walker, D., and Allain, L. Profiling of metal ions leached from pharmaceutical packaging materials. *PDA J Pharm Sci Technol* 2006, *60*(6), pp. 337–342.

36 Wang, Q. Selection of analytical techniques for pharmaceutical Leachables studies. *Am Pharm Rev* 2005, *8*(6), pp. 38–44.

37 Council of European Committees Directive: 2007/47/EC. September 2007.

38 Norwood, D.L., Prime, D., Downey, B.P., Creasey, J., Sethi, S.K., and Haywood, P. Analysis of polycyclic aromatic hydrocarbons in metered dose inhaler drug formulations by isotope dilution gas chromatography/mass spectrometry. *J Pharm Biomed Anal* 1995, *13*(3), pp. 293–304.

39 Simpson, N.J.K. *Solid-Phase Extraction: Principles, Techniques and Applications*. Taylor & Francis Group, LLC, Boca Raton, FL, 2000.

40 Akapo, S.O. and McCrea, C.M. SPME-GC determination of potential volatile organic leachables in aqueous-based pharmaceutical formulations packaged in overwrapped LDPE vials. *J Pharm Biomed Anal* 2008, *47*, pp. 526–534.

41 Rezaee, M., Yamini, Y., and Faraji, M.J. Evolution of dispersive liquid-liquid microextraction method. *J Chromatogr A* 2010, *1217*(16), pp. 2342–2357.

42 Xiao, B., Gozo, S.K., and Herz, L. Development and validation of HPLC methods for the determination of potential extractables from elastomeric stoppers in the presence of a complex surfactant vehicle used in the preparation of parenteral drug products. *J Pharm Biomed Anal* 2007, *43*, pp. 558–565.

43 Castner, J., Anderson, J., and Benites, P. Strategy for development and characterization of HPLC methods to investigate extractables and leachables (part II). *Am Pharm Rev* 2007, *10*(3), pp. 10–18.

44 Browne, D.C. Reference-standard material qualification. *Pharm Technol* 2009, *33*(4), pp. 66–73.

45 Crowther, J.B., Salomons, P., and Callaghan, C. Analytical method development for assay and impurity determination in drug substances and drug products. In: Miller, J.M. and Crowther, J.B., eds. *Analytical Chemistry in a GMP Environment: A Practical Guide*. John Wiley & Sons, New York, 2000; 332, 345.

46 Borman, P., Nethercote, P., Chatfield, M., Thompson, D., and Truman, K. The application of quality by design to analytical methods. *Pharm Technol* 2007, *31*(10), pp. 142–152.

47 Krull, I.S., Swartz, M.E., Turpin, J., Lukulay, P.H., and Verseput, R.A. Quality-by-design methodology for rapid LC method development, part I. Chromatography Online.com, December 1, 2008.

48 Montgomery, D.C. *Design and Analysis of Experiments*, 6th ed. John Wiley & Sons, New York, 2005.

49 Ryan, T.P. *Statistical Methods for Quality Improvement*, 2nd ed. John Wiley & Sons, New York, 2000.

50 Mahesan, B. and Lai, W. Optimization of selected chromatographic responses using a designed experiment at the fine-tuning stage in reversed-phase high-performance liquid chromatographic method development. *Drug Dev Int Pharm* 2001, *27*(6), pp. 585–590.

51 Gavin, P.F. and Olsen, B.A. A quality by design approach to impurity method development for atomoxetine hydrochloride (LY139603). *J Pharm Biomed Anal* 2008, *46*, pp. 431–441.

52 Rignall, A., Christopher, D., Crumpton, A., Hawkins, K., Lyapustina, S., Memmesheimer, H., Parkinson, A., Smith, M.A., Wyka, B., and Kaerger, S. Quality by design for analytical methods for use with orally inhaled and nasal drug products. *Pharm Technol Eur* 2008, *20*(10), pp. 24–31.

53 Kleinschmidt, G. Robustness. In: Ermer, J. and Miller, J.H.M., eds. *Method Validation in Pharmaceutical Analysis: A Guide to Best Practice*. Wiley-VCH Verlag GmbH & Co., Darmstadt, 2005; 120–169.

54 Eudy, T.L. Designing experiments for validation of quantitative methods. In: Torbeck, L.D., ed. *Pharmaceutical and Medical Device Validation by Experimental Design*. Informa Healthcare USA, New York, 2007; 1–46.

55 Krull, I.S., Swartz, M.E., Turpin, J., Lukulay, P.H., and Verseput, R. A quality-by-design methodology for rapid LC method development, part II. Chromatography Online.com, 2009.

56 Krull, I.S., Swartz, M.E., Turpin, J., Lukulay, P.H., and Verseput, R. A quality-by-design methodology for rapid LC method development, part III. Chromatography Online.com, 2009.

57 Murphy, T.D. Response surface methodology for validation of oral dosage forms. In: Torbeck, L.D., ed. *Pharmaceutical and Medical Device Validation by Experimental Design.* Informa Healthcare USA, New York, 2007; 141–168.

58 Riley, C.M. and Rosanske, T.W., eds. *Development and Validation of Analytical Methods.* Elsevier Science, Tarrytown, NY, 1996.

59 Krause, S.O. Analytical method validation for biopharmaceuticals: A practical guide. *BioPharm Int* 2005, *18*(10), pp. 52–59.

60 Standard operating procedures for validation and compliance. Available at: http://www.labcompliance.com/solutions/sops/default.aspx?sm=b_b#7 (accessed January 21, 2011).

61 Jenke, D.R. Guidelines for the design, implementation, and interpretation of validations for chromatographic methods used to quantitate leachables/extractables in pharmaceutical solutions. *J Liq Chromatogr Relat Technol* 2004, *27*(20), pp. 3141–3176.

62 Swartz, M.E. and Krull, I.S. Validation, qualification, or verification? *LC/GC N Am* 2005, *23*(10), pp. 1100–1109.

63 Porter, D.A. Qualification, validation, and verification. *Pharm Technol* 2007, *31*(4), pp. 146–154.

64 Jenke, D., Garber, M.J., and Zietlow, D. Validation of a liquid chromatographic method for quantitation of organic compounds leached from a plastic container into a pharmaceutical formulation. *J Liq Chromatogr Relat Technol* 2005, *28*(2), pp. 199–222.

65 Jenke, D., Poss, M., Story, J., Odufu, A., Zietlow, D., and Tsilipetros, T. Development and validation of chromatographic methods for the identification and quantitation of organic compounds leached from a laminated polyolefin material. *J Chromatogr Sci* 2004, *42*, pp. 388–395.

66 Boudreau, S.P., McElvain, J.S., Martin, L.D., Dowling, T., and Fields, S.M. Method validation by phase of development: An acceptable analytical practice. *Pharm Technol* 2004, *28*(11), pp. 54–66.

67 Bloch, M. Validation during drug product development—Considerations as a function of the stage of drug development. In: Ermer, J. and Miller, J.H.M., eds. *Method Validation in Pharmaceutical Analysis: A Guide to Best Practice.* Wiley-VCH Verlag GmbH & Co., Darmstadt, 2005; 243–264.

68 Ermer, J. Analytical validation within the pharmaceutical environment. In: Ermer, J. and Miller, J.H.M., eds. *Method Validation in Pharmaceutical Analysis: A Guide to Best Practice.* Wiley-VCH Verlag GmbH & Co., Darmstadt, 2005; 3–19.

69 Ermer, J. 2.4 Precision. In: Ermer, J. and Miller, J.H.M., eds. *Method Validation in Pharmaceutical Analysis: A Guide to Best Practice.* Wiley-VCH Verlag GmbH & Co., Darmstadt, 2005; 21–51.

70 Horwitz, W. The certainty of uncertainty. *J Assoc Off Anal Chem* 2003, *86*, pp. 109–111.

71 Riley, C.M. Statistical parameters and analytical figures of merit. In: Riley, C.M. and Rosanske, T.W., eds. *Development and Validation of Analytical Methods.* Elsevier Science, Tarrytown, NY, 1996; 64.

72 Massart, D.L., Smeyers-Verbeke, J., and Vander Heyden, Y. Benchmarking for analytical methods: The Horwitz curve. *LC/GC Eur* 2005, *18*(10), pp. 528–531.

73 Miller, J.H.M. 2.8 System suitability tests. In: Ermer, J. and Miller, J.H.M., eds. *Method Validation in Pharmaceutical Analysis: A Guide to Best Practice.* Wiley-VCH Verlag GmbH & Co., Darmstadt, 2005; 190.

74 Ermer, J. 2.4 Linearity. In: Ermer, J. and Miller, J.H.M., eds. *Method Validation in Pharmaceutical Analysis: A Guide to Best Practice.* Wiley-VCH Verlag GmbH & Co., Darmstadt, 2005; 80–100.

75 Hokanson, G.C. A life cycle approach to the validation of analytical methods during pharmaceutical product development, part II: Changes and the need for additional validation. *Pharm Technol* 1994, *18*(10), pp. 92–100.

76 Cox, R.A. Analytical procedures in a quality control environment. In: Ermer, J. and Miller, J.H.M., eds. *Method Validation in Pharmaceutical Analysis: A Guide to Best Practice.* Wiley-VCH Verlag GmbH & Co, Darmstadt, 2005; 354.

77 Paulson, D. Systems and modular approach: A proposed alternative to methods validation. *Inside Laboratory Management*, 2007, p. 9.

78 Norwood, D.L. and Ball, D. Product Quality Research Institute: Safety thresholds and best practices for extractables and leachables in orally inhaled and nasal drug products. Submitted to the PQRI

Drug Product Technical Committee, PQRI Steering Committee, and U.S. Food and Drug Administration by the PQRI Leachables and Extractables Working Group, 2006, *3*(3), pp. 135–151.

79   Norwood, D.L., Paskiet, D., Ruberto, M., Feinberg, T., Schroeder, A., Poochikian, G., Wang, Q., Deng, T.J., DeGrazio, F., Munos, M.K., and Nagao, L.M. Best practices for extractables and leachables in orally inhaled and nasal drug products: An overview of the PQRI recommendations. *Pharm Res* 2008, *25*(4), pp. 727–739.

80   Skoog, D.A., West, D.M., Holler, F.J., and Crouch, S.R. *Fundamentals of Analytical Chemistry*, 8th ed. Thompson Learning, Belmont, CA, 2004.

81   Ewing, G.W. (ed). *Analytical Instrumentation Handbook*. Marcel Dekker, New York, 1997.

APPENDIX: CASE STUDY 18.1

DEVELOPMENT OF A LIQUID CHROMATOGRAPHY/ MASS SPECTROMETRY (LC/MS) METHOD TO DETERMINE LEACHABLES FROM PLASTIC VALVE COMPONENTS

An LC/MS leachables method was created for the study of antioxidants retained on plastic valve components. The following activities are illustrations of method development activities that one might expect to complete.

Availability of Standard Compounds

In this example, the antioxidants are commercially produced so a standard is available. A check might be needed to ensure purity is acceptable since if a technical grade is used for plastic manufacture, it may not be sufficiently pure.

Solubility of Standards in Solvents Selected for Analysis

The solubility of the antioxidants was known to be low in water and methanol. They were freely soluble in hexane and dichloromethane; however, these solvents are not particularly suitable for use in reversed-phase high-performance liquid chromatography (HPLC). Acetonitrile was explored and found to solubilize the antioxidants at the required concentrations.

Defining Chromatographic System

A reversed-phase HPLC system based on an octadecyl silane (ODS) column, 5-μm particle size was explored. It was noted that the retention times on this initial 15 cm × 4.6 mm ID (inner diameter) column were quite long, and there was a large separation between the antioxidants under study. Given that the method was going to be used to process a large number of samples, a solution was sought to reduce the overall cycle time for the method analysis. This led to a reduction in the length of the column—The length was reduced to 1 cm and still gave adequate resolution between analytes. In addition, selectivity was ensured through the use of a mass spectrometer. The mobile phase employed was an isocratic system using a methanol:water mix with a small amount of ammonia to enhance ionization.

Optimization of Detection Parameters on Mass Spectrometer

The phenolic-based antioxidants were expected to produce negative ions, which can be used to monitor these leachables. Both electrospray ionization (ESI) and atmospheric pressure chemical ionization (APCI) were screened as detection methods. APCI was determined to give the best results under the conditions of analysis and was thus selected for further optimization. Selected ion monitoring (SIM) was chosen to enhance the selectivity and sensitivity of the method. The enhancement of the signal using this method also improved the linear response.

Optimization consisted of studying the effect of varying several of the mass spectrometer parameters (vaporizer temperature, fragmentor voltage, capillary current and voltage, and drying gas flow and temperature). These parameters formed the design of experiments (DOEs), and software on the instrument was able to automate the optimization experiments. The experimental output indicated the best combination of parameters to give a robust response.

Optimization of Chromatography

Further optimization of the chromatography was performed using the optimized mass spectrometer parameters to achieve a final HPLC/MS method. This work also included an evaluation of the sample matrix to ensure that the sample solutions did not interfere with detection of target analytes or show any other matrix effects. Introduced at this stage was some consideration around instrument performance. A prerun check with signal-to-noise ratio measurement was included along with an instruction to clean the spray chamber of the mass spectrometer with an acetone rinse to ensure continued performance.

Development of Procedure to Prepare Sample Solutions for Analysis

Sample preparation was based on other methods previously developed to study leachables in metered dose inhalers (MDIs). MDIs contain propellants, which liquefy under pressure or at low temperatures. Therefore, the propellants can be sampled by cooling the MDI then allowing the propellant to evaporate and leave behind a residue that contains the leachables of interest. Since the antioxidants are not volatile, the evaporation step does not reduce the antioxidant levels. The residue is redissolved to prepare a sample for analysis. A range of solvents were investigated for the new method to select the most suitable one based on its ability to solubilize the selected antioxidants.

Development of Study Design Protocol to Produce Suitable Samples for Experimental Study

During the development of the method, samples were placed on storage and subsequently used to confirm that the method was suitable for use. These samples were also used during formal validation.

APPENDIX: CASE STUDY 18.2

DEVELOPMENT OF A HIGH-PERFORMANCE LIQUID CHROMATOGRAPHY/ULTRAVIOLET (HPLC/UV) METHOD FOR THE DETERMINATION OF EXTRACTABLES IN POLYPROPYLENE COMPONENTS

Problem

A quality control (QC) release method was required for the determination of extractables in a polypropylene material used to mold multiple components of varying size and shape. The method had to be sensitive enough to detect known compounds at their required analytical evaluation thresholds (AETs), and to detect and separate unknown compounds. Additionally, the method needed to be developed in as short a time as possible to support an unexpected change in material supply.

Procedure

The first step was to prepare a method development requirements summary as follows:

Material	Replacement device material—polypropylene		
Deliverable	Analytical test method (ready for validation) and development report		
Delivery date	October 31, 2006 for final method and report		
Project team	Analytical lead		
	L&E Group leader		
	Project manager		
Key constraint(s)	Compressed timeline for development		
Method inputs			
Method scope	Routine extraction method for raw material and mouthpiece component		
	Quantitative method for known targets/qualitative profile method		
Potential targets	Irganox 1076	Octadecyl-3-(3,5-di-tert-butyl-4-hydroxyphenyl)-propionate (OBHP)	2082-79-3
	Irganox 1010	Pentaerythritol tetrakis (3,5-di-tert-butyl-4-hydroxyhydrocinnamate) (PBHC)	6683-19-8
	Irgafos 168	Tris(2,4-di-*tert*-butylphenyl)phosphite (TBPP)	31570-04-4
	Irgafos 126	Bis (2.4-di-t-butylphenyl) pentaerythritol diphosphite (BBPP)	26741-53-7
Peaks of interest	Irganox 3114	[1,3,5-Tris(3,5-di-(tert)-butyl-4-hydroxybenzyl)-1,3,5-triazine-2,4,6 (1H, 3H, 5H)-trione (THBT)]	27676-62-6
	Irganox PS 802		693-36-7
	Dibutyl phthalate (DBPH)		
	Trisphosphate		
	Unknowns ≥10 µg/g		
	(To be determined)		

Technical constraints	Instrumentation—none
	Materials—none
	Other—none
Method requirements	Limit of quantitation (LOQ)—12 μg/g (20 μg/g customer AET corrected for repeatability, accuracy)
	System precision—%RSD ≤ 2.0 for targets in standard injections
	Linearity—r^2 ≥ 0.995
	Specificity—purity threshold >995 by diode array detector (DAD), no interfering peaks in blank, extraction blank
	Accuracy—70%–130% recovery from spiked solutions at 50%, 100%, and 150% of nominal concentration
	Repeatability—%RSD ≤ 10 or replicate preps
	Intermediate precision—≤20% difference between analysts
	Chromatographic profile—similar profile for replicate sample preps within
	Retention time window (to be determined)
	Stability—85%–115% initial for at least 3 days at controlled room temperature
	Run time less than 45 minutes
	Sample prep time less than 60 minutes
Separation goals	R ≥ 2.0 for target peaks
	T ≤ 1.5 for target peaks
	k between 3 and 20
	Run time ≤30 minutes
	LC/MS compatible
Reference documents	
References	Protocol—polypropylene materials extractables method development
	Replacement material schedule

The second step was to progress chromatography and sample extraction method development simultaneously; see Section 8.5 above for extraction design of experiment (DOE) details.

Procedure for Chromatographic Method Development

Initial Assessment Initial experiments were performed to develop a suitable separation for five known extractables compounds in the material (BBPP, OBHP, PBHC, THBT, TBPP). Standard and sample solutions were assayed using water/acetonitrile and water/methanol scouting gradients on Phenomenex (Torrance, CA) Luna phenyl-hexyl and Technologies (Santa Clara, CA), Zorbax Eclipse SB-C$_8$, and Zorbax Eclipse XDB-C$_8$ analytical columns. The water/acetonitrile gradient on an Zorbax Eclipse XDB-C$_8$ (3.5 μm, 4.6 × 100 mm) best separated all targets and peaks of interest (Fig. CS 18.2.1).

Method Optimization Based on the initial results, methanol was added as a minor component to slightly alter the selectivity of the mobile phase in order to

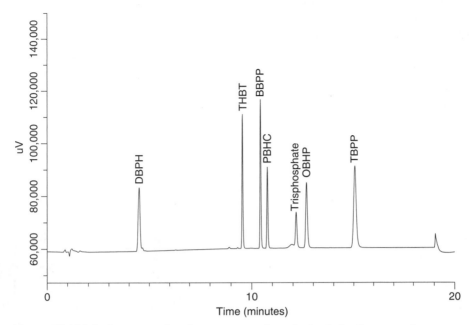

Figure CS 18.2.1   Representative chromatogram of standard solution from water/ acetonitrile scouting gradient on Zorbax Eclipse XDB-C$_8$ column.

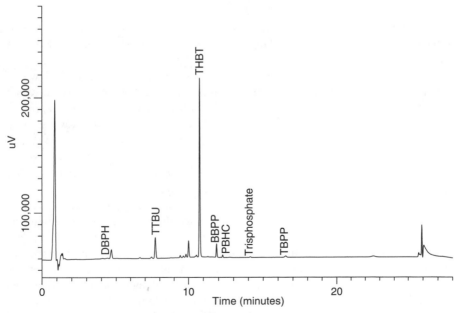

Figure CS 18.2.2   Representative chromatogram of extracted material, final method (full scale).

UV Apex spectrum of Peak 10.585 of 09260604.D

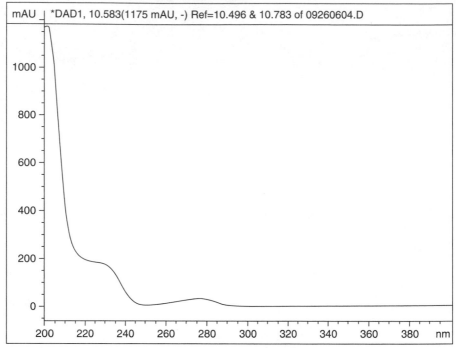

Figure CS 18.2.3   UV spectrum (200–400 nm) of THBT peak.

increase the separation of THBT and a closely eluting unknown compound. A hold at the final condition was extended to wash several late-eluting compounds from the column (Fig. CS 18.2.2).

Target UV spectra were evaluated to determine a suitable analytical wavelength. Most of the targets were aromatic compounds with long alkyl chains, which are weak chromophores. Although most targets had discrete maxima at approximately 280 nm, that wavelength did not provide adequate sensitivity. A wavelength of 230 nm was selected because it corresponded to higher absorbance maxima in the target compounds (Fig. CS 18.2.3) and internal standard compound (Fig. CS 18.2.4). This wavelength was low enough to detect poorly absorbing unknowns but was well above the UV cutoff of the mobile-phase system.

Because samples were extracted with isopropanol (IPA) using an automated solvent extractor, an internal standard was used to allow quantification without the need for time-consuming blowdowns and volumetric dilutions. IPA is a relatively viscous, nonpolar solvent compared with the mobile phase, so neat injections of samples extracted in IPA impacted the chromatographic profile. Because of the solvent mismatch with the mobile phase, injection volumes above 15 μL caused severe fronting of early-eluting peaks. An injection volume of 10 μL was selected to avoid peak fronting, along with a flow rate of 1.0 mL/min. An internal standard compound (2,4,6 Tri-tert-butylphenol [TTBU]) with a structure and UV response similar to the target peaks was evaluated to confirm that it eluted in a region of the chromatogram that is well separated from the targets and free from interfering peaks.

UV Apex spectrum of Peak 7.632 of BDA10008.D

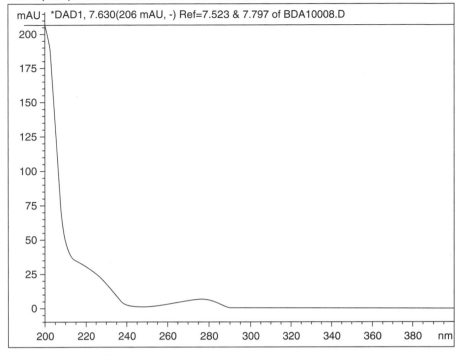

Figure CS 18.2.4   UV spectrum (200–400 nm) of internal standard peak.

Peak Purity   Specificity was assessed using peak purity analysis by diode array (liquid chromatography [LC]/DAD) and LC/mass spectrometry (MS) for all target peaks in a working standard, raw material sample, and mouthpiece sample. Spectral data collected from 215 to 400 nm were used to evaluate peak purity. Purity factors were above 995 for all targets except BBPP, which indicated that those peaks were spectrally pure. The purity factors for BBPP in the standard and samples were below 995, which indicated that the peak was not spectrally pure. Further analysis showed there was a small peak that co-eluted with BBPP present in the standards, samples, and extraction blank, but not in the diluent blank. This indicated that the small peak was contributed from the internal standard. Extracted chromatograms at 400 nm showed that the relative response of the peak was consistent across samples and blanks. The lack of a corresponding peak in the extraction blank chromatogram at 230 nm indicated that this peak did not have a significant UV absorbance at the analytical wavelength, so that the peak, when present, would not affect quantitation.

Because UV peak purity only identifies the absence of co-eluting peaks with dissimilar spectra, a peak purity analysis using MS (LC/MS) was performed for the main target peak (THBT) in sample extracts. Samples were analyzed by LC/MS using positive and negative atmospheric pressure chemical ionization (APCI). Target peaks were assessed for peak purity by comparing their mass spectra against those obtained from authentic reference standards. The mass spectra of the THBT peaks in the samples were identical, within instrumental uncertainty, to the mass spectra obtained from peaks in the standard. This indicated that the peaks were pure by LC/MS.

Chromatographic Conditions

HPLC system	Agilent 1100 HPLC	
Mobile phase	A: 50:40:10 acetonitrile:water:methanol	
	B: acetonitrile	
Column	Agilent Zorbax Eclipse XDB-C_8, 3.5 μm, 4.6 mm × 100 mm	
Column temperature	50°C	
Flow rate	1.0 mL/min	
Injection volume	10 μL	
Detection wavelength	230 nm	
Gradient		
Time (minutes)	% A	% B
0.0	90.0	10.0
10.0	0.0	100.0
24.0	0.0	100.0
24.1	90.0	10.0
30.0	90.0	10.0

CHAPTER 19

CRITICAL COMPONENT QUALITY CONTROL AND SPECIFICATION STRATEGIES

Terrence Tougas, Suzette Roan, and Barbara Falco

19.1 OVERVIEW

The intent of this chapter is to explore the "quality" aspects of pharmaceuticals as they relate to leachables and extractables (L&E). To start, it is important to define the term quality. The most common definition of quality is the extent to which a product or service meets customer expectations.[1] For a pharmaceutical product, the key quality attributes are primarily related to safety and efficacy. With respect to L&E, the clear emphasis is on the safety aspects. In other words, the quality control (QC) strategy should assure that leachables present in the drug product are at levels well below any level of safety concern. Previous chapters have dealt with characterizing the L&E profiles of drug products and establishing the thresholds of concern related to L&E. This chapter deals first with converting this information into a workable QC strategy including the product specifications that relate to L&E, and second with an overall approach to enhancing quality in the supply chain for orally inhaled and nasal drug product (OINDP) container closure systems and devices.

While a well-executed QC program is important to assuring quality, it is fundamental to understand that the quality of a commercial pharmaceutical product is established by a well-designed, -understood, and -executed manufacturing process using high-quality raw materials and components. No amount of testing can alter the inherent quality of a product. At best, QC testing can verify the quality of a product or detect when there is a quality issue. This chapter addresses approaches to the design and understanding of manufacturing processes, and ensuring quality of raw materials and components.

Leachables and Extractables Handbook: Safety Evaluation, Qualification, and Best Practices Applied to Inhalation Drug Products, First Edition. Edited by Douglas J. Ball, Daniel L. Norwood, Cheryl L.M. Stults, Lee M. Nagao.
© 2012 John Wiley & Sons, Inc. Published 2012 by John Wiley & Sons, Inc.

19.2 QUALITY BY DESIGN (QbD)

Much has been written over the last few years about the concept of "QbD." Regulatory bodies have encouraged the pharmaceutical industry to embrace these more modern quality concepts as a means to improve the overall quality of pharmaceutical products, and increase the efficiency of drug development and commercialization. The United States Food and Drug Administration (FDA) has commented in a guidance document on this topic as follows[2]:

> *The overarching philosophy articulated in both the CGMP regulations and in robust modern quality systems is:*
>
> *Quality should be built into the product, and testing alone cannot be relied on to ensure product quality.*
>
> Quality by design *means designing and developing manufacturing processes during the product development stage to consistently ensure a predefined quality at the end of the manufacturing process.*

This leads to the realization that L&E must be considered as an integral part of the overall product development. Material selection and manufacturing process design must balance performance characteristics with the L&E profile. This suggests that the common QbD tools (e.g., design of experiments [DOE s], analysis of variance [ANOVA], risk assessment, surface mapping, optimization strategies) should be applicable and advantageous for optimizing the product and process design with respect to these factors.

19.3 SPECIFICATION SETTING STRATEGIES

The fundamental goal of an L&E QC strategy associated with an OINDP is to assure that only safe levels of potential leachables are present in the drug product throughout its shelf life. However, any L&E QC strategy is predicated on a robust product and manufacturing process that has been developed with risk reduction of potential leachables as part of the design considerations. Key parts of the overall product design strategy should include material selection, process development including process controls, and verification of the leachables profile through stability studies conducted on the representative product.

The direct means of monitoring leachables is through laboratory investigation into the qualitative and quantitative nature of a particular OINDP leachables profile(s) over the proposed shelf life of the product (*leachables study*). Such studies are first conducted during product development to support leachables acceptance criteria and as part of the information required to establish an extractables/leachables correlation.

Whether or not these studies become a routine part of the QC strategy for a commercial product depends on the body of leachables/extractables information submitted in the marketing application, the quality strategy developed by the applicant and the acceptance by regulatory authorities of the proposed quality strategy. In principle, it should always be possible to make a sound technical argument that this particular quality attribute (leachables) can be controlled through proper material and process selection and, if necessary, monitored by appropriate extractables testing. This is in keeping with QbD principles of eliminating redundant testing, avoiding where possible end-product testing, and controlling quality attributes at the earliest appropriate point in the manufacturing process. Translating principle into practice depends primarily on the quality and thoroughness of the L&E studies conducted by the applicant.

Besides a regulatory expectation for maintaining controls on vendors supplying components for OINDP manufacture, business perspectives drive the need for robust controls on extractables, as this reduces the risk of costly quality failures due to substandard components and raw materials. It is advantageous to assure that components to be used in the manufacture of OINDPs contain the expected extractables at levels that will not later lead to unacceptable levels of leachables in the final product. This is typically achieved through *routine extractables testing*. Routine extractables testing is the testing by which OINDP container closure system critical components are qualitatively and quantitatively profiled for extractables. Such testing is first conducted during product development to determine the variability of the extractables profile(s) of the selected components and subsequently to determine if the components and process are capable of meeting the established extractables acceptance criteria. Note that the acceptance criteria should be established on the basis of safety and other product performance considerations.

19.3.1 Establishing a Correlation between L&E

It can be argued that if a control strategy has been established to ensure appropriate extractables levels and that this in turn assures that leachables levels in the final product will never exceed unacceptable levels, then end-product testing for leachables is redundant, unnecessary, and adds no value. This forms the basis for a strategy articulated in the Product Quality Research Institute (PQRI) L&E recommendations[3] that potentially eliminates routine leachables testing on commercial OINDPs. It is predicated on establishing, during product development, a so-called leachables/extractables correlation.

There are several caveats associated with this strategy. First, the acceptance of this type of control strategy is a policy matter subject to the approval of the appropriate regulatory authorities. In the United States, it is a regulatory expectation that even if leachables testing is not routinely performed, the company will still establish testing methods and appropriate acceptance criteria for leachables. The expectation is that if a product is tested for leachables, it will comply with these acceptance criteria. Thus, the main use for leachables testing under this scenario may be as part of an internal or regulatory investigation.

19.3.2 Establishing a Quality Standard

Historically, the relationship between acceptance criteria and the level of quality that they ensure has often been ill-defined within the pharmaceutical industry. One of

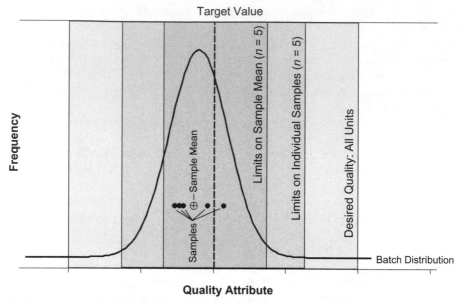

Figure 19.1   The relationship between quality limits (quality standard) and QC test acceptance criteria: distributional aspects of a batch versus QC samples used to make quality decisions.

the consequences of QbD is that the acceptance criteria are no longer established without a consideration of the corresponding quality level that they actually establish. As a result, a desired performance characteristic (e.g., essentially no leachables above an analytical evaluation threshold [AET]) should first be established and then used to derive criteria for a quality attribute that can be evaluated against the established criteria (e.g., "extractables levels").

The current practice in the pharmaceutical industry is to specify test acceptance criteria without defining an underlying "quality standard," or worse—to consider the acceptance criteria to be the quality standard. In a true "QbD" environment, one should first specify a quality standard suitable for the quality attribute in question and, based on this, determine a test structure (sample size, acceptance criteria, and tiers) appropriate to confirm whether the batch complies with the desired quality level or not.

This is illustrated in Figure 19.1. The desired quality is a boundary related to the acceptable performance of all units in a batch or population. Quality is achieved when virtually all units or a defined high percentage of all units produced lie within these boundaries. Typically, testing all units is impractical, and hence, QC testing attempts to determine adherence to a defined quality standard (with a certain confidence) for a specific batch by sampling a small number of units.

The results obtained are used to make quality decisions. This is done by comparing the results to acceptance criteria associated with the test. It is important to note that the acceptance criteria for the QC test are not the same limits used to define the desired quality (also referred to as quality standard or coverage) and that the former limits are inherently within the latter. Furthermore, these test acceptance criteria are also a function of the actual testing plan, that is, numbers of samples, averaged or individual results, and tiered testing.

It is also important to realize that it is impractical to consider the quality standard in terms of all or 100% of units being within the limiting quality. Hence, the practical alternative is to construct a quality standard that is cast in terms of a high percentage of units being within some quality boundaries. Within this construct, QC testing and the associated decisions can be cast in terms of conventional statistical hypothesis testing. It is then also possible to evaluate the ability of the QC testing to make good decisions through the construction of operating characteristics curves (OCCs) for the specific test. The construction and use of OCC will be discussed in more detail below. The selection of a quality standard is inherently not a statistically driven decision. It is instead a statement of the quality requirements for a given product.

In the case of L&E, the quality standard ultimately relates to the acceptable levels of leachables in the product. As previously discussed, the actual QC testing may consist primarily of extractables testing of critical components, but this may be performed as a surrogate for leachables testing. In either case (leachables or extractables testing), acceptable levels are established through a safety, not a statistical, consideration. Once these acceptable levels have been cast as quality standard statement, then statistical techniques can be used to derive test acceptance criteria based on a given test structure.

Once a suitable control strategy for L&E has been established including tests and acceptance criteria, then one can evaluate whether the developed processes are capable of producing a pharmaceutical product or critical component that consistently meets the desired quality standard. The statistical tools commonly employed for this purpose are process capability analysis and process performance analysis, and form the basis for a subsequent section of this chapter.

19.3.3 Statistics: Translating Quality Standard into Acceptance Criteria

As indicated previously, the OCCs used to characterize QC tests are derived from conventional statistical hypothesis testing. For a basic discussion of statistical testing of hypotheses, see Dudewicz's chapter in *Juran's Quality Handbook*.[4] An expanded discussion of OCCs is included in the American Society for Quality (ASQ) text on *Specifications for the Chemical and Process Industry*.[5]

If we consider the typical QC problem in the context of hypothesis testing, then the problem can be formulated as illustrated in Figure 19.2. Simply stated, a particular batch is either truly good or bad, and there are only two possible decisions to make about a batch, that is, to either release the batch as acceptable or reject it as unacceptable.

Recognizing that any measurement involves some degree of uncertainty, gives rise to the four scenarios illustrated in Figure 19.2. In two of the four resulting scenarios, the correct quality decision is made, but there are two other possibilities

Decision	True Situation: Batch is of acceptable quality.	True Situation: Batch is not of acceptable quality.
Accept batch	**Correct decision**	Type II error (β) ("consumer's risk")
Reject batch	Type I error (α) ("producer's risk")	**Correct decision**

Figure 19.2 Quality decisions: possible outcomes and consequences.

that involve making incorrect decisions. It is possible that one could reject a batch that is of suitable quality. The probability of this occurring is the type I error rate typically given the symbol "α." Conversely, one could accept a batch of unacceptable quality. The probability of this occurring is the type II error rate typically given the symbol "β." The probability of making these errors are also commonly referred to as "producer's risk" (α) and "consumer's risk" (β) for the obvious, but somewhat simplistic, reason. The goal in designing the QC test structure is to minimize both these risks of making incorrect decisions within the practical constraints of laboratory testing. Note that the risk tolerance for these two types of risk is typically unequal; that is, generally there is less risk tolerance for the so-called consumer's risk than for producer's risk.

An OCC consists of a plot of the probability of acceptance (or rejection) versus the true value of the quality attribute in question. Note that true value in this context refers to a population parameter that is typically estimated by a sample statistic. For example, consider a batch of tablets. A quality attribute of that tablet batch is its mean assay. One can never know absolutely the true mean assay of the batch without testing all tablets in the batch and even then there is measurement uncertainty. Instead, one samples and tests representative units in the batch and estimates the true population mean (with some level of uncertainty) via the sample mean (a statistic).

Typical forms of an OCC are

- P(accept) versus true batch mean,
- P(accept) versus true batch standard deviation, and
- P(accept) versus true % defects.

In the ideal situation, types I and II errors would be absent, implying that correct decisions are always made. This is graphically reflected in the ideal OCCs shown in Figure 19.3 where these curves become step functions. In real-world situations, there is always some uncertainty in the decision-making process when sampling is involved due to inherent variabilities. The impact on the OCCs is to introduce curvature to these curves (Fig. 19.4). The quality of the decision-making process is then reflected in how close the OCC approaches the ideal step functions for a given expenditure of effort in sampling and testing.

OCCs are used to characterize the statistical qualities of the decision-making process associated with a particular QC test's structure/form. Test structure/form in

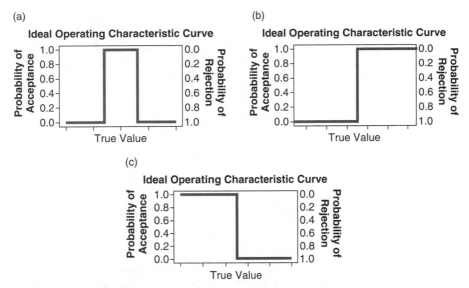

Figure 19.3   Ideal operating characteristic curves. (a) Two-sided limit; (b) one-sided lower limit; (c) one-sided upper limit.

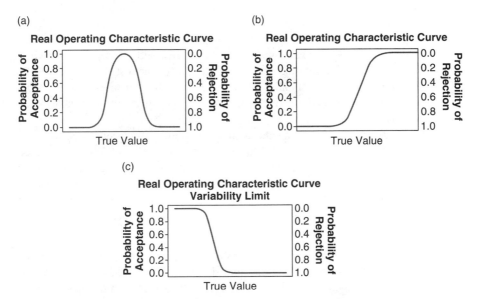

Figure 19.4   Real operating characteristic curves. (a) Two-sided limit; (b) one-sided lower limit; (c) one-sided upper limit.

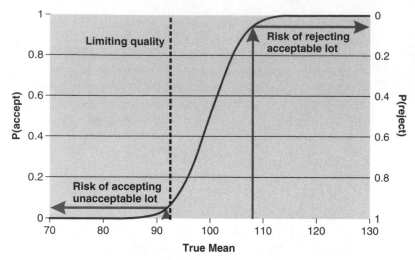

Figure 19.5 Example operating characteristic curve for a one-sided lower limit illustrating limiting quality and type 1 and 2 error rates. The figure shows the risks associated with testing in relationship to the operating characterisitic curve.

this context includes numbers of samples, limits, tiers, decision process flow, and quantities compared with limit. From the OCC, one can comment on the ability of the test structure to discriminate between acceptable and unacceptable "batches." Qualitatively, the closer the OCC approaches the corresponding step function (Fig. 19.3), then the better the decision making of the particular test. Furthermore, the OCC (Fig. 19.5) allows the estimation of the types I and II error rates: risk of failing an acceptable batch (producer's risk) and risk of passing an unacceptable batch (consumer's risk).

OCCs are constructed using the appropriate cumulative density probability distribution for the data in question. One of the most common OCCs is associated with a two-sided limit on the mean of a quality attribute. An example of this is the chemical assay of the active pharmaceutical ingredient in a drug product. In the case of L&E, the most common OCC is a single-sided upper limit on the leachable or extractable. The general steps for constructing these types of OCCs are as follows:

1. Start by selecting the appropriate probability distribution for the individual measurements. For this case, a normal distribution is often assumed. Where enough data is available, statistical tests may be used to aid this selection.

2. Obtain an estimate or assume a value for the standard deviation of the results. Note that an OCC on the mean is always at a specified standard deviation. If the standard deviation is not well known, then it may be necessary to construct a family of OCCs that vary the standard deviation.

3. Calculate the probability to accept for a given value of the true mean from the appropriate cumulative density probability distribution based on the test

construct. This is equivalent to obtaining areas under the curve of the associated probability distribution.

4. Repeat this process over the range of interest. With modern spreadsheet and statistical programs, it is a relatively simple task to use this numerical approach to construct OCCs. In all cases, it may be helpful to consult with a statistician to construct these curves.

Consider the simple hypothetical case of a tablet assay that is required to be between 95 and 100 mg/tablet. Suppose that the requirement is that the mean of three replicate determinations must be between those limits. Furthermore, we know from experience that the standard deviation of these assay means is 2 mg/tablet, and we are assuming that the means are normally distributed.

To calculate a point on the corresponding OCC for this test structure, we assume a value of the true mean (99 mg/tablet). The next step is to convert the limits (95 and 105) relative to the assumed mean (99) into standardized z-units (this allows us to use tabulations of the standard normal distribution [SND] to obtain areas under the distribution and determine acceptance probabilities). For $\mu = 99$ and $\sigma = 2$, the lower limit corresponds to $(95 - 99)/2 = -2$ z-units and the upper limit corresponds to $(105 - 99)/2 = +3$ z-units. Using these z-values, areas under the SND can be obtained. Note that we seek the area under the SND between -2 and $+3$ z-units. This corresponds to the area between 95 and 105 mg/tablet for a normal distribution with $\mu = 99$ mg/tablet and $\sigma = 2$ mg/tablet and represents the probability of accepting a batch with a true mean of 99 mg/tablet. Note that caution should be exercised to understand how areas under the SND are tabulated in the particular table used (i.e., area below or above the z-value). Sometimes only positive values are tabulated, and the symmetry of the SND must be used to obtain areas for negative z-values. For this example, the area under the SND below $z = -2$ is 0.02275, and the area above $z = 3$ is 0.00135. The area between $z = -2$ and $z = 3$ is

$$1 - 0.02275 - 0.00135 = 0.97590.$$

The interpretation of this value is that for the conditions in this example, there is a 97.59% probability of accepting a batch when the true mean of the batch is 99 mg/tablet.

In practice, to actually construct the OCC, these calculations would be set up for incremental values of the true mean within a spreadsheet. Many available spreadsheets or statistical programs contain statistical functions that can call the areas under the needed probability distribution. For more complicated test structures, the aid of a competent statistician is advised.

One of the main utilities of OCCs is that they allow comparisons of different test structures relative to the quality of the decision-making process. Through OCCs, the influence of sample size, tiered test structures, or other decision-making structures can be evaluated. In the next section, we will make some general observations concerning the influence of some of the more common factors.

Probably the most common consideration in a test structure is the sample size. Intuitively, one would expect that as the sample size increases, the ability to make correct QC decisions should improve. In practice, this depends on how the additional

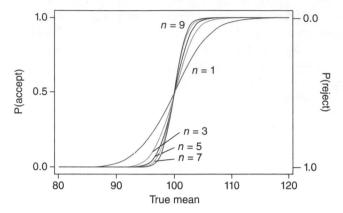

Figure 19.6   Operating characteristic curves showing influence of increasing sample size when mean of replicates is compared with a limit. One-sided limit: sample mean >100. Number of replicate samples, $n = 1, 3, 5, 7, 9$.

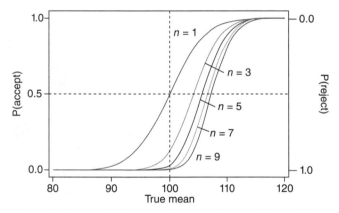

Figure 19.7   Operating characteristic curves showing influence of increasing sample size when individual replicates are compared with a limit (i.e., n of n replicates must meet the criterion). One-sided limit: sample mean >100. Number of replicate samples, $n = 1, 3, 5, 7, 9$.

results are used. In the case where the sample mean of a number of replicates is evaluated against one- or two-sided limits, this can be shown to be the case through OCCs (Fig. 19.6). However, in the case where individual replicates are compared with acceptance limits, it is a more complicated situation (Fig. 19.7).

In the former case where the sample mean is compared with the limit, the trend in the OCCs as "n" increases is that the curves sharpen and approach the ideal step function. In addition, the changes are symmetrical about the midpoint of the curves. This can be interpreted as an improvement in the ability to make good decisions and the simultaneous improvement in types I and II error rates.

In the second case where each individual replicate is compared with the limit and all must conform to the limit, a different trend is revealed in the OCCs.

While the curves do sharpen as "*n*" increases, there is a shift of the curves toward higher true means. In contrast to the former case, while the type II (consumer's risk) error rate is improving both due to the increased sharpness of the curves and the shift to the right, the type I (producer's risk) error rate is increasing dramatically due to the shift in the position of the OCC. One way to interpret this is that the true mean (or value of critical quality attribute) of batches must be larger as the number of replicates required increases in order to consistently meet the acceptance criteria for this one-sided lower limit. In other words, the quality standard of the product evaluated must increase to consistently meet this requirement. This is contrary to the original premise of establishing a quality standard and then designing a test that assures conformance to that quality standard.

Another view of these two schemes is that the former (acceptance criteria based on sample mean) rewards the producer for testing more samples, while the latter punishes the producer when *n* is increased without changing the acceptance criteria (increased producer's risk).

19.3.4   Determine Whether Process Is Capable of Meeting the Acceptance Criteria

Once critical quality attributes have been identified and a control strategy has been established, it is appropriate to address the question "Are the developed processes capable of producing product meeting the established quality standard?" Throughout the product life cycle, the manufacturer is interested in knowing how well the process is actually performing. These two concepts form the basis of process capability and performance analyses. Again for L&E, the assumption is that in most cases, a suitable correlation has been established between L&E. This places the focus on extractables testing of critical components. One advantage of the emphasis on extractables testing is that this generally allows acquisition of results on a larger number of batches than in the case of leachables testing on the final pharmaceutical dosage form.

Since both process capability and performance analyses are studies of the variability of processes, they both rely on estimates of the standard deviations of quality attributes. Reliable estimates of standard deviations and variances inherently require relatively large numbers of values. It can be easily demonstrated that the typical numbers of representative batches of the final drug product supporting most new drug submissions are inadequate to provide a reliable estimate of process capability. In contrast, if the L&E control strategy is based on extractives testing, obtaining sufficient numbers of batches of critical components to assess the initial process capability is much more feasible.

It is useful to make a distinction between the terms "process capability" and "process performance." Process capability can be defined as the statistical comparison of the inherent process variability (common cause variability only) of a process to some limits. Typical experimental designs rely on replicates within batches that allow estimation of the within-batch variability. It is a measure of the best possible performance of a given process.

Process performance is then the statistical comparison of the total observed variability of a process typically over a long period of time to some limits. It may include the so-called special cause variability, that is, sources of variability that can potentially be identified and controlled or eliminated. It represents an assessment of the current performance of a process including any special cause sources of variability.

Several different "capability indexes" exist to measure process capability. Among these are C_p and C_{pk}. They are defined as

$$C_p = (\text{USL} - \text{LSL})/6s_w \qquad\qquad (19.1)$$

and

$$C_{pk} = \min\{[(\text{USL} - \text{Avg})/3s_w], [(\text{Avg} - \text{LSL})/3s_w]\}, \qquad\qquad (19.2)$$

where

USL = upper specification limit,

LSL = lower specification limit,

s_w = within-batch sample standard deviation, and

Avg = sample average.

These indexes measure process capability relative to process requirements, that is, the specification limits. Since they are a dimensionless metric, in principle, this means that they can be used to compare the capability of different processes. This assumes however that a similar experimental design was used to estimate the within-batch variability.

A generally accepted minimum expectation for a new process is a C_p or C_{pk} value of 1.33. Values of these indexes below 1.33 indicate that the process as designed may not be capable of consistently delivering product of suitable quality. Note that for a process centered within its specification limits, this translates into the limits coinciding with ±4 standard deviations of the within-batch variability of the process.

Since the C_p index requires a two-sided limit, it is not generally applicable to L&E testing, which is predominately a one-sided limit. For a one-sided limit, the C_{pk} is interpreted with respect to only that part where there is an actual limit, that is, only the USL (common for L&E), where the limit is a maximum value, or only the LSL, where there is a minimum requirement. For a two-sided limit where both indexes can be determined, a comparison of the C_p and the C_{pk} is an indication of how well "centered" the process is. Comparable values of the C_p and the C_{pk} are an indication that the process is well centered between the specification limits.

The analogous process performance indexes to C_p and the C_{pk} are P_p and the P_{pk}. Again, the purpose of these indexes is to provide a measure of actual process performance relative to the specification limits. The definition of these two indexes is very similar in form to the process capability indexes. The important difference is that the total process standard deviation is used in the calculation:

$$P_p = (\text{USL} - \text{LSL})/6s_t, \qquad\qquad (19.3)$$

$$P_{pk} = \min\{[(USL - Avg)/3s_t], [(Avg - LSL)/3s_t]\}, \tag{19.4}$$

where

USL = upper specification limit,

LSL = lower specification limit,

s_t = total process sample standard deviation, and

Avg = overall sample average.

These indexes measure process performance relative to process requirements, that is, the specification limits. The same caveats apply as with capability indexes relative to comparing processes. In addition, the validity of these indexes depends on a process that is stable and has been studied over a sufficient period of time (at least 10–15 independent instances). Besides these simple metrics, it is recommended that process control charts be employed to aid in understanding the performance of any particular process.[6] It should be recognized that the discussion here is only a small part of what should be considered in the context of an overall quality system that potentially employs a variety of statistical tools.

A generally accepted minimum expectation for a new process is a P_p or P_{pk} value of 1.33. Values of these indexes below 1.33 indicate that the process is not performing in a manner that consistently delivers product of suitable quality.

Similar to the C_p index, the P_p index requires a two-sided limit and is not generally applicable to L&E testing, which is predominately a one-sided limit. For a one-sided limit, the P_{pk} is interpreted with respect to only that part where there is an actual limit, that is, only the USL (common for L&E), where the limit is a maximum value, or only the LSL, where there is a minimum requirement. For a two-sided limit where both indexes can be determined, a comparison of the P_p and the P_{pk} is an indication of how well "centered" the process is. In addition, comparable values of the P_{pk} and the C_{pk} are an indication of the absence of special cause variability.

The application and concepts connected with both process capability and performance indexes can be illustrated through the following example. Consider a set of extractables data collected as follows:

- Five gaskets were sampled from each of 10 batches of gaskets.
- The level of extractable "A" was determined for all 50 samples.
- Based on toxicology information and a leachables correlation, an upper limit of 120 ppm was established.
- To monitor material consistency, one may choose to establish a lower limit of, for example, 80 ppm (based on the information from the supplier).

The extractables results obtained are as follows along with the basic statistics needed to compute indexes.

The individual batch standard deviation (s_i) is a simple sample standard deviation computed from the five samples for a given batch (A–J). The within-batch standard deviation (s_w) is a pooled standard deviation computed as a weighted average of individual batch standard deviations. In this case, since the sample size is equal for all batches, s_w is the simple arithmetic average of the 10 individual batch standard deviations. Finally, the total standard deviation (s_t) is the grand standard deviation computed with all 50 individual results.

The basic questions that process capability and performance analyses are intended to address are the following:

- Is the gasket manufacturing process capable of producing gaskets conforming to the limits? (process capability)

- Is the gasket manufacturing process actually performing at a level that consistently produces gaskets of suitable quality, that is, conforming to the limits? (process performance)

- Are there opportunities to improve the performance of the process, that is, is there evidence of special cause variability? (process performance)

It is always a good practice to start by plotting the results to examine the data visually. Figure 19.8 illustrates the extractables results in Table 19.1.

Substituting the appropriate values into Equations 19.1–19.4 given the above results in the following values for the respective capability and performance indexes:

$$C_p = (120-80)/(6)(5.25) = 1.27 \text{ or } 1.3,$$
$$C_{pk} = \min[(120-99.1)/(3)(5.25)], [(99.1-80)/(3)(5.25)] = 1.21 \text{ or } 1.2,$$
$$P_p = (120-80)/(6)(8.72) = 0.76 \text{ or } 0.8,$$
$$P_{pk} = \min[(120-99.1)/(3)(8.72)], [(99.1-80)/(3)(8.72)] = 0.73 \text{ or } 0.7.$$

If we consider the process capability indexes, we can conclude that the process is reasonably centered within the specification window since $C_p \sim C_{pk}$ (1.3 vs. 1.2). However, these indexes fall slightly below the 1.33 criterion indicating that the process as designed is marginal at best in terms of being capable of producing a product of the desired quality.

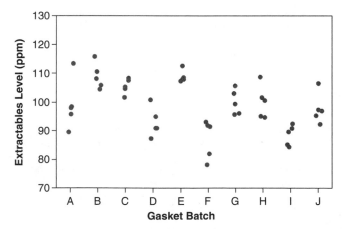

Figure 19.8   Plot of extractables results for example calculations of process capability and performance indexes.

TABLE 19.1. Example Extractables Results for Calculation of Process Capability and Performance Indexes

Gasket sample	Gasket batch									
	Batch A	Batch B	Batch C	Batch D	Batch E	Batch F	Batch G	Batch H	Batch I	Batch J
#1	98.7	104.6	108.6	91.0	112.8	82.1	99.5	108.8	89.8	92.5
#2	96.0	110.7	105.0	87.5	108.2	93.3	96.3	95.4	92.8	97.1
#3	98.3	106.0	101.9	95.1	107.4	78.2	106.0	100.8	91.2	95.4
#4	89.6	116.1	107.6	91.1	108.7	92.2	95.9	94.9	85.5	106.8
#5	113.5	108.4	105.5	101.1	108.1	91.7	103.2	101.8	84.5	97.5
s_i	8.8	4.5	2.6	5.2	2.2	6.9	4.4	5.7	3.6	5.4

$AVG = 99.1$

$s_t = 8.7$

$s_w = 5.3$

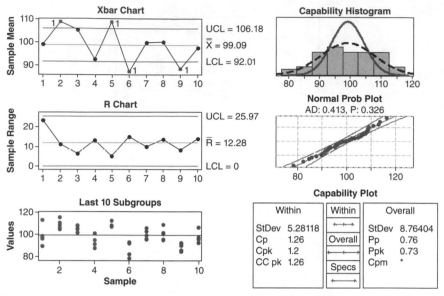

Figure 19.9 Process capability/performance analyses of example extractables results performed using MINITAB version.

Both process performance indexes (0.76 and 0.73) are well below the 1.33 criterion indicating the process is not performing at a level that will produce a product consistently meeting quality expectations. Furthermore, the P_{pk} of 0.73 is significantly lower than the C_{pk} of 1.21. This indicates that the process exhibits significant special cause variability that should be addressed. Note that the presence of special cause variability is manifested in the plot of the results (Fig. 19.8) as the clustering of results within a batch that appear less variable than the batch-to-batch variability. This is a signal that there may be an opportunity to improve the process performance by investigating the source of the special cause variability.

While the calculation of these capability and performance indexes is a relatively simple and straightforward task, several statistical software packages are capable of performing these same calculations. The typical advantages of utilizing these packages are that beyond calculating the capability and performance indexes, they also provide various related plots and additional related statistics, and do so in a fast, convenient manner. One example is the quality tools within MINITAB®. The example extractables results were analyzed using this statistical package. Figure 19.9 contains the resulting process capability and process analyses along with a plot of all results, conventional process control charts, tests for normality, and a histogram of the extractables results. While this software package is relatively easy to use, it does not substitute for the expertise of a qualified statistician. To ensure the appropriate use of control charts and these related statistics, the reader is advised to consult a statistician when designing and implementing a statistical process control program.

19.3.5   Conclusion

The preceding sections provide insights regarding a QbD approach to managing quality with respect to leachables. The emphasis is on establishing a leachables quality standard based on safety, developing a leachables/extractables correlation, and focusing QC testing on the extractables profile through a strong understanding of the component manufacturing process. With this approach, a limit for a given extractable is proposed based on safe exposure limits. Statistical tools (OCC) were described for translating a quality standard based on safety considerations into a testing scheme and related acceptance criteria that are the basis of a QC program for extractables.

The management of the extractables levels is fundamentally addressed through the ability of the manufacturing process to reliably produce components with extractables concentrations within the proposed limits. Extractables testing provides continuing assurance that the process is performing as expected. Statistical tools for evaluating the actual (control charts and process performance analysis) and potential (process capability analysis) performance of a manufacturing process were explored. The manufacturing process needs to be managed by the supplier of the component(s), if possible in collaboration with the OINDP manufacturer.

Indeed, important aspects of a QbD approach to extractables and leachables in OINDP lie in the management of extractables throughout the supply chain for OINDP container closure system/device components. Such management relies fundamentally on adherence to the concepts of current good manufacturing practices in the OINDP component supply chain.

19.4   CURRENT GOOD MANUFACTURING PRACTICES (cGMP) AND COMPLIANCE

19.4.1   Background

The pharmaceutical container closure system and device supply chain is often described using n, n-x nomenclature, where n is the pharmaceutical company, n-1 is the supplier that provides the packaging or device components, and n-2 is the supplier of materials and subcomponents procured by the n-1 supplier, and so forth.[7]

The compliance drivers related to L&E for OINDP naturally focus on components and the interface between the pharmaceutical manufacturer and the component supplier. This places added emphasis on the quality systems of the pharmaceutical manufacturer with respect to supplier quality and on the quality systems of the supplier (typically the material compounder or component fabricator). One of the challenges in this area is that critical L&E quality attributes may trace back to factors further back in the supply chain to material or additive suppliers. The net result is that consistent high quality for container closure systems correlates with strong healthy quality systems and relationships between the pharmaceutical manufacturer and its supply chain.

19.4.2   International Pharmaceutical Aerosol Consortium for Regulation and Science (IPAC-RS) Supplier Quality Guideline

Building on the importance of the role of the component supplier in driving quality with respect to L&E, the IPAC-RS, a consortium of OINDP manufacturers, under-took efforts to improve in general the quality relationship between pharmaceutical manufacturers and their component suppliers. A working group composed of rep-resentatives from both OINDP manufacturers and a variety of n-1 and n-2 suppliers came together to share quality concerns and generate recommendations related to quality topics. In 2006, the IPAC-RS "Good Manufacturing Practices Guideline for Suppliers of Components for Orally Inhaled and Nasal Drug Products" was pub-lished.[8] The goal of this guideline is to provide suppliers of OINDP components and OINDP manufacturers with recommendations and standards that harmonize quality expectations between both parties. This is a comprehensive guideline that, in addition to IPAC-RS recommendations, incorporates the requirements of both the IQA/PQG PS 9000 Application Standard for Pharmaceutical Packaging, ISO 9001:2000 Quality Management Systems—Requirements, and the ISO 9004:2000 Quality Management Systems—Guidelines for Performance. Four years after its inception, this guideline remains a frequently cited document and a recognized resource related to OINDP component quality. Furthermore, the OINDP recom-mendations and requirements from the guideline are being incorporated into the next version of PS 9000.[9]

19.4.3   Supply Chain and Supplier Management

The supply chain for OINDP delivery systems can be very complex. It is common for there to be contract manufacturers and multiple component suppliers, each manufacturing one or more components that are made from multiple materials. The complete supply chain for an OINDP can include dozens of suppliers. Additionally, the raw material or additive suppliers may produce materials that have variable compositions and are derived from several different manufacturing processes and may not be pharmaceutical or medical grade.[10]

The relationships between the OINDP manufacturer and its suppliers are even more complicated, especially at the n-2 and n-3 level, because often the identity of a supplier for an ingredient in a raw material is not disclosed to the OINDP manufac-turer. In cases such as this, the OINDP manufacturer has to rely on the supply chain management program of its supplier to ensure quality at the upstream supplier and to gain understanding of potential extractables and leachables. A change or impurity in one of the ingredients at an n-2 or n-3 supplier may be detected by the extractables methods for the final component or found in leachables studies on the OINDP. As a result, the understanding of the ingredients and the materials that enter into the final component is critical in developing appropriate analytical methodology.

A representation of a typical supply chain for a metered dose inhaler (MDI) valve is shown in Figure 19.10. This one valve has seven materials, which contain more than 50 total ingredients.

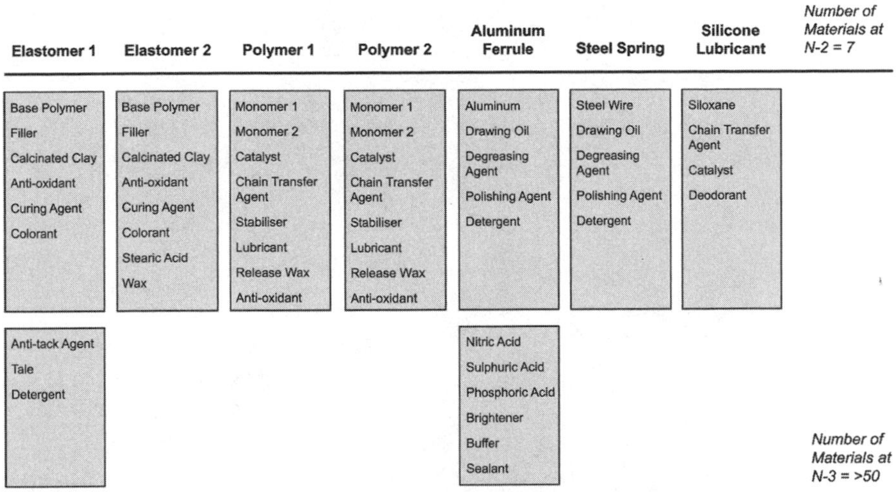

Elastomer 1	Elastomer 2	Polymer 1	Polymer 2	Aluminum Ferrule	Steel Spring	Silicone Lubricant	Number of Materials at N-2 = 7
Base Polymer	Base Polymer	Monomer 1	Monomer 1	Aluminum	Steel Wire	Siloxane	
Filler	Filler	Monomer 2	Monomer 2	Drawing Oil	Drawing Oil	Chain Transfer Agent	
Calcinated Clay	Calcinated Clay	Catalyst	Catalyst	Degreasing Agent	Degreasing Agent	Catalyst	
Anti-oxidant	Anti-oxidant	Chain Transfer Agent	Chain Transfer Agent	Polishing Agent	Polishing Agent	Deodorant	
Curing Agent	Curing Agent	Stabiliser	Stabiliser	Detergent	Detergent		
Colorant	Colorant	Lubricant	Lubricant				
	Stearic Acid	Release Wax	Release Wax				
	Wax	Anti-oxidant	Anti-oxidant				
Anti-tack Agent				Nitric Acid			
Talc				Sulphuric Acid			
Detergent				Phosphoric Acid			
				Brightener			Number of Materials at N-3 = >50
				Buffer			
				Sealant			

Figure 19.10   Example of a typical supply chain for an MDI valve.

The supplier and pharmaceutical relationship is inherently complicated. There is a need to balance the proprietary nature of the supplier's formulation against the pharmaceutical company's desire for further information in order to enhance its product and process understanding.

During OINDP product development, supplier development activities should include working closely with material and component suppliers, ensuring that quality is designed into the container closure system material and respective manufacturing processes as early as possible. This partnership needs to be based on open communication and sharing of information. Communication should be initiated prior to selecting the material for use in the product and should continue throughout the product development life cycle. Communication should include peers from the supplier and the pharmaceutical company working together, that is, engineers/staff scientists working with their counterparts and regulatory/quality assurance (QA) working with their counterparts. There are mutual benefits when the supplier and the pharmaceutical companies work together. Efficient exchange of information regarding expectations and capabilities for manufacturing and control leads to higher quality and more appropriate controls established early in the development process (Fig. 19.11).[11]

For materials and components subjected to routine extraction controls, discussions with the suppliers should focus on the technical aspects of the drug product that the container closure system materials are used in and the planned extractables controls. A goal of these meetings is to increase the sharing of formulation details suppliers are willing to disclose. Additionally, the importance of formulation control should be underscored, including management of supply at upstream suppliers, in addition to supplier-based formulation controls. This approach is consistent with QbD and risk management.[12]

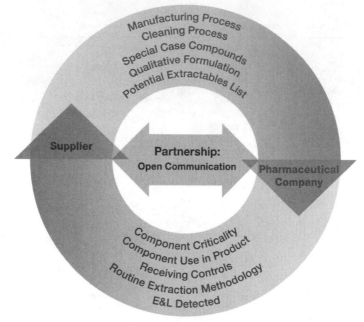

Figure 19.11   Information exchange between pharmaceutical manufacturer and supplier.

There are several ways that the parties may protect the information that they will share with each other. Quality agreements may be used to describe the requirements and expectations relating to quality and regulatory systems. Quality agreements are discussed in a later section in this chapter. Nondisclosure agreements are legally binding confidentiality agreements between the companies. These nondisclosure agreements can also be established between multiple parties, that is, three- or four-way nondisclosure agreements. Drug master files (DMFs) are confidential submissions made by the supplier to the FDA. The pharmaceutical company references the DMF in its regulatory submissions. DMFs are discussed in a later section in this chapter.

As there are many ways that extractables can make their way into the product, the processes at the supplier should be evaluated for sources of potential extractables. The recommended information to be shared with the pharmaceutical company by the supplier includes the following:

Shared information	Rationale for sharing
Manufacturing process	How ingredients are measured and added to the batch may impact extractables levels, and the type of manufacturing process used can provide information on potential presence of special case compounds

Shared information	Rationale for sharing
Cleaning process and whether the equipment is dedicated to the process	How the equipment is cleaned between batches may have an impact on extractables
Qualitative formulation, at a minimum. In some cases, the supplier may share the quantitative formulation, which provides additional understanding. This includes: ◦ base polymer, including any known additives ◦ additives (antioxidants, plasticizers, stabilizers) ◦ pigment packages	The components of the formulation are important to understand because they are useful in developing the potential extractables list
Special case compounds, specifically, nitrosamines, polynuclear aromatic hydrocarbons, or 2-mercaptobenzothiazole	The potential presence of these compounds require specialized testing to verify presence and levels, if seen
Potential extractables list—based on previous testing of the component/formulation, or theoretically based on: ◦ formulation ◦ component manufacturing, including any washing/rinsing processes ◦ equipment cleaning processes ◦ storage conditions for component after manufacture prior to assembly	The potential extractables list is a tool that can direct the experiments for controlled extractables and developing extractables methodology

The recommended information to be shared with the supplier by the pharmaceutical company includes the following:

Shared information	Rationale for sharing
L&E detected, including: ◦ results from controlled extraction and leachables experiments ◦ experimental conditions used for controlled extractables and leachables experiments	The supplier may be able to identify the source of the L&E
Routine extractables methodology, including description of routine extraction method details, if method is already developed, and description of planned routine extraction methodology, if method is in development	Provides greater understanding of the testing methodology at the pharmaceutical company
Receiving controls—description of tests that the pharmaceutical company will employ in order to accept the material/component as well as limits, if established	Provides greater understanding of the testing and controls at the pharmaceutical company, as well as the specification that the material/component will have to comply with

Shared information	Rationale for sharing
Component use in the product, including: 　○ the role of the component in the product 　○ design/functional requirements for the component 　○ expectations of component performance and use life 　○ contact with formulation or patient	Knowledge of the end use of the material/ component by the supplier provides the supplier the background to combine with the knowledge of the material/component, and the supplier may share suggestions and insights that provide for a resulting product that meets these requirements
Component criticality	Level of criticality drives testing, including whether extractables are required
Details of the delivery system, including expectations for the assembly environment, duration, and type of patient contact, for example, limited duration/surface and mucosal contact and anticipated product life cycle	Knowledge of the details of the delivery system by the supplier provides the supplier the background to combine with the knowledge of the material/ component, and the supplier may share suggestions and insights that provide for a resulting product that meets these requirements

This relationship between the material/component supplier and the pharmaceutical company can facilitate the teams' ability to evaluate product L&E. In best-case scenarios, the following are true. For leachables evaluations,

- source for each leachable detected in drug product is known;
- each leachable above threshold is correlated (qualitatively and quantitatively) with an extractable (or extractables);
- identified leachables at or above the safety concern threshold (SCT) are reviewed by pharmaceutical development toxicologists to identify any potential safety concerns; and
- control of the source of each correlated extractable is within the control of the supplier (directly or through controls with n-2 suppliers).

For establishing control of extractables,

- known source for each extractable (quantitative and qualitative) is specified;
- control of the source of each extractable is within the control of the supplier (directly or through controls with n-2 suppliers);
- where appropriate, robust extractables methods are established for routine testing;
- limits are established based on historical data, safety (for correlated leachables), and DOEs (enhanced product knowledge); and
- assurance for supplier and pharmaceutical company that the component will meet these limits is increased, as the extractables are understood and controlled.

19.4.4   Component and Procurement

The approach to procurement should be consistent with the FDA cGMP and quality system regulation requirements and with the principles of the ISO 9000 Quality System Standard, EN 46001, and ISO 13485. The procurement system should function within the framework of the site quality system and in concert with the other elements of assurance and control (design control, material qualification, and routine control), to ensure a consistent supply of goods that meet specification and are fit for their intended use. All raw materials, components, manufacturing materials, and processes used in the OINDP should be subject to the applicable elements of this procurement approach.

The key elements addressed by the procurement system are supplier selection, assessment, performance, and agreements. A brief summary of each follows:

Supplier selection. Suppliers are initially assessed to determine capabilities, capacities, availability of equipment and processes, resident skills and expertise, experience with medical and pharmaceutical requirements, suitability of the physical plant, and other factors. These initial evaluations also serve to establish the foundation of the intercompany relationship that will help facilitate communication and problem resolution, should one arise.

Supplier assessment. Suppliers are subject to a formal initial audit and subsequent periodic follow-up audits in accordance with appropriate due diligence. This audit usually includes a review of the organization and quality and manufacturing procedures; a tour of the manufacturing and/or testing facilities; and review of key studies and documents. A formal report of the audit is written with identification of the findings to which the supplier needs to respond. If the findings are significant, the company may re-audit the remediation. Otherwise, the resolution of the findings is verified at the next follow-up audit.

Supplier performance. The results of incoming inspection tests and analyses, as well as other sources of information on a supplier's ability to consistently meet specification, are tracked. Periodic feedback is provided to suppliers.

Supplier agreements. Suppliers are made aware of requirements and specifications in a formal, documented manner. Written purchase orders, quality agreements, detailed blueprints and drawings, and other documents formally communicate information to suppliers. All suppliers are formally alerted that changes to materials and components are not permitted without prior notification, review, and, if permissible, formal agreement.

19.4.5   Material Selection

The selection of appropriate materials is predicated primarily on the attributes of the drug product and its intended use and administration as defined in the drug product design input requirements.[13] During the process of selecting the material for a component, it is important to gather and analyze critical information about the component manufacturing process, supplier, and materials of construction. The goal is to ensure that there are no significant issues with compatibility or safety. When multiple

CCS or device materials options are available, use these data to compare options and to select the most appropriate CCS or device material.

Consider the following information:

- source of material and material manufacturing process;
- materials of construction including coatings, processing aids, and fillers;
- potential extractables lists (from supplier);
- in-house extractables and leachables data;
- all available biocompatibility data (from supplier);
- confirm bovine spongiform encephalopathy/transmissible spongiform encephalopathy (BSE/TSE) statements are available;
- other literature/data on the CCS system or device material: the supplier, the raw materials, or the suppliers of the raw materials (e.g., USP/EP/JP status and material grade); and
- for a rapid, qualitative assessment of volatile and semivolatile extractables, analysis by thermal desorption gas chromatography mass spectrometry (TD/GC/MS) is recommended.

Note: In many cases, a nondisclosure agreement will be needed before the supplier will share this information.

Perform a preliminary safety assessment of the identified extractables and leachables to identify those extractables that could be of high risk if detected as leachables in the final drug product. This should be performed by a toxicologist.

Perform *in vitro* cytotoxicity testing in conformance with ISO 10993-5/USP <87> on critical device components or raw material (if not available from supplier). Additional *in vivo* biocompatibility testing (i.e., sensitization, irritation) should be considered on a case-by-case basis, in consultation with a toxicologist. (Note: This testing is required for devices used in U.K. clinical studies—MHRA—Guidance on the Biological Safety Assessment—January 2006.)

Select the final material based on a full analysis and understanding of the above information. Include any potential risks associated with the final material in the overall project risk assessment. The rationale for the choice should be documented.

19.4.6   Component Qualification and Control

The OINDP component supplier and the customer should establish a formal agreement that describes product specifications, testing requirements, and acceptance criteria. These may be documented in the quality agreement. The supplier of the component is the expert on the manufacture of the component, and the customer is the expert on the manufacture of the final product; therefore, it is critical that they work together to develop the specification and testing regime needed to ensure that the critical attributes of the component are understood.

One important step in defining these criteria is categorizing the components. Categorization is performed by separating out the components based on whether they contact the drug or mucosal membranes of the patient. These would be

TABLE 19.2. Component Categorization

Criticality	Category	Category definition
Critical	Mucosal membrane and drug contacting components	All components that contact the patient's mouth and/or the drug
Critical	Mechanical/functional components	All components that are not included in the above category and affect the mechanics of the overall performance of the product
Noncritical	Noncritical components	All remaining components that are not considered to be critical components

considered the critical components. Increased control of this category of critical components is warranted due to the potential patient impact. Of the remaining nondrug or mucosal membrane contacting components, these should be further evaluated to determine which components affect the mechanics of the overall performance of the device and which components do not affect the mechanics of the overall performance of the device. An example grid of component categorization is provided in Table 19.2.

The determination of which components are mechanical/functional is specific to the device technology being used for the product. The draft guidance published by the FDA defined critical components of a dry powder inhaler (DPI) as "those that contact either the patient (i.e., the mouthpiece) or the formulation, components that affect the mechanics of the overall performance of the device, or any necessary protective packaging."[14] The mechanical/functional components that are not in the drug path fall under the classification of "those that affect the mechanics of the overall performance of the device."[15]

19.4.7 Qualification, Characterization, and Verification

19.4.7.1 Qualification Qualification of a component involves demonstrating that the component and the material used to fabricate the component are safe and appropriate for its intended use. This process involves a combination of obtaining information from the suppliers and a testing regime designed based on the product being evaluated.

Qualification and routine control requirements for the materials and components may be implemented using a tool such as a material qualification and control plan (MQCP). An MQCP is a version controlled document. The purpose of an MQCP is to define the requirements for material/component suitability for use in the product and ensure the continuing quality and integrity of the materials and components in the OINDP.

Components may be qualified for use in the product based on their assigned category. Typically, there are two aspects to qualification—functional/mechanical and safety. The functional/mechanical aspects are generally addressed by formal studies performed under a system of design controls. The safety aspects may be addressed by a combination of certifications to compendial requirements and performance of compendial tests. This qualification activity should consider the following tests, which may be relevant for mucosal membrane and/or drug contacting components:

- *Biocompatibility.* These tests may be performed by taking into consideration the recommendations of ISO 10993,[16] FDA Blue Book Memorandum G95-1,[17] and USP Monographs <87>,[18] <88>,[19] and <1031>.[20]

- *Physicochemical characterization.* As there is no physicochemical monograph test available for plastic and elastomeric components for pulmonary inhalers, the USP <381>[21] and USP <661>[22] physicochemical tests designed for containers may be applied as surrogate tests. These tests are intended to evaluate containers that will have long-term contact with the drug product. In contrast, the components that comprise the drug path of OINDP may have only transient contact with the drug product, so the acceptance criteria for these tests should be carefully considered.

19.4.7.2 Characterization The components of the product may be further characterized through evaluation of a number of characterization requirements. The purpose of these evaluations is to gain further knowledge regarding the components and their properties. Additionally, the information can help to provide additional assurance that the components are acceptable for their intended use. In some instances, the knowledge may be utilized to develop analytical methods and acceptance criteria for routine testing. As part of the characterization program, the components in the mucosal membrane/drug contacting category, the following tests may be relevant:

- *Controlled extraction studies.* The purpose of the controlled extraction studies is to qualitatively identify the extractables to facilitate leachables and routine extractables method development.

- *Leachables testing.* This test is performed to ensure that the components are constructed of materials that do not leach harmful of undesirable amounts of leachables into the drug product. Leachables are the compounds that migrate from the delivery systems of the drug product under normal conditions of use.

- *Infrared (IR) spectra interpretation.* The purpose is to confirm that IR spectra that are generated from the material are in accordance with its known composition.

19.4.7.3 Verification The remaining component testing comprises verification testing. The studies may be performed as part of other programs, such as tooling qualification and design verification testing. Successful completion of these activities further verifies that the components of the product are qualified for their intended use.

A summary of this example testing program is presented in Table 19.3. In a program such as this, one ensures that the materials are appropriate for their intended use and provide additional knowledge on the components and their properties.

TABLE 19.3.   Summary of Example Qualification, Characterization, and Verification Tests Based on the Component Categorization

Qualification, characterization, and verification tests		Category		
		Critical		Noncritical
		Mucosal membrane/drug contacting	Functional/ mechanical	
Qualification	Biocompatibility testing	X	–	–
	Physicochemical testing	X	X[a]	–
Characterization	Controlled extraction study	X	X[a]	–
	Leachables evaluation	X	–	–
	Physical characterization: FTIR spectra interpretation[b]	X	X	X
Verification	Dimensional measurement evaluation	X	X	X
	Performance/functional test	X	X	X
	Dimensional measurement evaluation	X	X	X

[a] Physicochemical testing and controlled extraction may be considered for functional/mechanical components based on risk assessments.

[b] Metal components are exempted from FTIR spectra interpretation.

Any future replacement or second source materials should consider following the same qualification, characterization, and verification as the materials that they are replacing, taking into consideration the compositional differences between the original material and the replacement material.

19.4.8   Control Strategy

The importance of having appropriate controls should be underscored, including management of supply at upstream suppliers, in addition to supplier-based formulation controls. Quality is built into the product from the start of development. OINDP quality starts with quality of the materials/components. The foundation for ensuring appropriate quality is dependent on constructive communication between suppliers and customers early in the design process in conjunction with robust quality systems at suppliers and manufacturers for quality throughout the life cycle. Upstream controls should be implemented to ensure consistency of the materials and processes used to manufacture these components, which may be linked to product performance. This approach is consistent with QbD and risk management.[12] It is also consistent with the recommendations of the Inhalation Technology Focus Group (ITFG)/IPAC-RS collaboration,[23] wherein the testing is proposed for the earliest point possible.

The material and component control approach for the manufacture of delivery systems components should incorporate the concept that the performance of a component is a function of both the material selected for use and its designed geometry. The control of the material quality, design of the component, and the control of the manufacturing processes are the foundations for controlling the quality of the final product.

A successful component design is one that will perform satisfactorily and is achieved with a balance of material properties and robust geometry. Typically, changes in the geometry, for example, length and/or diameter, have a squared or cubic effect on component performance, whereas changes in raw material (resin) properties; that is, strength, toughness, and hardness, have a linear effect. Therefore, the components used in the delivery system need to be designed to account for the typical specified ranges of physical and chemical properties associated with resins supplied by the plastics industries. The vendors' material (resin) specifications are typically used in the component design process to guide the selection of component geometry.

Individual lots of materials and components that are received or molded for ongoing production activities should be routinely tested to verify their identity and/or conformance to qualified characteristics. Where possible, testing should be performed upstream in the supply chain. The testing matrix for the materials and components may include some tests that are performed by the supplier with the results verified upon receipt.

Upon receipt, defined tests and inspections should be carried out. These may include the following:

- review and verification of certificates and other paperwork accompanying the shipment, which may include ISO or American Society of Testing and Materials (ASTM) Physical/Mechanical Testing, such as Melt Flow Rate, Izod Notched Impact, Flexural Strength, and Routine Extractables;
- visual inspection;
- identification by technologies such as Fourier transform infrared (FTIR) spectroscopy and/or differential scanning calorimetry (DSC); and
- dimensional inspections.

The controls, assigned to each material and component, should be based on the category assigned. As a result, the actual test regimen is specific to the material or component being evaluated; when a material is used in multiple components, the controls for the material should be based on the most stringent component category.

19.5 ESTABLISHING CONTROLS

The quality of OINDP products is dependent on the quality of the incoming components. Approximately 60% of the final product tests on an OINDP are directly related to the components. The regulatory and compliance obligation is that these specifications are formally established, justified, approved, and controlled according to the applicable cGMP requirements for all countries where the pharmaceutical product is marketed.

The routine control program should be developed using guidance provided in the Draft FDA Guidance for MDI/DPI Drug Products[14] in concert with the requirements of 21CFR §§ 210, 211, and 820. Routine controls for extractables should be but one part of the overall routine control program for the materials and components of OINDPs. The routine control program for the materials and components of these products should build on and complement qualification, characterization, and verification. The approach to routine control should encompass controls at the material and component suppliers, controls at contract manufacturers, controls at the pharmaceutical manufacturer, and final product testing. In addition to process controls, some tests may be performed prior to shipment, whereas the manufacturers may perform the other tests during incoming inspection. The combination of these controls assures a consistent quality product.

The manufacturer should consider the following factors when developing routine controls: function of material/component in the product; composition of the material/component; supplier quality systems; and results of risk analyses.[10] Identification of attributes to control for a component can be accomplished through a review of its design intent, and risk analysis of component failure through the use of the design failure modes and effects analysis (FMEA). Functional/mechanical controls on a resin material in addition to those of the supplier may be required and should be established by considering the end use of the component. As a result, the functional/mechanical controls will ensure that the material will perform in the component as intended.

The category of the component is an important consideration when developing controls. The MDI/DPI draft guidance recommends routine extractables controls on drug path components, mucosal contact components, and those that affect the mechanics of the device. Per this draft guidance, the impact of a component being defined as critical is that routine extractables testing is expected. However, extractables is not always the most appropriate test to be performed. The purpose of extractables testing is to provide a fingerprint identity of the composition of the material and quantitative control-specific extractables. Frequently, alternate controls are more appropriate for the components. The control strategy for the components should assure component functionality by establishing comprehensive controls over the entire process of manufacturing the resin through assembling the finished component into the delivery system.

A risk-based approach can be employed to develop the routine control strategy. Extractables testing may be applied to critical components, by routinely testing the materials used in the components of the drug path, as well as the components in other categories based on the outputs from the risk assessments, such as FMEA. FMEA is a systematic approach to identify and evaluate the relative risk of each potential failure mode, followed by improvements to control areas with the greatest risk. "To reduce risk, it is essential to identify hazards, evaluate the associated potential consequences and their likelihood, and then estimate the risk."[24] The FMEA technique is one of the analysis techniques used to identify risk and is described in the "FMEA" section.

The controls, assigned to each material and component, should be based on the category assigned to the component per the MQCP. As a result, the actual test regimen is specific to the material or component being evaluated. When a material is used in multiple components, the controls for the material should be based on the most stringent category assigned. Individual lots of materials and components received or molded for ongoing production activities should be routinely tested to verify their identity and/or conformance to the defined quality standard for the material/component.

The specification for the materials and components of the product should integrate the attributes to be controlled along with the purposes of test techniques selected. Routine extractables tests are designed for "indirect control of composition,"[25] "to ensure that the extractables profiles of the components used for the commercial drug product manufacture remain consistent with the profiles of the components evaluated as part of the development controlled extraction study,"[23] and is the process by which OINDP container closure system critical components are qualitatively and quantitatively profiled for extractables, either for purposes of establishing extractables acceptance criteria or for release according to already established acceptance criteria.[3] These purposes are ultimately focused on control of potential leachables in the drug product.

To achieve these purposes, routine extractables testing is typically applied to the materials used in the components of the drug path. The drug path includes all of those components in the OINDP that contact the drug formulation from the point of actuation in the delivery system to the contact with the patient. The drug path components should be subjected to both controlled extraction characterization studies and routine extractables control as these materials and components have the potential to affect the leachables profile. The leachables are determined by testing the drug product that has been in contact with the delivery system under normal use conditions. Routine evaluation of the extractables from the materials used in components of the drug path or the components themselves ensures that the profiles of the materials/components remain consistent and provides indirect control of composition.

But extractables may only be one of several attributes that require routine control. Once all appropriate attributes to control are identified and established, the risk of material inconsistency can be assessed in an FMEA specific to the material in question—a material FMEA or mFMEA. This risk is measured by determining the potential effect(s) of material inconsistency, the severity of that effect, the likelihood of that effect resulting from the inconsistency, and what controls are in place to detect that inconsistency and/or end effect. If the risk is determined to be acceptable, no further controls would be implemented. If the risk is determined to be too high, then additional controls should be identified, which may include routine extractables or other appropriate controls.

Based on the nature of the failure mode identified, additional or alternate control measures may include functional/mechanical testing, additive (e.g., polytetrafluoroethylene [PTFE]) content, and routine extractables. The mFMEA should also be used as a tool to ensure that the risks are mitigated with the control measures that are put in place. Where the conclusions from the mFMEA are that the non-drug-path components are appropriately controlled and the risk of failure due to material inconsistency is low, routine extractables testing for compositional control of the non-drug-path components would not be appropriate.

The testing matrix for the materials and components may include some tests that are performed by the supplier and verified by the manufacturer upon receipt and at periodic intervals. Where possible, testing is performed upstream in the supply chain. This approach is consistent with the recommendations of the ITFG/IPAC-RS collaboration and other recent IPAC-RS-related publications,[3,23,26] wherein the testing is proposed for the earliest point possible.

The following information relating to product quality should be considered for inclusion in the component specification:

Product name	Extractables testing (when applicable)
Product reference code	Sampling plans
Product dimensions and tolerances	Testing methods
Product performance (e.g., functionality), for example, leakage and spray pattern	Acceptance criteria (acceptable quality levels/ limits [AQLs])
Raw material (when applicable):	Product certification
Chemical requirements	Packaging details
Mechanical properties	Label details
Biological properties	Pallet details
Storage	Storage conditions and shelf-life setting
Grade	Product use by date/requalification date
Supplier	Limitations on use of processing aids
Product cleanliness standard: foreign particulates and microbial levels	Postprocessing treatments (e.g., passivation)
Visual standards	In-process checks

Once a testing regime is established and agreed upon, incoming materials should be subject to full testing. The testing ideally will be conducted by the customer upon receipt, or alternatively may be conducted by a qualified contract laboratory. The latter requires formal transfer of testing methods to the contract facility.

Once a suitable number of lots have undergone full analysis, a reduced testing scheme can be considered subject to regulatory approval. A risk analysis should be performed to determine how many lots would undergo full analysis prior to moving to a skip lot-testing plan. The risk analysis should also consider which, if any, tests can be accepted on supplier certification. In the event of a component lot failure, a clear procedure should be in place to revert to full analysis as well as a path to bring the supplier back to reduced testing.

19.5.1   Use of Alternate Controls[10]

The primary purpose of extractables for routine control is indirect control of material/ component composition and testing to ensure that the extractables profiles of components used commercially remain consistent with those used during development.

Extractables testing should be applied in those cases where these purposes are achievable.

In contrast, there are instances where the purposes of extractables testing may not be achievable and other approaches may be more appropriate. For metal drug path components, extractables results only provide information regarding the exterior of the component. An alternate control is to obtain a certification of chemical composition of the metal. A few case studies of the use of alternate controls follow in Table 19.4.

19.5.2 Routine Verification of Results Provided on Certificates of Analyses

The supplier of the material or component may perform the functional/mechanical testing, with results reported on the certificate of analysis provided to the manufacturers. Validation of the results needs to be performed in order to comply with the requirements of 21CRF 211.84.[27] A risk-based approach may be used, for example, by conducting validation for purchased items that are, or are used to make, critical components. Independent validation would need to be performed for attributes tested and reported by the supplier on the certificate of analysis. An independent laboratory periodically performs the validation. The manufacturers may also audit the performance of the test method in lieu of independent analysis. Using the risk-based approach, materials and components in the noncritical category may not be subjected to this verification.

Some tests are performed on-line during the manufacture of the component or material, and the analysis cannot be easily validated by an independent laboratory. In lieu of validation, the supplier of the component or material can be periodically audited. Additionally, there are methods that are developed and validated under the guidance of the manufacturer and controlled by the manufacturer, and the manufacturer is directly involved in the testing. In these cases, validation of the results on the certificate of analysis may not need to be performed, provided the manufacturer performs periodic audits of the testing facilities.

19.5.3 Quality Agreement

It has become current best practice to establish a quality agreement between the supplier and the customer. These documents are often requested by and reviewed during regulatory inspections. Quality agreements are contracts between a supplier and customer describing each party's expectations for how to manage the quality of the supplier's products. A detailed quality agreement, which outlines the responsibilities of both the supplier and the customer, forms a strong foundation on which a business relationship can be built. This strong foundation will inspire confidence in each party to the agreement as well as the regulators. Through a quality agreement, a pharmaceutical company and its suppliers can identify the types of information that could be shared. Sensitive information would be included in confidentiality agreements. Quality agreements should be established at all links in

TABLE 19.4. Examples of Alternate Component Controls

Example	FMEA	Regulatory guidance[a]	Function in product	Composition	Supplier quality systems	Approach	Final controls
Ink used for indicia on drug contact component	Low risk rating	Extractables expected because of drug contacting	Provides visual information for user	Ink	Successfully passed audits	Extractables would not detect changes in the ink due to low levels of ink present on the component in relation to the rest of the product	Certification of formulation composition, FTIR on incoming ink, visual inspection
Functional nondrug contact rubber component	High risk rating (reduced by adding extractables)	Extractables expected because of critical component	Holds pressure during inhaler use	Rubber	Successfully passed audits	Extractables added to monitor formulation composition	Extractables, FTIR, visual, dimensional, cure level (R-value)
Functional non-drug-path polymer component	High risk rating (reduced by adding PTFE content control)	Extractables expected because of critical component	Mechanical link between assemblies during inhaler use	Liquid crystal polymer with PTFE	Successfully passed audits	Control of PTFE content at compounding added to ensure composition	Material: PTFE content, FTIR, melt viscosity, visual component: visual, dimensional

[a] Draft guidance for industry: Metered dose inhaler (MDI) and dry powder inhaler (DPI) drug products. Department of Health and Human Services, Food and Drug Administration Center for Drug Evaluation and Research (CDER), 1998.

the supply chain (e.g., between n and n-1; between n-1 and n-2; and between n-2 and n-3).[7,8]

Quality agreements are used to outline and document the terms relating to key quality and regulatory systems. The specification referenced in the preceding section would typically be a controlled appendix of the quality agreement.

The following items should be considered for inclusion in quality agreements, where applicable:

Batch records	Process validation
Change control and notification	Qualification and/or validation of equipment
Cleanliness and hygiene	Recalls
Complaints and impact on commercial supply	Reference to applicable guidelines or guidance documents
Component testing	Regulatory compliance
Customer audits	Regulatory contacts and audits
Definitions	Requirements for raw materials and subcomponents
Document retention	Resolution of quality issues
Lot approval and product release	Retained samples
Manufacturing environment	Rework and reprocess
Material suppliers	Subcontractor management
OINDP manufacturer and component supplier responsibility matrix	Supply agreements

Quality agreements have the potential to increase communication and strengthen the partnership between the pharmaceutical company and supplier, thereby decreasing the amount and types of strictly confidential information that would go into the DMF, which would not be shared with the customers. Specific items that could be shared through business and quality agreements are presented in Figure 19.12.

The IPAC-RS Supplier Quality Working Group has developed a quality agreement template,[28] which can be used as a starting point to develop a product-specific quality agreement. The template provides guidance on elements of a quality agreement between OINDP manufacturers and their device/component/packaging suppliers (n-1 suppliers), and n-1 suppliers and their suppliers (n-2), and so on. The template links specific elements of the quality agreement with GMP recommendations made in the Working Group's GMP Guideline for Suppliers of Components for OINDP.

19.5.4 Change Control[29]

It is important that quality agreements address expectations regarding change control. The quality of a pharmaceutical product can be affected by what a supplier may consider an insignificant change. Changes can happen at any time during a product's life cycle and may be initiated by the OINDP manufacturer. The impact of a planned change needs to be balanced against the cost of making the change (safety, time,

Increased Communication and Sharing of Component Information

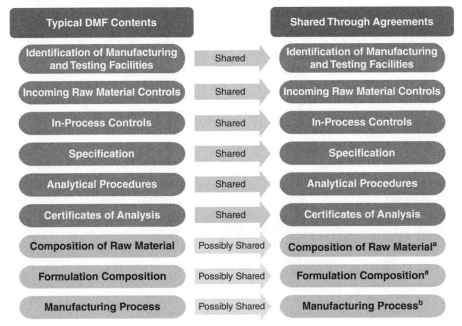

^aDepending on the formulation and business agreements, this information may be shared qualitatively and/or quantitatively.
^bSpecific portions of the Manufacturing Process may be trade secrets and therefore kept in DMF, but the rest of the process should be shared.

Figure 19.12   Communication and sharing of component information through various means.

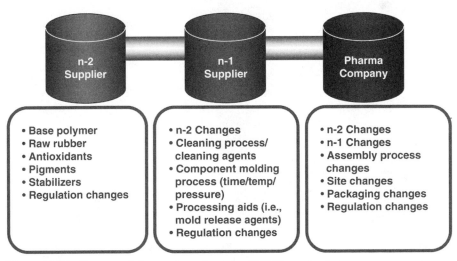

Figure 19.13   Sources of change throughout the OINDP supply chain.

money, and potential amendments to registered details). Change control is of critical importance to OINDPs due to the pulmonary route of administration, which requires controls as stringent as the intravenous route. An added complexity is that OINDP devices are comprised of multiple components typically fabricated from different materials and possibly sourced from different suppliers (Fig. 19.14).

When an unapproved or an unanticipated change occurs, the following are the best- and worst-case consequences:

- *Best case.* Unapproved change is detected upon receipt by the customer. This results in supplier investigation costing time and resources, material scrap, and customer production delays.

- *Worst case.* Unapproved changes to the material/component causes final product failure in the field. In addition to the investigations required, the OINDP manufacturer may need to recall the product from the market, resulting in significant cost, impacts to reputation and loss of trust from patients, and ultimately may compromise patient well-being.

Change control is defined as a process that ensures that changes to materials, methods, equipment, and software are properly documented, validated, approved, and traceable.[30] The IPAC-RS guideline[8] adds the additional requirement that the

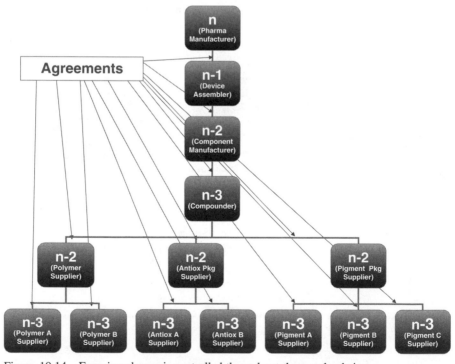

Figure 19.14   Ensuring change is controlled throughout the supply chain.

change control process includes evaluation to determine whether validation is required and the level of validation required. OINDP suppliers should establish written procedures for the identification, documentation, appropriate review, and approval of changes affecting the quality of products and/or processes, equipment, systems, and methods. Procedures should ensure changes will be implemented in a controlled manner. An independent group (i.e., quality unit) should have the responsibility and authority for the management/approval of changes.

Anticipated changes should be evaluated to determine the impact on component quality and validation status. This review is recommended to be performed as part of a documented risk assessment and should include an evaluation of whether there is an impact on the materials, an impact on the manufacturing process, or an impact on the finished product performance or design. *It is important to note that meeting product specification does not mean that the product has not changed and has not been impacted.*

For each product, the supplier and pharmaceutical company need to decide on what changes need prior approval. Agreement should be reached regarding what constitutes a change, which types of changes require notification/approval, and how changes will be implemented. It may be useful to include examples of changes that normally would not require prior approval by the customer. If the supplier is in doubt, they should always consult with their customer.

Additionally, the pharmaceutical company and suppliers should ensure that their suppliers have adequate change control programs in place, through audits of the change control programs at the suppliers and formalized in the agreements between the parties.

A typical change control process encompasses identification of the change, qualification of the change, and implementation. When a proposed change is identified, a change proposal may be used to notify the customer and document a multidisciplinary endorsement of that proposed change. The proposal should describe the change(s) in detail. A technical review and documented risk assessment need to be performed to assess the potential impact of the proposed change on the quality, safety, and efficacy of the resulting product. A regulatory review and documented risk assessment should be performed to assess the regulatory impact of the proposed change based on the affected registration(s) and regulatory guidance. Following approval of the change proposal, studies needed to support the change are performed. It is a good practice for the approved change proposal to define the documentation required to support the regulatory filing of the change.

Following completion of the supportive studies (qualification of the change), the results from these studies must be reviewed and risk assessments must be performed to determine whether they justify the change. A change request may be used to document the multidisciplinary endorsement of the request to implement the proposed change. Approval of the change request does not necessarily allow implementation of the change. When applicable, the supporting document in the approved change request is used to prepare the regulatory filing of the change. The change is implemented only after the approval of the regulatory filing of the change by the applicable health authority.

19.5.5   Material Retest Periods

To ensure that materials or components do not change during storage, appropriate expiry or reevaluation dates are assigned. Expiry dates may be provided by the supplier or determined by the manufacturer after reevaluation. The process for extending the review period for materials and components includes sampling and testing material from inventory against the reevaluation criteria. These reevaluation criteria are selected to demonstrate the suitability of the material for its application and typically consist of evaluating the material/component against the release tests. When a material/component is retested and passes the acceptance criteria, the material is deemed acceptable for use until the next retest date (calculated as test date plus review period). The material/component may be retested multiple times, but may not be used past the designated expiry date.

19.5.6   DMFs[7]

DMFs are a mechanism that material or component suppliers can use to share confidential information with health authorities without disclosing that information to the pharmaceutical company. In the United States, a "Drug Master File (DMF) is a submission to the Food and Drug Administration (FDA) that may be used to provide confidential detailed information about facilities, processes, or articles used in the manufacturing, processing, packaging, and storing of one or more human drugs."[31] While a DMF permits the sponsor to incorporate the information by reference when submitting an application or an amendment or supplement to the application, or to authorize other individuals to rely on the information to support a submission to the FDA without having to disclose the information, in the case of OINDPs, it is recommended to use DMFs for the actual trade secret information, and sharing the remainder of the non-trade-secret processing and control information with the pharmaceutical manufacturer.

19.5.7   FMEA

An mFMEA can be utilized to identify potential failure modes in the plastic and elastomeric components of the product caused by material inconsistency. A cross-functional team should perform this evaluation. The output of the mFMEA is the assignment of a rating to the component. The FMEA process helps to evaluate the relative risk of each potential failure mode, which can then be used as a tool to improve and control areas with the greatest risk. The relative risk contribution of each potential failure mode is characterized in terms of a rating called the risk priority number (RPN). The RPN is calculated by multiplying together the rank (1–10 scale) assigned to each of the three indexes for each failure mode. The three indexes, severity, occurrence, and detection, are defined in Table 19.5. Example scales for the three FMEA indexes are presented in Tables 19.6–19.8.

The mFMEA is a living document[32] and is updated periodically throughout the lifespan of the product. An overall "action required" limit should be defined by the manufacturer. When the RPN can be reduced through the addition of a control measure, the RPN should be reassessed based on the updated control measures.

TABLE 19.5. Material FMEA Indexes and Definitions

Index	Definition
Severity (S)	An assessment of the seriousness of the effect of the potential failure mode
Occurrence (O)	The probability that a specific failure mode will occur
Detection (D)	An assessment of the design controls to detect cause/mechanism of the potential failure mode or the failure mode itself, prior to release for production

TABLE 19.6. Severity (S) Ranking and Criteria

Rank	Severity criteria
1	No effect on user or patient.
2	Dissatisfaction experienced by patient or user.
3	Defect is noticeable. Device can be used but requires additional operations.
4	Device not operable; patient cannot receive treatment.
5	Device operable at reduced level of performance. Patient experiences some discomfort.
6	Device operable but patient experiences discomfort.
7	Device operable but not as effective, reduced level of performance.
8	Device causes injury or harm not of a permanent nature.
9	Very high severity ranking when a potential failure mode affects safe device operation and/or involves a permanent injury to patient or user.
10	Very high severity ranking when a potential failure mode affects safe device operation and/or involves death to patient or user without warning.

TABLE 19.7. Occurrence (O) Ranking and Criteria

Probability of failure	Possible failure rate	Rank
Very high: Failure is almost inevitable	1 in 2	10
	1 in 3	9
High: Repeated failures	1 in 8	8
	1 in 20	7
Moderate: Occasional failures	1 in 80	6
	1 in 400	5
Low: Relatively few failures	1 in 2000	4
	1 in 15,000	3
Remote: Failure is unlikely	1 in 150,000	2
	1 in 1,500,000	1

TABLE 19.8. Detection (D) Ranking and Criteria

Detection	Criteria: likelihood of detection by quality control	Ranking
Not possible	Quality control will not and/or cannot detect a potential cause/failure mechanism and subsequent failure mode, or there is no quality control.	10
Very remote	Very remote chance quality control will detect a potential cause/failure mechanism and subsequent failure mode.	9
Remote	Remote chance quality control will detect a potential cause/failure mechanism and subsequent failure mode.	8
Very low	Very low chance quality control will detect a potential cause/failure mechanism and subsequent failure mode.	7
Low	Low chance quality control will detect a potential cause/failure mechanism and subsequent failure mode.	6
Moderate	Moderate chance quality control will detect a potential cause/failure mechanism and subsequent failure mode.	5
Moderately high	Moderately high chance quality control will detect a potential cause and subsequent failure mode.	4
High	High chance quality control will detect a potential cause and subsequent failure mode.	3
Very high	Very high chance quality control will detect a potential cause and subsequent failure mode.	2
Almost certain	Quality control will almost certainly detect a potential cause and subsequent failure mode.	1

19.6 CONCLUSION

The preceding sections provide insights regarding a QbD approach to managing quality with respect to leachables. The emphasis is on establishing a leachables quality standard based on safety, developing a leachables/extractables correlation, and focusing QC testing on the extractables profile through a strong understanding of the component manufacturing process. With this approach, a limit for a given extractable is proposed, based on safe exposure limits. Statistical tools (OCC) were described for translating a quality standard based on safety considerations into a testing scheme and related acceptance criteria that are the basis of a QC program for extractables.

The management of the extractables levels is fundamentally addressed through the ability of the manufacturing process to reliably produce components with extractables concentrations within the proposed limits. Extractables testing provides continuing assurance that the process is performing as expected. Statistical tools for evaluating the actual (control charts and process performance analysis) and potential (process capability analysis) performance of a manufacturing process were explored. The manufacturing process would need to be managed by the supplier of the component(s), if possible in collaboration with the OINDP manufacturer.

Indeed, important aspects of a QbD approach to extractables and leachables in OINDP lie in the management of extractables throughout the supply chain for OINDP container closure system/device components. Such management relies fundamentally on adherence to the concepts of current good manufacturing practices in the OINDP component supply chain. Since the OINDP supply chain has several layers and can include multiple components, it is key that the management of extractables begin with the selection and qualification of materials so that quality is built into the product. A risk-based approach can be taken to ensure that appropriate controls are put in place. Such controls may include process controls or testing to a quality standard. Finally, change control procedures and agreements with suppliers are critical to ensure the quality of the OINDP throughout the life cycle of the product.

REFERENCES

1 Juran, J.M. How to think about quality. In: Juran, J.M. and Godfrey, A.B., eds. *Juran's Quality Handbook*, 5th ed. McGraw-Hill, New York, 1998; 2.1–2.3.

2 Guidance for industry: Quality systems approach to pharmaceutical CGMP regulations. U.S. Food and Drug Administration, September 2006.

3 Norwood, D.L. and Ball, D. Product Quality Research Institute: Safety thresholds and best practices for extractables and leachables in orally inhaled and nasal drug products. Submitted to the PQRI Drug Product Technical Committee, PQRI Steering Committee, and U.S. Food and Drug Administration by the PQRI Leachables and Extractables Working Group, 2006.

4 Schilling, E.G. Acceptance testing. In: Juran, J.M. and Godfrey, A.B., eds. *Juran's Quality Handbook*, 5th ed. McGraw-Hill, New York, 1998; 46.1–46.87.

5 ASQC Chemical and Process Industries Division, Chemical Interest Committee. *Specifications for the Chemical and Process Industries: A Manual for Development and Use*. ASQC Quality Press, Milwaukee, WI, 1996; 70–74.

6 Wheeler, D.J. and Chambers, D.S. *Understanding Statistical Process Control*, 3rd ed. SPC Press, Knoxville, TN, 2010.

7 Roan, S.M. Use of type III drug master files in product registrations. *Regulatory Focus* 2009, *14*(12), p. 40.

8 IPAC-RS. Good manufacturing practices guideline for suppliers of components for orally inhaled and nasal drug products, 2006.

9 The Chartered Quality Institute, Pharmaceutical Quality Group and IPAC-RS. PS9000:2011, A Standard for Pharmaceutical Packaging Materials Incorporating Good Manufacturing Practice (GMP).

10 Roan, S.M., Stults, C., Walawalker, A., Wood, J., Lloyd, R., and Schumacher, J. A risk based approach to setting acceptance criteria for materials and components in a novel pulmonary inhaler. Presented at the *PDA Conference*, September 7, 2007.

11 Roan, S.M. Use of type III drug master files in product registrations. *Regulatory Focus* 2009, *14*(12),p. 41.

12 Lostritto, R.T. Quality by design, risk management, and new review paradigms for inhalation drug products. Presented at *RDD Europe*, 2005; pp. 105–114.

13 Guidance for industry: Design control guidance for medical device manufacturers. Center for Devices and Radiological Health, Food and Drug Administration, March 11, 1997.

14 Guidance for industry: Metered dose inhaler (MDI) and dry powder inhaler (DPI) drug products. Department of Health and Human Services, Food and Drug Administration Center for Drug Evaluation and Research (CDER), 1998.

15 Guidance for industry: Metered dose inhaler (MDI) and dry powder inhaler (DPI) drug products; chemistry, manufacturing, and controls documentation. U.S. Food and Drug Administration, *Fed Regist* 1998, 63(*223*), p. 35, lines 1160–1161.

16 International Organization for Standardization (ISO). ISO 10993-1:2009. Biological evaluation of medical devices. Part 1: Evaluation and testing within a risk management process. Published in 2009. Available at www.iso.org (accessed August 25, 2011).

17 FDA blue book memorandum G95-1, use of international standard ISO 10993, Biological Evaluation of Medical Inhalers, May 1995. Available at http://www.fda.gov/MedicalDevices/DeviceRegulation-andGuidance/GuidanceDocuments/ucm080735.htm (accessed August 24, 2011).

18 U.S. Pharmacopeia (USP). XXXIV. Chapter 87. Biological Reactivity Tests, In Vitro, 2011.

19 U.S. Pharmacopeia (USP). XXXIV. Chapter 88. Biological Reactivity Tests, In Vivo, 2011.

20 U.S. Pharmacopeia (USP). XXXIV. Chapter 1031. The Biocompatibility of Materials used in Drug Containers, Medical Inhalers, and Implants, 2011.

21 U.S. Pharmacopeia (USP). XXXIV. Chapter 381. Elastomeric Closures for Injections, 2011.

22 U.S. Pharmacopeia (USP). XXXIV. Chapter 661. Containers—Plastic, 2011.

23 ITFG/IPAC-RS Collaboration Leachables and Extractables Technical Team: Leachables and Extractables testing: Points to consider. March 27, 2001, p. 13.

24 Anonymous. *Introduction in: Guidelines for Failure Mode and Effects Analysis for Medical Devices.* Dyadem Press, Ontario, Canada, 2003.

25 Guidance for industry: Metered dose inhaler (MDI) and dry powder inhaler (DPI) drug products; chemistry, manufacturing, and controls documentation. U.S. Food and Drug Administration, *Fed Regist* 1998 *63*(223), p. 37, line 1202.

26 Dohmeier, D.M., Norwood, D.L., Reckzuegel, G., Stults, C.L.M., and Nagao, L.M. Use of polymeric materials in orally inhaled and nasal drug products. *Med Device Technol* 2009, *20*(2), pp. 32–38.

27 Code of Federal Regulations. Current good manufacturing practice for finished pharmaceuticals: Testing and approval or rejection of components, drug product containers, and closures. U.S. Food and Drug Administration, 21(4), April 1, 2002, pp. 123–124.

28 IPAC-RS. IPAC-RS quality agreement template, 2009.

29 Roan, S. and Riddell, J. Change control—Perspective from a pharmaceutical company. IPAC-RS Supplier QC Training Course: Maidenhead, UK, 2008.

30 The Institute of Quality Assurance, Pharmaceutical Quality Group. PS 9000:2001, Pharmaceutical packaging materials, 2001, § 3.7.

31 Center for Drug Evaluation and Research, US Food and Drug Administration, Department of Health and Human Services. Guideline for drug master files, September 1989. Available at: http://www.fda.gov/Drugs/DevelopmentApprovalProcess/FormsSubmissionRequirements/DrugMasterFilesDMFs/ucm073164.htm (accessed August 25, 2011).

32 Anonymous. Follow-up of FMEA. In: *Guidelines for Failure Mode and Effects Analysis for Medical Devices.* Dyadem Press, Ontario, Canada, 2003; 6–17.

CHAPTER 20

INORGANIC LEACHABLES

Diane Paskiet, Ernest L. Lippert, Brian D. Mitchell,
and Diego Zurbriggen

20.1 INTRODUCTION

The preceding chapters focused on organic leachables. This chapter will highlight the exposure of drug products to inorganic leachables. Components of pharmaceutical container closure and delivery systems (CC/DSs) typically contain some inorganic elements or their compounds. These compounds may be extracted from the materials and remain soluble in the drug product matrix, or subsequently form fine precipitates upon reaction with the drug product or excipients. These insoluble materials are more commonly referred to as foreign particulates and further discussed in Chapter 21. Understanding the elemental form to target is key to developing an appropriate extractables and/or leachables study. Inorganic leachables can originate from glass, metal, or additives in polymer systems. Potential inorganic leachables may be limited to perhaps half of the periodic table (Fig. 20.1), but their detection and identification have unique challenges that depend in part on their speciation and solubility characteristics. This chapter will outline possible sources of inorganic contamination and common approaches to evaluation and measurement of inorganic extractables and leachables. No claim is made for inclusiveness; however, the information presented should provide insight to pharmaceutical development teams and regulators, who are responsible for the safety of the packaged product.

20.2 SOURCES OF INORGANIC LEACHABLES

Exposure of drug products to surfaces of CC/DS may be short term as in the case of intermediate storage, or as long as the intended shelf life. Throughout the drug product–CC/DS exposure, history of various temperature or pressure extremes during manufacture, shipping, or storage can facilitate migration of constituents from the CC/DS material. Glass, metals, plastics, and elastomers are common materials that can have direct or indirect contact with pharmaceuticals and biopharmaceuticals. All have the potential to contribute inorganic constituents to a leachables

Leachables and Extractables Handbook: Safety Evaluation, Qualification, and Best Practices Applied to Inhalation Drug Products, First Edition. Edited by Douglas J. Ball, Daniel L. Norwood, Cheryl L.M. Stults, Lee M. Nagao.

Figure 20.1   Periodic table of the elements.

Metals

Nonmetal

	IA	IIA	IIIB	IVB	VB	VIB	VIIB	VII			IB	IIB	IIIA	IVA	VA	VIA	VIIA	0
1	1 H																	2 He
2	3 Li	4 Be											5 B	6 C	7 N	8 O	9 F	10 Ne
3	11 Na	12 Mg											13 Al	14 Si	15 P	16 S	17 Cl	18 Ar
4	19 K	20 Ca	21 Sc	22 Ti	23 V	24 Cr	25 Mn	26 Fe	27 Co	28 Ni	29 Cu	30 Zn	31 Ga	32 Ge	33 As	34 Se	35 Br	36 Kr
5	37 Rb	38 Sr	39 Y	40 Zr	41 Nb	42 Mo	43 Tc	44 Ru	45 Rh	46 Pd	47 Ag	48 Cd	49 In	50 Sn	51 Sb	52 Te	53 I	54 Xe
6	55 Cs	56 Ba	57-71*	72 Hf	73 Ta	74 W	75 Re	76 Os	77 Ir	78 Pt	79 Au	80 Hg	81 Tl	82 Pb	83 Bi	84 Po	85 At	86 Rn
7	87 Fr	88 Ra	89-103**	104 Rf	105 Db	106 Sg	107 Bh	108 Hs	109 Mt	110 Ds	111 Rg	112 Uub	113 Uut	114 Uuq	115 Uup	116 Uuh	117 Uus	118 Uuo

Lanthanides

57* La	58 Ce	59 Pr	60 Nd	61 Pm	62 Sm	63 Eu	64 Gd	65 Tb	66 Dy	67 Ho	68 Er	69 Tm	70 Yb	71 Lu

Actinides

89** Ac	90 Th	91 Pa	92 U	93 Np	94 Pu	95 Am	96 Cm	97 Bk	98 Cf	99 Es	100 Fm	101 Md	102 No	103 Lr

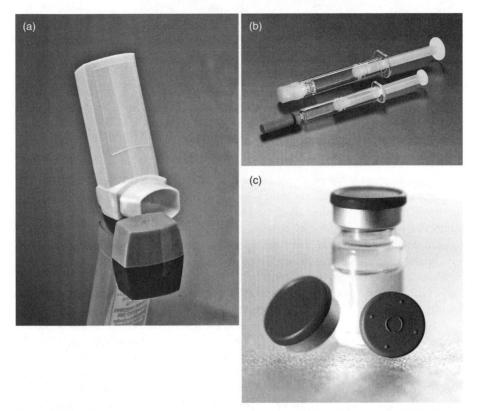

Figure 20.2   Examples of container closure systems used in the pharmaceutical industry, for example, (a) metered dose inhaler (showing mouthpiece, actuator, and canister), (b) prefilled syringe, and (c) glass vial with rubber closure and aluminum seal.

profile. Figure 20.2 shows examples of container closure systems that are used in the pharmaceutical industry as primary packaging and administration devices.

20.2.1   Glass

Silica, also known as quartz (SiO_2), the major component of sand and a precursor of glass, has a number of metastable phases.[1] Fused silica or quartz (SiO_2) has a softening point* of 1667°C.[2] At temperatures above 1705°C, the melt is very viscous. Upon cooling the melt, vitreous (glassy) silica results.[†] Vitreous silica is a random (as opposed to the crystallographic ordered structure of α-quartz) network of interconnected tetrahedra. Each silicon atom is surrounded tetrahedrally by four oxygen

---

* The softening point is the temperature at which a glass will deform under its own weight (American Society of Testing and Materials [ASTM-C338]).
[†] Molten silica is quite viscous. The kinetics of conversion to the possible crystalline phases, cristobalite, tridymite, or quartz, is slow. Each crystalline phase has two forms, α or β.

atoms. Each of the four *bridging* oxygen atoms is covalently bonded to silicon atoms in adjoining tetrahedra in such a way that each oxygen atom belongs to two tetrahedra. Although *fused quartz* has superior chemical durability and can be fabricated into containers and other useful shapes, it is not economically feasible to mass-produce the variety of shapes required by the pharmaceutical industry.

Glass, however, can be economically manufactured in a variety of shapes with desirable properties including mechanical strength, clarity, a wide range of colors, and chemical durability. Indeed, glass manufacture has been known since ancient times—Perhaps the earliest known bottle is a small decorative cosmetic jar bearing the name of King Thothmes (or Tuthmosis) III (18th Dynasty, 1479–1424 bce).[3]

Silicate glass is a super cooled, amorphous network of broken silicon-oxygen tetrahedra in which each silicon atom is linked to statistically fewer than four *bridging* oxygen atoms. Univalent oxides added to the melt disrupt the silicon-oxygen tetrahedra while maintaining electroneutrality. These oxides bind covalently to the oxygen-deficient silicon to maintain the stable octet of bonding electrons at the silicon atom. The added oxides reside in "holes" in the broken tetrahedral network of silicon and *bridging* oxygen atoms.

Glass is made by mixing a glass former with fluxes to reduce the melting point, heating the mixture above this temperature and shaping the molten mixture as it cools. Sand (>99% SiO_2), the most common glass former, when mixed with fluxes such as sodium carbonate (Na_2CO_3; soda ash) and/or potassium carbonate (K_2CO_3; potash), forms glasses that will melt* at temperatures in the range of 1000 to >1500°C. During melting, CO_2 is liberated. The Na_2O and K_2O become covalently bonded to the silicon atoms, modifying the silicon-oxygen network and, hence, the properties of the glass. This type of glass is readily attacked by water. Resistance to aqueous attack is achieved by adding structural modifiers in the form of multivalent metal carbonates: calcium oxide (CaO), magnesium oxide (MgO), and aluminum oxide (Al_2O_3). These become incorporated as multivalent metal oxides. The result is *soda lime* glass that melts at about 1500°C. Some typical glass compositions are shown in Table 20.1 along with the type classifications for chemical durability.

Compositions are carefully tailored to meet specific requirements of economical manufacture such as melting and forming temperatures, temperature range over which the forming operation can operate, and annealing properties. Annealing reduces any residual stress and results in a stronger container. Coupled with the requirements of chemical durability, these restrictions dictate the acceptable composition ranges.

Color in glass is produced when any number of metal oxides is dissolved in silicate glasses. The following are examples: cobalt produces blue; uranium produces green; manganese, particularly when exposed to X-rays, produces purple; ferrous iron produces greenish blue; ferric iron produces yellow; chromium produces green; and nickel may produce yellow to purple depending on the composition of the base glass. For amber (iron and sulfur) and green (chromium and sulfur) glasses, the hues depend on the extent of reduction of the contributing oxides. When an oxide

---

* The melting temperature of a glass is defined as the temperature corresponding to a viscosity of 100 poise. For comparison, at room temperature honey has a lower viscosity, about 70 poise.

TABLE 20.1.   Typical Glass Compositions[5,a]

Source	Oxide	Approximate composition (%, w/w)				
		Soda lime			Aluminoborosilicate	Borosilicate
		Colorless	Amber	Green	Amber	Colorless
Sand	SiO_2	71–75	70–74	70–74	70–74	70–82
Soda ash	Na_2O^d	12–16	13	13	7	4–8
Potash	K_2O^d	2	0.3	0.3	1	
Limestone	CaO^e	10–15	11	10	–	–
Dolomite	MgO^e	3	0.4	0.4	–	–
Feldspar	$Al_2O_3^e$	1.5–2.5	1.4	2	6	2–4
Borax	$B_2O_3^e$	–	–	–	7	7–13
Common	Fe_2O_3	0.01–0.05	0.2–0.5	0.1–0.3	–	–
sources[b]	SO_3	0.1–0.4	0.03–0.14	0.01–0.26	–	–
	Cr_2O_3	–	0.03	0.05–0.2	–	–
	Total other[c]	<2	<2	<2	5 (TiO_2)	<1
Classification by type						
	ASTM[6,7]	II or III[f]	II or III	II or III	I	I
					Class B	Class A
	USP[8]	II or III	II or III	II or III	I	I
	EP[9]	II or III	II or III	II or III	I	I
	ISO[10–13]	HC 3	HC 3	HC 3	HC 1	HC 1
	JP[14]	No unique designation				

[a] The ranges reported in Table 20.1 are based on the cumulative experience beginning in the early 1970s of two of the authors, Lippert and Mitchell, and their colleagues.

[b] Common sources of minor components intentionally added are hematite (Fe); iron chromite (Fe, Cr); iron pyrite (Fe, S); iron scale (Fe); salt cake (Na, S); and gypsum (Ca, S).

[c] F, Cl, BaO, MnO, PbO, SrO, ZnO, TiO_2, ZrO_2, As_2O_3, Sb_2O_3, and other additives or unintentional impurities.

[d] Tend to decrease chemical durability.

[e] Tend to increase chemical durability.

[f] Type II is derived from type III as noted in Section 20.2.1.3 "Glass Treatments."

or metal that does not dissolve is dispersed as a colloid, the color is caused by light scattering. The color depends on the size and concentration of the particles. Gold and copper produce ruby red, selenium produces red, and silver produces a yellow color. For a more extensive discussion of colored glasses, see Phillips.[4]

20.2.1.1   Enhancement of Glass Properties

Chemical durability, also called hydrolytic resistance, is measured by determining the increase in alkalinity of the attacking solution under prescribed conditions of temperature and surface area of the glass being evaluated. Borosilicate glass, also known by the trade names Pyrex®, Kimax®, or Duran®, has the best chemical durability (type I) of commercially available glasses. See Section 20.2.1.3 "Glass Treatments" for the description of a new, highly inert, glass surface.

At room temperature, B_2O_3 has a polymeric structure with the boron surrounded by six electrons. At high temperatures in the glass melt, the B–O network can assume a tetrahedral configuration that allows it to accept a lone pair of electrons (derived from Na_2O or K_2O) to complete the more stable octet of bonding electrons around boron. Because of this covalent bonding, the ease of alkali dissolution is hindered. This, along with the reduced amount of Na_2O, K_2O, and added Al_2O_3 as compared with soda lime glass, results in type I durability. Al_2O_3 has an effect analogous to B_2O_3. Amber aluminoboroslicate glass, which has higher levels of Al_2O_3 than borosilicate glass, also exhibits type I chemical durability.

SiO_2 is more soluble in basic than it is in acidic solutions. Under reaction with an alkaline environment, the silica-rich surface layer is continually removed by solvation so that fresh sodium and potassium ions are continually available to the solution. For the same glass composition, chemical durability is greater toward neutral or slightly acidic solutions than toward basic solutions. The type of glass chosen for a specific application depends on the chemical durability required. The European Pharmacopoeia provides general recommendations concerning the type of glass container that may be used for different types of pharmaceutical preparations. Table 20.2 lists general guidelines for pharmaceutical containers and extractables and leachables studies.

TABLE 20.2. **Recommendation of Glass Types for Pharmaceutical Use Abstracted from Section 3.2.1 of the European Pharmacopoeia 7.0[9] and USP <660>[8]**

	Classification of glass containers	Use of glass containers	Type of test
Type I	Neutral glass, with a high hydrolytic resistance due to the chemical composition of the glass itself	Suitable for most preparations whether or not for parenteral use	Powder (glass grains)
Type II	Usually of soda lime–silica glass with a high hydrolytic resistance resulting from suitable treatment of the glass surface	Suitable for most acidic and neutral aqueous preparations whether or not for parenteral use	Surface
Type III	Usually of soda lime–silica glass with only moderate hydrolytic resistance	[In general] Suitable for nonaqueous preparations for parenteral, for powders for parenteral use (except for freeze-dried preparations), and for preparations not for parenteral use	Powder (glass grains)

20.2.1.2   Glass Breakage   As is any other crystalline or amorphous solid, glass is quite strong in compression. This is because compression acts against the opposing atomic forces that resist decreasing interatomic distances. In contrast, tension forces act to increase the interatomic distances; hence, one would expect the tensile strength of glass to be very large. This is only true in the special case where the surface of the glass is pristine and without flaws.

Glass can only fail in tension. Microscopic *Griffith flaws*[15] (microscopic cracks) at the surface act as stress concentrators. An applied strain (whether pure tension or tension caused by bending) causes the crack to propagate, releasing the stress and leading to fracture. Griffith flaws result when the surface is abraded, impacted, or otherwise damaged.

It is important for the integrity of the container for the pharmaceutical packaging industry to treat glass containers gently during processing and delivery to avoid surface abrasion and hence maintain its intrinsic strength.

20.2.1.3   Glass Treatments   As the previous discussions have noted, glass is not inert and is particularly susceptible to chemical reactions at the surface. As the alkali (Na_2O and K_2O) fluxes are introduced, the tetrahedral structure of silica is broken. The added oxides then occupy "holes" in the broken silica network: The oxygen from the alkalies will weakly bond to silicon atom to preserve the stable octet of electrons about the silicon that has lost a network oxygen. Because these added oxides are not strongly chemically bonded to the network structure, they are available for solvation by the solution with which the glass surface is in contact. This solvation is more pronounced as the temperature and alkalinity of the solution increases. As solvation of the glass surface proceeds with the removal of sodium and potassium ions, the SiO_2-enriched layer becomes thick enough to spall from the surface. While basic solutions favor *siliceous flakes*, these flakes can also be observed in neutral or slightly acidic solutions that have been stored for some time, particularly if the containers have low chemical durability. The pharmaceutical literature often refers to the production of siliceous flakes as glass delamination. These flakes are a few or more nanometers thick. The width/length dimensions vary widely with lengths perhaps greater than 1000 μm. These flakes are best observed by swirling the liquid with light directed at 90° to the direction of observation. Interference colors may be observed. For an extensive study of this problem, see Iacocca and Allgeier.[16]

Weathering occurs when the glass surface is attacked by atmospheric moisture. It is most often observed with glass of low chemical durability. Atmospheric moisture promotes the migration of sodium oxide to the surface. This raises the pH of the condensed or adsorbed atmospheric moisture and will dissolve atmospheric carbon dioxide (CO_2), leading to the production of $NaHCO_3/Na_2CO_3$. Sodium oxide is effectively removed from the glass surface. The subsequent change in surface reflectivity leads to the appearance of dullness or fogging and the formation of siliceous flakes.

Sodium oxide depletion in the bottom of a vial will occur as a result of vaporization caused by overheating during manufacture. This leads to an increased propensity for the formation of siliceous flakes when the glass surface is in contact with a drug product or other aqueous solvent.

To protect the glass surface from chemical attack and minimize fogging or spallation of siliceous flakes, surface treatments may be utilized. The first type of treatment is dealkalization. After a container is formed, sulfur trioxide (SO_3) gas* is injected into the container at about 550°C as it begins the approximately hour-long journey through the annealing lehr. During annealing, residual stresses and probably surface defects are reduced. The immediate reaction of the gas with the hot glass surface removes the surface Na_2O by forming Na_2SO_4. The container is washed to remove the sodium sulfate and subsequently dried before filling with the pharmaceutical product. Such treatment produces a low-defect surface depleted in Na_2O, which results in type II durability.

A second type of surface treatment is the plasma impulse chemical vapor deposition (PICVD) process that was developed by Schott AG.[17] This deposits pure SiO_2 on the interior surface of a container. The PICVD process involves several steps. After washing and drying, the container is filled with oxygen and the surface is activated by a microwave-induced plasma. After activation, a coating gas is introduced and the low-pressure plasma is again ignited. This cracks the coating gas, generating SiO groups that diffuse to the container surface where they react with the excess oxygen, forming an SiO_2 layer. Oxygen is again introduced, and the procedure is repeated until a layer 0.1–0.2 μm thick has been deposited. No sodium oxide is leached from this quartz-like inner surface under autoclaving conditions that are more stringent than the treatments referenced in Table 20.2. These SCHOTT Type I plus® containers offer a highly inert surface with respect to ordinary type I surfaces from which leachables are commonly found.

A third type of surface treatment is applied to glassware that is used for trace analysis and may be employed in extractables or leachables studies. Chemical-resistant borosilicate glass is used for the determination of trace contaminants. It is obvious that this glassware (including platinum or polytetrafluoroethylene [PTFE] dishes) must be scrupulously clean. Common practice for glassware is to treat it with boiling nitric acid. Analytic reagent grade is sufficient unless the concentration of the analyte being sought is less than approximately three times the amount of analyte contributed by the acid. Otherwise, high purity acid is required. As an alternative, 10% KOH in ethanol can also be used. After cleaning, a thorough rinsing with high-quality deionized water is required.[†]

Volumetric flasks are cleaned by adding 10–15 mL HNO_3 and heating until the acid refluxes in the neck. Platinum dishes are cleaned by fusing with $NaHSO_3$ and rotating them during fusion to coat the inside walls. This is followed by water and high-quality deionized water rinses. Acid digestion may follow if necessary; for example, if trace sodium is being determined. PTFE dishes are usually cleaned by boiling them in the same acid that will be used in the analytical procedure.

---

* Other treatments include using SO_2 or ammonium sulfate.

[†] The requirements for high-purity water are given in 2009 USP, General Chapters: <600> Containers-Glass.

20.2.2 Polymers

The source of inorganic constituents in polymers systems will be described here. Colorants, catalysts, fillers, and other additives have inorganic origins with the potential to contaminate the drug product with soluble inorganic constituents or foreign particulate matter.

20.2.2.1 *Colorants for Polymers*[18]
The selection of an agent to color a plastic or elastomer is controlled by the physical and chemical characteristics both of the colorant and material being colored. The pigments or dyes may be organic or inorganic in nature. Pigments are essentially insoluble and are generally of micron size. They can be dispersed in a medium and form suspensions as opposed to dyes that are soluble in most solvents and polymers. Colors extended into an inorganic base to form an insoluble pigment are referred to as "lakes," which involve a chemical interaction of a dye with a metal compound or sorption of the dye onto an inorganic substrate. Pigment colors are classified due to chemical structure as well as differences in applications. The colorant properties must be compatible with the processing technique and desired color outcome. Properties of colorants critical to the final product are heat stability, dispersability, light fastness, chemical inertness, solubility, opacity/transparency, tinctorial properties, and electrical characteristics. The suitability of a colorant relies not only on the performance properties but resistance to heat, light, migration, and toxicity. Pigments are often used instead of dyes due to process requirement and cost. Because of their solubility, dyes may have a tendency to migrate. Many colorants contain heavy metals that may preclude their use in pharmaceutical products.[19]

Inorganic pigments tend to be opaque, while organic pigments and dyes are more likely to be transparent. Yellows and oranges and can be derived from lead chromate and can be co-precipitated with molybdate or sulfate. Cadmium sulfide, when co-precipitated with selenium, zinc, or mercury, also produces yellow hues, while nickel titanate produces pale yellow. Iron oxides, depending on the oxidation state, can produce a range of colors from yellows, reds, browns, and black. Chromium oxide produces a green color, and titanium dioxide is white. Aluminosilicate, known as ultramarine blue, has a red/blue hue and is often added to hide yellowing, while calcined cobalt aluminate produces a bright blue color.

Azo organic pigments are used to formulate shades of reds and yellow. Calcium, barium, strontium, or nickel can be used to precipitate the dye and form an insoluble pigment. These can be laked onto a zinc oxide aluminum substrate. Greens and blues are found in copper phthalocyanine and are halogenated to form green. Scarlets, reds, and violets can be derived from quinacridones. Examples of sanctioned colorants can be found in 21CFR parts 170–189, direct/food additives, along with uses and limitations.[20]

Dyes or pigments are incorporated into polymer systems using dry powders, concentrates (master batches), liquid carriers, paste, or other vehicles. Color can also be compounded into pellets using extruders with the addition of appropriate dyes.

The potential for migration, acceleration/retardation of cure, polymer degradation/ stabilization, bleeding, strength, and chroma are all factors that would impact the color choice. An extensive discussion of colorants for thermoplastics is given in the *Plastics Additive Handbook.*[21]

Pigments are also used for printing and decorating packaging systems. A typical printing ink will incorporate a resinous component or vehicle to bind pigment; solvents to dissolve and make fluid; and other additives such as wetting agents, dryers, antioxidants, viscosity agents, and tackifiers. Inks can be dried by different techniques such as evaporation, chemical reaction, or oxidation. Most inks solidify by a combination of several mechanisms and are formulated for specific substrates to achieve the desired outcome. Printing can be accomplished using various methods that include flexography, lithography, and gravure; common decorating methods are screen printing, hot foil stamping, embossing, and pad printing. Descriptions of printing and decorating methods are beyond the scope of this chapter, but there are many variations to be considered.[22]

20.2.2.2 *Polymer Additives, Fillers, Reinforcements, Stabilizers, and Catalysts*   The performance and function of polymers is enhanced by any of a number of additives used in the polymer system.[23] Many of the elements in the periodic table (Fig. 20.1) have potential to be found in a leachables profile because of the broad range of additives available.

A filler is a relatively inert material added to a plastic to modify its strength, permanence, working properties, or other qualities. A reinforcement is a high-strength fiber (inorganic or organic) that imparts strength properties far superior to those of the base resin.[24] There is no clear demarcation between fillers and reinforcements.

Fillers are solid materials added to the polymer to modify its physical and mechanical properties. The main classes of fillers are (1) thermoplastic fibers/ fabrics used to impart durability to brittle polymers; (2) glass fibers used to improve impact resistance and hollow glass or carbon microspheres used to reduce density; and (3) minerals used to extend or reinforce the resin. Calcium carbonate, clays, talcs, micas, silica/silicates, and aluminum trihydrate are examples of mineral fillers.[25]

Thermal stability during processing is achieved using heat stabilizers, and polymers can be protected from degradation by using antioxidants. Inorganic elements associated with stabilization are barium, cadmium, tin, and lead. Sulfur and phosphites act as antioxidants. Citrates, carbonates, or stearates of calcium, magnesium, zinc, potassium, and sodium are often used as stabilizers.

The catalyst system initiates and controls the propagation of chain growth during polymerization. Catalysts are often complexes produced by interaction of alkyls of metals in groups I–III of the periodic table (Fig. 20.1) with halide and other derivatives of groups IV–VIII metals. Soluble catalysts exist but solid or catalysts absorbed on a carrier particle are more common forms.[26] Residual inorganic species from polymer catalysts may be present at trace levels. The analytical procedure should have appropriate sensitivity and recognize any solubility problems of the elements being sought.

20.2.3   Metals

Metal containers can be used for a wide range of pharmaceutical applications and are processed by a number of different metal working operations. Usage in the pharmaceutical industry can range from rigid containers that can withstand high pressures, components that are part of a piercing mechanism, tubes that are collapsible, or a layer in a blister laminate that will have low tensile and tear strength. Large noncorrosive mixing vessels and storage drums may also be in direct contact with drug substances for an extended or intermediate period. Specialized applications can require components made of custom alloys, and secondary components such as pouches may be used to protect a product from moisture or gas permeation.

Metal containers are comprised of a mixture of elements of varying proportions (alloys) to achieve the desired properties and may be coated with organic or inorganic materials to protect them from corrosion. This section is intended to indicate the different types of elements that may be present in metal containers and may have potential to leach into a pharmaceutical product.

Tinplate is widely used to fabricate containers and closures. It is a low carbon steel coated on both sides with tin at various coating weights. Tin-free steel can be produced and coated with chromium/chromium oxide. Steel, such as black plate, is uncoated and susceptible to corrosion, but it can be galvanized by electroplating with zinc to reduce this susceptibility. Lacquers are often used to coat the containers to protect them from corrosion.

Aluminum and its alloys are used to fabricate tubes and rigid containers, especially for aerosols. Aluminum foil is used to provide an impervious layer in a blister laminate or protective sachet for a product. Aluminum oxide can also be deposited onto paper or plastic materials to provide protection for the enclosed product.[26]

Stainless steel is commonly used for mixing vessels, bulk containers, sampling tools, or delivery system components such as springs, screws, or piercing elements. Many grades are available for high-temperature application, usage requirements, and level of corrosion resistance. It is an iron-based alloy with at least 10.5% chromium; certain grades can contain the same amount of nickel. Various amounts of molybdenum, cobalt, copper, sulfur, manganese, titanium, phosphorus, niobium, and silicon may be used in alloys or super alloys depending on the application.[27] The 300 series of stainless steel is typically used in the pharmaceutical industry, and some example compositions are given in Table 20.3. Grade 304 is the most versatile commonly used stainless steel. Grade 316 or 316L has better overall corrosion resistance than grade 304. It is particularly resistant to pitting in a chloride environment. Varying combinations and levels of elements may be present in a given drug product, and the challenge in an extractables study is to ensure solubility and to be representative, sensitive, and specific.

TABLE 20.3.   Examples of Stainless Steel Compositions

	% Composition (wt/wt) Designation: ASTM/EN		
Element	304/X3CRNI18-10	316/X3CRNIMO17-12-2	316L/X2CRNIMO17-12-2
Fe	Major	Major	Major
C	≤0.07	≤0.07	≤0.030 (lower than 316)[a]
Si	≤1.00	≤1.00	≤1.00
Mn	≤2.00	≤2.00	≤2.00
P, max	0.045	0.045	0.045
S	≤0.015	≤0.015	≤0.015
N	≤0.11	≤0.11	≤0.11
Cr	17.5–19.5	16.5–18.5	16.5–18.5
Mo		2.00–2.50	2.00–2.50
Ni	8–10.50	10.00–13.00	10.00–13.00

[a] Grade 316L stainless steel is also easier to machine compared with grade 316 stainless steel due to its lower carbon content.

Source: Euro Inox, Grand Duchy of Luxembourg, 2011, p. 4. Available at: http://www.euro-inox.org/pdf/map/Tables_TechnicalProperties_EN.pdf (accessed August 25, 2011).

TABLE 20.4.   Compendial Tests for Glass

Compendial test	Description	Attribute	Type of test
EP 6.0,[9] USP <660>[8]	Classification of glass	Type I	Powder (glass grains)
		Type II	Surface
		Type III	Powder (glass grains)
EP 6.0[9]		Arsenic	
USP<231>[28]		Heavy metals	Water extract

20.3   METHODS FOR EXTRACTION AND ANALYSIS OF INORGANIC COMPOUNDS

20.3.1   Compendial Testing

There are compendial inorganic test procedures published for analysis of glass, elastomers, and plastic materials. The three compendia, USP, EP, and JP[8,9,14] in combination, provide general heavy metal data for glass, plastic, and elastomeric material. The USP and EP compendia classify glass for intended use based on leaching by water (chemical durability) shown in Table 20.4. Although a regulatory requirement, compendial testing (whether organic or inorganic) is not a substitute for conducting a controlled extraction or leachables study.

20.3.2   Inorganic Methods of Analysis

One factor critical to the identification and measurement of inorganic analytes is the sample introduction into the instrument and detector used in the method of analysis. In the case of an extraction study, the choice of solvent may or may not be suitable to represent the inorganic constituents in a given component, which may have the potential to contaminate the drug product. For instance, as a result of tungsten pins used during the manufacturing of prefilled syringes, soluble and insoluble tungsten species can be deposited in the glass barrel. The risk of tungsten-induced protein particle formation can be better controlled if tungsten can be reliably measured.[29]

Another similar problem of potential underreporting leachables or extractables is related to the evaluation of residual aluminum oxide, a polishing agent used in cleaning processes. The presence of aluminum oxide may not be detected if only soluble aluminum species are sought and a specialized sample preparation would need to be developed. These are among the many issues that need to be considered when assessing inorganic leachables and extractables. Attempts to provide an inorganic profile should consider the risk to oversight key trace elements due to choice of extraction medium, combination of elements, and speciation. Drug products can involve simple or complex matrices to put into solution in a way that will preserve the active ingredient and excipients. The extracting propensity will vary depending on the use of buffers, surfactants, chelating agents, organic solvents, alkaline, or acidic systems. The typical approach of a controlled extraction study, to identify organic extractables, may not be sufficient to indicate the inorganic extractables or migration behavior of the actual material; therefore, it would seem that a logical first step is to carry out a basic elemental scan to provide targets of what is known to exist, and a second step could be to perform a migration study using a surrogate solution in an attempt to predict migration. However, if it is found necessary to monitor for the presence of inorganic leachables, methods will need to be developed and validated to provide reliable data indicating the presence or absence of those inorganic constituents in the actual drug product.

The most commonly employed techniques for detection and quantification of inorganic analytes include the spectroscopic methods: atomic absorption (AA), atomic emission (AE), inductively coupled plasma (ICP), and X-ray fluorescence (XRF), as well as ion chromatography (IC). It is noteworthy that AA, ICP, and XRF provide elemental information, while IC is capable of providing information of the actual species present in a sample or sample extract. See Section 20.3.2.1.2 "Detection Methods" (2b) in the discussion on ICP for information on how IC–ICP–mass spectrometry (MS) can provide information on the species present.

XRF deserves particular attention as it allows for a nondestructive sample analysis. The following will describe instrumental techniques employed to detect and measure the majority of elements on the periodic table (Fig. 20.1).

20.3.2.1   ICP   ICP is an elemental technique that allows for the rapid, quantitative determination of metals, metalloids, and select nonmetals. Most modern optical ICP instrumentation is equipped with detectors capable of simultaneous detection and quantization of about 75 elements with a dynamic linear range of five to six orders of magnitude.[30] The sample solution is introduced into an argon plasma. Temperatures at the center of the argon plasma reach 6000–9000 K. As the aerosolized sample passes through the center channel of the plasma, the sample is first desolvated, followed by matrix decomposition and excitation of the electronic energy levels of the elements present in the sample.

20.3.2.1.1   ICP Sample Preparation   The preparation of the samples to be analyzed by ICP deserves special attention. The traditional approach to extractables and leachables screening has been to analyze the product in contact with the packaging system or to extract the packaging system, or its components, with one or more solvent systems. This approach carries the risk that if any of the constituents of the material tested is not soluble in the chosen solvent system, it will not be detected. Additionally, if the chosen solvent system is not representative of the interaction of the drug matrix with the packaging system, potential extractables and leachables will be missed.

Guidelines and regulations require that ever-increasing numbers of elements be monitored and quantified. Simple solvent extraction followed by ICP analysis may not be able to meet the increasingly demanding criteria applied to screening procedures. New approaches have been proposed that recommend initially performing a comprehensive analysis of the materials to be screened.[31] This comprehensive analysis can then be used to select the analytes to be monitored on a routine basis as well as to design a solvent system suited to the extraction and or leaching of the analytes of concern.

A comprehensive evaluation can be achieved by digestion of a material for elemental analysis. An appropriate digestion procedure is necessary to capture the analytes of interest. The composition of the material greatly affects the choice of digestion procedure. Organic-based samples can simply be ashed in a furnace, assuming no volatile analytes are to be analyzed. Ashing is not a suitable sample preparation method for inorganic materials such as glass or metals or when volatile elements such as mercury are to be analyzed.

Acid digestion is the most widely used technique for elemental analysis; however, care must be taken to ensure all analytes are soluble in the acid or acid mixture chosen for the sample digestion. Hydrofluoric acid and fluoride ions will lead to the loss of silicon as a gaseous SiF_4 while precipitating analytes such as calcium and magnesium. Chloride ions will precipitate silver, unless ammonium hydroxide is added to complex the silver chloride and bring it back into solution. Sulfuric acid will precipitate barium sulfate. Nitric acid is commonly used since nitrates are generally soluble. Nitric acid cannot be called a "universal acid" since tin will convert to metastanic acid, which is insoluble in nitric acid. Zirconium may also present a problem, particularly for long-term stability of standards prepared in nitric acid unless hydrofluoric acid (HF) is added to complex the Zr ions and prevent precipitation. Diligence is required to validate a particular method of sample preparation since unsuspected interactions may lead to unintended consequences.

Depending on the material to be tested, acid digestions can be performed on a hot plate. However, care must be taken, as some analytes can easily be lost due to their high volatility in an acid matrix; these include, among others, arsenic, mercury, ruthenium, osmium, and silicon (in the presence of HF). Another limiting factor of open-vessel digestion is the boiling point of digestion mixture. Samples may be lost due to evaporation. Sample size depends on the solubility of the analytes in the digestion mixture and is limited by the vessel size.

An alternative to open-vessel digestion is closed-vessel digestion in a microwave oven. Depending on the specific vessel design, there is little to no loss of volatile analytes during digestion if samples are allowed to cool to room temperature before opening the vessels. Microwave-assisted sample digestion also has the advantage that temperatures significantly higher than in an open vessel can be achieved, which has the potential for faster and more complete digestions.[32] It is important to note that microwave digestion is limited by sample size due to the pressure buildup caused by the sample. For organic materials, typical sample size is 250 mg or less. For inorganic samples, sample size is typically 500 mg.

Samples presented to the plasma must have all analytes of interest in soluble form. The residence time in the plasma is not long enough to decompose and atomize undissolved particles. It is desirable to keep the sample solution concentration to about 2% total dissolved solids to reduce plasma loading, which can negatively affect the energy available to create excited analyte species. ICP analysis is generally performed on aqueous solutions containing the analytes of interest. Organic sample solutions can be analyzed employing membrane desolvation to remove the solvent before introduction to the plasma. Alternatives to desolvation include analysis using either solvents with low vapor pressure (e.g., kerosene) or chilled sample introduction systems to reduce the solvent load on the plasma. Analysis of organic solvents tends to result in carbon deposits that degrade instrument performance. Carbon buildup can be prevented by introducing a small amount of oxygen into the plasma. When organic sample solutions are analyzed by ICP, great care must be taken to match closely calibration standards to sample solutions to account for variation in the volatility of different soluble organometallic compounds based on their respective oxidation states or the chemical form of the analyte.

20.3.2.1.2 Detection Methods Two different types of detection techniques are commonly used for ICP: visible spectroscopy or MS.

1. ICP–atomic emission spectroscopy *(AES)*. This technique is also referred to as ICP–optical emission spectroscopy (OES). When ICP is coupled with emission spectroscopy, the light emitted by the excited species as they return to the ground state is introduced into the optical part of the instrument and the emitted light is resolved into an emission spectrum using either a prism or an echelle grating. The resolved, element characteristic wavelengths are detected, and the intensity of the light emitted by the excited species is measured using a charge-coupled device or photomultiplier tube. The measured intensity is converted into a concentration by comparison with solutions of known concentrations of the analytes of interest.

   Detection limits are about 10 times lower for axially viewed ICPs than for radially viewed systems. Plasma viewed end-on in an axial system allows observation over the entire length of the center channel of the plasma containing the excited species. Since the length of the center channel is greater than the radial distance across the plasma, the total signal axially is greater than that obtained by observing only a cross section in the radial view. Typical detection limits are in the parts per billion range for an optical ICP.

2. *ICP-MS.* This utilizes the same argon plasma to desolvate and decompose the sample. The plasma settings are optimized to produce singly charged analyte ions. Using ion optics, the analyte ions generated in the plasma are extracted from the central channel of the plasma and presented to a mass selective analyzer, which separates the analytes based on their respective masses. Detection is most commonly accomplished using an electron multiplier. Typical detection limits are in the parts per trillion to parts per billion range for ICP-MS.

When sample extracts are to be analyzed by ICP-MS, consideration must be given to potential polyatomic interferences created in the plasma from the sample to be analyzed. Different approaches have been developed to eliminate isobaric polyatomic interferences such as $^{40}Ar^{35}Cl$ (75 amu) versus ^{75}As. The simplest approach is to analyze alternate isotopes free from isobaric interferences. Instruments equipped with a quadrupole mass analyzer employ algorithms to mathematically correct the observed results.

A different approach is the use of a cell gas to remove interferences before the analytes are presented to the mass analyzer. At present, two competing technologies are available:

2a. A reaction cell using a reactive gas such as ammonia (NH_3) or oxygen (O_2). Consider the example of $^{40}Ar^{35}Cl$ (75 amu) causing an interference with ^{75}As (75 amu). When O_2 is introduced into the reaction cell, it reacts with ^{75}As (75 amu) to form $^{75}As^{16}O$ (91 amu), thereby making it possible to analyze for As in a matrix-containing chloride.

2b. A collision cell using a nonreactive gas such as helium. In a collision cell, polyatomic ions generated in the plasma collide with an inert gas, such as helium. Ions with larger radii have a higher probability of colliding with helium than with ions with smaller radii. Each collision results in a reduction of the kinetic energy. This results in a kinetic energy distribution that allows for selective discrimination of the ions entering the mass analyzer. For example, $^{40}Ar^{23}Na$ (63 amu) has a larger radius than the analyte ^{63}Cu (63 amu). The larger $^{40}Ar^{23}Na$ has more frequent collisions with helium and looses kinetic energy faster than the smaller ^{63}Cr. This causes the arrival times of the isobaric ions at the detector to be different.

ICP is a quantitative, elemental technique. When interfaced with IC, liquid chromatography, or gas chromatography, an ICP can be used to perform speciation analysis.[33] For example, ICP alone only gives elemental information such as elemental chromium (Cr) in wastewater. In contrast, IC-ICP-MS can selectively analyze for Cr(VI) in addition to chromium.

20.3.2.2 IC IC is a liquid chromatography technique that employs ionic interactions with the stationary phase. Analyte ions are selectively separated. IC can be applied to a wide number of charged analytes such as inorganic ions, organic ions, charged molecules, and proteins. Detection of the analytes is accomplished by several methods including conductivity, electrochemical, and UV/Vis spectral analysis. To enhance the signal for inorganic analytes, chemical or electrochemical suppression is employed to greatly reduce the background conductivity introduced by the buffered mobile phase used to elute the analytes.

Samples to be analyzed by IC must be aqueous. Aqueous solutions may be taken from extracts or solvent exchange preparations. While small amounts of solvent are generally tolerated by most stationary phases, organic solvents cannot be analyzed neat (in their original form) and need to be solvent-exchanged into water. IC is capable of separating different species of the same element based on the respective charges. However, anions and cations cannot be separated at the same time. They require different stationary and mobile phases. Detection limits for IC are typically in the parts per billion range. With sample pretreatment or large-volume injections, parts per trillion levels are achievable.

20.3.2.3 AA Spectroscopy (AAS) This technique is based on the Beer–Lambert law relating the intensity of a specific spectral line attenuated by absorption to the concentration of the element. The instrumentation consists of (1) a source that emits a spectral line specific to the element sought; (2) a nebulizing/mixing chamber where the sample solution is nebulized and mixed with the fuel gas before entering the long, narrow sheet flame where it is atomized to free atoms in their ground states; (3) the spectral line from the source is attenuated as it passes through the long sheet flame where it loses energy (intensity) by exciting the analyte atoms to a higher energy level; (4) the spectral line from the source passes through a monochromator and then to an intensity measuring device such as a photomultiplier or charge coupled detector (CCD).

The concentration of the analyte is determined from a working curve prepared from standards of known concentration and matched to the matrix of the analyte solution.

Sources include hollow cathode lamps. These may be of single or multiple element construction. A high-resolution continuum source AAS (HR-CS AAS) based on a xenon arc lamp provides a spectral range of 185–900 nm.

Fuel gases may be acetylene air or acetylene-nitrous oxide. The flame temperature may range from 2100 to 2800°C. The effective path length through the sheet flame is 5 or 10 cm. As an alternative, a graphite furnace may be used to atomize the sample.

20.3.2.4 X-Ray Techniques: XRF Many forms of XRF analysis exist. These techniques are used mainly for elemental identification in micro to large samples. Semiquantitative analysis based on standardless analysis is often possible. Precise quantization is best provided by the techniques discussed above. XRF techniques depend on the excitation of the element by high-energy photons with energy above 1 keV. An inner shell (K or L) electron is ejected from the atom. When electrons replace the ejected electron, x-radiation is emitted. This fluorescent radiation has a wavelength (energy) characteristic of the atom. Two different detection methods are used. A Li-drifted silicon detector operating at liquid nitrogen temperature can provide both wavelength and intensity information. At room temperature, a crystal monochromator (determines wavelength) directs the fluorescent radiation to a photon counter to determine its intensity. An empirical calibration curve of intensity versus concentration allows the quantitation of the analyte.

Five of the many techniques available are the following:

1. *Energy-dispersive X-ray (EDX).* Excitation of the sample can be by high-energy X-rays or radioactive isotopes. Usually, the sample is placed in a vacuum chamber. Sample size is in the range of centimeters to millimeters.

2. *Electron probe microanalysis (EPMA).* Excitation of the sample is by a focused beam of electron in high vacuum. Sample size is in the range of millimeters to microscopic.

3. *Proton-induced X-ray emission (PIXE).* Excitation of the sample is by a beam of protons. Samples can be microscopic. Large samples (tens of centimeters) can be accommodated by placing them outside the vacuum chamber and enclosed in a helium sheath.

4. *Scanning electron microscopy (SEM).* Excitation of the sample is by a beam of electrons in vacuum. Sample size is in the range of millimeters to microscopic. The sample itself can be at a higher pressure.

5. *Wavelength dispersive X-ray (WDX).* Excitation of the sample in a vacuum is by radiation from an X-ray tube. Sample size is in the centimeter range and must be carefully prepared for quantitation against standards.

20.4   CONSIDERATIONS FOR INORGANIC LEACHABLES AND CASE STUDIES

This section illustrates the importance of understanding the chemistry of the system under study. Detection of inorganic elements that have leached into drug products can be challenging because some of these elements may already be part of the drug product matrix or exist as species that are not readily detected or are present at trace concentrations. In addition, combinations of elements may produce multiple spectral lines, creating interferences and masking trace ions in the presence of high concentration of competing ions, incompatible solubility characteristics, or interaction with the drug product to form a reaction product. Determinations for extractables or leachables are not as simple as submitting solutions for elemental analysis by ICP or another technique. Examples of these challenges are described in the following case studies regarding methods to measure titanium in the presence of phosphate and detection of nitrate and sulfate in buffered saline.

20.4.1   Determining Titanium in the Presence of Phosphate

An ICP method to detect and quantitate potential leachables in a drug product was required. Titanium (Ti) was among the several elements to be determined. During the preparation of the standard solutions, it was observed that when a small quantity of the Ti standard (100 μg/mL) was added to 5 mL of the drug product matrix (the components of the drug product without the active ingredient), cloudiness developed within 30 minutes. This precipitate was confirmed as a form of titanium phosphate using the EPMA X-ray technique. This simple method for the preparation of the standards could not be used because of the presence of phosphate in the matrix.

An alternative method for the preparation of standards was developed. A known quantity of EDTA (ethylene di-nitrilo tetra-acetic acid disodium salt dihydrate) and pH 10 buffer (NH_4Cl and NH_4OH), when mixed with the drug product matrix standards, resulted in clear solutions, because the Ti was complexed by EDTA, thus preventing the formation of insoluble titanium phosphate. At pH 10, the stability constant of the soluble titanium–EDTA complex is greater than that of insoluble titanium phosphate; therefore, titanium remains in solution as the more stable titanium–EDTA complex. The method quantitation limit (MQL) for Ti in the standards was 0.030 μg/mL.

20.4.2   Nitrate and Sulfate in Buffered Saline by IC

A leachables method was developed for the detection and quantization of nitrate and sulfate in phosphate buffered saline. Due to the large quantities of chloride and phosphate present in the sample, it was not possible to separate the chloride/nitrate and phosphate/sulfate peaks using isocratic elution with the standard carbonate/bicarbonate buffer. Instead, a new method was developed using gradient elution with a hydroxide eluent. To cope with the high loading of chloride and phosphate, a high capacity analytical column was chosen. Complete separation of all sample constituents was accomplished in 20 minutes. Sample preparation consisted of a simple fivefold dilution and direct injection into the IC system. Detection and quantification of the analytes were performed via conductivity monitoring of the column effluent after electrochemical suppression.

20.4.3   Trace Tungsten Determination

An extraction method was developed to detect and quantify residual water-soluble tungsten species in fillable syringes. The channel of the syringe is formed using a tungsten pin. While tungsten is very resistant, due to the high temperatures required, the tungsten pin may react with oxygen from the air and form tungsten oxides, which are then deposited on the inside surface of the syringe.[34]

The goal for this method was an extraction targeting all tungsten species, which may included W as well as WO_x, soluble in an aqueous solvent system. The method was to be aggressive enough to extract all potentially aqueous solvent system soluble species without attacking the glass surface. Tungsten species immobilized in the glass surface were not to be extracted. Ammonia was found to be the most aggressive solvent targeting potential tungsten species while benign enough not to attack the glass. This was a general method giving the worst-case scenario (using the most aggressive solvent).

Syringes were filled with an aliquot of 5% NH_4OH; a vacuum was applied to ensure wetting of the annular space where the needle is glued into the syringe. The filled syringes were stored for a week at room temperature after which the contents of the syringes were collected and analyzed using ICP-MS.

A calibration curve was established from matrix-matched W standards prepared from commercially available metals standards. Detection and quantitation was performed by ICP-MS, monitoring isotopes ^{182}W, ^{184}W, and ^{186}W with rhenium (^{187}Re) used as an internal standard. The MQL for W was set at 5 ng/mL.

20.4.4   Consideration of Inorganic Leachables with Respect to Biologic Products

The impact of inorganic leachables is especially apparent in biologic products; proteins can have significant interaction with metal ions. *In vivo*, about one-third of the known proteins require metal ions to perform their functions. In general, the metal ions either bind to catalytic sites to regulate protein functions, bind to structural sites to stabilize protein structure, or induce conformational changes. Alkali and alkaline earth metals bind proteins predominately through electrostatic interactions, while transition metals such as iron(II/III), cobalt(II), copper(II), aluminum(III), and manganese(II) covalently bind to proteins. Proteins can also function as chelates if a number of carboxyl groups in close proximity exist so as to form complexes analogous to EDTA. Trace levels of metal ions as contaminants in a biologic product can cause degradation via different mechanisms, such as protein oxidation, fragmentation, aggregation, or formation of insoluble particles; typical metal ions of interest are iron, cobalt, nickel, zinc, and manganese.[35] Case studies have shown the criticality of inorganic contaminants in therapeutic biologic proteins as cited in the following two cases discussed by Markovic:

1. A therapeutic protein product was changed from a lyophilized to a liquid presentation. Due to this change, a divalent metal cation migrated from the rubber stopper into the drug product vehicle. The released metal cation activated a metalloprotease (a process-related impurity that collated with the active pharmaceutical ingredient [API]) causing N-terminal degradation of the product. The problem was uncovered during stability studies under inverted conditions and was resolved by adding a chelator (e.g., EDTA) to the drug product formulation. Unfortunately, the new formulation was associated with an adverse safety outcome recognized by an increase in cardiovascular events as well as changes in the pharmacokinetic properties of the drug. This formulation was consequently withdrawn and replaced with the original. The leaching of the divalent metal cations was mitigated by implementing a modification in the elastomeric closure, which is now coated with PTFE.[36]

2. A change in a primary container closure vendor for glass vials led to barium leaching from the glass into the final drug product. As a result, visible particles were observed at the 18-month stability time point with no out-of-specification results for other physicochemical parameters tested. Upon

further examination, the particles were identified as barium sulfate crystals formed as a result of barium leachable from the glass vials reacting with the sodium sulfate excipient in the drug product formulation. While the resolution of the issue is in progress, the sponsor established low acceptance limit for barium in new glass vials and proposed to generate additional stability data.[37]

20.5 CONCLUSION

Inorganic leachables can originate from packaging components (primary, secondary, intermediate) during storage or manufacturing and may pose a safety risk to patients by causing toxicity, carcinogenicity, immunogenicity, and/or endocrine problems. They may also alter the physicochemical properties of the drug product by direct interaction with the API or indirectly by interaction with excipients in the drug product vehicle and adversely affecting drug product quality.[35] Metal ions and trace elements can affect the stability of the formulation and could catalyze the degradation of the API. The degradation products may also pose a toxicity threat. Metal ions from packaging, such as aluminum, cadmium, chromium, copper, lead, manganese, and zinc, have been shown to have both acute and chronic health effects, and repeated exposure to toxic metals leaching from the package may cause adverse effects in the patient.[38]

At trace levels, potential inorganic leachables may not be readily soluble in the extracting solution and therefore are not detected. Interferences from other species can mask trace levels of elements of interest even when they are present. There are several analytical techniques suitable for leachables and extractables evaluations, each having unique benefits and limitations. A combination of methods is often needed to be representative of potential and actual leachables however; a critical factor to be considered in these evaluations is the solubility of the inorganic species. A well-designed extractables evaluation starting with a basic understanding of materials and inorganic species in a given material will provide a baseline of targets to be referenced in order to determine the appropriate action for the assessment of inorganic leachables. Undetected inorganic leachables species can have a significant impact on clinical safety and efficacy of a drug product, and the approach to verifying the presence or absence of such leachables has distinctive challenges to recognize and overcome.

ACKNOWLEDGMENTS

The authors thank West Pharmaceutical Services, Inc., for providing the photographs.

REFERENCES

1 Sosman, R.B. New and old phases of silica. *Trans Br Ceram Soc* 1955, *54*, p. 655–670.
2 Shand, E.B. *Glass Engineering Handbook*. McGraw-Hill, New York, 1958.

3 Wilkinson, J.G. *The Manners and Customs of the Ancient Egyptians*, New edition, revised and corrected by Samuel Birch. John Murray, London, 1878: Vol. 2, Chap. 7, p. 11 n. 2; Chap. 9, p. 142.

4 Phillips, C.J. *Glass, the Miracle Maker*, 2nd ed. Pitman Publishing Corporation, New York, 1948.

5 Tooley, F.V., ed. *Handbook of Glass Manufacture*. Ogden Publishing Company, New York, 1953.

6 ASTM C225–85. Standard test methods for resistance of glass containers to chemical attack, 2004.

7 ASTM E438–92. Standard specifications for glasses in laboratory apparatus, 2006.

8 U.S. Pharmacopeia (USP). Chapter 660. Containers-glass, 2007.

9 European Pharmacopoeia 7.0. Section 3.2.1, 2011.

10 International Organization for Standardization (ISO). ISO 4802-1:2010(E) Determination by titration and method and classification. TC 76: Glassware—Hydrolytic resistance of the interior surfaces of glass containers, 2010. Available at: http://www.iso.org/iso/iso_catalogue/ and search for ISO 4802-1:2010 (accessed August 24, 2011).

11 International Organization for Standardization (ISO). ISO 4802-2:2010(E) Determination by flame spectrometry and classification. TC 76: Glassware—Hydrolytic resistance of the interior surfaces of glass containers, 2010. Available at: http://www.iso.org/iso/iso_catalogue/ and search for ISO 4802-2:2010 (accessed August 24, 2011).

12 International Organization for Standardization (ISO). ISO 719-1985(E) Method of test and classification. TC 48/SC 5: Glass—Hydrolytic resistance of glass grains at 98°C, 1985. Available at: http://www.iso.org/iso/iso_catalogue/ and search for ISO 719-1985 (accessed August 24, 2011).

13 International Organization for Standardization (ISO). ISO 720-1985(E) Method of test and classification. TC 48/SC 5: Glass—Hydrolytic resistance of glass grains at 121°C, 1985. Available at: http://www.iso.org/iso/iso_catalogue/ and search for ISO 720-1985 (accessed August 24, 2011).

14 Japan Pharmacopoeia, JP XV. Section 7.01, 2006.

15 Griffith, A.A. The phenomena of rupture and flow in solids. *Philos Trans R Soc Lond A* 1921, *221*, pp. 163–168.

16 Iacocca, R.G. and Allgeier, M. Corrosive attack of glass by a pharmaceutical compound. *J Mater Sci* 2007, *42*, pp. 801–811.

17 Schott, A.G. SCHOTT Type 1 plus container with quartz-like inner surface. Available at: http://www.schott.com/korea/korean/download/type1plus_eng.pdf (accessed August 24, 2011).

18 Simonds, H.R. and Church, J.M. *The Encyclopedia of Basic Materials for Plastics*. Reinhold Publishing Corporation, New York, 1967.

19 General chapter on inorganic impurities: Heavy metals. USP Ad Hoc Advisory Panel on Inorganic Impurities and Heavy Metals and USP Staff, USPNF, 2008.

20 Code of Federal Regulations. Indirect food additives: Adjuvants, production aids and sanitizers: Colorants for polymers. Title 21 Part 178, Section 3297.

21 Gächter, R., Müller, H., and Klemchuck, P.P. Stabilizers, processing aids, plasticizers, fillers, reinforcements, colorants for thermoplastics. In: *Plastics Additive Handbook*, 4th ed. Hansen Gardner Publishers, 1993.

22 Soroka, W. *Fundamentals of Packaging Technology*. Institute of Packaging Professionals, Naperville, IL, 1999.

23 Seymour, R.B. The role of fillers and reinforcements in plastics technology. *Polym Plast Technol Eng* 1976, *7*(1), pp. 49–79.

24 ASTM D-338-96. Standard terminology relating to plastics, 1996.

25 Wright, R.E. Reinforced plastics and composites. In: Harper, C.A., ed. *Modern Plastics Handbook*. McGraw-Hill, New York, 2004.

26 Bauer, E.J. *Pharmaceutical Packaging Handbook*. Informa Health Care, New York, 2009.

27 Deans, D.A., Evans, E.R., and Hall, H. *Pharmaceutical Packaging Technology*. Taylor & Francis, London and New York, 1996.

28 U.S. Pharmacopeia (USP). Chapter 231. Heavy metals, 2008.

29 Wei, L., Swift, R., and Torraca, G. Root cause analysis of tungsten: Induced protein aggregation in prefilled syringes. *PDA J Pharm Sci Technol* 2010, *64*(1), pp. 11–19.

30 Bolshakov, A.A., Ganeev, A.A., and Nemets, V.M. Prospects in analytical atomic spectrometry. *Russ Chem Rev* 2006, *75*, pp. 289–302.

31 Zuccarello, D.J. and Murphy, M.A. Comprehensive approach for the determination of extractable and leachable metals in pharmaceutical products by inductively-coupled plasma. *PDA J Pharm Sci Technol* 2009, *63*(4), pp. 339–352.

32 Kingston, H.M. and Haswell, S.J. *Microwave-Enhanced Chemistry—Fundamentals, Sample Preparation, and Applications*. American Chemical Society, Washington, DC, 1997.

33 Montaser, A. and Golightly, D.W. *Inductively Coupled Plasmas in Analytical Atomic Spectrometry*, 2nd Revised and Enlarged Edition. Wiley-VCH, New York, 1992.

34 Fries, A. Drug delivery of sensitive biopharmaceuticals with prefilled syringes. *Drug Deliv Technol* 2009, *9*(5), pp. 22–27.

35 Zhou, S., Lewis, L.M., and Singh, S.K. Metal leachables in therapeutic biologic products: Origin, impact and detection. *Am Pharm Rev* 2010, *May–June*, pp. 11–12.

36 Markovic, I. Risk management strategies for safety qualification of extractable and leachable substances in therapeutic biologic protein products. *Am Pharm Rev* 2009, *May–June*, pp. 96–101.

37 Markovic, I. Challenges associated with extractable and/or leachable substances in therapeutic biologic protein products. *Am Pharm Rev* 2006, *September–October*, pp. 20–27.

38 Fliszr, K.A., Walker, D., and Allain, L. Profile of metal ions leached from pharmaceutical packing materials. *PDA J Pharm Sci Technol* 2006, *60*(6), pp. 337–342.

CHAPTER 21

FOREIGN PARTICULATE MATTER: CHARACTERIZATION AND CONTROL IN A QUALITY-BY-DESIGN ENVIRONMENT

James R. Coleman, John A. Robson, John A. Smoliga, and Cornelia B. Field

21.1 INTRODUCTION: FROM QUALITY BY TESTING (QbT) TO QUALITY BY DESIGN (QbD)

The effects of foreign particulate materials (FPMs, or foreign particulate matter [PM]) in parenteral pharmaceutical products have been recognized for some time, and the testing of solutions for FPMs has received careful and extensive evaluation.[1,2] For parenterals and ophthalmic solutions, only particle size was considered important because it is the physical size of the particles that influences where in the circulatory system or in which tissues the particles are trapped and could cause damage. The composition of the particles was of interest only if they posed a toxicity hazard. The United States Pharmacopeia (USP) <788>[3] addresses "PM in injections" and recommends 100% inspection with prescribed lighting conditions for large (i.e., ≥50 μm) particles. According to Barber,[1] visual examination of solutions under standard lighting conditions would detect most particles of size ≥50 μm (unless the particle is transparent or similar in color, or color and refractive index, to the medium in which it is found).

Historically, counting the number of particles in particular size ranges constituted a test for acceptance or rejection of individual (transparent) drug product bottles or vials. One hundred percent inspection could be carried out, and individual bottles or vials containing excessive numbers of particles could be rejected. The quality of the product was achieved by discarding any unacceptable individual units. In this way, quality would be "tested in" resulting in "QbT." This approach attempted

Leachables and Extractables Handbook: Safety Evaluation, Qualification, and Best Practices Applied to Inhalation Drug Products, First Edition. Edited by Douglas J. Ball, Daniel L. Norwood, Cheryl L.M. Stults, Lee M. Nagao.

to compensate for manufacturing processes that yielded products highly variable in particle content.

Smaller particles in injection solutions are also of concern. Particles ≤50 μm are termed "subvisible" and must be analyzed by methods that can resolve particles in this size regime. Microscopy and light obscuration are the most frequently employed technologies for this task, and these technologies cannot analyze particles in solutions *in situ*. Samples of solutions must be removed from the container for analysis, and this eliminates the option of 100% inspection. Batches of product will be released, or not, based on the numbers of particles in representative samples. Interest in small particles has also been influenced by studies of the United States Environmental Protection Agency (EPA) on small- (less than 2.5 μm, and referred to as $PM_{2.5}$) and medium-size (less than 10 μm, and referred to as PM_{10}) airborne particles. These airborne particles are identified by aerodynamic properties and not identified by composition, but EPA studies have discovered that elevated levels of these particles are associated with a variety of undesirable health conditions.

Respiratory conditions are often treated by inhaled medicines that must have aerodynamic particle sizes in the same small range (2.5–10 μm). Increasing interest in inhalation as a drug delivery route[4,5] has resulted in greater interest in foreign particulates in orally inhaled and nasal drug products (OINDPs). Since OINDPs are designed to deposit active pharmaceutical ingredient (API)-containing particles or droplets in the respiratory tract, foreign particulates in the "respirable range" might be delivered to the same sites as the API. Consequently, regulatory agencies have shown interest in knowing the physical properties and chemical compositions of the contaminating foreign particles. This interest derives from the possibilities that the particles, as a consequence of their chemical compositions as well as physical properties, could exert effects either on or through the respiratory tract. For example, asbestos particles could lead to asbestosis or mesothelioma, particles of animal origin could trigger an immune response, and metals such as nickel could result in an allergic response or malignancy.[6]

QbT with respect to foreign particles in OINDPs is generally not practical for a process that produces batches with highly variable FPMs as is sometimes the case with OINDPs. It is difficult to confidently and consistently assure, as well as improve, the quality of OINDPs produced by such a manufacturing process when analysis is performed on only a few samples per batch. Furthermore, batch testing for foreign particles, the practice of accepting or rejecting a batch of product based on analysis of representative samples, does not provide a means to improve the manufacturing process. However, the concept of QbD, which was introduced in the International Conference on Harmonisation (ICH) Q8 and Q9 and has been advocated by the United States Food and Drug Administration (FDA), could alter the approach to control FPMs.[7–9]

QbD is aimed at promoting improved products and product manufacture. Ideally, the new processes will result in products of consistently high quality that require testing only to assure that the manufacturing processes are operating as designed. To achieve the state described in the FDA QbD initiative, pharmaceutical manufacturers will have to

- develop and accumulate extensive knowledge about critical product and process parameters during the development stage of a new drug product;
- design controls and testing based on the limits of scientific understanding using this knowledge; and
- utilize knowledge gained over the life cycle of the product to operate in an environment of continuous improvement.[7-9]

In contrast to "testing in," the QbD approach requires that products and manufacturing processes be designed to operate in a manner that consistently yields products within the appropriate acceptance limits, and in this way, quality is "designed in." Consequently, the capabilities of the production process and its operational limits must be well understood in terms of science and engineering to meet the goal of consistently producing acceptable products. In relation to FPMs, the production process must be sufficiently well understood so that the numbers of particles in various size categories and the numbers of particles of various compositions will be within control limits. To design in this type of control, it is necessary to know what sizes and what kinds of FPM occur during the production process. Thus, during the product development phase, the production process must be characterized in terms of numbers and compositions of FPMs. Efforts to control and reduce FPMs will require knowledge of the sources from which these FPMs originate.

To "design quality in," with respect to foreign PM, it is necessary to gain sufficient understanding of FPM types and sources so that the manufacturing process will operate within control parameters in a way that minimizes the introduction of FPMs. A fundamental aspect of this philosophy is the requirement that tests will be implemented to indicate whether the process is operating within the design parameters. If the process shows an excursion outside the control limits, there should be sufficient understanding of the process that it can be brought back into control rapidly and directly.

One goal of the FDA QbD initiative is for the industry to "design controls and testing based on the limits of scientific understanding at development stage."[8] In other words, it would no longer be sufficient to simply reject a batch of product that does not meet a specification for FPM content. QbD requires that the manufacturing process must be reviewed and brought into control through the application of corrective actions and/or preventive actions.

This chapter will (1) describe the best practices for understanding the number, composition, and sources of FPM, and suggest ways to exert control over the characteristics of the manufacturing processes that significantly impact FPM variability; (2) discuss the critical aspects of FPMs—particle size, particle number, particle composition, and particle sources—that must be assessed and understood to develop a pharmaceutical production process that effectively controls FPM; and (3) describe considerations for selection and application of analytical approaches that may be routinely required to evaluate these critical aspects of FPMs.

21.2 INTERNATIONAL PHARMACEUTICAL AEROSOL CONSORTIUM FOR REGULATION AND SCIENCE (IPAC-RS) OINDP "BUILDING QUALITY IN" APPROACH

Foreign PM can impact the functionality and safety of parenteral and inhalation products. For the control of FPMs, both quality and safety must be considered. Quality considerations for parenteral and OINDPs require clean manufacturing environments; aseptic handling procedures for specific products; and control of device/container closure system fabrication, manufacturing, packaging, and storage processes that generate PM. Consistent with the QbD concept, control of foreign PM is best attained through a "rational design" process that requires knowledge of the number, size, and kinds of particles that are present during normal operation of the manufacturing process. This information can be used to set specifications that will signal an out-of-control condition.[10] In essence, this is the process recommended by the IPAC-RS Working Group on Foreign Particulate Matter in their best practices papers.[6,11]

The rationale pursued by the IPAC-RS Working Group, though developed for OINDPs, is applicable to parenteral products as well. For OINDPs, IPAC-RS recommends that initial investigations begin with the drug product during the development phase. Parenteral products should be studied in the same way. The object of the initial focus of investigation should be the dose received by the patient. For parenterals, this is the solution to be injected or, if ophthalmic, the solution to be applied, and for OINDPs, the emitted or delivered dose. In some cases, alternative sampling procedures may also be appropriate.

In the development phase, the identity of the foreign particles must be established. Identity must be determined for individual particles and involves determining the size of a particle, and some chemical information about it. The identification must be sufficient to establish the source of the particle. Identification in terms of general material class, such as elastomer or synthetic fiber, may be sufficient, while some investigations may require greater specificity such as identifying the material as polypropylene or "type 304" stainless steel, for example, to establish the source.

The population of these particles is characterized by the numbers of each type of particle in specific size ranges. The size ranges for parenterals extend from ≥ 0.1 to ≤ 100 μm, or greater.[12,13] For OINDPs, the three ranges from ≥ 2 to ≤ 10 μm, >10 to ≤ 25 μm, and >25 to 100 μm are appropriate.[6] Particles smaller than approximately 2 μm are generally believed to be exhaled before they can settle.

One of the first steps after gathering the information mentioned above is to conduct a safety evaluation of the materials identified. Special consideration must be given to eliminating particles composed of or containing harmful or potentially harmful materials. During this phase of the overall strategy, the development team must be aware of the sources of foreign particulates so that appropriate specifications may be formulated for eventual manufacture. For a risk-based approach, these steps are the following:

- Identify potential sources of particulate contamination.
- Predict the likely effects of the PM.
- Indicate which steps of the overall manufacture must be closely monitored.

This experience-based information will provide the strategy for an effective manufacturing process that will control foreign particulates.

21.3   WHAT IS A FOREIGN PARTICLE?

21.3.1   What Is a Particle?

Although "everyone" thinks they know what a particle is, consensus agreement on a formal definition can be difficult to obtain. Barber,[1] referring to parenteral solutions, stated that "a particle is simply a physical object that is small in relation to the total volume in which it resides," a statement that provides a sufficiently comprehensive definition to include several other definitions. A particle must have other characteristics to be considered of interest in the pharmaceutical field. The small physical object must be insoluble in its drug product environment, must have distinct boundaries, must be three-dimensional, and must have a defined shape (in contrast to a droplet or amorphous smear of a liquid material). For our purposes, size limits will be determined by the proposed use of the particle-containing product. However, defining the concept of "size" will be dealt with in a later section.

It is important to recognize that the measurements of particle size and number are measurements of physical characteristics, and thus must be treated differently than chemical analyses. This difference will be cited repeatedly when relevant in this chapter. The question of chemical analysis of particles is a specific sort of chemical measurement, a microanalysis, restricted to small individual particles usually in the presence of other particles with differing compositions.

21.3.2   How to Define "Foreign"

"Foreign" in relation to pharmaceutical products has several working definitions that may not be mutually exclusive, but are appropriate to particular situations. The definitions provided in USP General Test Chapters <788> (Particulate Matter in Injections)[3] and <789> (Particulate Matter in Ophthalmic Solutions)[14] state the following:

Particulate matter consists of mobile, randomly sourced, extraneous substances, other than gas bubbles, that cannot be quantitated by chemical analysis because of the small amount of material they represent and because of their heterogeneous composition.

The broadest interpretation of this statement is to call particulate material foreign if it is material that was not deliberately designed into the formulation. If material generated during device/container closure system fabrication (e.g., metal flakes or glass shards) remained in, or on, the device/container closure system, that material would be considered foreign as cleaning prior to filling should remove such particles. Particulate material entering the product during drug product vehicle, API, or excipient manufacture would also be foreign. If it was known that an elastomer gasket would normally shed elastomer particles as a result of use, these would not be considered foreign. They would be considered a normal component of the drug product in the same sense that organic and inorganic leachables would be considered "normal." This is not intended to suggest that such particles should not be characterized and controlled in the manner similar to the way extractables/leachables are controlled. Should some component of the formulation precipitate and form particles, these would not be considered foreign, but could still be a problem. For example, there is concern that protein molecules used in protein therapeutics may aggregate to form particles of sufficient size to occlude capillaries, or may provoke immune responses.[12,13] Material that existed as a particle in the medium of the drug delivery system, but was soluble in the aqueous medium of the airways, would meet the criteria to be considered a particulate for this discussion, especially if the physical characteristics of the material as a particle determined its distribution in the pulmonary system.

Many products (e.g., paints, coatings, abrasives, and pigments), as well as many pharmaceutical products, are manufactured from materials in a powdered form, and remarkable progress has been made in developing methods to characterize powdered or particulate materials. However, it is important to note that the analysis of FPMs differs in several significant ways from the characterization of bulk particulate materials used in manufacture. Notably, as contaminants, foreign particulates exist as a minority component in a drug product. The fact that they are a minor particulate component has implications for sampling as will be discussed later.

Unlike particles used in manufacture that share common chemical and physical properties, foreign particulates are usually heterogeneous and can be of almost any composition. They must be analyzed individually, or, in some cases, they may be separated into classes (e.g., all ferromagnetic particles or all particles with a density greater than 2.5 g/cm^3) and then analyzed as individuals. The mean properties of a collection of foreign PM are unlikely to provide the information necessary for controlling their presence in a drug product. Knowledge of the distribution of sizes and compositions are crucial for control. Individual particles may differ significantly in size, shape, density, or composition from the medium in which they are found, and if so, may be readily identified. However, if they are similar in density, color, shape, or size to the bulk particulate material in which they reside, they may not be obvious. As an additional challenge, since they tend to be heterogeneous, some may be similar to the medium in which they occur, while others are quite different. Consequently, it can be misleading to identify only the most obvious particles without considering that other, less obvious, particles may also be present.

21.3.3 Foreign Particulate versus Airborne Particulate Pollution

There exists some confusion about the similarities and differences between "airborne particulate pollution" as defined by the EPA, and what is defined as foreign PM in OINDPs. According to the EPA,[15] PM is a complex and variable mixture of extremely small *airborne* particles and liquid droplets, also known as "particle pollution." Particle pollution is made up of a number of components, including acids (such as nitrates and sulfates), organic chemicals, metals, and soil or dust particles. They may come from combustion (e.g., forest fires, power plants, internal combustion engines), agitation of preexisting particulates such as stone dust and sand, or agricultural tilling and harvesting of trees. They may serve as nuclei that pick up constituents such as oxides of sulfur (SOx), and nitrogen (NOx) and/or hydrocarbons from the atmosphere.

In contrast, foreign particles in pharmaceutical products are likely to be derived from a relatively small number of sources such as the APIs, excipients, device/container closure system components, other formulation ingredients, or the manufacturing/filling environment. As stated by Blanchard et al.,[11] "foreign particles in orally inhaled and nasal drug products (OINDPs) are contaminant particles that may be derived from the active, excipients, container/closure components, formulation, environment, and/or the process of actuating the drug product device."

21.4 PARTICLE SIZE

21.4.1 Particle Sizes of Interest

The potential for causing respiratory health problems is strongly linked to particle size. For OINDPs, the categories of airborne small particles that are of concern are particles that are ≤ 10 μm in diameter and ≥ 2.5 μm (defined by the EPA[15] as PM_{10}). "Fine particles," labeled by the EPA[15] as "$PM_{2.5}$," such as those found in smoke and haze, are less than 2.5 μm in diameter. These particles can be directly emitted from sources such as forest fires, or they can form when gases emitted from power plants, industries, and automobiles react in the air. $PM_{2.5}$ particles are thought to be too light to settle in the respiratory system, and most are likely to be expired if inhaled. Recently, Phelan[16] reviewed the health effects of these particles. The PM_{10} particles are of interest because they constitute the "inhalable coarse particles," the particles that generally pass through the throat and nose and enter the lungs.

It is generally accepted that particles greater than 10 μm in diameter are deposited in the nose and pharynx, and thus are not carried into the lungs. Particles that dissolve in the aqueous medium of the nose and pharynx may be absorbed and, depending on the composition, could be toxic. On the other hand, insoluble large particles are likely to be carried into the gastrointestinal tract and again, depending on composition and amount, could be benign or toxic.[6] Airborne particulates originate from a wide variety of sources: agriculture, forestry, power plants, construction, combustion, and deterioration of such items as automobile tires, roads, and buildings. The smallest class of airborne particulates, $PM_{2.5}$, tends to originate from the coagulation of smaller particles from photochemical smog and gas-to-particle conversion, as well as directly from combustion. PM_{10} particles tend to originate directly from mechanical dispersion and the resuspension of particulate materials.[17] Consequently, the chemical composition of particles in these two different size classes is likely to be different. In products manufactured under conditions in which the number of particulates is controlled (e.g., "clean rooms" or sanitary conditions), the most likely origin for contaminant particulates will be the same (as discussed at length in Section 21.7.3 "Sources of FPMs"), and so the composition of the particle ensemble ≤10 and ≤2.5 μm will be the same.

21.4.2   Particle Size Definitions

Analogous to the definition of particle, "everyone" thinks they know what "size" means, but for more rigorous usage appropriate for analytical purposes, there are many definitions, each useful in certain circumstances (see Table 21.1). When dealing with bulk materials such as APIs and excipients, physical particle size may have different useful definitions depending on the technology used to measure size.

TABLE 21.1.   Some Generally Accepted Definitions of Particle Size

Name	Definition
Aerodynamic diameter	Diameter of a sphere having a density of 1 g/cm^3 that has the same inertial properties as the particle
Volume diameter	Diameter of a sphere having the same volume as the particle
Surface diameter	Diameter of a sphere having the same surface area as the particle
Surface volume diameter	Diameter of a sphere having the same surface area to volume ratio as the sphere
Drag diameter	Diameter of a sphere having the same resistance to motion in a fluid of the same viscosity and at the same velocity
Free-falling diameter	Diameter of a sphere having the same density and the same free falling velocity as the particle in a fluid of the same density and viscosity
Stokes diameter	Diameter of a spherical particle with the same density and settling velocity as the particle
Projected area diameter resting	Diameter of a circle having the same area as the projected area of the particle in a stable resting condition
Projected area diameter random	Diameter of a circle having the same area as the projected area of the particle in random orientation
Perimeter diameter	Diameter of a circle having the same perimeter as the projected outline of the particle
Sieve diameter	The width of the minimum square aperture through which the particle will pass
Feret diameter	The mean value of the distance between pairs of parallel tangents to the projected outline of the particle
Maximum feret diameter	The maximum distance between pairs of parallel tangents to the projected outline of the particle
Minimum feret diameter	The minimum distance between pairs of parallel tangents to the projected outline of the particle
Martin diameter	The chord dividing the projection area of the particle into two equal halves

Two general types of size analysis are typically employed: ensemble methods, which depend on the radiation scattering properties of a suspension of particles, and individual measurements, in which the sizes of individual particles are determined one by one to provide a particle size distribution. Ensemble methods, such as laser light diffraction and light scattering are widely used when determining the sizes of particles with similar composition. Results of these measurements are usually expressed in terms of a calculated equivalent circular or spherical diameter. The most widely used individual methods are light obscuration, electric zone sensing (Coulter principle), and microscopy. Light obscuration derives an equivalent size for particles from the quantity of light deflected by individual particles as they pass one by one between a light source and light sensor. Electric zone sensing derives an equivalent size by the change in conductance as an individual particle moves through an aperture separating two volumes of suspending medium. Microscopy provides two-dimensional projections of individual particle area, which may be extrapolated to equivalent circular diameter, or to one of the many other expressions of size (Table 21.1).[18,19]

When dealing with foreign particulates, particles usually have optical and physical properties that differ not only from the bulk material but also from each other so that ensemble size distribution measurements are difficult to interpret, even in those cases where the foreign particles can be separated from the bulk material. When dealing with inhaled or airborne particles, true physical size is only one aspect of what is referred to as "size." Aerodynamic size, usually determined using a cascade impactor, is a concept that takes into account shape and density. In many cases, it is useful to relate particle size measurements made with one technology to those made with another. Although no one measure is perfect, the concepts of aerodynamic diameter and Stokes diameter are useful when comparing values from different particle sizing technologies.

The aerodynamic diameter of a particle is the diameter of a particle of density 1 g/cm^3 that would have the same aerodynamic properties as the particle of interest, which has a *different* density. The Stokes diameter is the diameter of a sphere of the *same* density that would have the same aerodynamic diameter as the particle of interest. This relationship is expressed in Equation 21.1:

$$d_{pa} = d_{ps} \times (\rho_p)^{1/2},$$
$$d_{ps} = d_{pa} / (\rho_p)^{1/2},$$

(21.1)

where

d_{pa} = aerodynamic particle diameter (μm),

d_{ps} = Stokes particle diameter (μm), and

ρ_p = particle density (g/cm^3).

For smooth spherical particles, the Stokes diameter is equal to the physical diameter of the particle. For other shapes and for particles with surface features that will influence the movement of a particle through air or liquid, the Stokes diameter will differ from the actual physical diameter of the particle. Examples of this relationship are seen in Figures 21.1–21.3.

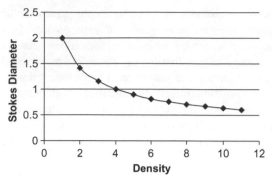

Figure 21.1   Stokes diameter versus density for particles with aerodynamic diameter = 2 μm.

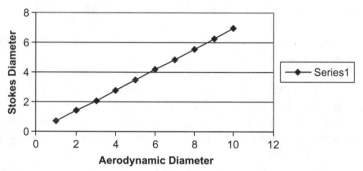

Figure 21.2   Stokes diameter versus aerodynamic diameter for particles with density = 2 g/cm³.

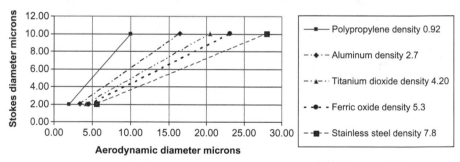

Figure 21.3   Stokes diameter versus aerodynamic diameter for particles of different densities.

Figure 21.1 shows how the Stokes diameter of a particle diminishes rapidly as the inverse of the square root of the density. Figure 21.2 shows how the aerodynamic diameter is linearly proportional to the Stokes diameter for particles of uniform density. The significance of this relationship is that the aerodynamic diameters of particles having higher density will have smaller Stokes diameters than those of lower density. A graphic demonstration of these relationships is given in Figure 21.3. Particle densities are chosen as representative of materials likely to be encountered as FPMs. For example, as the graph shows, particles with a density of 0.92 g/cm^3 and aerodynamic diameters between 2 and 10 µm will have Stokes diameters of similar dimensions. However, particles with a density of 7.8 g/cm^3 and aerodynamic diameters between 5.6 and 28 µm will have Stokes diameters between 2 and 10 µm. This distinction is important to keep in mind since microscopic methods as well as light scattering and light obscuration methods are likely to provide dimensions closer to Stokes diameters than aerodynamic diameters.

21.5 ALLOWABLE LIMITS

21.5.1 Parenteral and Injectable Solutions

For parenteral and injectable solutions, USP <788>[3] distinguishes the following classes:

- preparations supplied with a nominal volume or more than 100 mL;
- preparations supplied with a nominal volume of 100 mL; and
- preparations supplied with a nominal volume of less than 100 mL.

The USP criteria for acceptance are as follows:

When measured by light obscuration

Test 1.A

Supplied as a solution with volume >100 mL:

Not more than 25 particles/mL ≥ 10 µm and

Not more than 3 particles/mL ≥ 25 µm

Test 1.B

Supplied as a solution with volume <100 mL:

Not more than 6000 particles/container ≥ 10 µm and

Not more than 600/container ≥ 25 µm

Preparations supplied with a nominal volume of 100 mL will comply by satisfying either test 1.A or 1.B.

USP <789>[14] for ophthalmic solutions has similar size ranges of interest, with the exception that the limits are not more than 50 particles ≥ 10 µm per mL and not more than 5 particles ≥ 25 µm per mL.

When measured by microscopy

Test 2.A

Supplied as a solution with volume >100 mL:

Not more than 12 particles/mL ≥ 10 µm and

Not more than 2 particles/mL ≥ 25 µm

Test 2.B

Supplied as a solution with volume <100 mL:

Not more than 3000 particles/container ≥ 10 μm and

Not more than 300/container ≥ 25 μm

Preparations supplied with a nominal volume of 100 mL will comply by satisfying either test 2.A or 2.B.

21.5.2 OINDPs

Unlike the situation with parenteral products where the USP has established limits, as of this writing, the FDA has not published limits for FPMs in OINDPs. However, because particles ≤10 μm can be found to have deposited in the airways and alveoli of the lungs, particles in this size category must be considered for the safety of patients. The EPA has established the National Ambient Air Quality Standard (NAAQS)[20] that proposes standards for inhalation of particles that will be safe even for sensitive subpopulations such as children, the aged, and those suffering from diminished pulmonary function. These standards are based only on the mass of particles, and do not refer to chemical composition. Consequently, the standards are based on the premise that the harmful effects of particles are based only on the mass inhaled, and not dependent on chemical composition, or physical properties (other than aerodynamic diameter).

In 2004, IPAC-RS suggested that a limit of ≤1%–5% of the NAAQS[20] for foreign PM having an aerodynamic diameter ≥2 and ≤10 μm be used as a safety limit.[11] Limits may also be set for other size ranges, but these will be based on quality concerns rather than safety concerns. In a subsequent publication,[6] health effects of specific particulate materials were considered in terms of eliciting an immunologic response, a nonimmunologic response, or being "clinically" inert. The report states that to understand the clinical ramifications of foreign particulates in OINDPs, a full characterization of the particles is essential to estimate the potential for injury. However, for nondescript particles, assumed to have no specific toxicological effects, it is possible to estimate the maximum allowable dose of particles classed as foreign PMs. Table 21.2 shows the calculated allowable dose depending on density and

TABLE 21.2. Maximum Allowable Number of Particles of a Given Aerodynamic Diameter and Density to Meet the Criterion of ≤50 μg Additional Particulate Matter per Day

Diameter (μm)	Density (g/cm³)			
	1	2	4	8
2	1.2×10^7	6×10^6	3×10^6	1.5×10^6
3	3.5×10^6	1.8×10^6	8.8×10^5	6×10^5
4	1.5×10^6	7.5×10^5	3.7×10^5	1.9×10^5
6	64.4×10^5	2.2×10^5	1.1×10^5	5.5×10^4
8	1.9×10^5	69.3×10^4	4.7×10^4	2.3×10^4
10	9.6×10^4	4.86×10^4	2.4×10^4	1.2×10^4

aerodynamic size.[6] Since the NAAQS has relaxed its standard to not more than 150 µg/day, the values in the table provide an even greater safety limit than was originally proposed.

21.6 PARTICLE SIZING TECHNIQUES

As mentioned previously, two "ensemble" methods, laser diffraction and light scattering, are widely used for the sizing of particulate materials. These are termed "ensemble" methods to indicate that they are based on the collective properties of a suspension, rather than the analysis of particles individually. Light scattering and laser diffraction employ different interactions of light with particles. Light scattering depends on the form, size, and composition of the particles, and is useful for homogeneous particles with similar physical characteristics, such as those likely to be encountered in APIs and excipients, but is not recommended for heterogeneous particles such as those in contamination situations. Laser diffraction depends on the diffraction phenomenon that occurs at the edges of particles, and thus is largely independent of composition effects. However, both technologies are "ensemble" techniques and depend on the combined contributions from many particles analyzed simultaneously; thus, they are not suited for small numbers of particles of widely varying sizes and compositions.[2,21-23]

Electrozone technologies analyze one particle at a time and calculate a size measure of "equivalent spherical volume" or "equivalent spherical diameter," based on the current pulse when a single particle passes through the aperture. For contamination studies, the variation in shape and electrical resistance of heterogeneous particles, and the possibility of more than a single particle simultaneously passing through the aperture, makes this process subject to error. Light obscuration is designed to measure the amount of light blocked by a single particle as it passes between a light source and a light sensor. There are some limitations imposed by this arrangement: There is the potential for error from more than one particle simultaneously passing in front of the sensor, and variations in transparency, reflectivity, and color, may bias results. Comparisons of these technologies can be found in the literature.[24,25]

Both light and electron microscopy are widely employed for particle size analysis. These technologies, when carried out manually, require trained analysts and can be costly in time required to analyze a statistically significant number of particles. Fortunately, various forms of automation have been introduced to minimize the requirement for trained analysts. Automated systems that can operate unattended greatly reduce the elapsed time required for analysis. However, the process is critically dependent on sample preparation to avoid particle overlap, and on threshold settings to distinguish particles from the substrate.

The resolution limit for light or optical microscopy is usually taken to be the Rayleigh diffraction limit of about 0.25 µm in green light; however, this refers to the required distance between two particles in order to distinguish the particles from each other. A more practical limit in order to measure a diameter and to distinguish form is about 1 µm. The resolution of the scanning electron microscope is considerably greater and, depending on the configuration and the quality of the optics, may be about 20 nm for conventional tungsten electron sources or as small as 2 nm for cold field emission sources. As a result, accurate particle size analysis of 200 nm may be expected for a tungsten source and about 20 nm for a cold field emission source. Additionally, with electron beam technologies, it is possible to use backscattered electron (BSE) imaging to provide atomic number information and X-ray microanalysis by energy-dispersive X-ray spectrometry (EDS) for elemental information. Both technologies are readily adapted to scanning electron microscopes. Transmission electron microscopy is mostly used to identify asbestos fibers, and only rarely for FPMs in pharmaceuticals. The major disadvantage of any microscopic method is that the measurement is made of a two-dimensional projection of the particle, leaving the third dimension to be estimated.

21.7 PARTICLE COMPOSITION ANALYSIS

21.7.1 General

Identifying the composition of a foreign particle is important for two reasons: first, the composition tells the analyst that the particle is different from the materials designed to be in a device or solution; and second, the composition can be used to establish the source of the particle. Since a particle suspected to be foreign can be almost anything, some attention must be given to strategies that will lead efficiently to identification. Recently, Schearer[26] provided an excellent discussion of methods in use for the identification of individual particles found in pharmaceutical products. One aim of this chapter is to examine the need to quantify and identify foreign contaminant particles in relation to QbD principles. Suggested strategies are discussed in later sections. Automated particle composition analysis by scanning electron microscopy (SEM) coupled with EDS[27] and Raman microscopy[28,29] has been reported.

21.7.2 Which Particle Compositions Are of Interest to OINDPs?

Table 21.3 shows FPM compositions reported recently in intravenous solutions and OINDPs (data adapted from References 1 and 8). It is worth noting that the characterizations of these particles range from the very general (e.g., clay) to the highly specific (e.g., acrylonitrile butadiene styrene). Clearly, the analyst must be prepared to encounter a great variety of particle types, but in general, the composition of most of the FPMs will be the materials that the drug product is exposed to during synthesis and manufacture. The criterion for suitability of identification should be whether the level of identification or "speciation" is sufficient to indicate a source of the contaminant.

In an OINDP employing a stainless steel container, the presence of iron may be sufficient to point to the container as the likely source, but additional information indicating the type of stainless steel may be necessary to distinguish steel from the container versus steel from a stainless steel spring or processing machinery.

TABLE 21.3. Identities of Foreign Particulate Matter Reported in Various Sources

Acrylonitrile butadiene styrene	Polybutylene terephthalate
Aluminum	Polycarbonate
Bacterial fragments	Polychlorotrifluoroethylene
Bromobutyl rubber	Polyester
Clay	Polyethylene
Copper	Polyimide
Elastomer	Polyoxymethylene
Ethylene propylene diene polymer	Polypropylene
Glass	Polystyrene
Insect parts	Skin cells
Iron	Stainless steel
Kaolin	Starch
Metal	Talc
Paper/cardboard/cellulose	Transparent synthetic fibers
Polytetrafluoroethylene	Trichomes
Polyacetal	Unspecified inorganic
Polyamide	Zinc oxide

When a drug product contains more than one polymeric component, a method that can distinguish each of the polymer components in the drug product involved may be required. It may be necessary to enlist methods such as Raman microscopy to identify the entire list of components such as elastomer fillers or polymer pigments.

The presence of materials of biological origin as indicated by bacterial fragments, or skin cells such as mentioned in Table 21.3, represents special cases and requires specific identification methods. The presence of material of animal, plant, or bacterial origin will probably require the attention of a toxicologist or microbiologist.

21.7.3 Sources of FPMs

Several different ways to group contamination sources have been employed. Barber[1] suggested using six source classes, each with a different "fingerprint" of contaminant materials so that when a particular pattern of foreign particles appears, attention may be directed to certain likely sources. The source classes identified by Barber[1] are as follows:

1. *Workstation.* Site where product is modified, such as welding on a medical device.
2. *Equipment.* Devices used to transform material (e.g., tableting).
3. *Materials.* Items that may contact the product, such as sampling devices or packaging.

4. *Process flow.* Sites such as filters or mixers that may contribute contamination as the product passes through.

5. *Air handling.* Heating, ventilation, and air conditioning (HVAC) and filtered air anywhere that the product is exposed to environmental conditions.

6. *Personnel and clothing.* The worker and the activity of workers.

Barber[1] also used a different system of four classes, arranged from least likely to most likely as

1. *diffuse or environmental,* for example, ambient air in the process room;

2. *localized,* a small source but one producing a higher concentration than a diffuse source (e.g., a vial unscrambler that produces glass particles within a vial);

3. *point related,* involves both particle source and its proximity to the product (e.g., a stoppering punch that builds up particles during operation); and

4. *product related,* particles resulting from the instability of the product or chemical reactions of the product itself.

It has also proven useful to classify particle sources as likely to originate from either the components of the product itself or from the environment.

From components, the likely sources are

1. *drug product components,* such as APIs, excipients, and delivery devices;

2. *container components* that come into contact with the container contents such as valves, gaskets, and springs; and

3. *packaging.*

From the environment, likely sources are

1. *fabrication* of the device or container;

2. *filling,* where airborne PM may enter a container during filling, even in a clean environment; and

3. *packaging,* where materials may come into contact with the contents of the container during activation or use.

Elastomeric components are frequently the sources of FPM. Incoming elastomer components may carry particles picked up during manufacture or fabrication. Cutting or punching of elastomer sheets may generate particulates. The elastomeric particle may be primarily elastomer but may also contain nonrubber components (e.g., sulfur-containing or metal oxide cross-linking agents, sulfur-containing or peroxide accelerators, or pigments and fillers such as clay or talc), which are added to uncured rubber to control physical properties.

Polymeric FPMs can originate from particles produced by polymer components during fabrication, packaging, shipping, and assembly. Assembly of components may generate PM. Mechanical components may be shipped in polymer bags coated with materials such as mold release agents, slip agents, or extrusion aids. Waxes, for example, can migrate and segregate to form particles. Rigid polymer components can contribute PM in the form of whiskers, or flashings that are associated with molding or forming. The fabrication process can also add metal particles as a result of tool wear.

Even though components are cleaned before being employed, and filling/closure occurs in clean conditions, or more properly controlled environment areas (CEAs), personnel can be expected to be a significant source of a wide variety of particles from clothing, skin, and hair. Abrasion/wear of the clothing materials in the clean environment may also contribute to the overall numbers of FPMs. Wear of processing devices and machine parts in the clean environment may also contribute FPMs. Materials of the CEA may be a significant source of particulate materials due to continuous wear/abrasion as well as episodic release due to cleaning, repair, equipment maintenance, material import, or product export.

Contamination by microorganisms is not usually considered under the topic of FPMs because of the potential for harm from the presence of such materials. Microbial contamination is properly the area of expertise of a microbiologist and is not discussed here, except to say that if "bacterial fragment" PM is identified, as in Table 21.3, microbiological consultation is necessary not only because of potential pathogens but also because of the possibility of immune responses to bacterial components.

21.8 HOW TO ANALYZE OINDPS FOR FPMS

21.8.1 General Comments on Sample Preparation

The most challenging aspect of sample preparation is keeping the sample uncontaminated by environmental material. The presence of extraneous material can not only confuse the search for a source, but can also bias the particle counting process. Diligent attention to cleanliness of all items involved in isolation, capture, and analysis is essential. All reagents or suspending media must be filtered and demonstrated to meet established cleanliness standards. Deionized, distilled water is preferred for preparation procedures. Washing and rinsing with copious amounts of water is good practice for achieving clean glassware. Cleaned glassware should be dried in a laminar flow hood to remain particle free. It is wise to give some consideration to the number of particles likely to be required for the levels of confidence desired and to have sampling criteria in place before investigation (e.g., $\pm 3\%$ or $\pm 30\%$ for the largest or smallest category).

Clearly, the importance of the FPM will be derived from the nature and frequency of the particulates. For benign particles, wider confidence limits can be tolerated, while those materials that may have toxic effects will call for more stringent confidence intervals. An analyst may choose to consider only the total number of particles collected to set confidence intervals, but if one class of particle is of greater significance than others, then the collected number of particles in this class may be used to calculate confidence intervals.

Barber[1] also provided advice on setting alert limits based on extended experience with a product under development. He pointed out that numerous measurements during the development of the number of particles per drug product, for example, metered dose inhaler (MDI), should provide a reasonably good estimate of the "true" mean number and standard deviation of particles per drug product. Based on this information from actual measurements, it is then possible to calculate what number of items will have to be tested and to calculate an upper control limit for a sample of this size. Experienced analysts know that during development, when variations in processing are being evaluated, it is necessary to know the history of the samples analyzed in order to identify sources of contamination and to decide which variations may have a beneficial effect on contaminant particle content.

21.8.2   Sampling

Two types of sampling must be considered: sampling from product in order to identify and quantify the FPM burden, and sampling from suspected sources of FPMs to minimize contamination.

For product contamination analysis, when the number of particles per unit may be quite small, and the particles may be diverse, it is good practice to obtain a sufficiently large number of particles to establish some statistical confidence in the relative importance of each category of the diverse particles. It may be necessary to isolate FPMs from large volumes of API or excipient, or to combine the contents of several delivery devices in order to obtain a sufficient number of particles. It is possible to estimate the number of particles necessary to obtain a desired level of confidence, using the approximation that the standard deviation is roughly equivalent to the square root of the number of items counted (Table 21.4):

$$\text{standard deviation} \approx \sqrt{\text{number particles counted}}.$$

It is wise to give some consideration of the number of particles likely to be required for the levels of confidence desired and to have sampling criteria in place before investigation (e.g., ±3% or ±30% for the largest or smallest category).

Clearly, the importance of the FPM will be derived from the nature and frequency of the particulates. For benign particles, wider confidence limits can be tolerated, while those materials that may have toxic effects will call for more stringent confidence intervals. An analyst may choose to consider only the total number of particles collected to set confidence intervals, but if one class of particle is of greater significance than others, then the collected number of particles in this class may be used to calculate confidence intervals.

TABLE 21.4. Estimated %RSD as a Function of Number of Particles Counted

Number of particles collected (N)	$\sqrt{N}$	%RSD
10	3.16	31.6
100	10.0	10.0
1000	31.62	3.2

21.8.3   Methods of Isolation

When dealing with particles ≤10 μm, it is important to recognize that as physical size diminishes, chemical identification technologies become less reliable and less able to distinguish among similar substances. There are four categories of drug product that may have to be analyzed for FPMs: (1) solution, (2) dry powder, (3) suspension of particles, and (4) "suslution," a mixture of a solution with a suspension of particles. The defining characteristic of FPMs as objects that are insoluble in the aqueous environment of the lungs means that water-based isolation procedures are not likely to dissolve the particles.

To isolate FPM from a solution, it may be a simple matter to capture the particles by using a membrane filter of known pore size, usually 0.2 or 0.4 μm. An example of an apparatus used to isolate FPM by filtration from an OINDP device is shown in Figures 21.4–21.6. The solution may flow freely through the filter or flow may be aided by gentle vacuum. The use of pressure-aided filtration must be handled with caution: It is possible to deform soft particles and drive them into the pores of the filter where they may not be detected. It is possible that excessive pressure differences may also deform the pores and permit large particles to pass. If a large amount of material is trapped by the filter, the filter may be "blinded," and no longer pass fluid. If analysis will be by microscopy, then it will be necessary to capture particles as individuals and not as aggregates that have been crowded together in a way that boundaries between particles are obscured. If light obscuration is employed, the concentration of particles must be adjusted to fall within the quantitation range of the instrument.

Isolating FPM from a dry powder can be an especially challenging task. The most convenient method is usually to take advantage of selective dissolution. Since

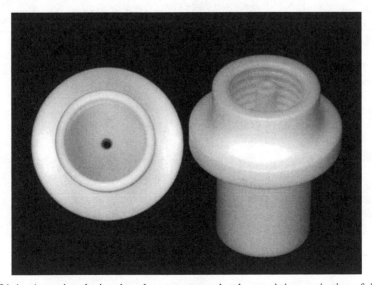

Figure 21.4   Actuation device threads onto capture bottle permitting expiration of drug product directly into solvent.

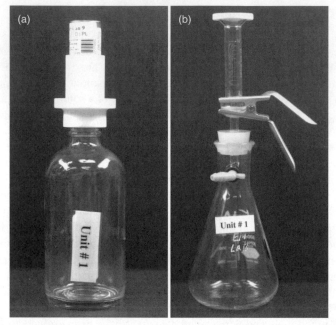

Figure 21.5   (a) MDI drug product in firing position atop capture bottle containing appropriate solvent. (b) Assembled 25-mm filter apparatus used to isolate suspended FPM from capture bottle suspension.

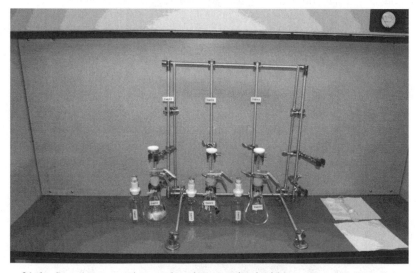

Figure 21.6   Sample preparation workstation contained within a class 100 (ISO 5) laminar flow HEPA-filtered hood.

the dry powder is likely to be water soluble in the lungs, and the FPM must be insoluble in water to remain as particles in the lungs, then simply suspending the powder in an aqueous medium may permit the capture of the FPM on a filter, as described above.

If the FPM is in a suspension of particles, then selective dissolution may offer a path to isolation of the FPM. It may be possible to identify a solvent that will dissolve the API without affecting the FPM. Detergents may offer advantages in this regard, but use of nonpolar and strongly polar solvents must be avoided to prevent dissolving foreign PM. Exposure to selective dissolution may be carried out after both FPM and API particles have been captured on a filter.

If the drug product to be analyzed contains a "suslution," a mix of material in solution and particles in suspension, a combination of selective dissolution and filtration is likely to permit isolation of the FPM

In cases where filtration is impractical, for example, if the product contains a viscous material that impedes filtration, dilution may decrease the viscosity of the interfering material, but if not, then it may be possible to separate FPM though centrifugation. Recovering all the particles from a centrifugation may be difficult. Manual isolation of individual particles, especially for particles ≤10 μm, is possible, but hardly convenient. When employing separation based on density differences, foreign PM may rise to the surface or fall to the bottom of a centrifuge tube or both. Harvesting particles from the surface or from the bottom of a centrifuge tube requires some degree of skill. Care must be taken to prevent the particles that have sedimented to the base of the tube from aggregating so tightly that they will not separate into individual particles when trapped on a filter. If removing the particles that have risen to the surface with a pipette, one must observe whether some of the particles adhere to the surface of the pipette. If so, then treating the pipette to make it more hydrophobic or more hydrophilic may help.

21.8.4   General Comments Concerning Filtration

Some words of caution are appropriate when using filtration to capture foreign PM. Choice of membrane filter material may affect the capture of FPM if some of the particles are strongly hydrophobic or hydrophilic. Choice of pore size is a compromise: Larger pores speed up filtration but may permit elongated and soft materials to pass through or become trapped in the pores. Most filtration chambers are of small volume and difficult to examine by light microscopy, so that retention of particles on chamber walls is difficult to detect. Copious washing of the filter apparatus walls with suspending medium is necessary to minimize particle adhesion to the walls of the filtration apparatus. If some particles have a greater propensity than others to adhere to the walls of the chamber, it is likely that a selective loss of this class of particles will bias the analysis. In many filtration apparatus chambers, the last small volume of fluid to pass through the filter collects as a ring around the edge of the filter where the filter touches the filter chamber. If this fluid leaves the chamber only slowly, a large number of particles may be deposited around this part of the filter resulting in a nonuniform distribution of particles on the filter. Dilution of the suspending medium may help by increasing filtration rate, but if this phenomenon cannot be avoided, some comment about the possible bias of the measurements is appropriate. A useful exercise to assess the uniformity of particle deposition is to count the number of particles in a line across the diameter of the filter. Even greater confidence can be attained if two lines across the filter diameters at 90° to each other are analyzed. If the relative standard deviation (RSD) of particles per field is less than a particular value, this may be taken as a criterion to justify analysis of a filter. If the RSD is greater than the chosen value, the filter should be discarded for nonacceptable sample preparation. The analyst must look for any tendency of particles to associate or aggregate, or if large particulates are present, that they do not fragment during sample preparation or obscure smaller particles. Finally, if examining filters with optical microscopy, the use of low-angle illumination is helpful in visualizing any particles that might be thin, transparent, or similar in color to the filter. The use of SEM can be helpful for verifying that thin or transparent particles will be counted.

21.8.5   General Comments Concerning Manual Identification and Removal

When all else fails and it is necessary to capture individual particles manually using PLM imaging, grouping particles by optical properties, such as color, opacity, reflectivity, and even shape, can make sizing, enumeration, and analysis more efficient. However, it is important to keep in mind that those particles of similar color, opacity, or reflectivity need not be of identical composition. Generally, this operation must be carried out under dust-free conditions. The process is usually slow and prone to loss of individual particles. Furthermore, because of the labor-intensive nature of the process, only a small sample of the total number of particles is likely to be analyzed, and the total number of particles will be extrapolated.

21.9   METHOD VALIDATION

21.9.1   General Comments

ICH, USP, the European Pharmacopoeia and regulatory authorities require that analytical methods be validated, but the nature of particulate materials imposes different validation criteria than those appropriate for nonparticulates such as solutions and homogeneous liquids and solids. The European-based Pharmaceutical Analytical Sciences Group (PASG) has presented a white paper that addresses the special needs of sample preparation and validation for particle characterization.[30] The recommendations in this chapter are based on the arguments in this white paper.

The analysis of FPMs differs significantly from the analysis of homogeneous materials such as solutions, and also of homogeneous powders and suspensions encountered in drug products. The primary difference is that FPMs are likely to be both heterogeneous and of unpredictable compositions. These conditions impose stringent requirements on the analytical procedures employed. The analyst must be able to gather sufficient qualitative information to identify the nature of each particle or class of particle. An experienced analyst may begin with some insight, based on experience, about which material classes are likely to be present, but in fact the analyst must deal with the probability that many different materials will be encountered.

For foreign PM characterization, three aspects of a validatable analytical procedure must be developed:

1. sample harvest and preparation;

2. particle detection, sizing, and quantification; and

3. particle speciation.

Sample harvest and preparation. Particle loss during this process should be minimized. No one class or type of particle must be selectively lost. The particles must not be altered by the preparation process. Brittle particles should not be fractured during preparation, and soft particles should not be deformed so that size and shape characteristics would be altered. The physical state of the particles must be preserved so that the preparation does not separate particles that are associated, and loosely associated particles should not aggregate. It is difficult to verify that these criteria have been met, but logical modifications of the preparation method can provide some defensible estimate of its effects.

Particle detection, sizing, and quantification. It is necessary to estimate the efficiency with which particles in a heterogeneous population are identified. For example, if some particles are transparent, they may not be detected in optical microscopes or light obscuration instruments. Very thin particles may not be detected by BSE imaging in SEM, but secondary electron (SE) imaging can often increase the contrast between a thin particle and the substrate (see Fig. 21.7). Particle detection efficiency may be estimated from repeated analyses of the same portions of the

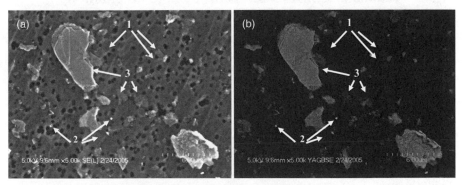

Figure 21.7   A mixture of talc and calcium carbonate FPM particles on a membrane filter. Comparison of (a) secondary electron (SE) detection and (b) backscattered electron (BSE) detection images illustrates challenges of automated analysis using BSE imaging. Notice that the very thin talc particles (1's) observed using SE detection are poorly imaged using the BSE signal. Conversely, the smaller calcium carbonate particles (2's) are clearly visible in both images. Notice that thicker particles of talc are visible in both SE and BSE images (3's).

same sample, with the proviso that the sample is not altered by the analysis method. Repeated exposure to an energetic electron beam may lead to degradation or movement. Particle sizing capability can be estimated by analysis of latex or glass sphere reference materials, but the physical attributes of these materials may be significantly different from the heterogeneous particles recovered from a drug product.

Particle speciation. Preliminary investigations to characterize the populations of particles likely to be encountered are essential during drug product development. With this characterization information in hand, qualitative analytical technologies suitable for the materials expected can be chosen. It should be possible to estimate the minimum size of particles that can be identified, and to make some judgment about the likelihood of interferences occurring. The question of interferences, especially with spectroscopic methods, should also be considered when dealing with composite materials.

Analytical procedures for what may be termed bulk analysis (analysis of solutions or homogeneous particles either in suspension or powder form) are validated by establishing characteristics of the procedure such as precision, specificity, detection limit, quantification limit, linearity, range, and robustness. In contrast, for FPMs, these characteristics of analysis must be redefined to accommodate the properties of individual particle analysis. Additionally, a fundamental difference between the analysis of FPMs and more conventional analytical chemical procedures arises from the fact that there are no reference or standard materials for FPMs that can mimic the heterogeneity of size, shape, and composition of actual FPMs. Consequently, validation of FPM analysis frequently requires indirect measures of capability. The discussions in the following two subsections address validation characteristics with respect to sizing and speciation. The recommendations provided are intended to be consistent with the perspective provided in the PASG white paper referenced above.

21.9.2 Validation Considerations for Sample Preparation, Particle Sizing, and Quantification

21.9.2.1 General The harvesting and preparation of particulate material for analysis is affected by the affinity of many particulate materials for surfaces. Once in contact with a suitable surface, small particles may be extremely difficult to remove. This characteristic of particles has been a source of extensive activity in the manufacture of semiconductors where microscopic particles have proven a major problem in making clean surfaces. For the purposes of FPM analysis, it is wise to keep in mind that during sample preparation, particles may deposit on every surface that they contact, leading to a loss of particles for the analysis. It is prudent to minimize the surface involved with preparation and to rinse containers repeatedly to minimize losses.

21.9.2.2 Precision *Repeatability.* The repeatability of the entire analytical procedure, including the sample preparation, must be characterized. The number of times the procedure can be repeated will be influenced by the amount of sample available, and the uniformity of the number and size of the FPM being studied. If for example, it is possible to obtain several samples from the same drug product, it is possible to determine the extent of variation among samples. If it is necessary to compare samples from different batches, one may expect greater variability in number and size. Calculation of an RSD will provide an estimate of the variability that will be characteristic of the materials or products analyzed. Sample collection and preparation can affect results significantly and can be a major influence on measured repeatability. Sampling must be carried out in a "clean" environment, preferably a high efficiency particulate air (HEPA)-filtered laminar flow hood or, if available, a "cleanroom." All materials employed must be scrupulously clean and particle free. Cleanliness must be documented by preparing "blank" samples.

Intermediate precision. This is an extended measure of repeatability within the same lab when the entire method is performed with different analysts and equipment, and on different days. At a minimum, results should be compared when at least two different analysts collect and prepare samples and use the same instrument. When available, comparison of results using different instruments should be performed. This should be carried out on a number of samples and for each variation in the procedure.

Reproducibility. This criterion measures the variability between laboratories when a method is to be transferred. It is recommended to consult a statistician to establish equivalence criteria.

21.9.2.3 Specificity In classical terms, this is the ability of the method to measure the substance of interest without interferences from other substances known or expected to be present in the sample. For particle sizing and quantification, this parameter is similar to accuracy but is not strictly applicable in the sense it is used when dealing with the analysis of a nonparticulate matter. In some FPM methodologies, it can be interpreted as the ability of the technique to count specific particle sizes within specified limits of variation. As will be mentioned later, some particle detection technologies may not recognize certain types of particles.

21.9.2.4 Detection Limit The detection limit is that level below which foreign PM is not confirmed to be present. This may be addressed in terms of count and size. The ability to detect foreign PM is governed by the quality of the sample preparation. A sample cannot be expected to have lower counts than a blank preparation. The size detection limit of the instrumentation should be determined using standard or reference particles.

21.9.2.5 Quantification Limit *Quantification limit.* A useful measure of this criterion is the number of particles that can be detected above the background, where the background is the number of particles per unit volume in blanks. Cleanliness in sample preparation has a significant influence on this parameter. Equations 21.2 through 21.4 demonstrate the importance of clean blank samples. For example, if the mean number of particles on a blank is 5 per microscopic field, one may estimate that the standard deviation will be $\sqrt{5}$ or ± 2.24. In that case, 99% of the blanks will fall in the range of $\pm 3 \times$ (standard deviation) or 0.0–11.71 particles per field. The number of particles counted must be greater than 12 to be significantly different from the background with a confidence level of 99%. Table 21.5 also shows the

TABLE 21.5. Relative Standard Deviations for Blanks Containing Increasing Numbers of Background Particles

Mean number of particles on a blank sample (P)	Estimated standard deviation (SD)	3 x (estimated SD) $CI_{99\%}$	Maximum mean number of background particles based on SD (R_{SD})	t-Value	Maximum mean number of background particles based on t-values (tCI)
5	2.24	6.71	11.71	4.00	9.00
10	3.16	9.49	19.49	3.17	13.17
20	4.47	13.42	33.42	2.85	22.85
30	5.48	16.43	46.43	2.75	32.75
40	6.32	18.97	58.97	2.70	42.70
50	7.07	21.21	71.21	2.68	52.68
100	10.00	30.00	130.00	2.63	102.63

relation of the 99% confidence interval, based on Student's t-values. Student's t-values are better for predicting confidence intervals for small numbers of particles. This estimate is provided to illustrate the fact that as the number of background particles increases, the range of the 99% confidence intervals standard deviations decreases. Paradoxically, the smallest number of background particles, $N = 5$, is associated with a wide confidence interval range, while the largest number, $N = 100$, provides the smallest relative range of the number of particles on the blank:

$$P_b = \text{mean number of particles on blank}$$
$$SD_{Pb} = \text{estimated standard deviation of number of particles on blank}$$
$$SDCI_{99\%} = 99\% \text{ confidence interval of the number of particles}$$
$$\text{on blank based on estimated standard deviation}$$
$$SDCI_{99\%} = 3 \times SD_{Pb} \tag{21.2}$$
$$R_{SD} = \text{maximum mean number of particles on blank} + \text{mean}$$
$$\text{number of particles on blank} = CI_{99\%} + P_b \tag{21.3}$$
$$tCI = 99\% \text{ confidence interval based on Student's } t\text{-table.} \tag{21.4}$$

A different approach to this criterion is to measure several samples or samples from several batches to establish the normal variation of FPM burden. Once it is established that the standard deviation of the normal FPM burden may vary by as much as 10% (for example), then it is possible to conclude that 99% of the normal values will fall within the range of ±40%. Consequently, when a sample differs by more than 40% of the mean value, it is likely to represent a significant difference.

21.9.2.6 Linearity Generally, the concept of linearity is interpreted to mean that, within certain limits, there is a consistent relationship between the actual number of particles present and the number of particles reported or counted. The limits are generally stated as the range of particle numbers, particle sizes, or particle types.

This criterion is influenced by the conditions that may affect the accuracy of sizing or quantification. Examples of these effects are the following:

1. *Number of particles per unit volume.* For example, in a microscope or light obscuration instrument when the number of particles is large, two particles may be so close as to be detected as one particle causing a bias toward larger size and smaller numbers of particles. If the sizes of the particles are large in relation to the microscope field, then large particles may be partly outside of the field and selectively excluded from the count.

2. *Particle transparency.* When some particles are essentially invisible because they are the same refractive index as the medium in which they are suspended, this may lead to selective underestimation of one type of particle.

3. *Particle shape or particle aspect ratio.* When some particles are so thin that if seen on the edge they will be invisible, this may also lead to underestimation of one type of particle.

21.9.2.7 Range This parameter should indicate both upper and lower size or count limits at which measurement becomes unreliable or nonlinear. The upper limit reflects the ability to accurately size and count large particles. Large particles or numbers of particles may exceed the capability of the instrument to detect the particle as a single particle, or to indicate accurately the dimensions of the particle. The lower limit indicates the size or counts at which detection as well as quantification becomes nonlinear. These may be from the manufacturer's specifications or from actual measurement using standard or reference particles of the appropriate sizes.

21.9.2.8 Robustness This criterion refers to the amount of variation resulting from changes of conditions or sample preparation procedure. Where sufficient sample material exists, this can be assessed by analyzing the same material after varying the sample preparation or varying other conditions likely to be encountered.

21.9.3 Validation Considerations for Particle Speciation

21.9.3.1 General When speciation of particulate material is required, method validation involves different criteria. Unless single classes of FPM can be isolated (e.g., by sedimentation), then individual particles must be analyzed. A survey of a representative sample, for example, a pooled sample from several drug product batches, should be performed to determine whether the particles are organic or inorganic. This can be accomplished by a combination of analytical microscopy technologies, including PLM, Raman, infrared (IR), and X-ray microanalysis (see Table 21.6).

21.9.3.2 Specificity This parameter is similar to accuracy but is not strictly applicable in the same sense as it is used when dealing with analysis of nonparticulate matter. This term implies identification, purity, and content of the analyte

TABLE 21.6. Analytical Technology Useful in Determining the Composition of Individual Foreign Particles

Technology	Uses	Comments
Polarized light microscopy	Size (two dimensions)	Excellent automated systems available
	Number	Particle properties extensively documented for reference
	Identification by:	2 μm lower limit for speciation
	Morphology (fibers)	Labor intensive unless automated
	Birefringence (crystals)	
	Refractive index	
	Color (pigments)	
SEM/EDS	Number	Submicron resolution
	Size (two dimension)	Automated systems exist
	Elemental composition	Requires vacuum
		Electron beam may cause damage
		Does not provide molecular level ID
		Based on atomic number differences, backscattered electron imaging can distinguish small thin particles from the background
Fourier transform infrared microscopy	Organic, polymer identification	12 μm lower practical limit
	Some inorganic identification	Labor intensive
Raman microprobe	Organic, polymer identification	Fluorescence interference
Raman microscope	Some inorganic identification	Does not identify metals
		<1 μm lower limit claimed
Light obscuration	Particle numbers in size ranges	2 μm lower limit
		No chemical information
		May not measure very thin or translucent/transparent particles
Electrozone sensing	Particle numbers in size ranges	0.4 μm lower limit
	Particle volume	No chemical information
		Sizing may be inaccurate for materials with different conductivities
		Sizing may be inaccurate for materials with differing conductivities
Micro X-ray fluorescence	Elemental composition	10 μm lower limit

according to ICH Q2(R1). Testing for content and for the presence of impurities is not generally applicable when the analyte, a particle, is an impurity. Identification refers to the ability to assess unequivocally the analyte in the presence of components that may be expected to be present. The type of analysis appropriate for speciation will be influenced by the concept of "fitness for use"; that is, understanding what level of information is necessary to assign a likely origin to a particle. For example, when dealing with a product that has been processed in stainless steel vessels and machinery, and then packaged in an aluminum container equipped with a valve having elastomer gaskets and polymer parts, it is probably sufficient in many, if not most, instances to simply determine that a particle has come from one of the four classes of materials mentioned. If investigation indicates that the source is unlikely to be from the manufacture of the product, then additional analyses will probably be necessary to learn if the stainless steel or aluminum or elastomer or polymer found in the particle is the material that should be in the product. If the particle composition does not fall into one of those classes, then it is probably necessary to obtain additional information in order to assign a source. The choice of analytical technology to accomplish a complete analysis will be determined by the type of material to be analyzed: Raman analysis might be chosen for polymer and elastomer, while X-ray microanalysis might be chosen for inorganics. The capability to elucidate chemical composition of various materials will be characteristic of the technology/instrumentation employed. The capacity of the technology to identify a specific material unequivocally can be demonstrated by analysis of reference or standard materials. The limits of detection and quantification then can also be determined by analyzing appropriate reference or standard materials with the same technology. Each analytical technology will have its own measures of specificity. Known limitations should be considered (see Table 21.6).

21.9.3.3 *Precision*

Repeatability. The repeatability of the entire analytical procedure, including the sample preparation, should be characterized. The number of times the procedure can be repeated will be influenced by the amount of sample available, but up to six times is recommended if sample availability permits. This is important because of the likely heterogeneity of the FPM, so that one sample is rarely identical with another. This exercise can be used to predict the normal variability among samples. Using nondestructive analytical methods will permit repeated analyses of the same sample. Some methods may not be entirely nondestructive (such as X-ray microanalysis) but may permit a limited number of repeated analyses to demonstrate this characteristic. Keep in mind that if the population of particles in a class of particles is very large, the RSD will be relatively small, and if particles of a different class occur only infrequently, the RSD associated with these particles will naturally be large.

Intermediate precision. At a minimum, comparison of results when different analysts use the same instrument with the same samples should be performed. This may be costly when it requires extensive training of a second person for sample preparation and instrument operation in order to accomplish this one criterion for evaluation. When available, comparison of results using different instruments should be performed. When practical, this should be carried out on a number of samples and for each variation in the procedure, or at a minimum by different analysts evaluating the same samples with the same instrument. Using nondestructive analytical methods will permit repeated analyses of the same sample. As noted above, some methods may not be entirely nondestructive, but may permit a limited number of repeated analyses to demonstrate this characteristic.

Reproducibility. This measures the variability between or among laboratories when a method is to be transferred. This may be best accomplished by analysis of a large pooled sample, with the caveat that the sample must remain stable during the study. Sample stability is an important consideration for liquid samples because FPM may settle out from suspension and form aggregates.

21.9.3.4 *Detection Limit*

The detection limit of the instrumentation should be specified based on manufacturer's specifications or on actual measurements using standard or reference particles. X-ray microanalysis, for example, is less sensitive to low atomic number elements than higher atomic number elements, but higher atomic number elements may require a more energetic electron beam, increasing the likelihood of radiation-induced degradation. For particles containing more than a single molecular species or element, detection limits will be important to specify (e.g., for X-ray microanalysis, detection limits of 10^{-12} g are common). Most FPM can be traced to a source with qualitative information, and it is rare that a quantitative microanalysis will be required. Often when describing the qualitative composition of a particle, it can be helpful to indicate whether a particular element is present as a "major component," "minor component," or simply as "present."

Minimum detectable change or sensitivity may be used to express the capability of automated systems. Gunshot residue (GSR) standards are often used to evaluate performance of automated particle characterization systems employing X-ray microanalysis. The standard consists of a known number of approximately spherical particles composed of lead, barium, and antimony. There are five size classes of particles, and the position of each particle is known. Upon completion of the analysis, the analyst manually compares the results obtained with a particle "map" for the specific standard. Because the size range, composition, and location of each particle are known, the system can be evaluated for accuracy of detection, sizing, and elemental analysis, although it can be cumbersome to use.

Recently, an improved standard, the Performance Grading System™ or PGS™ (ASPEX Corp., Delmont, PA), has been introduced. The standard consists of thousands of features with known locations and sizes that can be used to quickly probe the capabilities of the scanning electron microscope. Upon completion of an automated analysis using the PGS™, results are reviewed by the accompanying Map-Match™ program (ASPEX Corp.). MapMatch™ compares current results against known values of the analyzed features. Magnification and aspect ratio accuracy as well as particle recognition capabilities of the automated SEM system are quickly determined by MapMatch™. When coupled with a routine calibration of the X-ray microanalysis portion of the system, a robust verification of automated capabilities can be performed. Since no standard material that can mimic the heterogeneity of size, shape, and composition of an actual FPM exists, these indirect standards provide a practical estimate of instrument capability.

21.9.3.5 Quantification Limit This criterion is rarely applicable to qualitative analysis of a heterogeneous sample.

21.9.3.6 Linearity This criterion describes the conditions that may affect the accuracy of the particle composition analysis. These may be, for example, X-ray peak overlap in X-ray microanalysis or spectral overlap in Fourier transform infrared (FTIR) spectroscopy. Organic particles may degrade and loose volatile materials when exposed to the electron beam energies and vacuum conditions in a scanning electron microscope. It may not be practical to provide specifications for this property, but it will be important for the analyst to be aware of such potential problems.

21.9.3.7 Range This parameter should indicate both upper and lower composition limits at which measurement becomes unreliable or nonlinear. The lower limit describes the capability of the instrument to distinguish the composition of a small or thin particle from the background or medium, essentially a signal-to-noise consideration. The upper size limit describes the composition or concentration at which the signal overwhelms the detector, and may prevent recognition of minor components. These values may be from the manufacturer's specifications or from actual measurements using standard or reference particles of the appropriate sizes.

21.9.3.8 Robustness This parameter refers to the amount of variation resulting from changes of conditions or sample preparation procedure. Where sufficient sample material exists, this can be assessed by analyzing the same material after varying the sample preparation or other conditions likely to be encountered.

21.10 CURRENTLY AVAILABLE TECHNOLOGY FOR FPM ANALYSIS

21.10.1 General

Increasing interest in particle detection and identification in such fields as forensics, high performance aircraft, and cleanroom technology as well as contamination in food and pharmaceuticals has resulted in significant and continuing improvements of particle analytical technology. Table 21.6 lists currently available technologies useful in characterizing particulate material. Technology is developing rapidly in this area making a detailed comparison of available and soon to be available technology of limited value. However, it is possible to provide practical guidance on what must be done to validate the various classes of particle analytical technology as indicated in Table 21.7.

Analytical methods for *homogeneous materials* generally depend on validation by measurement of standards or reference materials. Analytical methods for *particulate materials* are validated by measuring particle characteristics, for example, size, shape, and composition, with several measurement technologies. It is important to keep in mind that instrumentation performance is based on the analysis of standard

TABLE 21.7. Practical Steps in FPM Method Validation

Precision	Sample preparation	Particle sizing and quantification	Particle speciation
Repeatability	Calculate RSD for one analyst analyzing multiple samples	Calculate RSD of each size class for one analyst analyzing the same sample multiple times	Prepare histogram of numbers of particles in each size class, sorted by composition class
Intermediate precision	Calculate RSD for samples prepared by multiple analysts	Calculate RSD for multiple analysts analyzing the same sample with different instruments[1]	Compare histograms of numbers of particles in each size class sorted by composition class generated by multiple analysts
Reproducibility	RSD between/among laboratories	Calculate RSD for the same sample analyzed in different laboratories with different instruments[1]	Compare histograms of numbers of particles in each size class sorted by composition class analyzed in different laboratories with different instruments employing the same technologies, for example, Raman, SEM/EDS
Specificity	Estimate if classes of FPM are lost during capture[2]	Determine if thin or transparent particles are seen by SEM or IR or Raman, but not detected by LO	Determine whether a composition class is not detected by speciation technology
Detection limit	Determined by a number of particles on blank preparations	Determine at which size class does recognition begin to fail	Elemental standard can be used to determine minimum detectable concentrations or atomic percentages for composition classes
Quantification limit	Increase sample size until FPM interacts or agglomerates	Determine what numbers of particles can be reliably detected	Determine molecular or atomic percent necessary for detection
Linearity	Test multiples of unit sampled	Test by analyzing samples of 12×, 2×, and 4× particle concentration	Employ performance grading system to test minimum size for detection
Range	Similar to linearity	Similar to linearity	Elemental standards can be used to demonstrate
Robustness	Alter conditions: temperature method of sampling, type or pore size of filter	Alter instrument parameters to determine stable analytical conditions	Alter instrument parameters to determine stable analytical conditions

LO, light obscuration.

or reference materials. Reference standards will rarely match the characteristics of all the materials in FPM. Thus, the results with actual FPM samples containing diverse materials of varying sizes, shapes, and composition may not match the performance obtained with well-characterized standards or reference materials of known sizes, shapes, and compositions. Consequently, validation activities must be tailored to the specific sample types and instrumentation used. The two most commonly employed analytical technologies for OINDPs are discussed below.

21.10.2 Light Obscuration

Basically, this technology employs various algorithms to derive equivalent size (e.g., equivalent circular diameter and equivalent spherical diameter) based on light attenuation. USP <788> provides detailed instructions for the calibration and verification of light obscuration instruments for the enumeration and sizing of PM.[3,14] Sample volume, sensor resolution, and sensor dynamic range must be demonstrated to be within stated limits. Practical experience indicates that $\leq \pm 10\%$ in particle count is to be expected.

Validation consists of measuring the accuracy, repeatability, and linearity of the technology using particle standards and/or particle reference materials. Sample preparation is a critical step in the measurement of FPMs, and demonstration of robust preparation methods in terms of reproducibility, repeatability, and precision is essential.

21.10.2.1 Standards
- *National Institute of Standards and Technology (NIST) traceable particle counter size standards.* Typically, particle standards of 2, 5, 10, 15, 25, and 30 μm are used to document particle size discrimination. Acceptable criteria for the mean and variation in the number of counts recorded with each size standard must be established and met before analyzing samples of FPM.
- *NIST particle size standard.* Dilute solution with very narrow particle size distribution used to demonstrate the instrument can accurately measure particle size.
- *USP particle count reference standard.* Solution with known counts per milliliter used to demonstrate the instrument can accurately count a large number of particles.
- *USP particle count reference blank.* This is used to establish the background count level of the instrument and demonstrate its ability to detect small numbers of particles.

The combination of the USP particle count standard and USP particle count reference blank defines the useful range in numbers of particles per milliliter for the instrument.

21.10.2.2 Validation Parameters
- *Accuracy.* Assessed by measuring the sizes and counts reported for particle standards and/or reference materials of known size and number.

- *Linearity.* Assessed by measuring the counts reported for serial dilutions of particle standards and/or reference materials.

- *Repeatability.* Assessed by repeated analyses of the same particle standards, reference materials, and samples prepared in the same way by the same person using the same person using the same instrument.

21.10.3 SEM/EDS

This combination of technologies is used to produce a measure of particle size based on a two-dimensional image of a particle and an elemental composition based on electron-beam-induced X-ray emission from the particle. The essential parameters of these technologies are for size measurements: sample suitability, accuracy, and repeatability; and for elemental analysis: accuracy of element identification. Imaging may use either SE imaging or BSE imaging or both. Because contrast in BSE imaging is dependent on the average atomic number (Z) of the material being analyzed, it is frequently applied to automated investigations. Appropriate selection of a low Z substrate, such as a polycarbonate microporous filter, can significantly improve one's ability to detect small foreign particles since the background signal intensity of the carbonaceous material is low in comparison to most FPM materials typically associated with pharmaceutical manufacturing. However, it is also important to recognize that the particle features may also play a significant role in the ability to use BSE detection. For example, when the interaction volume of the electron probe exceeds the particle thickness, a reduced portion of the BSE signal originates from the particle, regardless of the particle's elemental composition. At some point, the particle brightness drops below the threshold value, and the particle is either undersized or missed completely. It is a limiting factor that needs to be addressed during method development since it can have a dramatic effect on the results obtained using this type of analysis as illustrated in Figure 21.7. BSE imaging offers significant advantages in variable pressure electron beam columns that reduce the need for conductive coatings. EDS is employed to perform elemental analysis.

Sample suitability. The conditions of the sample capture, preparation, and analysis impose strict constraints. If particles aggregate, accurate size measurements may not be possible. If the analysis will only sample some of the particles, and extrapolate to determine the total content of a container, it is necessary to demonstrate that the sample preparation has produced a uniform distribution of particles, and did not produce areas of particle aggregation and/or regions devoid of particles. As mentioned previously, an acceptability criterion for particle distribution based on the RSD of several sampled areas will demonstrate the uniformity, or lack thereof, of particle distribution.

Accuracy. Size measurements are based on magnification calibration of the microscope using NIST traceable size standards. X-ray energy measurement is based on calibration using commercial elemental standards. Feature recognition, the capability of distinguishing a particle from the substrate on which it sits, determines how successfully the instrument can identify all particles or just a fraction of those present. This can be a problem, for example, when dealing with very thin particles of the sort that may occur in talc. In the case of very thin particles, the electron beam can penetrate the particle with the consequence that the X-ray signal may be almost entirely from the substrate.

Precision. The parameter of repeatability for these technologies should, whenever practical, include repeated sample preparation from the same batch of material and repeated analysis of the same preparation by one person. Intermediate precision should, whenever possible, include repeated sample preparation from the same batch of material and sample analysis by more than one person. Repeated analyses are not always feasible when the particle materials are affected by exposure to the electron beam. Analyses of different regions of the same preparation, with uniform distribution of particles, by different operators will provide a measure of precision. When using automated instrumentation, the parameter of precision will have to include instrument setup by different operators.

Resolution. Spatial resolution in SEM can be measured using reference materials such as sputtered gold particles on carbon; however, the resolution achieved in actual practice will be strongly influenced by such factors as particle thickness, composition, edge definition, and contrast.

Size discrimination. For OINDPs, the size classes of greatest interest are usually <2 μm, ≥2 to ≤10 μm, >10 to ≤25 μm, and >25 μm, and these can be confirmed with particle reference materials. If finer size discrimination is required (e.g., to distinguish 2-μm particles from 2.5-μm particles), it will be up to the analyst to demonstrate that such discrimination can be accomplished.

Automated counting, sizing, and speciation can offer significant savings in time and cost per analysis.[27] Additionally, operator fatigue is greatly reduced and operator-to-operator variance will improve. The number of particles that can be analyzed quickly will increase statistical confidence. Typically, a nonautomated analysis can evaluate, that is, size, enumerate, and speciate a few hundred particles per day, while an automated analysis can evaluate several thousand overnight, unattended. However, establishing the parameters of analysis will depend on experience in determining which elements should be reported and which combinations of elements are characteristic of which types of FPM. Prior to investing in automated speciation, it is necessary to involve nonautomated efforts in determining which types of particles will have to be analyzed.

Figures 21.8 and 21.9 show examples from an ASPEX Corporation P-SEM automated SEM/X-ray microanalysis system. The P-SEM system provides unattended particle recognition, sizing, and elemental analysis. Figure 21.8 shows the user interface for SEM/X-ray microanalysis control and automation, and Figure 21.9 shows the data analysis screen. The system methodically analyzes a sample by physically moving to preprogrammed stage locations (as defined by operator method). A low-resolution image of the stage location actively being analyzed is captured and displayed. The area is then divided into nine segments that will be analyzed at higher resolution. The electron beam is then rastered across the sample until BSE signal reaches the threshold value. At that point, the BSE signal, processed by one of several possible algorithms, is used to determine the particle size and shape. Upon completion of the operation, a representation of the particle is added to the particle map. The electron beam is then relocated to the particle and elemental characterization using X-ray microanalysis begins. Upon completion of

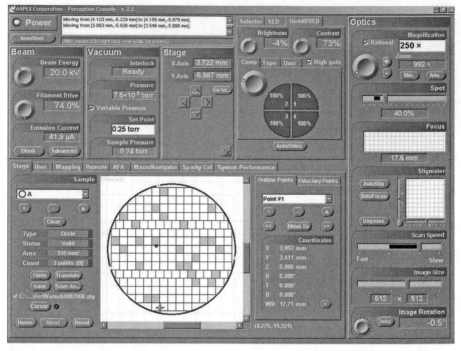

Figure 21.8 User interface to SEM/EDS control and data analysis functions. The complex series of steps required to automate the SEM/EDS analysis are organized into logical groupings using a tabbing layout.

the X-ray analysis, a thumbnail image of the particle is captured. The collected parameters are used to classify the particle accordingly, and the operation sequence continues to the next particle. The system analyzes each of the nine regions, then the stage moves to a new location and the process is repeated until all preprogrammed areas have been analyzed.

21.10.4 PLM

Optical microscopy, in particular PLM, utilizes the characteristic optical properties of solid substances for identification and as a means to provide the necessary contrast in order to distinguish different solid phases. The advantage of PLM is that samples can be analyzed in their native environment, whether a dry powder or liquid suspension.

Contrast may be achieved in a number of ways. Taking advantage of refractive index properties of materials, a sample may be mounted in an appropriate refractive index liquid such that the contrast of the particles of interest (i.e., FPM) is enhanced while other particles (i.e., API) are diminished, rendering them nearly invisible. The optical crystallographic property of birefringence may be utilized to easily distinguish between crystalline and amorphous phases. It may also be used as a means of providing the necessary contrast to distinguish various crystalline solid phases,

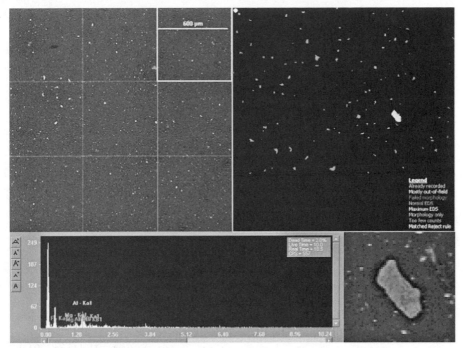

Figure 21.9   Screen view permits the analyst to monitor the analysis in real time. The top-left quadrant contains a survey image defining the nine areas to be analyzed at high resolution. The top-right quadrant documents particles, "on the fly," as they are identified. The lower-left quadrant displays the X-ray microanalysis acquisition in real time. Upon completion of the microanalysis identification, a thumbnail image of the particle being analyzed is captured and displayed in the lower-right quadrant.

whereby the polarization colors may be utilized to distinguish various phases. A heating/freezing stage may be employed for substance identification by determination of melting and/or crystallization temperatures. Additionally, a heating stage may be utilized to provide contrast whereby a sample containing a mixture of drug substance and foreign particulates is heated to a temperature above the melting point of the drug substance, leaving behind solid FPM. This requires that the drug substance has a melting point lower than that of the FPM, which in many instances is the case. Additional techniques that rely on optical properties of solid phases include phase contrast microscopy and differential interference contrast. Both of these techniques are particularly useful in instances when the refractive index of the FPM is relatively close to that of the medium in which they occur (i.e., FPM in parenteral or ophthalmic solutions). In the case of opaque phases, reflected light by means of a vertical illuminator may be used to distinguish particles; however, special sample preparation techniques such as mounting in a support epoxy and polishing may be required. Additionally, shape/morphology if characteristic of the particular sample being analyzed (i.e., fiber vs. nonfiber, crystal morphology) may be used to distinguish FPM.

Automation may be employed utilizing image analysis software on digitized images. Contrast between the particles of interest (FPM) and the remainder of the sample can be improved using various software filters; however, it is still best to achieve maximum contrast prior to image capture utilizing sample preparation or optical enhancement techniques. In any case, reproducibility and validation of imaging software can be extremely difficult and must be developed for each type of sample being analyzed. Size measurements can be performed manually using a calibrated eyepiece reticule or through automated analyses of digitized images. Calibration of the optical microscope is relatively easy using a NIST traceable stage micrometer.

21.11   SUGGESTED STRATEGIES FOR IMPLEMENTING QbD PRINCIPLES INTO FOREIGN PM CONTROL

The following is a suggested approach to implementing QbD principles during drug development so that the information and experience gained during development can be utilized during the manufacture of the drug product. The steps recommended in the text below are captured in the flowcharts (as shown in Figs. 21.10–21.12).

The IPAC-RS reports[6,11] recommended that investigations focus on the drug product, and for OINDPs, emitted dose. The same reports provide a high-level description of the recommended process (Fig. 21.10). Figures 21.11 and 21.12 present in graphic form the control process outlined by IPAC-RS as a best practice. The two charts describe two phases of control: the processes recommended for product development, and the processes appropriate for routine quality control in a production environment.

21.11.1   Steps in the Formulation and Process Development Batches (Fig. 21.11)

During the development of a new drug product, information is collected showing the level of FPM types, within specified size ranges, which occurs normally. If this number is not within the range regarded as safe, or acceptable for quality concerns, then an investigation is appropriate to identify sources of the FPM and to establish the root causes of the particles. Corrective and preventive actions (CAPAs) should be devised to bring the number of particles within the ranges considered safe or acceptable. These CAPAs should be applied to formulation batches still being produced. When the FPM burden is considered stable and acceptable, process development can proceed.

Steps in the formulation and process development batches include

1. count, size, and characterize particles; and
2. assess
   a. preliminary estimate of normal number and types of particles in size ranges,
   b. identify probable sources of particles, and
   c. devise and initiate corrective actions/preventive actions.

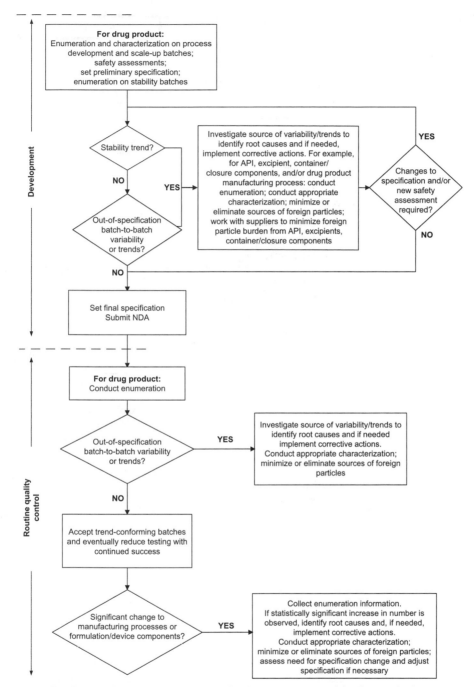

Figure 21.10   IPAC-RS recommendations for the management of foreign particulate matter in orally inhaled and nasal drug products. NDA, new drug application.

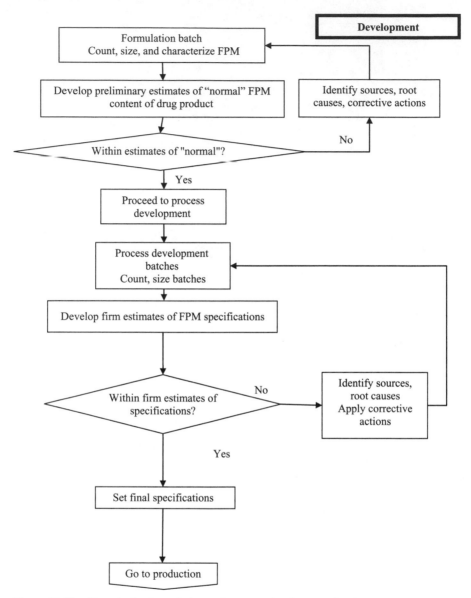

Figure 21.11   Steps in the formulation and process development batches.

During the process development phase, the number of particles in each size range is measured. The preliminary estimates of "normal" FPM content should be examined against the number and types of FPM encountered in formulation. If necessary, CAPAs will be initiated or devised and initiated, to bring the number of particles within the ranges considered to be normal for this stage of development.

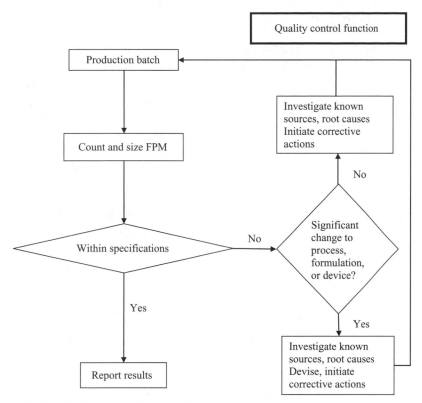

Figure 21.12   Quality control function during production.

The aim of this process of identifying root causes and applying CAPAs is to collect the knowledge gained from repeated analyses and corrections to improve the understanding of the eventual process chosen for production scale manufacturing. This knowledge will facilitate the development of specifications based on the understanding of the manufacturing process.

21.11.2   Steps in Quality Control Function during Production (Fig. 21.12)

1. Count and size FPMs.
2. If numbers and sizes comply with specifications, report results.
3. If numbers and sizes are "out of specification," initiate an investigation.
   - If there has been a significant change in process, formulation, or device, initiate an investigation.
   - If there has been no significant change, initiate an investigation of known root causes, apply CAPAs, and, if necessary, devise new CAPAs.

21.12   CONCLUSION

Control of foreign PM in a QbD environment requires solving several challenging problems. To properly understand and manage FPM content in OINDPs, it is necessary to analyze individual particles and therefore isolate them, count and determine the size of each particle, and identify the source or sources of each type of particle. There are no standard or reference materials that mimic the heterogeneity of the foreign particles that must be analyzed, and this imposes an especially stringent need to demonstrate that the capabilities of the entire analysis process are sufficient to meet the needs of providing safe and efficacious products through a QbD development and manufacturing approach. However, careful and thoughtful development of these capabilities early in the formulation and development phases of a product can pay significant dividends during product manufacture. These analysis considerations and approaches focus on the critical characteristics of FPM measurements, and are therefore necessary in developing and managing an appropriate production process with respect to FPM.

REFERENCES

1   Barber, T.A. *Control of Particulate Material Contamination in Healthcare Manufacturing*. Interpharm Press, Denver, CO, 2000.

2   Lich, B. and Willen, U. When size and shape matter. *Drug Discov Dev* 2009, *12*(2), pp. 26–28.

3   U.S. Pharmacopeia (USP). Chapter 788. Particulate matter in injections. Available at: http://www.pharmacopeia.cn/v29240/usp29nf24s0_c788.html (accessed April 1, 2010).

4   Ball, D., Blanchard, J., Jacobson-Kram, D., McClellan, R.O., McGovern, T., Norwood, D.L., Vogel, W.M., and Nagao, L.M. Development of safety qualification thresholds and their use in orally inhaled and nasal drug product evaluation. *Toxicol Sci* 2007, *97*(2), pp. 226–236.

5   Greb, E. Inhalable drugs on the launch pad: Will they take off? *Pharm Technol* 2008, *32*(4), p. 48.

6   Blanchard, J., Coleman, J., Crim, C., D'Abreu-Hayling, C., Fries, L., Ghaderi, R., Haeberlin, B., Malcolmson, R., Mittelman, S., Nagao, L., Saracovan, I., Shtohryn, L., Snodgrass-Pilla, C., Sundahl, M., and Wolff, R. Best practices for managing quality and safety of foreign particles in orally inhaled and nasal drug products, and an evaluation of clinical relevance. *Pharm Res* 2007, *24*(3), pp. 471–479.

7   Hussain, A.S. Quality by design (QbD)—Integration of prior knowledge and pharmaceutical development into CMC submission and review. Presented at the AAPS Workshop—Pharmaceutical Quality Assessment—A Science and Risk-Based CMC Approach in the 21st Century, North Bethesda, MD, October 5–7, 2005.

8   Nasr, M.M. Risk-based CMC review and quality assessment: What is quality by design (QbD)? Presented at the FDA/Industry Conference, School of Pharmacy, Temple University, March 29, 2006.

9   Woodcock, J. Pharmaceutical quality in the 21st century—An integrated systems approach. Presented at the AAPS Workshop—Pharmaceutical Quality Assessment—A Science and Risk-Based CMC Approach in the 21st Century, North Bethesda, MD, October 5–7, 2005.

10   Nagao, L.M., Lyapustina, S., Munos, M.K., and Capizzi, M.D. Aspects of particle science and regulation in pharmaceutical inhalation drug products. *Cryst Growth Des* 2005, *5*(6), pp. 2261–2267.

11   Blanchard, J., Coleman, J., D'Abreu Hayling, C., Ghaderi, R., Haeberlin, B., Hart, J., Jensen, S., Malcolmson, R., Mittelman, S., Nagao, L.M., Sekulic, S., Snodgrass-Pilla, C., Sundahl, M., Thompson, G., and Wolff, R. Foreign particles testing in orally inhaled and nasal drug products. *Pharm Res* 2004, *21*(12), pp. 2137–2147.

12   Singh, S.K., Afonina, N.A., Awwad, M., Bechtold-Peters, K., Blue, J.T., Chou, D., Cromwell, M., Krause, H.J., Mahler, H.C., Meyer, B.K., Narhi, L., Nesta, D.P., and Spitznagel, T. An industry

perspective on the monitoring of subvisible particles as a quality attribute for protein therapeutics. *J Pharm Sci* 2010, *99*(8), pp. 3302–3321.

13 Cao, S., Jiang, T., and Narhi, L. A light-obscuration method specific for quantifying subvisible particles in protein therapeutics. *Pharmacopeial Forum* 2010, *36*(3), pp. 824–834.

14 U.S. Pharmacopeia (USP). Chapter 789. Particulate matter in ophthalmic solutions. Available at: http://www.pharmacopeia.cn/v29240/usp29nf24s0_c789.html (accessed April 1, 2010).

15 United States Environmental Protection Agency. Particulate matter: PM standards. Available at: http://www.epa.gov/air/particlepollution/standards.html (accessed April 1, 2010).

16 Phelan, R.F. *The Particulate Air Pollution Controversy: A Case Study and Lessons Learned.* Kluwer Academic Publishers, Norwell, MA, 2002.

17 Whitby, K.T., Charlson, R.E., Wilson, W.E., and Stevens, R.K. The size of suspended particulate matter in air. *Science* 1974, *183*(129), pp. 1098–1099.

18 Allen, T. *Particle Size Measurement Vol. 2 (Particle Technology Series)*, 5th ed. Chapman and Hall, London, 1997.

19 Hirleman, E.D., Bachalo, W.D., and Felton, P.G., eds. *Liquid Particle Size Measurement Techniques: 2nd Volume.* American Society for Testing and Materials International, West Conshohocken, PA, ASTM Special Technical Publication Stp, 1990.

20 National ambient air quality standards for particulate matter. *Fed Regist* 1997, *62*(138), pp. 38651–38760.

21 Avilés, A.I., Ferraris, C.F., and Hackley, V.A. Measurement of particle size distribution in Portland cement powder: Analysis of ASTM round robin studies. *J Cement Concrete Aggr* 2004, *26*(2), pp. 1–11.

22 Iacocca, R.G. Particle characterization from and end user's perspective. *Am Pharm Rev* 2009, *12*(1), pp. 54–61.

23 Jillavenkatesa, A., Dapkunas, S.J., and Lum, L.-S.H., eds. Particle size characterization. NIST Spec Publ 960(1), 2001.

24 Burgess, D.J., Duffy, E., Etzler, F., and Hickey, A.J. Particle size analysis: AAPS Workshop Report, cosponsored by the Food and Drug Administration and the United States Pharmacopeia. *AAPS J* 2004, *6*(3), p. e20.

25 Etzler, F.M. and Deanne, R. Particle size analysis. A comparison of various methods. Part 2. *Part Part Syst Charact* 1997, *14*(6), pp. 278–282.

26 Schearer, G.L. Contaminant identification in pharmaceutical products. *Microscope* 2003, *51*(1), pp. 3–10.

27 Vicens, M.C. Foreign particle size distribution and characterization in pharmaceutical products using a high throughput electron beam analyzer. Pharmaceutical Processing, 2008. Available at: http://www.pharmpro.com/Articles/2008/11/Foreign-Particle-Size-Distribution-and-Characterization-in-Pharmaceutical-Products-Using-a-High-Throughput-Electron-Beam-Analyzer/ (accessed April 1, 2010).

28 Kreher, C., Bootz, W., Niemann, M., Scaffidi, L., and Spallek, M. Foreign particle characterization in inhalation drug products: A critical comparison to methods and techniques. *Respir Drug Deliv* 2004, *4*, pp. 373–376.

29 Lankers, M., Munhall, J., and Valet, O. Differentiation between foreign particulate matter and silicone oil induced protein aggregation in drug solutions by automated Raman spectroscopy. *Microsc Microanal* 2008, *14* (Suppl. 2), pp. 1612–1613.

30 Bell, R., Dennis, A., Hendriksen, B., North, N., and Sherwood, J. Position paper in particle sizing: Sample preparation, method validation and data presentation. *Pharm Technol Eur* 1999, *1*(3), pp. 1–3.

APPENDIXES*

APPENDIX 1   EXPERIMENTAL PROTOCOL FOR CONTROLLED EXTRACTION STUDIES ON ELASTOMERIC TEST ARTICLES

A1.1   INTRODUCTION

In November 1998 and May 1999, the United States Food and Drug Administration (FDA) issued two chemistry, manufacturing, and controls (CMC) draft guidances addressing orally inhaled and nasal drug products (OINDPs): (1) the draft *Metered Dose Inhaler (MDI) and Dry Powder Inhaler (DPI) Drug Products Chemistry, Manufacturing, and Controls Documentation*[2] (referred to here as the "MDI/DPI draft guidance"); and (2) the draft *Nasal Spray and Inhalation Solution, Suspension, and Spray Drug Products Chemistry, Manufacturing, and Controls Documentation* (referred to here as the "nasal spray draft guidance"). In July 2002, the Nasal Spray Guidance was finalized.[3]

Currently, both guidances recommend that the sponsor identifies, reports, and conducts toxicological analyses on all extractables found in the controlled extraction study (referred to in the guidances as a "control extraction study"). Examples of these recommendations are described in the draft MDI/DPI Guidance regarding MDI canisters, valves, and actuators (lines 883–884, 990–991, and 1073):

> ... *the profile of each extract should be evaluated both analytically and toxicologically.*

The Product Quality Research Institute (PQRI) Leachables and Extractables Working Group has developed this experimental protocol as an example of a controlled extraction study for elastomeric (i.e., rubber) test articles. Various experimental parameters will be investigated, test article extracts will be analyzed, and results will be evaluated within the context of the Working Group's approved work plan and experimental hypothesis.

This experimental protocol will be used by all laboratories and investigators participating in the study.

* These Appendixes are from the Product Quality Research Institute Safety Thresholds and Best Practices for Extractables and Leachables in Orally Inhaled and Nasal Drug Products, published by the PQRI Leachables and Extractables Working Group in 2006.

Leachables and Extractables Handbook: Safety Evaluation, Qualification, and Best Practices Applied to Inhalation Drug Products, First Edition. Edited by Douglas J. Ball, Daniel L. Norwood, Cheryl L.M. Stults, Lee M. Nagao.
© 2012 John Wiley & Sons, Inc. Published 2012 by John Wiley & Sons, Inc.

617

A1.2   PURPOSE AND SCOPE OF WORK

A1.2.1   Purpose

The purpose of the experiments outlined in this protocol is to generate data from controlled extraction studies that will contribute to a larger database, which the Working Group will use to investigate its hypotheses[4]:

1. Scientifically justifiable thresholds based on the best available data and industry practices can be developed for the

   a. reporting and safety qualification of leachables in OINDPs, and

   b. reporting of extractables from the critical components used in corresponding container/closure systems.

   Reporting thresholds for leachables and extractables will include associated identification and quantitation thresholds.

2. Safety qualification of extractables would be scientifically justified on a case-by-case basis.

A1.2.2   Scope

A1.2.2.1   Topics Addressed by This Protocol   This protocol covers only controlled extraction studies that would be applied to components from MDIs. The MDI represents the best example of "correlation" between extractables from components and leachables in drug products. Controlled extraction studies will be performed following the general outline described in the guidances. Test articles will be subjected to different extraction conditions to show how different experimentally controlled parameters affect resulting extractables profiles. Additionally, the Working Group will assess experimental results to identify reasonable approaches for sample preparation and analysis of extractables from container and closure components.

As no single analytical technique can be used to identify and quantify all unknown extractables, a variety of methods will be utilized in this protocol to maximize the likelihood that all extractables compounds associated with the test articles are evaluated analytically. Overlap between methods will supply corroborating data that the procedures are valid. To provide a full analytical survey of possible analytes, the following strategies will be employed:

1. direct injection gas chromatography/mass spectrometry (GC/MS) for identification and assessment of relatively volatile extractables;

2. high-performance liquid chromatography/diode array detection (HPLC/DAD) for identification and assessment of relatively polar/nonvolatile ultraviolet (UV)-active extractables;

3. high-performance liquid chromatography/mass spectrometry (LC/MS) for identification and assessment of relatively polar/nonvolatile extractables, which may or may not have UV activity;

4. inductively coupled plasma/mass spectrometry (ICP/MS), inductively coupled plasma/atomic emission spectroscopy (ICP/AES), and/or energy-dispersive X-ray (EDX)/wavelength-dispersive X-ray (WDX) to detect single elements in the extracts (i.e., metals); and

5. Fourier transform infrared (FTIR) spectroscopy for the characterization of major components in the nonvolatile extractables residues.

A1.2.2.2 Topics Not Addressed by This Protocol Studies designed to assess recovery (i.e., mass balance) for individual extractables relative to the known formulations of chemical additives in the elastomeric test articles, or reproducibility of extractables profiles for multiple "batches" of any particular test article are not within the scope of this test protocol.

The extraction procedures, analytical techniques/methods, and analysis conditions described in this experimental test protocol will not be validated as material control methods, since they will be performed in order to collect qualitative information. However, during the course of these experiments, the PQRI Leachables and Extractables Working Group will review the results and may initiate additional experimental work for quantitative assessment of extractables.

This protocol does not address system suitability tests for quantitative methods. Appropriate system suitability tests will be addressed later, and agreement on this issue will be reached with all of the participating laboratories.

Special case studies such as organic volatile impurities (OVIs), *N*-nitrosamines, or polynuclear aromatic hydrocarbons (PAHs or polynuclear aromatics [PNAs]) will not be considered in this study. These "special case" classes of extractables have defined and highly specific analytical methods, which are generally accepted and commonly used for their identification and quantitative assessment.

It should be noted that the outlined experimental procedures, analytical instrumentation parameters and conditions, and other details are intended as a guidance for laboratory studies. Details of actual experimental procedures, and so on, should be reviewed by the entire group of participating laboratories and investigators so that harmonization among laboratories working on the same test articles can be achieved.

A1.3 REGULATORY STATUS

This is a good manufacturing practice (GMP)[5] study. All experiments shall be performed under GMP conditions to the extent practical in a particular laboratory.* Any changes to these protocols shall be documented, following appropriate GMP change control procedures.

---

* These experiments are considered research projects to be conducted in research labs, which are not strictly GMP compliant. However, all participating labs will perform these experiments in the spirit of GMP, which means that they will implement appropriate documentation, sample handling, data traceability, and so on.

A1.4   SAFETY AND ENVIRONMENTAL IMPACT

Organic solvents are commonly used to enhance solubility of lipophilic targets and to increase transport of small molecules out of complex matrices. These solvents may be flammable and/or show short-term and long-term environmental health risks. Care must be exercised with their use. Consult the material safety and data sheet (MSDS) for appropriate personal protection and disposal.

A1.5   TEST ARTICLES

Elastomeric materials will be provided in sheet form for use as test articles. The additive formulations and manufacturing conditions for these test articles are known and will be provided to all laboratories participating in the study at the appropriate times.

Note that reference compounds and additive mixtures may be required for the completion of this test protocol and will be provided as appropriate.

A1.6   CHEMICALS AND EQUIPMENT

Extraction and analytical methods have been chosen and designed so as to utilize chemicals, apparatus, and instrumentation available in typical laboratories routinely involved with this type of study.

A1.6.1   Extraction Solvents

Extractions will be performed on each test article using three solvents representing a range of polarity selected from the list below. The solvents should be American Chemical Society (ACS) grade or better:

- methylene chloride (dichloromethane),
- 2-propanol (isopropanol), and
- hexane (*n*-hexane, not hexanes).

Depending on the behavior of the test articles in these particular solvent systems, additional solvents may be chosen. Changes in extracting solvent will be discussed by all study participants prior to change initiation by any particular study participant or laboratory.

A1.6.2   Extraction Apparatus

- Soxhlet apparatus with an Allihn condenser, flask (500 mL), and hot plate or heating mantle
- Sonicator
- Reflux apparatus consisting of an Erlenmeyer flask (125 mL or larger) and condenser with ground-glass joints, hot plate, or heating mantle

A1.6.3 Analytical Instrumentation

- Gas chromatograph equipped with a flame ionization detector (GC/FID)
- Gas chromatograph equipped with a mass spectrometer (GC/MS)
- Liquid chromatograph equipped with a photodiode array detector
- Liquid chromatograph equipped with an atmospheric pressure chemical ionization (APCI)-capable mass spectrometer (LC/MS)
- FTIR spectrometer
- EDX and/or WDX equipped with a microprobe or scanning electron microprobe
- ICP/MS and/or ICP/AES

A1.7 EXTRACTION PROCEDURES

For each extraction technique and solvent type, appropriate blanks (no test article sample) must be prepared. These must be prepared concurrently using a different extraction apparatus (same type) under the same conditions, or by using the same apparatus prior to charging with the sample.

Note that the extraction parameters and conditions outlined below are subject to modification, and the details of any particular extraction process must be agreed to among all laboratory study participants prior to initiation of experimental work in any particular laboratory.

A1.7.1 Soxhlet Extraction

A1.7.1.1 Sample Preparation Samples of each test article should be cut into strips appropriately sized to fit into pre-extracted Soxhlet cellulose thimbles. Sample amounts may be in the range of 1–3 g (2 g) using 200 mL of solvent. For quantitative measurements, extracts prepared by Soxhlet will have to be evaporated to dryness and the resulting residues redissolved to a known volume (25–50 mL). Alternatively, an internal standard can be used for quantitative measurements.

A1.7.1.2 Extraction Conditions Under normal laboratory conditions, three physical extraction parameters may be modified: turnover number, total extraction time, and temperature. Temperature is the most difficult of the three parameters to control as the sample holder is maintained above the vapor level (temperature may be above the boiling point), but will be continuously bathed in freshly distilled solvent (coil temperature). It is recommended that the coil temperature be kept as low as possible to avoid heating above the solvent flash point.

Turnover number is controlled by the heating rate and should be limited by safety concerns. At low turnover numbers, the extraction characteristics will resemble those of reflux and may be limited by equilibrium phenomena. It is recommended that turnover numbers to be at least 10 during the course of the extraction.

Extraction time should be in the range of 24 hours to guard against possible degradation of thermally labile or reactive compounds.

A1.7.2   Reflux

Reflux extraction is a common and easily implemented approach for the production of extractables (e.g., USP <381> "Elastomeric Closures—Physicochemical Tests"). Conditions are easily standardized as the temperature and pressure are at the defined boiling points of the extraction solvents. Unlike Soxhlet extraction, reflux extraction is an equilibrium phenomenon.

A1.7.2.1   Sample Preparation   Transport of extractables out of the complex matrix may be affected by the surface area and thickness of the test article. Test articles will be prepared by two methods: grinding and cutting into strips appropriately sized to fit into the reflux apparatus.

Sample amounts should be in the range of 2 g using 25–50 mL of solvent. For quantitative measurements, the solvent with the sample and the flask can be weighed and returned to original weight after extraction. Alternatively, an internal standard can be used for quantitative measurements.

In reflux extraction, the sample-to-solvent ratio may affect the completeness of the technique. This should be addressed when optimizing the method for the measurement of extractables.

A1.7.2.2   Extraction Conditions   The only adjustable physical parameter for reflux extraction is time. Extraction time should be 2–4 hours. The solvent reservoir level must be monitored and periodically recharged to provide the correct amount of solvent.

A1.7.3   Sonication

Sonication uses ultrasonic energy instead of thermal energy to increase the rate of diffusion of small analytes out of a solid matrix. Similar considerations as reflux extraction (equilibrium conditions) should be evaluated, but these cannot be calculated using thermodynamic parameters. Sonication equipment may be standardized by measuring the temperature rise after a set exposure time and evaluating the energy deposited into the solvent. Standardization of conditions should be accomplished after consultation between participating laboratories.

A1.7.3.1   Sample Preparation   Transport of extractables out of the complex matrix may be affected by surface area and thickness of the test article. Test articles will be prepared by two methods: grinding and cutting into strips appropriately sized to fit into the sonication apparatus.

In sonication, the sample-to-solvent ratio may affect the completeness of the technique. Therefore, a weight ratio of at least 20:1 solvent to sample should be maintained with sample amounts of 2 g.

A1.7.3.2   Extraction Conditions   The only adjustable physical parameter for sonication is time. Bath temperatures should be either standardized using ice water (0°C) or monitored by a calibrated thermometer. Extractions may be completed in as little as 15 minutes. Safety considerations are paramount as extractions are performed under normal atmosphere and the technique may provide easy ignition. The solvent reservoir level must be periodically recharged to provide the correct amount of solvent.

A1.8   ANALYTICAL METHODS

A1.8.1   Chromatographic Methods System Suitability for Extractables Profiling (Qualitative Analyses)

Standard reference materials will be used for qualitative chromatographic analytical techniques to ensure system suitability. The standard reference materials are selected to represent a range of common extractables compounds found in polymeric materials. No one analytical technique is suitable for detection of all targets. The following table presents a list of system suitability analytes for GC- and HPLC-based analytical techniques. The presence of these analytes should be verified at the recommended concentrations prior to analysis of test article extracts by any participating laboratory.

Note that the entire group of participating laboratories and scientists will judge whether a given participating laboratory has met system suitability for its analytical techniques prior to that laboratory analyzing test article extracts.

Compound name	Suggested techniques	Recommended target concentration (μg/mL)
Pyrene	GC and LC/UV	1
2-Mercaptobenzothiazole	GC or LC	50
Tetramethylthiuram disulfide	GC and LC/UV	50
Butylated hydroxytoluene (BHT)	GC or LC	50
Irganox 1010	LC	50
Diphenylamine	LC	50
Bis(2-ethylhexyl)phthalate	GC or LC	50
Bis(dodecyl)phthalate	GC or LC	50
Stearic acid	GC and LC/MS	100
2-Ethylhexanol	GC	50

A1.8.2   Nonvolatile Residue Analysis

The nonvolatile residue from the extracts will be qualitatively examined for inorganic and organic substances. For inorganic species, ICP/MS and EDX/WDX will be employed. For nonvolatile organic substances, infrared spectroscopy will be employed.

An aliquot of each appropriate extract (10–20 mL) will be transferred to a suitable weighing dish and evaporated to dryness using a hot water bath. Other drying methods can be used, but care should be taken to not degrade the residue.

Note that the choice of extracts submitted to these analyses will be made in consultation with all participating laboratories and investigators.

A1.8.2.1 *ICP/MS or EDX/WDX*   For ICP, samples must be digested to obtain a solution as required in the referenced analytical method.[6] Digestions should be performed using aqueous solutions (i.e., aqueous solution of nitric acid).

For EDX/WDX, the dried residues of the extracts are mounted for analysis. A scanning electron microprobe or other suitable analytical instrument is used to generate the X-ray spectrum showing the elements detected in the sample. The results are reported qualitatively.

A1.8.2.2 *Infrared Spectroscopic Analysis*   The residue from the extract can be transferred onto a KBr or KRS-5 crystal with the aid of a solvent if necessary. The sample should be scanned 100× from 4000 to 400 cm^{-1} having a resolution of at least 4 cm^{-1}. The spectra can be qualitatively evaluated by comparing with a spectral library or identification of major functional groups.

A1.8.3 GC/MS

Semivolatile compounds will be analyzed by GC/MS using a predominantly nonpolar capillary column with wide (40–300°C) temperature programming.[7] Each GC/MS analysis will produce an extractables "profile" in the form of a total ion chromatogram (TIC). As a first pass, identifications of individual extractables will be accomplished with manual interpretation of the electron ionization (EI) spectra assisted by computerized mass spectral library searching. Beyond this, more difficult identifications may require the collection of additional data (such as chemical ionization GC/MS for molecular weight confirmation and high-resolution MS for elemental composition), the purchase of reference compounds, and so on.

The following GC/MS conditions are provided as an example. Any nonpolar (100% dimethyl siloxane) or slightly polar (5% diphenyl siloxane) column can be used along with full temperature programming. Data cannot be collected while the injection solvent is in the ion source.

Note that additional identification work beyond the first-pass analysis will be accomplished only after consultation with all participating laboratories and investigators.

Also, note that the GC/MS instrumental conditions presented below are target conditions for all participating laboratories and investigators. The actual conditions employed by any participating laboratory should be reviewed by the entire group of participating investigators so that harmonization among laboratories can be preserved.

Gas chromatograph conditions	
Instrument	Hewlett-Packard 5890 Series II Plus, Agilent 6890, or equivalent
Injection mode	Cool on-column or splitless injection
Injection volume	1 μL
Injector temperature/program	40°C initial; oven track ON for on-column injection 280°C for splitless injection
Purge valve	On at 1.00 minute, off initially
Column	Restek Rtx-1, 30 m × 0.25 mm (0.1-μm film), or equivalent
Oven temperature	40°C for 1 minute, heated at 10°C/min to 300°C, and hold for 10 minutes
Pressure program	Constant flow (helium) at 1 mL/min
Transfer line	280°C

Mass spectrometer conditions	
Instrument	Hewlett-Packard 5972, Agilent 5973 MSD, or equivalent
Ionization mode	EI
Scan mode	Scanning; m/z 50–650
Scan cycle time	Approximately 2 s/scan

A1.8.4 HPLC/DAD

UV-active species will be identified in the extracts by retention time and UV spectral matches. Reversed-phase HPLC conditions will be employed using a gradient range from 50% to 100% solvent.[8] The chromatogram of the extracts will be compared with that of a library of compounds, and identification will be confirmed by obtaining the actual compound and analyzing with the sample.

Note that the HPLC/DAD instrumental conditions presented below are target conditions for all participating laboratories and investigators. The actual conditions employed by any participating laboratory should be reviewed by the entire group of participating investigators so that harmonization among laboratories can be preserved.

Liquid chromatograph conditions	
Instrument	Hewlett-Packard 1050, Agilent 1100, or equivalent
Flow rate	1 mL/min
Injection volume	10 μL
UV wavelength	200 nm
Column	Vydac (201tp5415) C18, 5-μm particles 15 cm × 4.6 mm or equivalent
Temperature	60°C

Liquid chromatograph conditions	
Mobile phase	Initial 50:50 acetonitrile (ACN)/water
	11-minute linear gradient
	Final 100% ACN
	Hold 8 minutes
	50:50 ACN/water at 1.5 mL/min for 5 minutes at 25 minutes return to 1.0 mL/min

A1.8.5 LC/MS

Compounds will be analyzed by LC/MS with in-line UV absorbance detection. The method will use reversed-phase chromatography with a wide (gradient) range of solvent strengths.[9] Each LC/MS analysis will produce two extractables "profiles" in the form of a TIC and a UV chromatogram. As a first pass, identifications of individual extractables will be accomplished with manual interpretation of the APCI spectra. Note that computerized mass spectral library searching is not available for APCI. Correlation with the GC/MS profiles will be attempted manually.

Beyond this, more difficult identifications may require the collection of additional data such as tandem mass spectrometry (MS/MS) for induced fragmentation and the purchase of reference compounds.

Note that additional identification work beyond the first-pass analysis will be accomplished only after consultation with all participating laboratories and investigators.

Also, note that the LC/MS instrumental conditions presented below are target conditions for all participating laboratories and investigators. The actual conditions (i.e., solvent strength, etc.) employed by any participating laboratory should be reviewed by the entire group of participating investigators so that harmonization among laboratories can be preserved.

Liquid chromatograph conditions	
Instrument	Hewlett-Packard 1050, Agilent 1100, or equivalent
Injection volume	10–50 µL, as appropriate
UV wavelength	280 nm
Column	Alltech Alltima C18, 4.6 mm × 25 cm
	5-µm particles, or equivalent
Mobile phase	A—75:25 ACN/water
	B—50:50 ACN/tetrahydrofuran

Gradient		
Time (minutes)	% A	% B
0	100	0
10	60	40

Gradient		
Time (minutes)	% A	% B
20	0	100
30	0	100
32	100	0
45	100	0

Mass spectrometer conditions	
Instrument	Micromass Platform II, Agilent 1100 MSD, or equivalent
Ionization mode	APCI
	(Both APCI+ and APCI– will be accomplished)
Scan mode	Scanning; *m/z* 50–1350
Scan cycle time	Approximately 5 s/scan

A1.9 ANALYTICAL PROCEDURES

A1.9.1 Qualitative Analysis Procedure

A1.9.1.1 Sample Extract Preparation The resulting extracts will usually contain low-level amounts of extractables. Sample concentration may be necessary as well as solvent switching to provide compatible samples for different analytical instrumentation. It is possible to manipulate extracts to provide very large concentration ratios, but this also has the effect of concentrating normal solvent impurities. For known targets in well-characterized matrices, this is possible. As this protocol is for characterization purposes, no analyte or matrix behavior will be presupposed. Therefore, extracts will be concentrated no more than 100× as can be considered reasonable given normal ACS reagent purities of 99+%.

Concentration may be affected by residue formation and reconstitution in a smaller volume or by concentration to a fixed volume. Solvents may be switched during these procedures as appropriate. Residues may be prepared using standard techniques, rotary evaporation, nitrogen blowdown, lyophilization, or centrifugal evaporation. Details of the sample preparation techniques will be based on good scientific reasoning and recorded in the laboratory notebook at time of analysis.

Note that the actual conditions employed by any participating laboratory should be reviewed by the entire group of participating investigators so that harmonization among laboratories can be preserved.

A1.9.1.2 Blank Solvent Extract Preparation The solvent blanks are extracted and prepared in the same manner as the sample and analyzed prior to sample extracts.

A1.9.1.3 Analysis The extracts are surveyed using appropriate analytical methodology described in Section A1.8.

A1.9.2 Quantitative Analysis Procedure (if Required)

A1.9.2.1 Sample Extract Preparation The sample extracts can be obtained from the qualitative solutions, or new extracts can be prepared to optimize for the extraction and analysis techniques.

A1.9.2.2 Blank Solvent Extract Preparation A blank solvent extract is prepared in the same manner as the sample and analyzed prior to sample analysis.

A1.9.2.3 Standard Reference Material Preparation Standard reference materials can be prepared at the appropriate concentrations as mixtures in a single solvent. Quantitative standardization will be performed using a single point relative to an internal or external standard.

A1.9.2.4 Analysis The extracts will be analyzed using methods that are optimized to detect the substances identified in the survey analysis.

Note that the actual conditions and procedures employed by any participating laboratory should be reviewed by the entire group of participating investigators so that harmonization among laboratories can be preserved.

A1.10 DATA EVALUATION AND REPORTING

A1.10.1 Qualitative Analysis

- A list of all identified extractables for all techniques will be generated that were not detected in the corresponding blank
- A list of all unidentified peaks in chromatogram that were not detected in the corresponding blank at signal-to-noise ratios greater than 10
- Amount of nonvolatile residue relative toward blank
- Indication of presence of known materials and techniques used in detection
- For each extraction, the solvents, condition, and sample size
- For each analytical technique, the equipment, conditions, and calibration method
- Provide copies of chromatograms and spectra

A1.10.2 Quantitative Measurement (if Required)

- List of analytes and source of standard reference materials
- Extraction and analysis techniques needed to determine all analytes
- For each extraction, the solvents, condition, and sample size
- For each analytical technique, the equipment, conditions, and calibration method
- Report as microgram per gram sample
- Comparison to the known analytes/amounts
- Provide copies of sample and standard reference chromatograms and spectra

A1.11 GLOSSARY

Abbreviations	
GC/FID	Gas chromatography/flame ionization detection
OVIs	Organic volatile impurities
EDX	Energy-dispersive X-ray
WDX	Wavelength-dispersive X-ray
ICP/MS	Inductively coupled plasma/mass spectrometer
GC/MS	Gas chromatography/mass spectrometry
HPLC/DAD	High-performance liquid chromatography/diode array detection
LC/MS	Liquid chromatography/mass spectrometry
AES	Atomic emission spectroscopy
ELSD	Evaporative light-scattering detector
RI	Refractive index
TIC	Total ion chromatogram
APCI	Atmospheric pressure chemical ionization

Compounds	Chemical Abstracts Service (CAS) numbers
Pyrene	129-00-0
2-Mercaptobenzothiazole	149-30-4
Tetramethylthiuram disulfide	137-26-8
Butylated hydroxytoluene (BHT)	128-37-0
Diphenylamine	122-37-4
Bis(2-ethylhexyl)phthalate	117-81-7
Bis(dodecyl)phthalate	2432-90-8
Stearic acid	57-11-4
2-Ethylhexanol	104-76-7

APPENDIX 2 EXPERIMENTAL PROTOCOL FOR CONTROLLED EXTRACTION STUDIES ON PLASTIC TEST ARTICLES

A2.1 INTRODUCTION

In November 1998 and May 1999, the United States Food and Drug Administration (FDA) issued two chemistry, manufacturing, and controls (CMC) draft guidances addressing orally inhaled and nasal drug products (OINDPs): (1) the draft *Metered Dose Inhaler (MDI) and Dry Powder Inhaler (DPI) Drug Products Chemistry, Manufacturing, and Controls Documentation*[2] (referred to here as the "MDI/DPI draft guidance"); and (2) the draft *Nasal Spray and Inhalation Solution, Suspension, and Spray Drug Products Chemistry, Manufacturing, and Controls Documentation* (referred to here as the "nasal spray draft guidance"). In July 2002, the Nasal Spray Guidance was finalized.[3]

Currently, both guidances recommend that the sponsor identifies, reports, and conducts toxicological analyses on all extractables found in the controlled extraction study (referred to in the guidances as a "control extraction study"). Examples of these recommendations are described in the draft MDI/DPI Guidance regarding MDI canisters, valves, and actuators (lines 883–884, 990–991, and 1073):

> ... the profile of each extract should be evaluated both analytically and toxicologically.

The Product Quality Research Institute (PQRI) Leachables and Extractables Working Group has developed this experimental protocol as an example of a controlled extraction study for plastic test articles. Various experimental parameters will be investigated, test article extracts will be analyzed, and results will be evaluated within the context of the Working Group's approved work plan and experimental hypothesis.

This experimental protocol will be used by all laboratories and investigators participating in the study.

A2.2 PURPOSE AND SCOPE OF WORK

A2.2.1 Purpose

The purpose of the experiments outlined in this protocol is to generate data from controlled extraction studies that will contribute to a larger database, which the Working Group will use to investigate its hypotheses[4]:

1. Scientifically justifiable thresholds based on the best available data and industry practices can be developed for the

    a. reporting and safety qualification of leachables in OINDPs, and

    b. reporting of extractables from the critical components used in corresponding container/closure systems.

Reporting thresholds for leachables and extractables will include associated identification and quantitation thresholds.

2. Safety qualification of extractables would be scientifically justified on a case-by-case basis.

A2.2.2 Scope

A2.2.2.1 Topics Addressed by This Protocol
This protocol covers only controlled extraction studies that would be applied to components from MDIs. The MDI represents the best example of "correlation" between extractables from components and leachables in drug products. Controlled extraction studies will be performed following the general outline described in the guidances. Test articles will be subjected to different extraction conditions to show how different experimentally controlled parameters affect resulting extractables profiles. Additionally, the Working Group will assess experimental results to identify reasonable approaches for sample preparation and analysis of extractables from container and closure components.

As no single analytical technique can be used to identify and quantify all unknown extractables, a variety of methods will be utilized in this protocol to maximize the likelihood that all extractables compounds associated with the test articles are evaluated analytically. Overlap between methods will supply corroborating data that the procedures are valid. To provide a full analytical survey of possible analytes, the following strategy will be employed:

1. direct injection gas chromatography/mass spectrometry (GC/MS) for identification and assessment of relatively volatile extractables;

2. high-performance liquid chromatography/diode array detection (HPLC/DAD) for identification and assessment of relatively polar/nonvolatile ultraviolet (UV)-active extractables;

3. high-performance liquid chromatography/mass spectrometry (LC/MS) for identification and assessment of relatively polar/nonvolatile extractables, which may or may not have UV activity;

4. inductively coupled plasma/mass spectrometry (ICP/MS), inductively coupled plasma/atomic emission spectroscopy (ICP/AES), or energy-dispersive X-ray (EDX)/wavelength-dispersive X-ray (WDX) to detect single elements in the extracts (i.e., metals); and

5. Fourier transform infrared (FTIR) spectroscopy for the characterization of major components in the nonvolatile extractable residues.

A2.2.2.2 Topics Not Addressed by This Protocol
Studies designed to assess recovery (i.e., mass balance) for individual extractables relative to the known formulations of chemical additives in the plastic test articles, or reproducibility of extractables profiles for multiple "batches" of any particular test article are not within the scope of this test protocol.

The extraction procedures, analytical techniques/methods, and analysis conditions described in this experimental test protocol will not be validated as material control methods, since they will be performed in order to collect qualitative information. However, during the course of these experiments, the PQRI Leachables and Extractables Working Group will review the results and may initiate additional experimental work for quantitative assessment of extractables.

This protocol does not address system suitability tests for quantitative methods. Appropriate system suitability tests will be addressed later, and agreement on this issue will be reached with all of the participating laboratories.

Special case studies such as organic volatile impurities (OVIs), *N*-nitrosamines, or polynuclear aromatic hydrocarbons (PAHs or polynuclear aromatics [PNAs]) will not be considered in this study. These "special case" classes of extractables have defined highly specific analytical methods, which are generally accepted and commonly used for their identification and quantitative assessment.

It should be noted that the outlined experimental procedures, analytical instrumentation parameters and conditions, and other details are intended as a guidance for laboratory studies. Details of actual experimental procedures, and so on, should be reviewed by the entire group of participating laboratories and investigators so that harmonization among laboratories working on the same test articles can be achieved.

A2.3   REGULATORY STATUS

This is a good manufacturing practice (GMP)[5] study. All experiments shall be performed under GMP conditions to the extent practical in a particular laboratory.* Any changes to these protocols shall be documented, following appropriate GMP change control procedures.

A2.4   SAFETY AND ENVIRONMENTAL IMPACT

Organic solvents are commonly used to enhance solubility of lipophilic targets and to increase transport of small molecules out of complex matrices. These solvents may be flammable and/or show short-term and long-term environmental health risks. Care must be exercised with their use. Consult the material safety and data sheet (MSDS) for appropriate personal protection and disposal.

A2.5   TEST ARTICLES

Polypropylene and low-density polyethylene (LDPE) materials will be provided in disk form for use as test articles. The additive formulations and manufacturing

---

* These experiments are considered research projects to be conducted in research labs, which are not strictly GMP compliant. However, all participating labs will perform these experiments in the spirit of GMP, which means that they will implement appropriate documentation, sample handling, data traceability, and so on.

conditions for these test articles are known and will be provided to all laboratories participating in the study.

The following known formulation ingredients will be provided for use as identification and potentially quantitation reference compounds/mixtures:

- Irganox 1010,
- Ultranox 626,
- calcium stearate,
- Pationic 901, and
- Millad 3988.

Note that additional reference compounds and additive mixtures may be required for the completion of this test protocol and will be provided as appropriate.

A2.6   CHEMICALS AND EQUIPMENT

Extraction and analytical methods have been chosen and designed so as to utilize chemicals, apparatus, and instrumentation available in typical laboratories routinely involved with this type of study.

A2.6.1   Extraction Solvents

Extractions will be performed on each test article using three solvents representing a range of polarity selected from the list below. The solvents should be American Chemical Society (ACS) grade or better:

- methylene chloride (dichloromethane),
- 2-propanol (isopropanol), and
- hexane (*n*-hexane, not hexanes).

Depending on the behavior of the test articles in these particular solvent systems, additional solvents may be chosen. Changes in extracting solvent will be discussed by all study participants prior to change initiation by any particular study participant or laboratory.

A2.6.2   Extraction Apparatus

- Soxhlet apparatus with an Allihn condenser, flask (500 mL), and hot plate or heating mantle
- Sonicator
- Reflux apparatus consisting of an Erlenmeyer flask (125 mL or larger) and condenser with ground-glass joints, hot plate, or heating mantle

A2.6.3 Analytical Instrumentation

- Gas chromatograph equipped with a flame ionization detector (GC/FID)
- Gas chromatograph equipped with a mass spectrometer (GC/MS)
- Liquid chromatograph equipped with a photodiode array detector
- Liquid chromatograph equipped with an atmospheric pressure chemical ionization (APCI) -capable mass spectrometer (LC/MS)
- FTIR spectrometer
- EDX and/or WDX equipped with a microprobe or scanning electron microprobe
- ICP/MS and/or ICP/AES

A2.7 EXTRACTION PROCEDURES

For each extraction technique and solvent type, appropriate blanks (no test article sample) must be prepared. These must be prepared concurrently using a different extraction apparatus (same type) under the same conditions, or by using the same apparatus prior to charging with the sample.

Note that the extraction parameters and conditions outlined below are subject to modification, and the details of any particular extraction process must be agreed to among all laboratory study participants prior to initiation of experimental work in any particular laboratory.

A2.7.1 Soxhlet Extraction

A2.7.1.1 Sample Preparation   Samples of each test article should be cut into strips appropriately sized to fit into pre-extracted Soxhlet cellulose thimbles. Sample amounts may be in the range of 1–3 g (2 g) using 200 mL of solvent. For quantitative measurements, extracts prepared by Soxhlet will have to be evaporated to dryness and the resulting residues redissolved to a known volume (25–50 mL). Alternatively, an internal standard can be used for quantitative measurements.

A2.7.1.2 Extraction Conditions   Under normal laboratory conditions, three physical extraction parameters may be modified: turnover number, total extraction time, and temperature. Temperature is the most difficult of the three parameters to control as the sample holder is maintained above the vapor level (temperature may be above the boiling point), but will be continuously bathed in freshly distilled solvent (coil temperature). It is recommended that the coil temperature be kept as low as possible to avoid heating above the solvent flash point.

Turnover number is controlled by the heating rate and should be limited by safety concerns. At low turnover numbers, the extraction characteristics will resemble those of reflux and may be limited by equilibrium phenomena. It is recommended that turnover numbers to be at least 10 during the course of the extraction.

Extraction time should be in the range of 24 hours to guard against possible degradation of thermally labile or reactive compounds.

A2.7.2 Reflux

Reflux extraction is a common and easily implemented approach for the production of extractables (e.g., USP <381> "Elastomeric Closures—Physicochemical Tests"). Conditions are easily standardized as the temperature and pressure are at the defined boiling points of the extraction solvents. Unlike Soxhlet extraction, reflux extraction is an equilibrium phenomenon.

A2.7.2.1 Sample Preparation Transport of extractables out of the complex matrix may be affected by the surface area and thickness of the test article. Test articles will be prepared by three methods: pressing, grinding, and cutting into strips appropriately sized to fit into the reflux apparatus.

Sample amounts should be in the range of 2 g using 25–50 mL of solvent. For quantitative measurements, the solvent with the sample and the flask can be weighed and returned to original weight after extraction. Alternatively, an internal standard can be used for quantitative measurements.

In reflux extraction, the sample-to-solvent ratio may affect the completeness of the technique. This should be addressed when optimizing the method for the measurement of extractables.

A2.7.2.2 Extraction Conditions The only adjustable physical parameter for reflux extraction is time. Extraction time should be 2–4 hours. The solvent reservoir level must be monitored and periodically recharged to provide the correct amount of solvent.

A2.7.3 Sonication

Sonication uses ultrasonic energy instead of thermal energy to increase the rate of diffusion of small analytes out of a solid matrix. Similar considerations as reflux extraction (equilibrium conditions) should be evaluated, but these cannot be calculated using thermodynamic parameters. Sonication equipment may be standardized by measuring the temperature rise after a set exposure time and evaluating the energy deposited into the solvent. Standardization of conditions should be accomplished after consultation between participating laboratories.

A2.7.3.1 Sample Preparation Transport of extractables out of the complex matrix may be affected by surface area and thickness of the test article. Test articles will be prepared by three methods: pressing, grinding, and cutting into strips appropriately sized to fit into the sonication apparatus.

In sonication, the sample-to-solvent ratio may affect the completeness of the technique. Therefore, a weight ratio of at least 20:1 solvent to sample should be maintained with sample amounts of 2 g.

A2.7.3.2   *Extraction Conditions*   The only adjustable physical parameter for sonication is time. Bath temperatures should be either standardized using ice water (0°C) or monitored by a calibrated thermometer. Extractions may be completed in as little as 15 minutes. Safety considerations are paramount as extractions are performed under normal atmosphere and the technique may provide easy ignition. The solvent reservoir level must be periodically recharged to provide the correct amount of solvent.

A2.8   ANALYTICAL METHODS

A2.8.1   Chromatographic Methods System Suitability for Extractables Profiling (Qualitative Analyses)

Standard reference materials will be used for qualitative chromatographic analytical techniques to ensure system suitability. The standard reference materials are selected to represent a range of common extractables compounds found in polymeric materials. No one analytical technique is suitable for detection of all targets. The following table presents a list of system suitability analytes for GC- and HPLC-based analytical techniques. The presence of these analytes should be verified at the recommended concentrations prior to analysis of test article extracts by any participating laboratory.

   Note that the entire group of participating laboratories and scientists will judge whether a given participating laboratory has met system suitability for its analytical techniques prior to that laboratory analyzing test article extracts.

Compound name	Suggested techniques	Recommended target concentration (μg/mL)
Pyrene	GC and LC/UV	1 ppm
2-Mercaptobenzothiazole	GC or LC	50 ppm
Tetramethylthiuram disulfide	GC and LC/UV	50 ppm
Butylated hydroxytoluene (BHT)	GC or LC	50 ppm
Irganox 1010	LC	50 ppm
Diphenylamine	LC	50 ppm
Bis(2-ethylhexyl)phthalate	GC or LC	50 ppm
Bis(dodecyl)phthalate	GC or LC	50 ppm
Stearic acid	GC and LC/MS	100 ppm
2-Ethylhexanol	GC	50 ppm

A2.8.2   Nonvolatile Residue Analysis

The nonvolatile residue from the extracts will be qualitatively examined for inorganic and organic substances. For inorganic species, ICP/MS and EDX/WDX will be employed. For nonvolatile organic substances, infrared spectroscopy will be employed.

An aliquot of each appropriate extract (10–20 mL) will be transferred to a suitable weighing dish and evaporated to dryness using a hot water bath. Other drying methods can be used, but care should be taken to not degrade the residue.

Note that the choice of extracts submitted to these analyses will be made in consultation with all participating laboratories and investigators.

A2.8.2.1 *ICP/MS or EDX/WDX*

For ICP, samples must be digested to obtain a solution as required in the referenced analytical method.[6] Digestions should be performed using aqueous solutions (i.e., aqueous solution of nitric acid).

For EDX/WDX, the dried residues of the extracts are mounted for analysis. A scanning electron microprobe or other suitable analytical instrument is used to generate the X-ray spectrum showing the elements detected in the sample. The results are reported qualitatively.

A2.8.2.2 *Infrared Spectroscopic Analysis*

The residue from the extract can be transferred onto a KBr or KRS-5 crystal with the aid of a solvent if necessary. The sample should be scanned 100× from 4000 to 400 cm^{-1} having a resolution of at least 4 cm^{-1}. The spectra can be qualitatively evaluated by comparing with a spectral library or identification of major functional groups.

A2.8.3 GC/MS

Semivolatile compounds will be analyzed by GC/MS using a predominantly nonpolar capillary column with wide (40–300°C) temperature programming.[7] Each GC/MS analysis will produce an extractables "profile" in the form of a total ion chromatogram (TIC). As a first pass, identifications of individual extractables will be accomplished with manual interpretation of the electron ionization (EI) spectra assisted by computerized mass spectral library searching. Beyond this, more difficult identifications may require the collection of additional data (such as chemical ionization GC/MS for molecular weight confirmation and high-resolution MS for elemental composition), the purchase of reference compounds, and so on.

The following GC/MS conditions are provided as an example. Any nonpolar (100% dimethyl siloxane) or slightly polar (5% diphenyl siloxane) column can be used along with full temperature programming. Data cannot be collected while the injection solvent is in the ion source.

Note that additional identification work beyond the first-pass analysis will be accomplished only after consultation with all participating laboratories and investigators.

Also, note that the GC/MS instrumental conditions presented below are target conditions for all participating laboratories and investigators. The actual conditions employed by any participating laboratory should be reviewed by the entire group of participating investigators so that harmonization between laboratories can be preserved.

Gas chromatograph conditions	
Instrument	Hewlett-Packard 5890 Series II Plus, Agilent 6890, or equivalent
Injection mode	Cool on-column or splitless injection
Injection volume	1 μL
Injector temperature/program	40°C initial; oven track ON for on-column injection 280°C for splitless injection
Purge valve	On at 1.00 minute, off initially
Column	Restek Rtx-1, 30 m × 0.25 mm (0.1-μm film) or equivalent
Oven temperature	40°C for 1 minute, heated at 10°C/min to 300°C, and hold for 10 minutes
Pressure program	Constant flow (helium) at 1 mL/min
Transfer line	280°C

Mass spectrometer conditions	
Instrument	Hewlett-Packard 5972, Agilent 5973 MSD, or equivalent
Ionization mode	EI
Scan mode	Scanning; m/z 50–650
Scan cycle time	Approximately 2 s/scan

A2.8.4 HPLC/DAD

UV-active species will be identified in the extracts by retention time and UV spectral matches. Reversed-phase HPLC conditions will be employed using a gradient range from 50% to 100% solvent.[8] The chromatogram of the extracts will be compared with that of a library of compounds and identification confirmed by obtaining the actual compound and analyzing with the sample.

Note that the HPLC/DAD instrumental conditions presented below are target conditions for all participating laboratories and investigators. The actual conditions (i.e., solvent strength, etc.) employed by any participating laboratory should be reviewed by the entire group of participating investigators so that harmonization among laboratories can be preserved.

Liquid chromatograph conditions	
Instrument	Hewlett-Packard 1050, Agilent 1100, or equivalent
Flow rate	1 mL/min
Injection volume	10 μL
UV wavelength	200 nm
Column	Vydac (201tp5415) C18, 5-μm particles 15 cm × 4.6 mm or equivalent
Temperature	60°C

Liquid chromatograph conditions	
Mobile phase	Initial 50:50 acetonitrile (ACN)/water
	11-minute linear gradient
	Final 100% ACN
	Hold 8 minutes
	50:50 ACN/water at 1.5 mL/min for 5 minutes at 25 minutes return to 1.0 mL/min

A2.8.5 LC/MS

Compounds will be analyzed by LC/MS with in-line UV absorbance detection. The method will use reversed-phase chromatography with a wide (gradient) range of solvent strengths.9 Each LC/MS analysis will produce two extractables "profiles" in the form of a TIC and a UV chromatogram. As a first pass, identifications of individual extractables will be accomplished with manual interpretation of the APCI spectra. Note that computerized mass spectral library searching is not available for APCI. Correlation with the GC/MS profiles will be attempted manually.

Beyond this, more difficult identifications may require the collection of additional data such as tandem mass spectrometry (MS/MS) for induced fragmentation and the purchase of reference compounds.

Note that additional identification work beyond the first-pass analysis will be accomplished only after consultation with all participating laboratories and investigators.

Also, note that the LC/MS instrumental conditions presented below are target conditions for all participating laboratories and investigators. The actual conditions employed by any participating laboratory should be reviewed by the entire group of participating investigators so that harmonization among laboratories can be preserved.

Liquid chromatograph conditions	
Instrument	Hewlett-Packard 1050, Agilent 1100, or equivalent
Injection volume	10–50 μL, as appropriate
UV wavelength	280 nm
Column	Alltech Alltima C18, 4.6 mm × 25 cm
	5-μm particles or equivalent
Mobile phase	A—75:25 ACN/water
	B—50:50 ACN/tetrahydrofuran

Gradient		
Time (minutes)	% A	% B
0	100	0
10	60	40
20	0	100
30	0	100
32	100	0
45	100	0

Mass spectrometer conditions	
Instrument	Micromass Platform II, Agilent 1100 MSD, or equivalent
Ionization mode	APCI
	(Both APCI+ and APCI– will be accomplished)
Scan mode	Scanning; m/z 50–1350
Scan cycle time	Approximately 5 s/scan

A2.9   ANALYTICAL PROCEDURES

A2.9.1   Qualitative Analysis Procedure

A2.9.1.1   Sample Extract Preparation   The resulting extracts will usually contain low-level amounts of extractables. Sample concentration may be necessary as well as solvent switching to provide compatible samples for different analytical instrumentation. It is possible to manipulate extracts to provide very large concentration ratios, but this also has the effect of concentrating normal solvent impurities. For known targets in well-characterized matrices, this is possible. As this protocol is for characterization purposes, no analyte or matrix behavior will be presupposed. Therefore, extracts will be concentrated no more than 100× as can be considered reasonable given normal ACS reagent purities of 99+%.

Concentration may be affected by residue formation and reconstitution in a smaller volume or by concentration to a fixed volume. Solvents may be switched during these procedures as appropriate. Residues may be prepared using standard techniques, rotary evaporation, nitrogen blowdown, lyophilization, or centrifugal evaporation. Details of the sample preparation techniques will be based on good scientific reasoning and recorded in the laboratory notebook at time of analysis.

Note that the actual conditions employed by any participating laboratory should be reviewed by the entire group of participating investigators so that harmonization among laboratories can be preserved.

A2.9.1.2   Blank Solvent Extract Preparation   The solvent blanks are extracted and prepared in the same manner as the sample and analyzed prior to sample extracts.

A2.9.1.3   Analysis   The extracts are surveyed using appropriate analytical methodology described in Section A2.8.

A2.9.2   Quantitative Analysis Procedure (if Required)

A2.9.2.1   Sample Extract Preparation   The sample extracts can be obtained from the qualitative solutions, or new extracts can be prepared to optimize for the extraction and analysis techniques.

A2.9.2.2   Blank Solvent Extract Preparation   A blank solvent extract is prepared in the same manner as the sample and analyzed prior to sample analysis.

A2.9.2.3   Standard Reference Material Preparation   Standard reference materials can be prepared at the appropriate concentrations as mixtures in a single solvent. Quantitative standardization will be performed using a single point relative to an internal or external standard.

A2.9.2.4   Analysis   The extracts will be analyzed using methods that are optimized to detect the substances identified in the survey analysis.

Note that the actual conditions and procedures employed by any participating laboratory should be reviewed by the entire group of participating investigators so that harmonization among laboratories can be preserved.

A2.10   DATA EVALUATION AND REPORTING

A2.10.1   Qualitative Analysis

- A list of all identified extractables for all techniques will be generated that were not detected in the corresponding blank
- A list of all unidentified peaks in chromatogram that were not detected in the corresponding blank at signal-to-noise ratios greater than 10
- Amount of nonvolatile residue relative toward blank
- Indication of presence of known materials and techniques used in detection
- For each extraction, the solvents, condition, and sample size
- For each analytical technique, the equipment, conditions, and calibration method
- Provide copies of chromatograms and spectra

A2.10.2   Quantitative Measurement (if Required)

- List of analytes and source of standard reference materials
- Extraction and analysis techniques needed to determine all analytes
- For each extraction, the solvents, condition, and sample size

- For each analytical technique, the equipment, conditions, and calibration method
- Report as microgram per gram sample
- Comparison to the known analytes/amounts
- Provide copies of sample and standard reference chromatograms and spectra

A2.11 GLOSSARY

Abbreviations	
GC/FID	Gas chromatography/flame ionization detection
OVIs	Organic volatile impurities
EDX	Energy-dispersive X-ray
WDX	Wavelength-dispersive X-ray
ICP/MS	Inductively coupled plasma/mass spectrometer
GC/MS	Gas chromatography/mass spectrometry
HPLC/DAD	High-performance liquid chromatography/diode array detection
LC/MS	Liquid chromatography/mass spectrometry
AES	Atomic emission spectroscopy
ELSD	Evaporative light-scattering detector
RI	Refractive index
TIC	Total ion chromatogram
APCI	Atmospheric pressure chemical ionization

Compounds	Chemical Abstracts Service (CAS) numbers
Pyrene	129-00-0
2-Mercaptobenzothiazole	149-30-4
Tetramethylthiuram disulfide	137-26-8
Butylated hydroxytoluene (BHT)	128-37-0
Diphenylamine	122-39-4
Bis(2-ethylhexyl)phthalate	117-81-7
Bis(dodecyl)phthalate	2432-90-8
Stearic acid	57-11-4
2-Ethylhexanol	104-76-7

APPENDIX 3 PROTOCOL ADDITION, PHASE 2 STUDIES: QUANTITATIVE CONTROLLED EXTRACTION STUDIES ON THE SULFUR-CURED ELASTOMER

Protocol for Validation of a Quantitative Extractables Profiling Method for a Sulfur-Cured Elastomer Using Soxhlet Extraction and Gas Chromatography/ Flame Ionization Detection

A3.1 INTRODUCTION AND BACKGROUND

Qualitative controlled extraction studies guided by a specific and detailed protocol have been accomplished on a sulfur-cured elastomeric test article of known additive composition. These qualitative studies produced extractables profiles by gas chromatography/mass spectrometry (GC/MS) and high-performance liquid chromatography/mass spectrometry (LC/MS), which exactly reflect the known additive composition of the elastomeric test article.

This protocol addition is designed to extend the qualitative controlled extraction study to a quantitative controlled extraction study, with appropriate method optimization and investigation of validation parameters. The analytical system chosen for validation is gas chromatography/flame ionization detection (GC/FID).

A3.2 TEST ARTICLE

The elastomer test article to be employed in this study is a sulfur-cured and carbon-black-containing rubber especially created for this Product Quality Research Institute (PQRI) project by the West Pharmaceutical Services. The qualitative extractables profile of this elastomeric material was fully characterized under a preceding test protocol.

A3.3 METHOD DEVELOPMENT

Based on the results of the qualitative controlled extraction studies, Soxhlet extraction in methylene chloride with quantitative GC analysis of extracts has been selected for optimization and validation. Internal standardization utilizing appropriate authentic reference materials will be employed for quantitative calibration of the analytical system. The known additives in the elastomeric test article that can be quantitated by this analytical technique include

- 2,2'-methylene-bis(6-*tert*-butyl-4-ethylphenol),
- coumarone-indene resin-related species,
- *n*-alkanes derived from paraffin/oils, and
- additional relatively minor extractables.

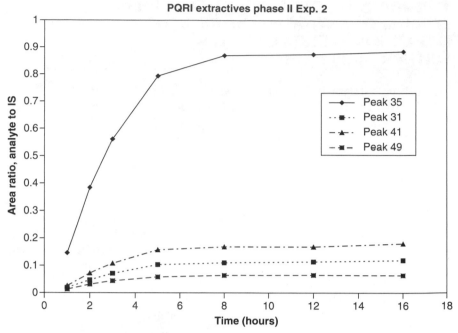

Figure A3.1   Model extraction optimization experiment (methylene chloride Soxhlet extraction; GC/MS analysis of extracts; internal standard [IS] added to extracting solution).

All details of the analytical method, including the extraction procedure and analysis system will be documented in laboratory notebooks and/or other appropriate documentation media.

Prior to method validation, the extraction procedure will be optimized to produce maximum quantities of target extractables (i.e., "asymptotic" levels; note the example experiment in Fig. A3.1). The optimized extraction conditions will then be employed for an initial examination of extraction method repeatability. Individual representative target extractables will be used to evaluate linearity and various chromatographic parameters, to establish appropriate dynamic ranges for quantitation, and to assess method accuracy. The optimized quantitative analytical method will then be taken to validation with acceptance criteria based on either the method development studies or the expected performance of such analytical methods.

A3.4   VALIDATION PARAMETERS AND ACCEPTANCE CRITERIA

The following validation parameters that include appropriate acceptance criteria will be investigated. When appropriate, the following representative target extractables will be employed:

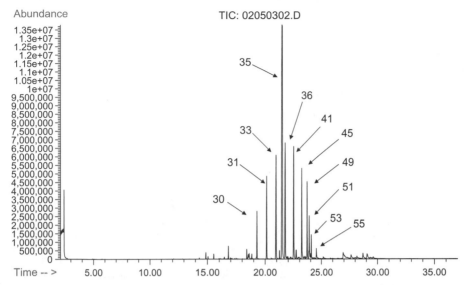

Figure A3.2   GC/MS extractables profile (total ion chromatogram [TIC]) of the West sulfur-cured elastomer (16-hour Soxhlet extraction with dichloromethane).

- 2,2′-methylene-bis(6-*tert*-butyl-4-ethylphenol),
- *n*-docosane,
- *n*-tricosane,
- *n*-tetracosane,
- *n*-pentacosane,
- *n*-hexacosane,
- *n*-octacosane, and
- internal standard: 2-fluorobiphenyl.

These target extractables include the primary phenolic antioxidant and several *n*-alkanes, which represent the bulk of the remaining extractables profile. The qualitative GC/MS extractables profiles of the sulfur-cured elastomeric test article are shown in Figures A3.2 and A3.3, with extractables identifications in Table A3.1. A representative GC/FID extractables profile is shown in Figure A3.4.

A3.4.1   System Suitability

A3.4.1.1 Instrument Precision   A test solution of target extractables with internal standard will be prepared at concentrations demonstrated not to produce adverse effects on chromatographic performance, and at levels determined to encompass the concentrations of target extractables determined in the method development phase of this study. Utilizing optimized chromatography conditions, six replicate

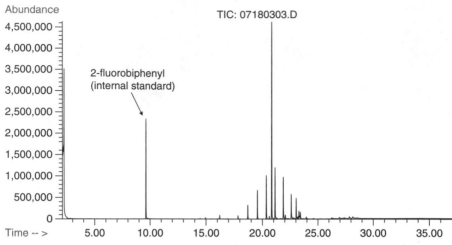

Figure A3.3   GC/MS extractables profile (total ion chromatogram [TIC]) of the West sulfur-cured elastomer (16-hour Soxhlet extraction with dichloromethane; internal standard added; optimized injection volume).

TABLE A3.1.   Identifications of Major Extractables from the West Sulfur-Cured Elastomer

Peak number	Retention time (minutes)	Identification	Comments
30	19.28	n-Docosane	Confirmed
31	20.12	Tricosane	Confirmed
33	20.94	Tetracosane	Confirmed
35	21.47	2,2′-Methylene-bis(6-tert-butyl-4-ethylphenol)	Confirmed (antioxidant)
36	21.73	Pentacosane	Confirmed
41	22.48	Hexacosane	Confirmed
45	23.20	Heptacosane	Confirmed
49	23.68	Trimer (two indenes with one α-methylstyrene)	Tentative (derived from the coumarone-indene resin)
51	23.88	Octacosane	Confirmed
53	24.06	Trimer (two indenes with one α-methylstyrene, containing one double bond)	Tentative (derived from the coumarone-indene resin)
55	24.54	Nonacosane	Confirmed

Notes: Peak numbers are taken from the controlled extraction study results in which a total of 66 major and minor extractables were identified. Confirmed implies a positive match with an authentic reference material, library mass spectrum, or both. Tentative implies a certain level of uncertainty in the exact molecular structure; however, the compound class is confirmed.

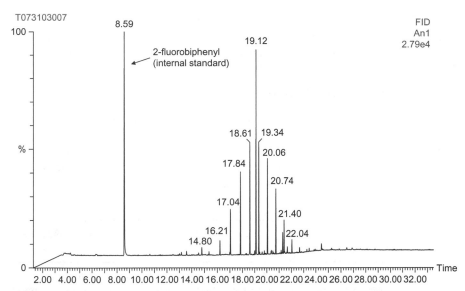

Figure A3.4   GC/FID extractables profile of the West sulfur-cured elastomer (test run from a preliminary GC/FID feasibility study; internal standard added).

injections of the test solution will be analyzed. Peak area and area ratio measurements of target extractables and the internal standard will be determined, and means and percent relative standard deviations (%RSDs) of area ratios and relative response factors (RRFs) will be calculated.

Acceptance criteria: %RSDs for area ratios and RRFs to be determined during method development.

Note: RRF is defined as

$$RRF = (A_a C_i)/(A_i C_a),$$

where

A_a = area of the analyte peak,

A_i = area of internal standard peak,

C_a = concentration of the analyte, and

C_i = concentration of the internal standard.

A3.4.1.2   *Chromatographic Resolution (USP)*   Utilizing the analyses accomplished for instrument precision, chromatographic resolution between appropriate peak pairs will be determined. Means and %RSDs will be calculated. Appropriate peak pairs will be selected during method development.

Acceptance criteria: To be determined during method development.

A3.4.1.3   *Chromatographic Tailing Factor (USP)*   Utilizing the analyses accomplished for instrument precision, chromatographic tailing factors for appropriate peaks will be determined. Means and %RSDs will be calculated. Appropriate peaks will be selected during method development.

Acceptance criteria: To be determined during method development.

A3.4.2 Linearity and Range

Linearity and range will be determined by analyzing selected target extractables at six different concentration levels (in duplicate), over a range established during the method development phase of this study. For each target extractable linearity experiment, a linear regression analysis will be accomplished on peak area ratios versus analyte concentration. Slopes, *y*-intercepts, and coefficients of determination (r^2) will be calculated.

Target extractables	*2,2′-Methylene-bis(6-tert-butyl-4-ethylphenol)*
	Pentacosane

Acceptance criteria: To be determined during method development.

In addition to the linearity study for selected target extractables, single-point RRFs will be determined for additional identified extractables for which authentic reference compounds are available. The list of extractables for which this will be accomplished and the concentration level at which the measurements will be made will be determined during the method development phase of the study. These additional extractables may or may not be limited to those listed in Table A3.1.

Acceptance criteria: Report results.

A3.4.3 Precision

A3.4.3.1 Repeatability Utilizing optimized extraction procedures, six separate extractions will be accomplished and target extractables will be quantitated with the analytical method. Means and %RSDs of individual target extractable amounts will be calculated.

Acceptance criteria: %RSD for each target extractable ≤10%.

A3.4.3.2 Intermediate Precision Intermediate precision will be evaluated by a second analyst accomplishing the repeatability study utilizing a different GC column and analytical instrument (if available).

Acceptance criteria	*1. %RSD for each target extractable ≤10%*
	2. %Difference between analyst means for each target extractable ≤25%

A3.4.4 Specificity

Specificity was demonstrated in the qualitative phase of the controlled extraction studies utilizing GC/MS.

Acceptance criteria: Confirms peak identifications and confirms no significant co-eluting peaks for each target extractable.

A3.4.5   Accuracy

Accuracy will be expressed as the percent recovery of known amounts of target extractables spiked into the extraction system.

Spiking solutions of appropriate target extractables will be prepared and spiked at three different levels (in triplicate). The individual spiking levels will be chosen to represent the appropriate range of analyte concentrations expected based on the method development experiments. Spiked samples will be analyzed by the optimized analytical method and individual mean recoveries determined for each spiking level.

Acceptance criteria: Mean recovery for each target extractable at each spiking level should be between 80% and 120% of known spiking level.

A3.4.6   Limit of Quantitation (LOQ)

A standard solution of target extractables designed to produce a response of approximately 10 times the LOQ; that is, a response that provides a signal-to-noise (RMS) ratio (S/N) of approximately 100:1, will be analyzed six times by the optimized analytical method. Based on the average S/N ratios for each target extractable, LOQs will be estimated by extrapolation (S/N 10:1). Based on these extrapolated LOQs, a solution of target extractables will be prepared and analyzed six times for LOQ confirmation.

Acceptance criteria: Report results based on extrapolated LOQs.

A3.4.7   Standard and Sample Stability

Standard and sample stability will be evaluated over a period of 5 days by analyzing on each day an appropriate mixed standard of target extractables (as in the "System Suitability" section), and an appropriate test article extract (as in the "Precision" section). Appropriate area ratios of target extractable to internal standard will be determined, and the solutions will be considered stable if:

Acceptance criteria: Area ratios for target extractables on each subsequent day should be ±10% of those determined on day 1.

A3.4.8   Robustness/Ruggedness

Robustness/ruggedness experiments will not be accomplished as a part of this validation protocol. However, this decision may be revisited and modified during the course of the validation exercise. Any robustness/ruggedness studies will be based on critical method parameters identified during the method development and validation phases of the study.

Method for Quantitative Extractables Profiling of a Sulfur-Cured Elastomer Using Soxhlet Extraction and Gas Chromatography/Flame Ionization Detection

A3.5   PURPOSE

The purpose of the method is to produce a quantitative extractables "profile" from a sulfur-cured elastomeric test article prepared for the PQRI Leachables and

Extractables Working Group by the West Pharmaceutical Services. The method employs a weighed sample of the elastomer test article, Soxhlet extraction of the test article with methylene chloride, an internal standard for quantitation of individual extractables via single-point response factors, and analysis of the resulting methylene chloride extract by GC/FID. The resulting chromatogram is considered to be an "extractables profile."

A3.6 APPARATUS

- 250-mL round-bottom boiling flasks, with ST 24/40 ground glass female joints
- Soxhlet extractors, to hold a 22 × 39 mm cellulose thimble, with a male ST 24/40 joint on the bottom and a female ST 45/50 joint on top
- Allihn condenser, male ST 45/50 joint on bottom
- Heating mantle, to accommodate a 250-mL round-bottom flask
- Variac or equivalent variable transformer
- 200-mL volumetric flasks
- 100-mL volumetric flasks
- 10-mL volumetric flasks for dilutions
- 250-mL graduated cylinders
- Volumetric pipettes (1, 2, 5, 10, 15, 20 mL, etc., as needed)

A3.7 REAGENTS AND STANDARDS

- EM Scientific HPLC Grade methylene chloride or equivalent
- 2-Fluorobiphenyl as the internal standard (Aldrich, 99%)
- 2,2'-Methylene-bis(6-*tert*-butyl-4-ethylphenol) (Chem Services)
- *n*-Docosane (Chem Services, 99.4%)
- *n*-Tricosane (Chem Services, 99.2%)
- *n*-Tetracosane (Chem Services, 99%)
- *n*-Pentacosane (Chem Services, 99.0%)
- *n*-Hexacosane (Chem Services, 99.2%)
- *n*-Octacosane (Chem Services, 99.5%)
- Ultra-high-purity helium
- Ultra-high-purity hydrogen
- Zero air

A3.8 PREPARATION OF STANDARDS AND CALIBRATION SOLUTIONS

A3.8.1 Internal-Standard-Spiked Extraction Solution/ Calibration Diluent

This methylene chloride solution spiked with internal standard (2-fluorobiphenyl) is used to extract the elastomer samples. It is also used as a diluent for the preparation of analyte calibration standards. This extraction solution/calibration diluent preparation may be scaled up as needed. The concentration of the internal standard in this preparation is nominally 100 µg/mL. This example is for a 500-mL preparation:

1. Accurately weigh approximately 50 mg of 2-fluorobiphenyl into a 500-mL volumetric flask.
2. Partially fill the flask with methylene chloride. Shake to dissolve.
3. Dilute to the mark with methylene chloride. Store at room temperature.

A3.8.2 Analyte Calibration Solution (for the Determination of RRFs)

1. Accurately weigh approximately 10 mg of each target analyte into a 100-mL volumetric flask.
2. Add about 40 mL of calibration diluent (containing internal standard) to the volumetric flask and agitate to dissolve the target analytes. Note that sonication may be required to completely dissolve some of the alkanes.
3. Dilute to the mark with calibration diluent (nominal concentration 100 µg/mL for each analyte and the internal standard).
4. Pipet 1.0 mL of solution in step 3 into a 10-mL volumetric flask. Dilute with pure methylene chloride.
5. Transfer approximately 2 mL to a GC vial for analysis.

A3.8.3 Linearity Solutions (for System Suitability)

Note that the actual levels and preparation procedure used for validation will be determined during method development. The following is an example:

1. Prepare a stock solution of 2,2'-methylene-bis(6-*tert*-butyl-4-ethylphenol) and *n*-pentacosane by accurately weighing 10 mg of each analyte into a 100-mL volumetric flask and bringing to volume with methylene chloride. Sonicate as required to dissolve the solid material.
2. Into individual 100-mL volumetric flasks, pipet 1.0, 2.0, 5.0, 10.0, 15.0, and 20 mL of the analyte stock solution. The levels of each analyte will be approximately 1, 2, 5, 10, 15, and 20 µg/mL.
3. Into each volumetric flask, pipet 10.0 mL of internal standard calibration diluent.

4. Dilute each solution to the mark with methylene chloride. The nominal concentration of internal standard is 10 μg/mL.

A3.9 SAMPLE PREPARATION

A3.9.1 Pre-extraction of Cellulose Thimbles

1. Place about 10 boiling chips into a 250-mL round-bottom flask and add approximately 200 mL of methylene chloride.
2. Place an empty cellulose thimble into a Soxhlet extractor.
3. Assemble the heating mantle, round-bottom flask, Soxhlet extractor, and condenser, and hook up to a Variac. Cap the unused neck of the round bottom with an ST 24/40 ground glass stopper.
4. Turn on water; observe that the water is flowing, there are no leaks, and the condenser is cold.
5. Turn on Variac to a setting between 40 and 50.
6. Pre-extract for 2 hours once boiling starts.
7. Allow extractor(s) to cool.
8. Properly discard the solvent.

A3.9.2 Preparation and Extraction of Elastomer Sample

1. Remove the protective material from a sheet of elastomer sample (Note: These elastomer samples were shipped in sheets from the West Pharmaceutical Services wrapped in a protective material, which must be removed prior to extraction.)
2. Accurately weigh 7 ± 0.2 g of rubber sample.
3. Cut the rubber into approximately 15–25 roughly square (approximately 5 mm) pieces to fit into the bottom of the thimble. The rubber swells considerably in methylene chloride; this is to prevent the swollen rubber from protruding above the siphon in the Soxhlet, preventing full extraction.
4. Load the pieces into the pre-extracted thimble. Put the thimble into the Soxhlet.
5. Place about 10 boiling chips into a 250-mL round-bottom flask.
6. Using a graduated cylinder, measure 200 mL of internal-standard-spiked methylene chloride into the flask.
7. Assemble the extraction apparatus as above. Turn on the water, and verify flow and that there are no leaks.
8. Turn the Variac to a setting of between 40 and 50.
9. Once boiling starts, observe the time it takes for the thimble to fill and siphon. This is the turnover time. Adjust the Variac power so that this time, the setting is between 18 and 22 minutes.

10. Once boiling starts, observe and record the clock time.

11. Extract under these conditions for 16 hours (Note: Extraction may be accomplished in 2- to 8-hour increments; i.e., the extraction may be stopped after 8 hours, the system is allowed to cool to room temperature, and the extraction is continued for a further 8 hours the next day.)

A3.9.3 Extraction Blank

Prepare an extraction blank in the same manner as the elastomer sample extract, but without the elastomer sample.

A3.9.4 Sample/Blank Collection

1. After the 16-hour extraction time, turn off the Variac at the power switch without disturbing the power level dial.

2. Allow the solvent to stop boiling. This will take about 10 minutes.

3. Siphon the solvent from the Soxhlet, and clip the thimble to the top of the extractor and allow to drain.

4. Siphon last remaining solvent into the boiling flask.

5. Quantitatively transfer solution into a 200-mL volumetric flask. Rinse the boiling flask with small amounts of pure methylene chloride (no internal standard) and add these to the volumetric flask. Fill to the mark with methylene chloride.

6. Pipet 1.0 mL of solution in step 5 into a 10-mL volumetric flask. Dilute with pure methylene chloride.

7. Transfer approximately 2 mL to a GC vial for analysis.

A3.10 GC CONDITIONS

Instrument: Hewlett-Packard 5890 Series II Plus, Agilent 6890, or equivalent

Column: Restek RTX-1, 30 m × 0.25 mm (0.1-μm film) or equivalent

Injection mode: Splitless

Injection volume: 1 μL

Injector temperature/program: 280°C for splitless injection

Purge valve: On at 1.00 minute; off initially

Oven temperature: 40°C for 1 minute
40–300°C at 10°C/min
300°C for 10 minutes

Pressure: Constant helium flow at 1.0 mL/min

Transfer line: 280°C

A3.11   INJECTION SEQUENCE

1. Six injections of the diluted analyte calibration solution (used for determining chromatographic resolution, chromatographic tailing factor, and RRF precision)
2. Two injections of the extraction blank
3. Two injections of each linearity solution (from low to high concentration; used for determining linearity and sensitivity)
4. Two injections of each sample extract

A3.12   SYSTEM SUITABILITY

A3.12.1   Linearity

Evaluate linearity by plotting area ratio for each analyte in each linearity solution versus individual analyte concentration.

Acceptance criteria: To be determined in method development.

A3.12.2   Sensitivity

For each analyte in the second injection of the lowest concentration of linearity solution, determine the S/N ratio (the term noise is taken to mean root mean square noise).

Acceptance criteria: To be determined in method development.

A3.12.3   Chromatographic Resolution

For the second injection of the analyte calibration solution, calculate the chromatographic resolution between 2,2′-methylene-bis(6-*tert*-butyl-4-ethylphenol) and *n*-pentacosane.

Acceptance criteria: To be determined in method development.

A3.12.4   Chromatographic Tailing Factor

For the second injection of the analyte calibration solution, calculate the chromatographic tailing factors for 2,2′-methylene-bis(6-*tert*-butyl-4-ethylphenol) and *n*-pentacosane.

Acceptance criteria: To be determined in method development.

A3.12.5   RRF Precision

Calculate RRFs for all individual analytes for each injection of the analyte calibration solution and then determine means and RSDs for RRFs for each individual analyte:

$$RRF = A_a \times C_i / A_i \times C_a,$$

where

A_a = peak area for an individual analyte,

A_i = peak area for the internal standard,

C_a = concentration of an individual analyte, and

C_i = concentration of the internal standard.

Acceptance criteria: To be determined in method development.

A3.13 CALCULATION OF ANALYTE LEVELS IN THE ELASTOMER SAMPLE

For each individual analyte, use the mean RRF determined in the "System Suitability" section (Section A3.12.5):

1. Calculate the concentration of each individual analyte in the extraction solution as follows:

$$C_a = A_a \times C_i / A_i \times RRF.$$

2. Calculate the total mass of each individual analyte in the solution as follows:

Total mass = concentration of the analyte in $\mu g/mL \times 200$ mL.

3. Calculate the amount of each individual analyte in the elastomer as follows:

Analyte ($\mu g/g$ elastomer) = total mass of an analyte (μg)/mass of elastomer (g).

APPENDIX 4 PROTOCOL ADDITION, PHASE 2 STUDIES: QUANTITATIVE EXTRACTABLES STUDIES ON SULFUR-CURED ELASTOMER AND POLYPROPYLENE

Validation of a Quantitative Gas Chromatography Method for Sulfur-Cured Elastomer Extractables

A4.1 INTRODUCTION AND BACKGROUND

Qualitative controlled extraction studies guided by a specific and detailed protocol have been accomplished on a sulfur-cured elastomeric test article of known additive composition. These qualitative studies produced extractables profiles by gas chromatography/mass spectrometry (GC/MS) and high-performance liquid chromatography/mass spectrometry (LC/MS), which exactly reflect the known additive composition of the elastomeric test article.

This protocol addition is designed to extend the qualitative controlled extraction study to a quantitative controlled extraction study, with appropriate method optimization and investigation of validation parameters.

A4.2 METHOD DEVELOPMENT

Based on the results of the qualitative controlled extraction studies, Soxhlet extraction in methylene chloride with quantitative GC analysis of extracts has been selected for optimization and validation. Internal standardization utilizing appropriate authentic reference materials will be employed for quantitative calibration of the analytical system. The known additives in the elastomeric test article that can be quantitated by this analytical technique include

- 2,2′-methylene-bis(6-*tert*-butyl-4-ethylphenol),
- coumarone-indene resin-related species,
- *n*-alkanes derived from paraffin, and
- additional relatively minor extractables.

All details of the analytical method, including the extraction procedure and analysis system will be documented in laboratory notebooks and/or other appropriate documentation media.

Prior to method validation, the extraction procedure will be optimized to produce maximum quantities of target extractables (i.e., "asymptotic" levels). The optimized extraction conditions will be documented and taken to method validation.

A4.3   VALIDATION PARAMETERS AND ACCEPTANCE CRITERIA

The following validation parameters that include appropriate acceptance criteria will be investigated. When appropriate, the following model extractables will be employed:

- 2,2′-methylene-bis(6-*tert*-butyl-4-ethyl phenol),
- docosane,
- hexacosane,
- nonacosane, and
- internal standard: 2-fluorobiphenyl.

A4.3.1   System Suitability

A4.3.1.1   Instrument Precision   A test solution of target extractables with internal standard will be prepared at concentrations demonstrated not to produce adverse effects on chromatographic performance. Utilizing optimized chromatography conditions, six replicate injections of the test solution will be analyzed. Peak area and area ratio measurements of target extractables and the internal standard will be determined, and means and percent relative standard deviations (%RSDs) of area ratios and relative response factors will be calculated.

Acceptance criteria: %RSDs for area ratios ≤10%.

A4.3.1.2   Chromatographic Resolution   Utilizing the analyses accomplished for instrument precision, chromatographic resolution between appropriate peak pairs will be determined. Means and %RSDs will be calculated.

Acceptance criteria: To be determined.

A4.3.1.3   Chromatographic Tailing Factor   Utilizing the analyses accomplished for instrument precision, chromatographic tailing factors for appropriate peaks will be determined. Means and %RSDs will be calculated.

Acceptance criteria: To be determined.

A4.3.2   Linearity and Range

Linearity and range will be determined by analyzing target extractables at six different concentration levels (in duplicate), over a range established during the qualitative phase of the controlled extraction study.

Acceptance criteria: To be determined.

A4.3.3   Precision

A4.3.3.1   Repeatability   Utilizing optimized extraction procedures, six separate extractions will be accomplished and target extractables will be quantitated with the analytical method. Means and %RSDs of individual target extractable amounts will be calculated.

Acceptance criteria: %RSD for each target extractable ≤10%.

A4.3.3.2   Intermediate Precision   Intermediate precision will be evaluated by a second analyst accomplishing the repeatability study utilizing a different chromatographic system (including mobile phase and GC column). A different analytical instrument will also be utilized if available.

Acceptance criteria: (1) %RSD for each target extractable ≤10%; (2) %difference between analyst means for each target extractable ≤25%.

A4.3.4   Specificity

Specificity was demonstrated in the qualitative phase of the controlled extraction studies utilizing GC/MS.

Acceptance criteria: Confirms peak identifications and confirms no co-eluting peaks for each target extractable.

A4.3.5   Accuracy

Accuracy will be expressed as the percent recovery of known amounts of target extractables spiked into the extraction system.

Spiking solutions of appropriate target extractables will be prepared and spiked at three different levels (in triplicate). The individual spiking levels will be chosen to represent the appropriate range of analyte concentrations expected based on the method development experiments. Spiked samples will be analyzed by the optimized analytical method and individual mean recoveries determined for each spiking level.

Acceptance criteria: Mean recovery for each target extractable at each spiking level should be between 80% and 120% of known spiking level.

A4.3.6   Limit of Quantitation (LOQ)

A standard solution of target extractables designed to produce a response of approximately 10 times the LOQ (i.e., a response that provides a signal-to-noise [RMS] ratio [S/N] of approximately 100:1) will be analyzed six times by the optimized analytical method. Based on the average S/N ratios for each target extractable, LOQs will be estimated by extrapolation (S/N 10:1). Based on these extrapolated LOQs, a solution of target extractables will be prepared and analyzed six times for LOQ confirmation.

Acceptance criteria: Report results based on extrapolated LOQs.

A4.3.7   Robustness

Since there is no intention to transfer this analytical method to other laboratories, robustness experiments will not be accomplished as a part of this validation protocol.

Quantification of Mercaptobenzothiazole Compounds from Sulfur-Cured Rubber

A4.4 PURPOSE

To quantify mercaptobenzothiazole (MBT) and 2,2′-dibenzothiazyl disulfide (MBTS) from the extracts of sulfur-cured rubber using both high-performance liquid chromatography (HPLC) and LC/MS. Two extraction procedures will be compared for the extraction efficiency.

A4.5 REFERENCE STANDARDS, SOLVENTS, AND SAMPLES

- MBT, Aldrich
- MBTS, Aldrich
- Methyl-*tert*-butyl ether (MTBE)
- Methylene chloride
- Sulfur-cured rubber

A4.6 INSTRUMENTATION

- Soxhlet extraction apparatus
- Ultrasonication bath
- Agilent 1100 series HPLC system equipped with ultraviolet detector
- PE Sciex API-2000 Triple-Quadrapole Mass Spectrometry equipped with atmospheric pressure chemical ionization (APCI) source

A4.7 EXTRACTION PROCEDURE

(Note: Extraction conditions can be modified to obtained better recovery.)

A4.7.1 Sonication

Approximately 1 g of rubber sample, cut into small pieces, and 10 mL of MTBE will be transferred into a suitable glass vial with screw caps. The vial will be sonicated for 30 minutes in an ultrasonication bath. Triplicate sample extraction will be performed.

A4.7.2 Soxhlet Extraction

Approximately 2 g of rubber sample, cut into small pieces, will be transferred into a cellulose thimble and extracted with methylene chloride in a Soxhlet extraction apparatus for 24 hours. Triplicate sample extraction will be performed.

A4.8 STANDARD AND SAMPLE PREPARATION

A4.8.1 Reference Standard Solutions

Mixture of MBT and MBTS will be prepared at five concentration levels between 0.1 and 10 µg/mL in acetonitrile.

A4.8.1.1 Sample Solution The MTBE extract from the sonication will be evaporated to dryness under nitrogen stream and reconstituted into 1 mL of acetontrilc. The methylene chloride extract from the Soxhlet extraction will be brought to 200 mL in volume, and 50 mL of the extract will be evaporated to dryness and reconstituted into 1 mL acetonitrile.

A4.9 ANALYTICAL METHODS

A4.9.1 HPLC/UV

Column: Symmetry C18, 2.1 × 50 mm, 3.5 µm

Column temperature: 40°C

Autosampler temperature: Ambient

Diluent: 60:40 acetonitrile : water, v/v

Detection wavelength: UV at 280, 325 nm

Flow rate: 0.4 mL/min

Injection volume: 20 µL

Run time: 35 minutes

Mobile phase: A: 0.02 M sodium acetate buffer, pH 3.5; B: acetonitrile

Gradient profile:

Time	MP(A)	MP(B)
0	80	20
10	20	80
20	20	80
21	80	20
35	80	20

A4.9.2 LC/MS

Column: Symmetry C18, 2.1 × 50 mm, 3.5 µm

Column temperature: 40°C

Autosampler temperature: Ambient

Diluent: 60:40 acetonitrile : water, v/v

Flow rate: 0.4 mL/min

Injection volume: 20 µL

Run time: 35 minutes

Mobile phase: A: 0.1% formic acid; B: acetonitrile

Gradient profile:

Time	MP(A)	MP(B)
0	80	20
10	20	80
20	20	80
21	80	20
35	80	20

Mass spectrometer

Ionization mode: Positive APCI

Detection mode: Selected ion monitoring (SIM) at m/z 168

A4.10   QUANTITATION

The area response of the working standard solutions will be plotted against their corresponding concentration. The concentration of the extract sample solution will be calculated against the curve and converted to microgram per gram of rubber (parts per million) based on the extraction solvent volume and concentration factors. If the area response of the sample is out of the working curve range, the sample solution will be diluted accordingly to fit into the working curve range.

A4.11   FURTHER READING

Hansson, C., et al. Extraction of mercaptobenzothiozole compounds from rubber products. *Contact Dermatitis* 1997, *36*, pp. 195–200.

Gaind, V.S. and Jerdrzejczak, K. HPLC determination of rubber septum contaminants in the iodinated intravenous contrast agent (sodium iothalamate). *J Anal Toxicol* 1993, *17*, pp. 34–37.

Validation of a Quantitative High-Performance Liquid Chromatography/ Ultraviolet Detection Method for Polypropylene Extractables

A4.12   INTRODUCTION AND BACKGROUND

Qualitative controlled extraction studies guided by a specific and detailed protocol have been accomplished on a polypropylene test article of known additive composition. These qualitative studies produced extractables profiles by GC/MS and high-performance liquid chromatography/diode array detection (HPLC/DAD), which exactly reflect the known additive composition of the polypropylene test article as well as showing oligomer patterns indicative of polypropylene.

This protocol addition is designed to extend the qualitative controlled extraction study to a quantitative controlled extraction study, with appropriate method optimization and investigation of validation parameters.

A4.13   METHOD DEVELOPMENT

Based on the results of the qualitative controlled extraction studies, reflux extraction in 2-propanol with quantitative HPLC/DAD analysis of extracts has been selected for optimization and validation. External standardization utilizing appropriate authentic reference materials will be employed for quantitative calibration of the analytical system. The known additives in the polypropylene test article that can be quantitated by this analytical technique include

- Millad 3988   1,3:2,4-bis(3,4-dimethylbenzylidene)sorbitol
- Ultranox 626   Bis(2,4-di-*tert*-butylphenyl)pentaerythritol diphosphite
- Irganox 1010   Tetrakis(methylene-3-(3′,5′-di-*tert*-butyl-4′-hydroxyphenyl)
                  propionate)methane

All details of the analytical method, including the extraction procedure and analysis system will be documented in laboratory notebooks and/or other appropriate documentation media.

Prior to method validation, the extraction procedure will be optimized to produce maximum quantities of target extractables (i.e., "asymptotic" levels). The optimized extraction conditions will be documented and taken to method validation.

A4.14   VALIDATION PARAMETERS AND ACCEPTANCE CRITERIA

The following validation parameters that include appropriate acceptance criteria will be investigated.

A4.14.1   System Suitability

A4.14.1.1   Instrument Precision   A test solution of target extractables will be prepared at concentrations demonstrated not to produce adverse effects on chromatographic performance. Utilizing optimized chromatography conditions, six replicate injections of the test solution will be analyzed. Peak area measurements of target extractables will be determined, and means and %RSDs of area ratios and relative response factors will be calculated.

Acceptance criteria: %RSD not more than (NMT) 5.

A4.14.1.2   Chromatographic Resolution   Utilizing the analyses accomplished for instrument precision, chromatographic resolution between appropriate peak pairs will be determined. Means and %RSDs will be calculated.

Acceptance criteria: Half-width resolution not less than (NLT) 2.

A4.14.1.3   Chromatographic Tailing Factor   Utilizing the analyses accomplished for instrument precision, chromatographic tailing factors for appropriate peaks will be determined. Means and %RSDs will be calculated.

Acceptance criteria: Tailing factor NMT 2.

A4.14.2 Linearity and Range

Linearity and range will be determined by analyzing target extractables at six different concentration levels (in duplicate), over a range established during the qualitative phase of the controlled extraction study.

Acceptance criteria: Correlation coefficient 0.99.

A4.14.3 Precision

A4.14.3.1 Repeatability Utilizing optimized extraction procedures, six separate extractions will be accomplished and target extractables will be quantitated with the analytical method. Means and %RSDs of individual target extractable amounts will be calculated.

Acceptance criteria: %RSD NMT 15.

A4.14.3.2 Intermediate Precision Intermediate precision will be evaluated by a second analyst accomplishing the repeatability study utilizing a different chromatographic system (including mobile phase and HPLC column). A different analytical instrument will also be utilized if available.

Acceptance criteria: %RSD NMT 15 and % absolute difference of the mean between analysts 1 and 2 is NMT 15.

A4.14.4 Specificity

Specificity was demonstrated in the qualitative phase of the controlled extraction studies utilizing HPLC/DAD and LC/MS.

Acceptance criteria: Confirms peak identifications and confirms no co-eluting peaks for each target extra extractable.

A4.14.5 Accuracy

Accuracy will be expressed as the percent recovery of known amounts of target extractables spiked into the extraction system.

Spiking solutions of appropriate target extractables will be prepared and spiked at three different levels (in triplicate). The individual spiking levels will be chosen to represent the appropriate range of analyte concentrations expected based on the method development experiments. Spiked samples will be analyzed by the optimized analytical method and individual mean recoveries determined for each spiking level.

Acceptance criteria: Mean recovery for each target extractable at each spiking level should be between 80% and 120% of known spiking level.

A4.14.6 LOQ

A standard solution of target extractables designed to produce a response of approximately 10 times the LOQ (i.e., a response that provides a signal-to-noise [RMS] ratio [S/N] of approximately 100:1) will be analyzed six times by the optimized analytical method. Based on the average S/N ratios for each target extractable, LOQs will be estimated by extrapolation (S/N 10:1). Based on these extrapolated LOQs, a solution of target extractables will be prepared and analyzed six times for LOQ confirmation.

Acceptance criteria: Report results based on extrapolated LOQs.

A4.14.7 Robustness

Since there is no intention to transfer this analytical method to other laboratories, robustness experiments will not be accomplished as a part of this validation protocol.

Appendix to Protocol Addition: Method for Extractables Profiling of a Sulfur-Cured Elastomer Using Soxhlet Extraction and Gas Chromatographic Analysis

A4.15 INTRODUCTION AND BACKGROUND

This extractables profiling method was developed in support of investigational studies undertaken by the Product Quality Research Institute (PQRI) Leachables and Extractables Working Group. The purpose of the method is to produce a quantitative extractables "profile" from a sulfur-cured elastomeric test article prepared for the Working Group by the West Pharmaceutical Services. The method employs Soxhlet extraction with methylene chloride of a weighed sample of the elastomer test article, followed by the analysis of the resulting extract by GC. The resulting chromatogram is considered to be an "extractables profile." An internal standard (2-fluorobiphenyl) is used for the quantitation of individual extractables.

A4.16 APPARATUS AND EQUIPMENT

- Analytical balance, capable of weighing to 0.00001 g
- Wax-coated weighing paper

  For each extraction:

- 250-mL round-bottom boiling flasks, with two ST 24/40 ground glass female joints
- Soxhlet extractors, to hold a 22×39 mm cellulose thimble, with a male ST 24/40 joint on the bottom and a female ST 45/50 joint on top
- Allihn condenser, male ST 45/50 joint on bottom
- ST 24/40 ground glass stoppers
- Teflon or glass boiling chips
- Cold tap or recirculated water

- Tygon tubing to connect condensers to tap and together
- Heating mantle, to accommodate a 250-mL round-bottom flask
- Variac or equivalent variable transformer
- Cellulose thimbles, 33 × 80 mm, Schleicher 7 Schuell or equivalent
- Glass volumetric pipettes, 0.5 mL
- Pipette bulbs or automatic pipettor
- Glass volumetric flasks with ground glass stoppers (5 mL)
- 250-mL glass graduated cylinder
- Ring stands, monkey bars, or equivalent to hold extractors
- Clamps and clamp holders
- Disposable 5¾ in. glass pipettes
- 2-mL rubber bulbs

  For GC/MS or GC/FID:

- Hewlett-Packard 5890 Series II Plus, Agilent 6890, or equivalent gas chromatograph, equipped with a mass selective detector (MSD) and/or an FID
- Restek RTX-1 30 m × 0.25 mm (0.1-μm film) GC column or equivalent
- 2-mL glass vials, caps, and fluoropolymer-lined septa

A4.17 CHEMICALS/REAGENTS

- EM Scientific HPLC Grade methylene chloride or equivalent
- 2-Fluorobiphenyl (Aldrich, 99%)
- Ultra-high-purity helium
- Ultra-high-purity hydrogen
- Zero air

A4.18 PREPARATION OF INTERNAL-STANDARD-SPIKED EXTRACTION SOLUTION

This may be scaled up as needed. The concentration of the internal standard is approximately 100 μg/mL. This example is for 500 mL of internal standard solution:

1. Accurately weigh approximately 50 mg of 2-fluorobiphenyl into a 500-mL volumetric flask.
2. Partially fill the flask with methylene chloride. Shake to dissolve.
3. Dilute to the mark with methylene chloride. Store at room temperature.

A4.19 PRE-EXTRACTION OF CELLULOSE THIMBLES

1. Place about 10 boiling chips into a 250-mL round-bottom flask and add approximately 200 mL of methylene chloride.

2. Place an empty cellulose thimble into a Soxhlet extractor.

3. Assemble the heating mantle, round-bottom flask, Soxhlet extractor, and condenser, and hook up to a Variac. Cap the unused neck of the round bottom with an ST 24/40 ground glass stopper.

4. Turn on water; observe that the water is flowing, there are no leaks, and the condenser is cold.

5. Turn on Variac to a setting between 40 and 50.

6. Pre-extract for 2 hours once boiling starts.

7. Allow extractor(s) to cool.

8. Properly discard the solvent.

A4.20 PREPARATION AND EXTRACTION OF RUBBER SAMPLE

1. Remove any release liner/coating from the rubber.

2. Tear a piece of wax weighing paper.

3. Cut the rubber so that it fits on the weighing paper. Add or remove portions to get to 7 ± 0.2 g; weigh to the nearest 0.00001 g.

4. Cut the rubber into approximately 15–25 roughly square pieces to fit into the bottom of the thimble. The rubber swells considerably in methylene chloride; this is to prevent the swollen rubber from protruding above the siphon in the Soxhlet, preventing full extraction.

5. Load the pieces into the pre-extracted thimble. Put the thimble into the Soxhlet.

6. Place about 10 boiling chips into a 250-mL two-neck round-bottom flask.

7. Using a graduated cylinder, measure 200 mL of internal-standard-spiked methylene chloride into the flask.

8. Assemble the extraction apparatus as above. Cap the unused port with an ST 24/40 ground glass stopper.

9. Turn on the water, and verify flow and that there are no leaks.

10. Turn the Variac to a setting of between 40 and 50.

11. Once boiling starts, observe the time it takes for the thimble to fill and siphon. This is the turnover time. Adjust the Variac power so that this time, the setting is between 18 and 22 minutes.

12. Once boiling starts, observe and record the clock time.

13. Extract under these conditions for 16 hours (Note: Extraction may be accomplished in 2- to 8-hour increments; i.e., the extraction may be stopped after 8 hours, the system allowed to cool to room temperature, and the extraction continued for a further 8 hours the next day.)

A4.21 SAMPLE COLLECTION

1. After the 16-hour extraction time, turn off the Variac at the power switch without disturbing the power level dial. Record the clock time.
2. Allow the bulk of the fluid to stop boiling. This will take about 10 minutes.
3. Remove the ground glass stopper.
4. Using a glass 0.5-mL glass volumetric pipette, remove 0.5 mL of extract and transfer it to a 5-mL volumetric flask.
5. Dilute the extract to the mark with pure methylene chloride. *Do not use the internal standard solution.* Shake to mix.
6. Using a glass disposable pipette, transfer a portion of the diluted extract to a 2-mL glass vial. Cap the vial with a fluoropolymer-lined septum and cap.
7. Collect GC/MS or GC/FID chromatogram.

A4.22 GC WITH MSD OR FID

GC conditions are the following:

Instrument: Hewlett-Packard 5890 Series II Plus, Agilent 6890, or equivalent

Column: Restek RTX-1, 30 m × 0.25 mm (0.1-μm film) or equivalent

Injection mode: Splitless

Injection volume: 1 μL

Injector temperature/program: 280°C for splitless injection

Purge valve: On at 1.00 minute; off initially

Oven temperature: 40°C for 1 minute
40–300°C at 10°C/min
300°C for 10 minutes

Pressure: Constant helium flow at 1.0 mL/min

Transfer line: 280°C

If a mass spectrometer is used:

Instrument: HP 5972, Agilent 5973 MSD, or equivalent

Ionization mode: Electron ionization (EI)

Scan mode: Scanning; m/z 50–650

Scan cycle time: Approximately 2 s/scan

A4.23 CALCULATIONS (FOR DATA COLLECTED BY MS)

1. Using the selected ion extraction menu, selections of m/z 172 (2-fluorobiphenyl); 191 (phenolic); 71 (hydrocarbons); and 233 (coumarone-indene).

2. Integrate each selected ion chromatograph.

3. Calculate the ratio between each analyte peak area and that of the internal standard. For the hydrocarbons, it is useful to select one well-resolved peak to either side of the phenolic peak. In this work, docosane (C22) and hexacosane (C26) are used.

4. Plot the ratio versus time for each analyte.

5. Select an extraction time well onto the asymptotic part of the curve.

REFERENCES

1 Norwood, D.L. and Ball, D. Product Quality Research Institute: Safety thresholds and best practices for extractables and leachables in orally inhaled and nasal drug products. Submitted to the PQRI Drug Product Technical Committee, PQRI Steering Committee, and U.S. Food and Drug Administration by the PQRI Leachables and Extractables Working Group, 2006.

2 Draft guidance for industry: Metered dose inhaler (MDI) and dry powder inhaler (DPI) drug products. Department of Health and Human Services, Food and Drug Administration, Center for Drug Evaluation and Research (CDER). 1998.

3 Guidance for industry: Nasal spray and inhalation solution, suspension, and spray drug products— Chemistry, manufacturing, and controls documentation. Department of Health and Human Services, Food and Drug Administration, Center for Drug Evaluation and Research (CDER), 2002.

4 Product Quality Research Institute (PQRI) and Leachables and Extractables Working Group. Reporting and qualification thresholds for leachables in orally inhaled and nasal drug products, 2002.

5 Code of Federal Regulations. Food and drugs, Part 210 and 211, Title 21, 2010.

6 Magellan Laboratories, Inc. Magellan analytical test method ATM-MAG-M0012. Determination of metals by inductively coupled plasma mass spectrometry.

7 Norwood, D.L., Prime, D., Downey, B., Creasey, J., Sethi, S., and Haywood, P. Analysis of polycyclic aromatic hydrocarbons in metered dose inhaler drug formulations by isotope dilution gas chromatography/mass spectrometry. *J Pharm Biomed Anal* 1995, *13*(3), pp. 293–304.

8 American Society of Testing and Materials. ASTM designation: D 5524-94. Standard test method for determination of phenolic antioxidants in high density polyethylene using liquid chromatography.

9 Vargo, J.D. and Olson, K.L. Characterization of additives in plastics by liquid chromatography-mass spectrometry. *J Chromatogr* 1986, *353*, pp. 215–224.

INDEX

Page numbers in *italics* refer to Figures; those in **bold** refer to Tables.

*Leachables and Extractables Handbook: Safety Evaluation, Qualification, and Best Practices Applied
to Inhalation Drug Products*, First Edition. Edited by Douglas J. Ball, Daniel L. Norwood,
Cheryl L.M. Stults, Lee M. Nagao.
© 2012 John Wiley & Sons, Inc. Published 2012 by John Wiley & Sons, Inc.